FLOOD FORECASTING

FLOOD FORECASTING
A Global Perspective

Edited by

THOMAS E. ADAMS, III
University Corporation for Atmospheric Research, Boulder, CO, United States

THOMAS C. PAGANO
Australian Bureau of Meteorology, Melbourne, Australia

AMSTERDAM • BOSTON • HEIDELBERG • LONDON
NEW YORK • OXFORD • PARIS • SAN DIEGO
SAN FRANCISCO • SINGAPORE • SYDNEY • TOKYO
Academic Press is an imprint of Elsevier

Academic Press is an imprint of Elsevier
125 London Wall, London EC2Y 5AS, UK
525 B Street, Suite 1800, San Diego, CA 92101-4495, USA
50 Hampshire Street, 5th Floor, Cambridge, MA 02139, USA
The Boulevard, Langford Lane, Kidlington, Oxford OX5 1GB, UK

Library of Congress Cataloging-in-Publication Data
A catalog record for this book is available from the Library of Congress

British Library Cataloguing-in-Publication Data
A catalogue record for this book is available from the British Library

ISBN: 978-0-12-801884-2

For information on all Academic Press publications visit our website at https://www.store.elsevier.com

DEDICATION

To our parents, Anne, Jess, Megan, Mary, Thomas, Soroosh Sorooshian, Phil Pasteris, Rob Vertessy, Florian Pappenberger, Andy Wood, Maria-Helena Ramos, Tanya Smith, and the global community of operational flood forecasters.

CONTENTS

Contributors xiii
Foreword xvii
Acknowledgments xxi
Flood Forecasting: A Global Perspective xxiii

Part 1 National and Regional Flood Forecasting Systems **1**

1. Australian Bureau of Meteorology Flood Forecasting and Warning **3**
T.C. Pagano, J.F. Elliott, B.G. Anderson, J.K. Perkins

1. Introduction 3
2. Demographics, Climate, and Floods of Australia 3
3. History of Flood Forecasting at the Bureau of Meteorology 10
4. Current Characteristics of the Bureau and Its Services 13
5. Operational Forecasting and Systems 18
6. Operational Challenges 32
7. Future Directions 37
8. Summary 38

Acknowledgments 38
References 39

2. Hydrological Forecasting Practices in Brazil **41**
F.M. Fan, R.C.D. Paiva, W. Collischonn

1. Introduction 41
2. Geography, Climate, and Floods in Brazil 41
3. General Overview of Hydrological Forecasts in Brazil 43
4. Examples of Operational and Preoperational Forecasting Systems 46
5. Conclusions 61

References 63

3. The Development and Recent Advances of Flood Forecasting Activities in China **67**
Z. Liu

1. Introduction 67
2. Evolution of Flood Control and Management in China 68

3. Development of Operational Hydrological Forecasting and Prediction in China 71
4. Conclusions and Outlook 85
References 86

4. A Regional Perceptive of Flood Forecasting and Disaster Management Systems for the Congo River Basin 87

R.M. Tshimanga, J.M. Tshitenge, P. Kabuya, D. Alsdorf, G. Mahe, G. Kibukusa, V. Lukanda

1. Introduction 87
2. Physiographic Setting and Physical Characteristics of the Congo River Basin 89
3. Flood-Bearing Processes in the Congo River Basin 92
4. Trends and Socio-Economic Impacts of Floods in the Congo River Basin 107
5. Current Status of Flood Forecasting and Disaster Management Systems in the Congo River Basin 111
6. Conclusion 120
References 121

5. Flood Forecasting in Germany — Challenges of a Federal Structure and Transboundary Cooperation 125

N. Demuth, S. Rademacher

1. Introduction 125
2. International Cooperation — The Rhine Basin 127
3. Measured Meteorological and Hydrological Data 130
4. Numerical Weather Predictions 133
5. Snowmelt Forecasts 136
6. Regional Organization and Transboundary Data Exchange 137
7. River Forecasting Models 138
8. Forecast Dissemination 145
9. Flood Partnerships 147
10. Summary 148
References 148

6. Operational Flood Forecasting in Israel 153

A. Givati, E. Fredj, M. Silver

1. Introduction 153
2. Climate and Hydrological Characteristics of Israel 154

3. The Flood Forecasting Modeling System in Israel 156
4. Summary 165
References 165

7. Operational Hydrologic Forecast System in Russia 169
S. Borsch, Y. Simonov

1. Introduction 169
2. Hydrological Practices in Russia 170
3. The Hydrologic Forecasting System of the Roshydromet 171
4. Hydrometeorological Data 173
5. Long-Range Spring Flood Predictions 176
6. Early Warning Flood Forecasting Systems 177
7. Conclusions 181
References 181

8. Increasing Early Warning Lead Time Through Improved Transboundary Flood Forecasting in the Gash River Basin, Horn of Africa 183
G. Amarnath, N. Alahacoon, Y. Gismalla, Y. Mohammed, B.R. Sharma, V. Smakhtin

1. Introduction 183
2. Building a Flood Forecast and Early Warning System 184
3. Geographical Setting of the Gash River Basin 187
4. The Modeling Approach 189
5. Results and Discussion 195
6. Conclusion 197
Acknowledgments 198
References 198

9. Flood Forecasting — A National Overview for Great Britain 201
C. Pilling, V. Dodds, M. Cranston, D. Price, T. Harrison, A. How

1. Background and Catalysts for Change 201
2. Countrywide Flood Forecasting Modeling Approach 211
3. Forecast Dissemination Protocols and Products 229
4. Measuring Performance of the Flood Forecasting and Warning Service 233
5. Future Forecasting Challenges 238
Disclaimer 244
Dedication 244
References 244

10. Flood Forecasting in the United States NOAA/National Weather Service **249**

T.E. Adams III

1. Introduction 249
2. History 251
3. A Brief History of Models and Modeling Systems Used by the NWS 254
4. Current Models and Modeling System 274
5. Current Operations 281
6. Additional Topics 301
7. Future Developments 303

References 307

Part 2 Continental Modeling and Monitoring — The Future? **311**

11. On the Operational Implementation of the European Flood Awareness System (EFAS) **313**

P.J. Smith, F. Pappenberger, F. Wetterhall, J. Thielen del Pozo, B. Krzeminski, P. Salamon, D. Muraro, M. Kalas, C. Baugh

1. Introduction 313
2. EFAS Structure 317
3. Data Acquisition 318
4. Model Components 320
5. Generating Forecasts 325
6. Forecast Products 327
7. Forecast Dissemination 332
8. Operational Performance 335
9. Case Studies 337
10. Conclusions 342

Acknowledgments 344

References 344

12. Developing Flood Forecasting Capabilities in Colombia (South America) **349**

M. Werner, J.C. Loaiza, M.C. Rosero Mesa, M. Faneca Sànchez, O. de Keizer, M.C. Sandoval

1. Introduction 349
2. Physiography and Demographics 351

3. Development of a Pilot for the National Hydrological Forecasting System 353
4. Design and Setup of the National Hydrological Forecasting System 355
5. Expansion of the Pilot System 362
6. Discussion and Outlook 365
References 366

Part 3 Challenges Facing Flood Forecasting 369

13. Streamflow Data 371

G.J. Wiche, R.R. Holmes Jr.

1. Introduction 371
2. History of Streamgaging in the United States 373
3. Streamflow Collection and Computation 376
4. Delivery of Streamflow Information 388
5. Opportunities for the Streamgage Program 391
References 395

14. A Simple Streamflow Forecasting Scheme for the Ganges Basin 399

Y. Jiang, W. Palash, A.S. Akanda, D.L. Small, S. Islam

1. Introduction 399
2. The Ganges Floods and Rainfall-Runoff Relationships 407
3. A Simple Q-Q and Q+P-Q Model for the Ganges 410
4. Results 411
5. Discussion 417
Acknowledgment 419
References 419

Index *421*

CONTRIBUTORS

T.E. Adams III
University Corporation for Atmospheric Research, Boulder, CO, United States

A.S. Akanda
University of Rhode Island, Kingston, RI, United States

N. Alahacoon
International Water Management Institute (IWMI), Colombo, Sri Lanka

D. Alsdorf
The Ohio State University, Columbus, OH, United States

G. Amarnath
International Water Management Institute (IWMI), Colombo, Sri Lanka

B.G. Anderson
Australian Bureau of Meteorology, Melbourne, Australia

C. Baugh
European Centre for Medium-Range Weather Forecasts, Reading, United Kingdom

S. Borsch
Hydrometeorological Research Centre of Russian Federation, Moscow, Russia

W. Collischonn
Federal University of Rio Grande do Sul (UFRGS), Porto Alegre, Brazil

M. Cranston
Scottish Flood Forecasting Service, SEPA, Perth, United Kingdom

O. de Keizer
Deltares, Delft, the Netherlands

N. Demuth
State Environmental Agency Rhineland-Palatinate, Mainz, Germany

V. Dodds
Bureau of Meteorology, Brisbane, QLD, Australia

J.F. Elliott
Australian Bureau of Meteorology, Melbourne, Australia (Retired)

F.M. Fan
Federal University of Rio Grande do Sul (UFRGS), Porto Alegre, Brazil

M. Faneca Sànchez
Deltares, Delft, the Netherlands

E. Fredj
The Jerusalem College of Technology, Jerusalem, Israel

Y. Gismalla
Hydraulic Research Station (HRS), Wad Madani, Sudan

A. Givati
Israeli Hydrological Service, Jerusalem, Israel

T. Harrison
Flood Incident Management, Environment Agency, Solihull, United Kingdom

R.R. Holmes Jr.
U.S. Geological Survey, Rolla, MO, United States

A. How
Flood and Operational Risk Management, Natural Resources Wales, Cardiff, United Kingdom

S. Islam
Tufts University, Medford, MA, United States

Y. Jiang
Tufts University, Medford, MA, United States

P. Kabuya
University of Kinshasa; CB-HYDRONET, Kinshasa, Democratic Republic of Congo

M. Kalas
European Commission Joint Research Centre, Ispra, Italy

G. Kibukusa
Ministère des Affaires Sociales, Actions Humanitaires et Solidarité Nationale, Kinshasa, Democratic Republic of Congo

B. Krzeminski
European Commission Joint Research Centre, Ispra, Italy

Z. Liu
Hydrological Forecast Center, The Ministry of Water Resources of China, Beijing, China

J.C. Loaiza
IDEAM, Bogotá, Colombia

V. Lukanda
University of Kinshasa, Kinshasa, Democratic Republic of Congo

G. Mahe
Laboratoire HydroScience, IRD, Montpellier, France

Y. Mohammed
Hydraulic Research Station (HRS), Wad Madani, Sudan

D. Muraro
European Commission Joint Research Centre, Ispra, Italy

T.C. Pagano
Australian Bureau of Meteorology, Melbourne, Australia

R.C.D. Paiva
Federal University of Rio Grande do Sul (UFRGS), Porto Alegre, Brazil

W. Palash
Tufts University, Medford, MA, United States

F. Pappenberger
European Centre for Medium-Range Weather Forecasts, Reading, United Kingdom

J.K. Perkins
Australian Bureau of Meteorology, Brisbane, QLD, Australia

C. Pilling
Flood Forecasting Centre, Met Office, Exeter, United Kingdom

D. Price
Flood Forecasting Centre, Met Office, Exeter, United Kingdom

S. Rademacher
German Federal Institute of Hydrology, Koblenz, Germany

M.C. Rosero Mesa
IDEAM, Bogotá, Colombia

P. Salamon
European Commission Joint Research Centre, Ispra, Italy

M.C. Sandoval
CVC, Cali, Colombia

B.R. Sharma
International Water Management Institute (IWMI), New Delhi, India

M. Silver
Ben Gurion University, Beersheba, Israel

Y. Simonov
Hydrometeorological Research Centre of Russian Federation, Moscow, Russia

V. Smakhtin
International Water Management Institute (IWMI), Colombo, Sri Lanka

D.L. Small
Tufts University, Medford, MA, United States

P.J. Smith
European Centre for Medium-Range Weather Forecasts, Reading, United Kingdom

J. Thielen del Pozo
European Commission Joint Research Centre, Ispra, Italy

R.M. Tshimanga
University of Kinshasa; CB-HYDRONET, Kinshasa, Democratic Republic of Congo

J.M. Tshitenge
University of Kinshasa; CB-HYDRONET, Kinshasa, Democratic Republic of Congo

M. Werner
UNESCO-IHE, Delft, Netherlands; Deltares, Delft, the Netherlands

F. Wetterhall
European Centre for Medium-Range Weather Forecasts, Reading, United Kingdom

G.J. Wiche
U.S. Geological Survey, Bismarck, ND, United States

FOREWORD

Why read this book? This book is about the development and implementation of operational systems for predicting future flood events in an effort to reduce flood damages and to enable flood risk management at regional and national scales. Flood forecasting methods and systems generally exploit the most current developments in meteorology, hydrology, remote sensing, computer science, and decision analysis, as well as risk communication and management. Such systems literally define the current state of knowledge in those fields because they tend to employ the most sophisticated state-of-the-art advances for the grand societal purpose of flood risk management. As two professors who are often charged with introducing and motivating future scientists and engineers to embark on careers in hydrology and hydrometeorology, we often use existing flood forecasting systems to describe the state of the art of our profession because they include climatic and weather forecasts, rainfall-runoff hydrologic watershed models, reservoir systems models, river hydraulic models, along with very sophisticated graphical user interfaces. Thus such systems integrate most of the topics and developments in the fields of hydrology and hydrometeorology. What better way to introduce and impress students about the extraordinary sophistication and societal value of our field to encourage future careers in this area?

This is not a book about the theory of flood forecasting; rather, it is about actual flood forecasting systems written and edited by two authors who have spent their entire careers creating such systems in the United States and Australia. This book provides a comprehensive review of developments in flood forecasting and flood risk management using experience written by experts of systems in the United States, China, Russia, Britain, Israel, Germany, Europe, Brazil, Australia, Congo, and the Sudan. It focuses on national and regional flood forecasting systems, which tend to be the most sophisticated among existing flood forecasting systems, though there are exceptions, and there are also a myriad of small flood forecasting systems, globally.

Flood forecasting systems can be developed over a wide range of temporal scales ranging from hours to days, or even months, depending on the spatial scale of interest and/or the availability of reliable weather forecasts. Rather than providing an overview of this book, which is given in the next introductory section, we use this foreword to emphasize a few aspects of flood forecasting, not commonly discussed, which could benefit from further research and development:

Future advances in flood forecasting and water management: The skill associated with weather forecasts has improved considerably over the last 3 decades. Improvements in weather forecasts, combined with advances in monitoring, remote sensing, data collection, and models, have led to concurrent improvements in our flood forecasting skill. Unfortunately, many reservoirs are still operated using fixed "rule curves" designed decades ago. Similarly, for most existing reservoirs, flood control storage requirements were determined from a long-range frequency analysis of *n*-day flood volumes. As a result, many existing reservoirs have never experienced spills! For such systems, and many others, there is vast potential for developing dynamic rule curves that provide dynamic changes in flood control storage, while concurrently allocating the additional water for the benefit of society, including human and ecological requirements, all conditioned on flood forecasts. Thus a basic and open research question involves an assessment of methods for exploiting improvements in flood forecasting skill for altering the flood control space allocation without increasing the current downstream flood risk, yet leading to concurrent improvements in human and ecological flow needs. What we suggest is a shift in flood control operations and management practices that effectively blends improvements in flood forecasts with operational guidelines for flood control systems. Such a shift may become increasingly important as society attempts to account for increasing concerns over nonstationary hydrologic behavior resulting from climate change, urbanization, and other anthropogenic influences on flood control systems.

Seamless integration of streamflow forecasts over multiple time scales: Short-term flood forecasting is useful if the flood event exceeds a critical threshold, particularly during tropical storms and hurricanes. Such forecasts are particularly useful for issuing flash flood warnings for watersheds with little or no flood storage. However, for watersheds with considerable flood storage capacity, such forecasts are of limited value because existing flood storage will tend to dampen flood impacts. For systems with significant flood storage capacity, longer-term streamflow forecasts at monthly to seasonal time scales may be more useful. Thus, what is really needed are continuous streamflow forecasting systems which span seamlessly from short-term meteorological to longer-term climate time scales.

Flood forecasts for run-of-the-river hydropower and other water resource systems: Currently, operational flood forecasts tend to target large watersheds.

Most run-of-the-river hydropower systems are on upstream reaches; thus such systems are not commonly included in operational flood forecasts. Continuous streamflow forecasts would be of immense value for such locations, since they provide useful information on available hydropower potential for meeting peak loads in addition to information on imminent floods. Similarly, the National Weather Service commonly develops flood forecasts for a variety of reservoir systems in their effort to delineate flood risks for flood-prone regions, yet remarkably, agencies charged with reservoir flood control operations (ie, the U.S. Army Corps of Engineers and the U.S. Bureau of Reclamation) do not usually employ such forecasts for their own reservoir operations. Clearly there is room for improvement.

Data assimilation: Assimilation of observations into models is a well-established practice in the meteorological community. However, it is still in its infancy in streamflow forecasting due to limited data availability on land-surface states. Recent advances in the availability of remotely sensed soil moisture and other variables provide opportunities; however, the space-time availability of such remotely sensed data does not usually match well with hydrologic model data requirements. An important opportunity exists for the use of observed streamflow and groundwater states to enable real-time updating of the land-surface states of the hydrologic model.

Mobile technologies for flood warning and risk management: A recent PhD student at the University of North Carolina in Chapel Hill, Maura Allaire, performed detailed surveys during the 2011 Bangkok flood. Interestingly, this was one of the first major flood disasters to affect a city connected to social media; thus her study was the first to examine the role of the Internet in reducing flood disaster losses. She provides evidence that social media led to reductions of flood damages on the order of 20–30% by providing localized and real-time updates on flood locations and depths, enabling residents to move their possessions to higher ground. Thus it is extremely important to develop streamflow forecasts that may be customized for access using easily and commonly used mobile-based maps and other applications. This has enormous value from many perspectives, including recreational and many other uses, when issuing flash flood warnings. Ideally such forecasts for any given location should be given with verification statistics discussing their past performance. Where available, it is important to integrate such verification statistics within the forecast portal.

A full economic analysis of flood damages: A recent paper in the journal *Water Resources Research* titled "Economic costs incurred by households in the 2011 Greater Bangkok flood" presents a comprehensive approach for evaluating both the financial and economic costs experienced by households during and after a flood. Prior to this study, a full economic accounting of flood damage costs may never have been achieved. Such an analysis includes, in addition to the usual infrastructure and other flood prevention costs, *ex post* flood losses, compensation payments received, and any new income generated during the flood. For example, in the Bangkok flood, the post damage costs were the largest component of the overall flood costs. Such findings could provide very important inputs for the evaluation of future flood control mitigation and preventive measures. Analogous to the need for a full economic accounting for environmental damages when performing benefit-cost analyses, there is a need to incorporate a full economic and financial accounting of flood damages to enable more efficient flood forecasting and flood risk management systems.

Summary: This book summarizes extensive experience with actual flood forecasting and risk management systems. Within the myriad of institutional settings discussed throughout this book, we see this book as both a summary of our past experience in flood forecasting, as well as an opportunity for the hydrometeorological community to extend existing practices to enable streamflow forecasts over a broad range of spatial and temporal scales for the benefit of society. In an era of increasing concern over climate change and variability, improvements in flood forecasting take on even greater importance. Providing such streamflow forecasts in an easily accessible platform (eg, mobile technology) and extending such forecasts to serve the reservoir operations community should provide benefits beyond flood warning and control, including benefits to energy, agriculture, water supply, and ecologic regimes. This book, *Flood Forecasting: A Global Perspective*, gives an excellent and balanced summary of the field, which should provide a tremendous resource for those interested in improving our ability to manage future flood risks.

Richard M. Vogel
Professor, Tufts University.
Email: Richard.Vogel@tufts.edu.

Sankarasubramanian Arumugam
Professor, North Carolina State University.
Email: sankar_arumugam@ncsu.edu.

ACKNOWLEDGMENTS

We want to thank the staff at Elsevier for their help, patience, and guidance through the process of getting this book completed, particularly Rowena Prasad. We are, of course, extremely appreciative of the work and dedication of our colleagues who have contributed to the book chapters. After all, it is their individual contributions that cause this book to have any value. We would all be remiss for not acknowledging the contributions of all the engineers and scientists developing the flood forecasting systems that aim at the protection of lives and property of those we serve.

Flood Forecasting: A Global Perspective

T.E. Adams III*, T.C. Pagano†
*University Corporation for Atmospheric Research, Boulder, CO, United States
†Australian Bureau of Meteorology, Melbourne, Australia

1 INTRODUCTION

Throughout history, civilizations have had to cope with the consequences of flooding. Accounts of flooding and their impacts along the Nile River in Egypt and in Mesopotamia from the Euphrates and Tigris Rivers have been documented by ancient Greek philosophers (Kirk, Raven, & Schofield, 2013). Passages in *Homeric* poems and writings by *Herodotus* (484–25 BCE) postulated the existence of *Okeanos*, a river surrounding the Earth, as the source of all waters and, by extension, tried to understand and explain flooding. Thales (624–546 BCE) proposed that flooding of the Nile was caused by the *Etesian* winds (strong, dry north winds originating over the *Aegean Sea* from about mid-May to mid-September), which prevented the river from discharging out to the Mediterranean Sea. Similar ancient accounts of great floods and their causes can be found in other cultures and civilizations. The underlying themes are the occasional devastating effects that flooding has on human populations and the ongoing need to understand and predict the occurrence of floods. To that end, nothing has changed since ancient times.

Flooding is a serious and expensive issue. In 2013, flooding was responsible for nearly half of all natural catastrophe-related losses (Swiss Re, 2013a). The 2011 Thailand flood caused US$48 billion in losses and left 5 million people homeless (Swiss Re, 2013b). Climate change may increase the intensity of floods in some regions (Hirabayashi et al., 2013). Flood forecasting and warning systems have been shown to reduce impacts and save lives, and these systems are an increasingly important tool for water managers and emergency response services.

The distinct advantage we have over our predecessors is the advancement of science and technology. Better scientific understanding of physical phenomena and processes allows us to produce better and more realistic models, improve observational measurements of hydrometeorological variables, and improve prediction of model inputs. Dramatic advancements in affordable technology have made the application of modern hydrometeorological science and hydraulics to flood prediction possible due to three factors: (1) the availability of very fast computers, with significantly more memory and data storage capability than was available even just 10 years ago; (2) widespread availability of high-resolution geophysical data sets from remote-sensing sites, which are needed for model parameter estimation and calibration; and (3) highly reliable telecommunications systems for data transmission from ground-based and satellite data collection platforms (DCPs). Furthermore, the expansion of the Internet and the proliferation of mobile phones have had a dramatic impact on the distribution of flood warnings.

The intent of this book is to catalog public, national, and regional-scale flood forecasting systems, providing a snapshot of current systems while illuminating the historical context from which the various systems arose. The emphasis is on the current state-of-practice in the field, understanding systems as they are now and the real-world challenges they face. This understanding helps us identify gaps and opportunities for forecasting improvements, some of which are being addressed by the emerging systems described near the end of this book.

The book is divided into four sections:

(1) National and regional flood forecasting systems
(2) Continental modeling and monitoring — the future?
(3) Challenges facing flood forecasting
(4) Forecast dissemination issues.

Summaries of these topics are below, with commentary on common themes distinguishing both similarities and differences between the different forecast systems.

An unexpected challenge in assembling this book was the lack of documentation of existing systems. It was sometimes difficult to identify who would be available within a given country to describe actual operational practices. Some waterways are forecasted independently by multiple agencies. This is particularly true in countries with both national and regional forecasting authorities with varying mandates and resources. Some national agencies have regional offices with practices that vary from one another, partly due to local climate and the needs of local customers, but also because of the legacy of independently developed systems.

At the opposite extreme, some waterways have no hydrological forecasting authorities at all, possibly because of a lack of resources and/or data. In some countries, meteorological services provide generalized flood advice without doing hydrologic analysis or using river data, and instead base the assessment solely on rainfall. In some cases, flood forecasting is considered a technical specialist activity to support the functions of government, with forecasters having limited or no direct contact with the public or the media. Furthermore, language barriers are a challenge for anyone attempting to assemble an international compilation.

Even when there is a common language, the challenge of finding documentation points to a general lack of international communication on the subject of national and large regional flood forecasting systems, particularly with the aim of sharing ideas and information. Historically, many forecasting systems were developed in isolation, developing custom solutions to address the challenges at hand. Forecasting agencies, like many large and small organizations, may sometimes have an attitude of "not invented here," which underpins a preference for in-house development over the work done by third parties such as researchers, companies, or other agencies. This attitude is exacerbated by the notion that "all hydrology is local," and that neighboring and distant countries do not have similar enough problems to make it worthwhile building external partnerships. The result can be a duplication of effort, reinvention of methods, and needless repetition of mistakes.

We hope the chapters following in this book demonstrate that some of the challenges facing flood forecasters are nearly universal and that some agencies are successfully collaborating to improve their systems and services. For example, we are encouraged by the adoption of common forecasting software by dozens of agencies. Furthermore, nations are pooling their resources to develop trans-national and global "top cover" forecasting systems. These affordable and adaptable technologies make the application of scientific advancements possible. Additionally, the benefits of exchange are not just limited to better models producing more accurate forecasts — there is much information to share about interfacing with end-users, distributing warnings, communicating certainty, and so on.

2 NATIONAL AND REGIONAL FLOOD FORECASTING SYSTEMS

The book chapters covering operational national and regional flood forecasting systems illustrate that no single system provides complete operational solutions that are applicable universally. We show that creative solutions to common issues, with many different approaches, can successfully solve difficulties inherent in the development, implementation, and operational success of national-scale flood forecasting systems. For example, data collection, transmission, quality control (QC), processing of data into needed formats for model use and data storage must be adequately handled by all flood forecasting systems. The World Meteorological Organization (WMO) has a manual that includes recommended features and methodologies for the development and implementation of flood forecast systems (World Meteorological Organization, 2011). However, the extent to which the WMO manual is used in constructing forecast systems is unknown.

To be sure, some issues in some countries do not exist. For example, frozen ground and river ice effects are not issues in tropical climates and, generally, nor are snow effects and glacial melt. Often simply the historical context, which could involve the effect of national or regional cultural norms, surrounding the development of a national or regional flood forecast system determines that one solution to a problem is applied to the exclusion of others. To this end, some differences between systems can result from a range of factors (eg, personal preferences, relationships) that are not strictly objective criteria.

The national-level systems described are not complete. Other national-level systems and regional centers exist in, for example, Canada (Environment Canada), Sweden (Swedish Meteorological and Hydrological Institute, Arheimer, Lindström, and Olsson, 2011), Romania (Romanian *Institutl National de Hidrologie si Gospardarire a Apelor* (IHNGA)), and others. Unfortunately, time constraints prevented potential authors from contributing to this book.

Globally, many local, primarily urban, flood forecasting systems have been developed and implemented by consultancies. While these systems provide tremendous value to communities, descriptions of these systems are not included to avoid the appearance of favoritism — due to the inevitability that some may be inadvertently excluded. Furthermore, this book necessarily focuses on riverine flooding and places less of an emphasis on coastal flooding from storm surge, tsunamis, or flooding due to sea-level rise as the consequence of a warming climate. Each of these is critically important and should be covered in separate volumes dedicated to those topics alone.

Consequently, we include chapters on:

(1) *Australian Bureau of Meteorology flood forecasting and warning* (Australia);
(2) *Hydrological forecasting practices in Brazil* (Brazil);
(3) *The development and recent advances of flood forecasting activities in China* (People's Republic of China);
(4) *A diagnostic analysis of flood forecasting systems for the Congo River Basin* (Congo);
(5) *Flood forecasting in Germany — challenges of a federal structure and transboundary cooperation* (Germany);
(6) *Operational flood forecasting in Israel* (Israel); *WRF-Hydro and Tel Aviv systems*;
(7) *Operational hydrological forecast systems in Russia* (Russia);
(8) *Transboundary river basin flood forecasting and early warning system experience in Gash River Basin, Horn of Africa*;
(9) *Flood forecasting — a national overview for Britain* (United Kingdom); *National and Regional systems*;
(10) *Flood forecasting in the United States NOAA/National Weather Service* (United States);
(11) *On the operational implementation of the European Flood Awareness System* (EFAS).

We have included discussion of European Flood Awareness System (EFAS) in this section as well as the subsequent section on Continental Modeling and Monitoring because of the robust operational standing of EFAS and utility among European Union (EU) member nations to supplement national flood forecast operations. There is representation for systems on each continent, with the exception of Antarctica.

Table 1 compares features of the national systems featured in the book. The intent is to show both the similarities and dissimilarities of the systems, which address common operational hydrologic concerns, albeit, with different approaches. We emphasize that what we present should *not* be interpreted as a "score card." Instead, we think the side-by-side comparisons demonstrate both strengths and weakness of the systems and how there can be multiple solutions to solving common problems, depending on local or regional perspectives.

The forecast systems for Colombia, the Congo and Ganges Basins are either pilot projects or not yet fully operational. Nonetheless, we include a discussion of the work being done in these regions to emphasize a largely global problem in many less-developed countries that relates to a lack of infrastructure and capital to fund the development of modern flood forecast systems.

Table 1 Comparison of National flood forecasting systems featured in this book. Some information is not applicable, so it is left blank intentionally

	Australia Flood	Australia Short-term	UK National	UK Regional[a]	EFAS	Colombia	Israel, WRF-Hydro	Israel, Tel Aviv	Russia flood[b,c]	Brazil[a]	China[a]	Germany[a]	United States	Congo
Domain: Regional (R), National (N), Continental (C), Global (G)	R[c]	R[c]	N	R	C	R[c]	N	R	R[2]	R	N	N	R,N	R,N
Climate: Tropical (Tr), Dry (D), Temperate (Te), Continental (C), Polar/Alpine (A)	Tr, D, Te	Tr, D, Te	Te	Te	Te, C, A	Tr, D, Te	D, Te	D	D, Te, C, A	Tr, D, Te	D, Te, C, A	Te, C, A	Te, C, A	Tr, Te
Observed rainfall forcing: In Situ (I), Radar (R), Satellite (S)	I	I	I,R	I,R	I	I	I	I	I	I,S	I	I,R	I,R,S	I, S
Rainfall-runoff models: Lumped (L) Semi-Distributed (S) or Gridded (G)	S	S	G	L,S	G	S	G	S		L,S	L,G	S	L,G	S
Streamflow routing: None (N), Hydrologic (H), Hydraulic (D)	H	H	H,D	H,D	D	D	H	D		H,D	H	H,D	H,D	H
Real-time flood inundation mapping?	No	No		Some	No	No	No	No		No	No	No	No	No
Reservoir simulation?	Yes	No		Some	Yes	Yes	No	Yes		Yes	Yes	Yes	Yes	No

Coastal forecasting: None (N) Tides (T) Storm Surge (S)	T	N	T,S	T,S	N		N			N	N	T,S	T	N
Snow modeling?	No	No	Yes	Some	Yes	No	No	No	Yes	No		Yes	Yes	No
Software: In-house development (I) or "Off the shelf" (O)	I, O	I, O	O	I,O	I	O	O	O	I	I,O		I,O	I,O	I, O
Maximum lead time	Variable[d]	7 days	6.5 days	10 days	46 days	6 days	4 days	Hours	3 days	14 days		7 days	10 days	
Forecast frequency	As needed	Daily	6 Hourly	4 Hourly or more frequent	12 Hourly	As needed	12 Hourly	12 Hourly	As needed	As needed		As needed, from hourly to daily	Twice daily[e]; As needed during flooding	None
Numerical weather guidance (spatial resolution, maximum leadtime)	40 km, 10 days 12 km, 3 days 4 km, 1.5 days	40 km, 7 days	1.5 km, 4 km, 5 days	4 km, 10 days 1.5 km 12 hours	16 km, 15 days 9 km, 10 days 7 km, 5.5 days	15 km, 3 days 5 km, 3 days 1.7 km, 3 days	27 km, 3 days 9 km, 3 days 3 km, 3 days	27 km, 3 days 9 km, 3 days 3 km, 3 days	14 km, 3 days 1 km, 1.5 days	15 km, 7 days		13 km, 7 days 7 km, 3 days 2.8 km, 27 hours	4 km, 3days 28 km, 16 days 70 km, 14 days	None
Precipitation forecast: Numerical Weather Prediction (NWP), Expert (E)	NWP,E	NWP	NWP	NWP,E	NWP	NWP	NWP	NWP	NWP	NWP,E		NWP, E	NWP, E	
QPF processing: Raw (R) downscaled (D), bias-corrected (B)	R	R	R, D	D	B	R				R		R	R	
Forecaster Interaction? Manual (M), Engaged (E), Automated (A)	M	A	A	M,A, E	A		A	A		M,A,E		M,E, A	M	

(Continued)

Table 1 Comparison of National flood forecasting systems featured in this book. Some information is not applicable, so it is left blank intentionally—cont'd

	Australia flood	Australia short-term	UK national	UK regional[a]	EFAS	Colombia	Israel, WRF-Hydro	Israel, Tel Aviv	Russia flood[b]	Brazil	China	Germany	United States	Congo
Verification: None (0), Some (+), Extensive (++)	+	++	++	+	++					+		+	++	
Publicly available forecasts on Internet?	Some	Yes	No	Planned	No, after 30 days	No				Some		Yes	Yes	No
Stakeholder product form: Warnings (W) Hydrographs (H) Maps (M)	W	H	M	W,M	W,H,M	H				H		W, H, M	W,H,M	
Stakeholder products: Categorical (C) Deterministic (D) Probabilistic (P) Ensemble (E)	C, D	D	C	D,P,E	C,P					D,P,E		C,D	C,D,P,E	
Internal guidance: Categorical (C) Deterministic (D) Probabilistic (P) Ensemble (E)	D	D, E	D,P,E	D,P,E	P,E	D,P				D,E		C, D,P, E	C,D,P,E	

[a] Multiple systems exist within a country
[b] Russian floods also happen on seasonal timescales associated with snowmelt, which is covered by a different forecasting system
[c] Regional: A national agency forecasts although geographic coverage is not complete within the country
[d] Variable: Days, but weeks for largest inland rivers
[e] Variable, depending on the River Forecast Center

The description of the various fields in Table 1 follows. Some entries are blank because of unavailable information:

Domain: How large is the geographic domain of the forecasting system? Does it cover individual catchments within a region, most of a nation or a continent? A number of systems operate at a federal/national scale but may have geographic diversity or not cover the entire nation (possibly because of the lack of significant population centers in certain areas, such as the interior of Australia).

Climate: What is the dominant Köppen climate classification for the area covered by the forecasting system? Some operational systems are shaped around the primary climate of the region, whereas some countries have to provide services in a variety of climates.

Observed rainfall forcing: Do the forecasters typically rely on direct measurements from raingauges, or make routine and quantitative use of estimates from radar or satellites? Although not listed in this table, some agencies use radar and satellite data qualitatively to develop situational awareness. However, in situ raingauge measurements remain the primary data source for all agencies surveyed.

Rainfall-runoff models: Is the rainfall-runoff model spatially lumped (ie, uses a single catchment average rainfall time series to drive a model to forecast at the catchment outlet) or spatially distributed/gridded? Semi-distributed models often have multiple irregularly shaped but hydrologically homogeneous regions, the results of which are routed through the river channel network to the outlet. Gridded models are similar to semi-distributed models but have a regular grid spacing (eg, every 5 km). Some gridded models produce runoff to be used as inputs to a separate channel routing model, whereas others have lateral subsurface interactions between soil moisture in adjacent grid cells. Nearly all agencies had some spatial representation in their rainfall-runoff models and many agencies ran both lumped and distributed models side-by-side.

Streamflow routing: What kind of model, if any, is used to convey runoff from upstream to the forecast locations of interest? Lumped models do not have a separate stream routing model. Hydrologic routing (eg, the Muskingum method) treats an entire stream reach as a single storage unit. More complicated hydraulic routing models solve partial differential equations of fluid flow — consequently, they require more data related to river geometry and morphology and take longer to run. When computing time was limited, hydrologic routing was the standard technique, but now over half of the agencies are operating hydraulic models in at least some locations.

Real-time flood inundation mapping: Does the agency produce real-time maps of flooded areas? Inundation maps are a standard tool for designing flood defenses and regulating building in the floodplain. They help local emergency services understand what structures are at risk and how deep the water will be in various places in the community (the alternative is a prediction of flood depth at the river gauging station). However, inundation models are currently extremely expensive to run in real time and almost no agencies do this. In a few cases, agencies provide predictions at river gauging stations and then rely on a pre-run library of multiple inundation scenarios to estimate the likely extent of the flooding.

Reservoir simulation: Do the forecasting agencies simulate the behavior of reservoirs so as to generate forecasts downstream? These may be simple models that assume that all the incoming streamflow is captured until the reservoir fills and begins spilling. Other models reflect more complicated operating rules. Only a few agencies do not simulate reservoirs, primarily in places where the flow at river forecast locations is not significantly regulated by dams.

Coastal forecasting: When rivers meet the coastline, how is the interaction with the ocean modeled, if at all? Is the forecast river height adjusted by predictable astronomical tides? Or is there a more dynamic model that accounts for storm surge, such as for the landfall of a hurricane/tropical cyclone? Some agencies do not forecast near coasts and therefore do not require coastal models. However, for the remaining countries, nearly all consider tides and a few consider storm surge.

Snow modeling: Does the hydrologic modeling system include the accumulation and melting of snowpack? Whereas rainfall-runoff models typically focus on rainfall and evaporation, snow models must also consider the energy balance at the surface, often requiring data about temperature. Those countries where snow is an important component of the hydrologic cycle typically model snow.

Software: Is the core of the model done using a widely used commercial or free "off the shelf" software system, or is it a custom system built by or for the agency? In many cases, the line between in-house and external development is blurring as agencies are increasingly using one or more custom modeling engines within widely available forecast production environments. For example, dozens of agencies use Deltares' Flood Early Warning System (FEWS) for data handling, workflow management, visualization, and product generation, but only a few agencies use the same the core modeling engine within FEWS (eg, Australia and the Mekong River Commission use the FEWS environment and share the URBS engine; Pagano, 2014).

Maximum lead time: If forecasts are produced regularly, is there a common maximum lead time used? Two to four days' lead time is typical of most flood forecasting systems, although some systems extended their lead time by using medium to long-range weather predictions. Furthermore, some countries have large basins that respond slowly, making long-lead time forecasts possible. This table does not include seasonal forecasting systems (ie, 1–9 months ahead).

Forecast frequency: Are the forecasts produced on an as-needed basis, or at a regular interval? About half of the systems surveyed produced forecasts only as needed during and in anticipation of flood conditions. Agencies that provide forecasts during fair-weather conditions are often serving a variety of customers beyond emergency responders, such as irrigators, hydropower generators, recreationalists, and others.

Numerical weather guidance: If the hydrologic models are driven by forecast variables (eg, precipitation), what is the spatial resolution and maximum lead time of the weather prediction? Impressively, all operational systems make quantitative use of weather forecasts and nearly all agencies rely on a variety of products of various resolutions. This ranges from relatively coarse resolution (>15 km) long-lead time (>5 days) to fine resolution (<2 km) short-lead time (<2 days) forecasts.

Precipitation forecast: Does the weather forecast come directly from one or more weather models? Or is an expert assessment of a meteorologist (often making an interpretation of one or more weather model outputs). While all agencies used weather model outputs, only a quarter of agencies emphasized the interface between hydrologists and meteorologists. There may be a variety of reasons for this — hydrologists and meteorologists may reside in different agencies, making it practically difficult to maintain a strong operational working relationship. However, some agencies use their weather forecasts in a way that precludes involving an expert meteorologist (eg, the weather model output is statistically post-processed using a method trained on an archive of past predictions from the same model).

Quantitative precipitation forecast processing: If numerical weather predictions (NWPs) are used to force the hydrologic model, what if any statistical processing is used to prepare the forcings? The weather prediction may be at a coarse spatial resolution and may be downscaled to resolve sub-gridscale features (such as rugged terrain). The weather prediction may also have known systematic biases that can be corrected. A few agencies post-process weather model outputs, but the majority rely on just the raw predictions. This is sometimes because the weather prediction models are rapidly changing and the necessary hindcasts are unavailable for training the processing method.

Forecaster interaction: Do the hydrologic models largely run on their own with complete automation (eg, on a scheduler at the same time every day)? Is a hydrologist required to make assumptions every time the model is run, manually tuning parameters in real time using expertise? Or is there a blend, with occasional manual corrections as needed to a mostly automated system? There is a relatively even number of agencies that emphasize hydrologist expertise in running the model iteratively to get the best model output compared to those agencies that look to the hydrologic model output as guidance to be interpreted for the end-user. In rare instances, agencies run an automated model and deliver the output to users unaltered as the final product. This is sometimes due to a lack of human resources to tend to all the detail of the model output, but can also be a statement of trust in the quality of the model output. However, many agencies label any "hands-off" publicly available hydrologic model output as experimental guidance or unofficial.

Verification: How much emphasis is put into regular and routine assessment of forecast accuracy? While nearly all agencies evaluate the quality of a model during its setup and calibration, some agencies have lacked rigorous verification of as-issued forecasts. In extreme cases, this could be due to a lack of archiving of past forecasts, which was regrettably common in a pre-digital era or when computer storage was prohibitively expensive. Some agencies have limited verification activities, such as occasional assessment of some forecasts for a specific high-profile event. Other agencies calculate myriad accuracy metrics and have sophisticated diagnostics of how the systems are performing.

Public availability: Are the agencies forecasts publicly available on the Internet? In many cases, agencies employ a funneled approach — the hydrologists are able to see the results a wide variety of tools and interpret them so as to generate a subset of products. They may produce an expanded set of products for registered users and the products are transmitted directly or available on a password-protected website. A number of agencies do not provide any public product, often because another agency (eg, emergency services) has responsibility for warning the public or a trans-national agency does not want to conflict with the messaging of local agencies.

Stakeholder product form: Is the primary method of communicating with stakeholders through warnings (text describing current and anticipated conditions, along with options for actions), hydrographs (time series charts describing river levels, perhaps with respect to critical thresholds), or maps (showing areas in danger)? Agencies use a variety of channels to communicate with stakeholders and sometimes an agency may use multiple channels. Warnings are often considered labor-intensive to produce but can be tailored to respond to the particular situation and include expert interpretation so as to inspire appropriate action by stakeholders. Hydrographs provide a rich source of information at a specific location, including the timing and magnitude of the flood. Maps also help in the assessment of how widespread the event will be or what particular communities will need the most attention.

Product communication style: Stakeholder products may describe flooding categorically (eg, minor, moderate, major flooding), deterministically (eg, the river will reach 4.5 m at 3 pm Sunday), probabilistically (eg, there is a 35% chance of flooding), or in an ensemble (ie, a set of many deterministic forecasts, each with some likelihood). Again, agencies use a variety of methods to communicate their products, with an even mix of agencies providing single-valued/single-category predictions, whereas others express certainty/uncertainty. Many agencies communicate in different ways simultaneously, which can be a communications challenge if the various products disagree (eg, the official deterministic forecast is different from the experimental ensemble forecast).

Internal communication style: Similar to the public communication style, but do the forecasters have access to a broader suite of guidance than the stakeholders? In many cases, agencies may generate ensemble or probabilistic forecasts for themselves, but not release the results to the public, sometimes because of a perception that such forecasts are too technically sophisticated and are prone to misinterpretation. In some cases, the hydrologists may use their expertise and situational awareness to convert the ensemble forecasts into deterministic or categorical forecasts of what they think will happen.

3 CONTINENTAL MODELING AND MONITORING

We examine the fast-growing development of global and continental-scale flood monitoring, modeling, and prediction systems. The chapter by Smith et al. (Chapter 11) describing the European Flood Awareness System (EFAS) provides an excellent overview of an operational continental-scale flood forecasting system. EFAS became operational in 2012 with several European organizations having responsibility for producing and providing the flood information to EU member countries. EFAS was developed as a complementary system for existing national and regional flood forecasting systems in EU countries. Similarly, the Global Flood Awareness System (GloFAS), which was developed jointly by the European Commission and the European Centre for Medium-Range Weather Forecasts (ECMWF), provides countries with information on upstream river conditions as well as continental and global overviews. GloFAS, found at http://www.globalfloods.eu, couples output from NWP model ensembles from the ECMWF *Ensemble Prediction System* with a hydrological model covering continental domains. Additional flood monitoring and prediction systems are in place operationally or as real-time demonstration or experimental projects that are not included herein as book chapters, but are mentioned below.

For example, the Global Flood Monitoring System (GFMS) (http://flood.umd.edu) is an experimental system, funded by the National Aeronautics and Space Administration (NASA), that uses real-time Tropical Rainfall Measuring Mission (TRMM) Multi-satellite Precipitation Analysis (TMPA) precipitation information as input to a quasi-global (50°N–50°S) hydrological runoff and routing model running on a 1/8th degree latitude/longitude grid. Consequently, since there is no use of forecasted precipitation, the utility of the GFMS is limited to flood predictions derived from observed precipitation only. Estimates of flood magnitudes are based on 13 years of retrospective model runs with TMPA input, with flood thresholds derived for each grid location using surface water storage statistics (95th percentile plus parameters related to basin hydrologic characteristics). Streamflow, surface water storage, and inundation estimates are made at a 1 km spatial resolution. The system makes available the latest maps of instantaneous precipitation and totals from the prior 1, 3, and 7 days. The underlying GFMS hydrologic model is based on the University of Washington Variable Infiltration Capacity (VIC) land surface model (Liang, Lettenmaier, Wood, & Burges, 1994). It is coupled with the University of Maryland Dominant River Tracing Routing (DRTR) model (Wu et al., 2011; Wu et al., 2014). The VIC/DRTR coupled model is referred to as the Dominant river tracing-Routing Integrated with VIC Environment (DRIVE) model. The flood detection algorithm is described in Wu, Adler, Hong, Tian, and Policelli (2012) and Wu, Kimball, et al. (2012). The real-time TMPA precipitation data product (Huffman, Adler, Bolvin, & Nelkin, 2010) is obtained from the NASA Goddard TRMM/GPM Precipitation Processing System. The new GFMS with the DRIVE model has been evaluated based on 15-year (1998–2012) retrospective simulation against more than 1000 gauge streamflow observations and more than 2000 reported flood events across the globe (Wu et al., 2014).

Although it provides more of an observational rather than predictive services, the Dartmouth Flood Observatory (http://floodobservatory.colorado.edu) uses satellite-based global remote sensing for fresh water flow measurement and inundation mapping in "near real time" (see, eg, Adhikari et al., 2010; Brakenridge et al., 2011; Cohen et al., 2012; Khan et al., 2011, 2012). Since 1998, the Flood Observatory has used orbital sensors to gather and store, in a permanent archive, globally consistent information concerning surface water flow and flood inundation. The Flood Observatory is engaged in hydrological research in the area of surface water variability, using both remote sensing and modeling for the estimation of river/stream flow, which is critically important in remote areas where streamgauging is sparse or nonexistent. The Flood Observatory collaborates with humanitarian and water organizations by providing critically needed river discharge and flood inundation information in remote regions and where data are difficult to collect. Processing of optical and Single Aperture Radar imaging and satellite-based systems, such as MODIS (late 1999) and VIIRS and their automated data distribution systems, permit assessment of how climate and land cover directly affects flood and drought frequency and severity through observed long-term means and extremes of river discharge and lake water storage. The impacts of climate change cannot be accurately predicted without adequate knowledge concerning present conditions and rate of change of hydrologic systems. With its reliance on remotely sensed observational data, the Flood Observatory can provide data required for the validation of hydrologic models that predict runoff and streamflow where none would exists otherwise.

Flash flooding has long been a concern (see, eg, American Meteorological Society, 2000; Gaume et al., 2009; Ruin et al., 2009). Historically, flash flood monitoring and forecasting have often been restricted to local-scale systems because of the detailed local data and knowledge necessary to set up effective systems. This is increasingly changing, with the emergence of national and continental flash flood systems. Many countries have used technological and scientific advances found with the availability of high spatial and temporal resolution radar rainfall estimates, digital terrain models, distributed hydrologic models, use of geographic information system based analyses, etc. to greatly improve flash flood forecasting capabilities. One such national-scale system is the United States, National Oceanic and Atmospheric Administration (NOAA)/National Severe Storms Laboratory (NSSL), Flooded Locations And Simulated Hydrographs Project (FLASH) (https://blog.nssl.noaa.gov/flash) system. Ntelekos, Georgakakos, and Krajewski (2009), Clark, Gourley, Flamig, Yang, and Clark (2014), and

Gourley, Flamig, Hong, and Howard (2014) found that NWS flash flood alert and warning systems based on Flash Flood Guidance (FFG) produced by River Forecast Centers (Chapter 10) performs relatively poorly in predicting flash flooding and that there is considerable uncertainty associated with the predictions. FLASH was developed, in part, to address the inadequacies of NWS FFG.

Key to the evaluation and verification of flash flood prediction systems is a comprehensive database of flash flood events. The US National Flash Flood database (Gourley et al., 2013) is a georeferenced, long-term, detailed characterization of flash flooding in terms of spatiotemporal occurrence and specific impacts. The database is composed of three primary sources: (1) the entire archive of automated discharge observations from the US Geological Survey that has been reprocessed to describe individual flooding events; (2) flash-flooding reports collected by the National Weather Service from 2006 to the present; and (3) direct reports from flash flood witnesses obtained directly from the public in the Severe Hazards Analysis and Verification Experiment during the summer months from 2008 to 2010. This database has served as a basis to evaluate existing NWS flash flood prediction systems, to identify strengths and weaknesses and to serve as validation of the FLASH system.

Similarly, the European HYDRATE (http://www.hydrate.tesaf.unipd.it) initiative has the stated objective to improve the scientific basis of flash flood forecasting by understanding past flash flood events. A European-wide flash flood observation methodology, using a network of existing hydrometeorological observatories located in high flash flood potential regions, is being combined with the development of technologies and tools to produce early warning systems. The observation strategy within the HYDRATE program is utilization of radar, traditional hydrometeorological monitoring data, and concurrent post-event surveys. These data will be stored in the HYDRATE European Flash Flood Database, which will be made freely accessible to the international research community.

Another example of continental-scale hydrometeorological observation system is the NOAA/NWS Snow Data Assimilation System (SNODAS) at National Operational Hydrologic Remote Sensing Center (NOHRSC). SNODAS provides daily, gridded estimates of snow depth, snow water equivalent (SWE), and related snow parameters at a 1-km^2 resolution for the conterminous United States (CONUS) (Carroll et al., 2001; Carroll, Cline, & Li, 2000). SNODAS ingests national, near real-time, ground-based hydrometeorological data sets, real-time, NWP model data sets, which are used as forcings to a physically based, snow-model as part of the snow-data-assimilation system. The near real-time data include ground-based, airborne, and satellite snow cover observations are assimilated into the gridded fields generated by the snow accumulation and ablation model (Clow et al., 2012). Major system components include (1) data ingest, QC, and downscaling procedures; (2) an energy- and mass-balance, spatially uncoupled, vertically distributed, multilayer snow model; and (3) snow-model-data assimilation and updating procedures. The model incorporates the mathematical methods of Tarboton and Luce (1996) to handle the snow surface temperature solution. The modeling approach of Jordan (1990) is used to address the snow thermal dynamics for energy and mass fluxes as represented in SNTHERM.89. The methods developed by Pomeroy, Gray, and Landine (1993) are used to account for net mass transport from the snow surface to the atmosphere through sublimation of the snow surface transport processes. The snow model, with three major-layer state variables of water content, internal energy, and thickness, is forced by CONUS, hourly, 1 km, gridded, meteorological input data downscaled from the mesoscale NOAA/NWS NCEP NWP model (RUC2 — http://ruc.noaa.gov) analyses. The model generates total SWE, snow pack thickness, and energy content of the pack, several energy and mass fluxes at the snow surface and between the snow and soil layers.

Recently, the NOAA/NWS National Water Center (NWC) has been established as the centerpiece of the hydrologic services program within the overall NWS reorganization and serves as the foundation of the NWS' Integrated Water Resources Science and Services (IWRSS) initiative (Office of Hydrologic Development, 2009). Key to IWRSS is the development of a comprehensive real-time national hydrologic modeling system, which, in time, aims to include coupling of all elements of the hydrologic cycle, including land surface, infiltration, subsurface and groundwater flow, channel routing, snow accumulation and melt, and estuarine and costal processes. The national model will include reservoir simulation and comprehensive handling of water resources related activities for irrigation, water supply, and, ultimately, water quality applications. The modeling system is being designed to provide "street-level" flood prediction to the public and other targeted end-users. The NWC will provide single-valued and ensemble hydrologic forecasts ranging in time scales from day 1 to months.

Overall, the current state of operational, large-scale flood forecasting systems is summarized by Emerton et al. (2016). They describe similarities and differences of six recently implemented operational large-scale systems, including an evaluation of the methodologies of the forecast systems used to forecasting floods at continental and global scales. The authors state, "Operational systems currently have the capability to produce coarse-scale discharge forecasts in the medium-range and disseminate forecasts and, in some cases, early warning products in real time across the globe, in support of national forecasting capabilities." The implication is that these continental and global systems have some utility by providing a longer lead time — on the order of 1 week or longer — flood risk outlooks, aiding more regionally based national systems, but lack sufficient detail and accuracy to supplant regionally based national flood forecasting systems.

3.1 Challenges Facing Flood Forecasting

In a survey of operational forecasting systems in developed and developing countries, Pagano et al. (2014) found four common challenges (with 18 subthemes): (1) making the most of available data; (2) making accurate predictions using models; (3) turning hydrometeorological forecasts into effective warnings; and (4) administering an operational service.

A major challenge facing flood forecasting is the lack of high-quality hydrometeorological data of sufficient spatial and temporal resolution for model parameter estimation and as real-time model forcings. Tshimanga et al. (Chapter 4) describe a 95% decline in streamflow gauges in the Congo in the past 50 years, where a significant amount of historical data storage is still on paper records. Large parts of Canada and northern Russia have lost around 40% of their streamflow gauges in recent decades.

The challenge is also significant in seemingly data-rich countries, as they attempt to run highly demanding models. A good example of this is the data needs for real-time flood inundation mapping utilizing an unsteady hydraulic flow model, as shown in Fig. 1, which are

1. tributary inflows, lateral inflows, upstream flow boundary, downstream water level boundary;
2. high-resolution (LiDAR) digital terrain model;
3. high-resolution channel bathymetry;
4. georeferenced detailed building and other structure locations (building footprints);
5. detailed levee location and heights; and
6. control structure details.

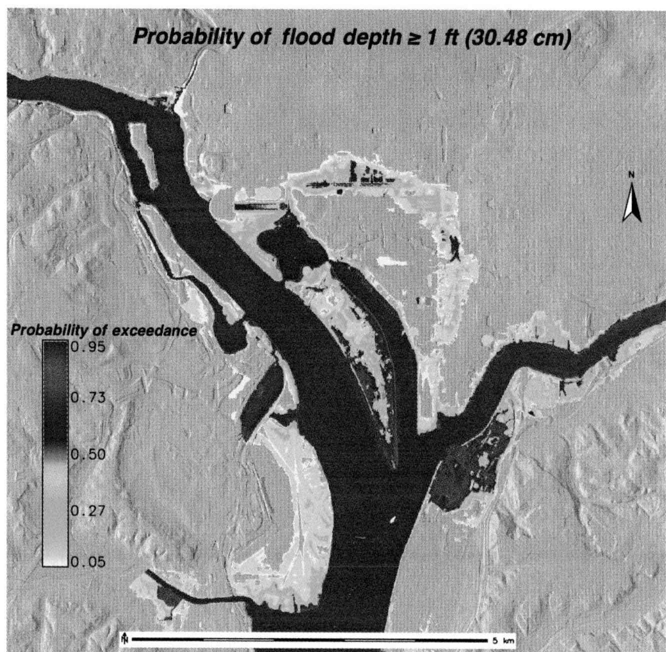

Fig. 1 Map showing the probability of exceedance for a flood event in the Washington, DC, USA area from the Potomac and Anacostia Rivers using the US Army Corps of Engineers (USACE) HEC-RAS 1-D, unsteady flow hydraulic model. HEC-RAS flow inputs are from the NOAA/NWS RDHM distributed hydrologic model. A depth of 1 ft is 30.48 cm.

In addition:

1. all data must be georeferenced with a common vertical and horizontal geodetic datum at the same distance-preserving map projection;
2. all data must be available for model development and calibration; and
3. forcing data must be available in real time.

Even where such data are available, the costs associated with obtaining the data and setting-up a modeling framework and forecast system to produce routine operational forecasts over broad geographic regions or on a national basis can be substantial. However, within the proper context and with sufficient resources detailed model implementations are highly feasible. For instance, Adams, Chen, Davis, Heim, and Young (2010) and Adams, Chen, Heim, Davis, and Young (2012) described the implementation of a detailed 1-D unsteady hydraulic flow model in an operational setting for routine daily forecasts in a highly complex setting with bridges, levees, off-channel storage areas, locks, and dams for navigation and upstream flood control reservoirs for the mainstem of the Ohio River and major tributaries. It is not that producing a forecast of the type shown in Fig. 1 is difficult, it is the associated costs that make doing so extremely difficult, particularly in developing or emerging nations.

Previously, many of the forecasting systems described in this book were limited by available computing power; this problem is lessening with technological improvements. With sufficient computational efficiency, operational modeling environments can produce multiple forecast updates daily, over large geographic regions. Computational efficiency is particularly important if, as is shown in Fig. 1, probabilistic flood inundation mapping is desired using hydrologic ensembles, which implies the need to combine flood inundation maps from each of the hydrologic/hydraulic model ensemble members. Further, mapping software must be capable of running efficiently within the forecast operations workflow. Additionally, maximum water level data must be gathered to verify model inundation predictions.

Wiche and Holmes (Chapter 13) detail the challenges faced with streamgauging in the United States by the US Geological Survey to obtain accurate records of flow, stage (water level) and rating curves. These data are critically important to assist in hydrologic and hydraulic model calibration, real-time hydrologic forecasting, and real-time model evaluation. Such streamgauging difficulties are universal.

Referring to Table 1, another significant concern is that most forecast systems rely on the use of QPF, which is known to have associated significant errors (see, eg, Diomede, Nerozzi, Paccagnella, & Todini, 2008; Ebert, Damrath, Wergen, & Baldwin, 2003) in terms of the placement and magnitude of precipitation, and this has implications for hydrologic forecasts (Danhelka, 2004; Nurmi et al., 2010; Xuan, Cluckie, & Wang, 2009). This leads to the overall problem of data QC and maintenance of DCPs, also known as automated weather stations. Many of the chapters in this book describe the large effort put into quality controlling in situ data for use in modeling. Hydrologists spend substantial time checking, cleaning, infilling, using, archiving, and redistributing hydroclimatic data. Poor-quality data affect the entire forecast enterprise, beginning with model parameter estimation and model calibration through real-time operations and forecast verification. Equally important is the physical maintenance of data transmission system from observation stations to forecast centers.

Another common theme throughout this book is balancing using proven modeling techniques and integrating the latest scientific advances.

It is interesting to consider that every agency relies on NWPs (which are often run on supercomputers, assimilating tens of millions of observations) and these are used in some countries to drive hydrologic models where the physics are based on decades-old concepts. The hydrologists are aware of the limitations of these simple models; however, they often provide sufficient predictions. Furthermore, there is a cost associated with changing systems — not only the effort in setting up software and deriving model parameters, but also in developing forecaster training and expertise in using the new system.

One of the wider gaps between science and practice has been in the spatial representation of hydrologic models. There has long been the promise that distributed hydrologic models would overtake lumped parameter hydrologic models with superior model performance. Kampf and Burges (2007), Pechlivanidis, Jackson, Mcintyre, and Wheater (2011), Spies, Franz, Hogue, and Bowman (2015), and Devia, Ganasri, and Dwarakish (2015) review many distributed hydrologic models in current use. A significant advantage of distributed models is their ability to make predictions of interior variables such as soil moisture and evapotranspiration. Unfortunately, results reported from the Distributed Model Inter-comparison Project, DMIP1 and DMIP2 (Reed et al., 2004; Smith, Georgakakos, & Liang, 2004; Smith & Gupta, 2012; Smith, Seo, et al., 2004; Smith et al., 2012), show that lumped parameter hydrologic models continue to generally outperform distributed models based on a range of verification measures at modeled catchment outlets. Other problems exist in terms of model parameter estimation, calibration, estimation of evapotranspiration, and characterization of land surface processes adequately (Bhatt, Kumar, & Duffy, 2014; Overgaard, Rosbjerg, & Butts, 2006; Song et al., 2015). Interesting, almost no operational systems described in this book use purely lumped models, with semi-distributed models being common and a few gridded models in use.

3.2 Forecast Dissemination Issues

The chapter by Jiang et al. (Chapter 14) describes a proposed flood forecast system for the Ganges River in India and some of the challenges and limitations due to limited availability of data. Relatively long-lead time forecasts are needed in order to have sufficient time to mobilize rural, poor populations to evacuate to areas of safety. This points to general problems human societies have in relation to our complex interrelationship with our environment, both in a dependent way, in terms of fishing, agriculture, transportation, water supply, etc., and in ways that we directly influence our environment, such as using waterways for waste disposal, withdrawing water for irrigation, constructing large multipurpose reservoirs, and so on (Carter et al., 2014; Liu, Dietz, Carpenter, Alberti, et al., 2007; Liu, Dietz, Carpenter, Folke, et al., 2007). Consequently, these interdependencies exist with our streams and rivers, yet too often individuals fail to learn from past events, take authoritative advice for safety, or heed alerts and warnings. Ashley and Ashley (2009) addresses issues related to flood-related fatalities in the United States, reporting that "people between the ages of 10 and 29 and ≥60 years of age are found to be more vulnerable to floods. Findings reveal that human behavior contributes to flood fatality occurrences." This suggests the need for continued and improved education on flood awareness and safety and that expanded and better methods for flood alerts and warnings are needed. Significant published literature can be found on efforts to promote public outreach for flood awareness and safety (Barnes, Schultz, Gruntfest, Benight, & Hayden, 2007; Benight, Gruntfest, Hayden, & Barnes, 2007; Demuth, Gruntfest, Drobot, Morss, & Lazo, 2007; Downton, Morss, Gruntfest, Wilhelmi, & Higgins, 2005; Drobot, Benight, & Gruntfest, 2007; Gruntfest, 2009; Gruntfest, Ruin, & League, 2009; Gruntfest, Showalter, & Ruin, 2008; Hayden et al., 2007; Kuhlman, Gruntfest, Stumpf, & Scharfenberg, 2009; Morss, Wilhelmi, Gruntfest, & Downton, 2005; Ruin et al., 2012). Much of this public outreach and awareness activity has been supported by NWS public outreach efforts for flood awareness and safety, which can be found at http://www.floodsafety.noaa.gov. One significant and visible result is the "Turn around, don't drown" program in flash flood-prone areas.

Details for the various regional and national flood forecast system dissemination policies and procedures can be found in this book's section titled "National and Regional Flood Forecasting Systems." An example of more advanced web-based forecast products comes from the joint US Geological Survey and NWS Flood Inundation Mapping (FIM) Program (http://water.usgs.gov/osw/flood_inundation). The web-based products make static flood inundation maps at NWS flood forecast point locations that are tied to the NWS Advanced Hydrologic Prediction Services (AHPS) website (http://water.weather.gov/ahps), available to the public. Increasingly, several agencies are reaching stakeholders through social media (eg, Facebook, Twitter) and the widespread availability of mobile phones makes geo-targeted warnings an attractive possibility, especially in places without much access to the Internet.

The importance of stakeholder engagement on the effectiveness of an operational forecasting and warning service cannot be overstated. There can be a tendency for operational agencies to focus on modeling technologies when developing their services. However, if stakeholders were customers in a restaurant, modeling technologies would be akin to the implements used to prepare the food in the kitchen. While well-prepared food is a necessary element of a successful restaurant, it gives an incomplete picture of the customers' total experience. Customers also care about presentation, service, responsiveness, and other factors. To this end, countries such as Australia (Chapter 10) have placed a heavy emphasis on recognizing that "monitoring and prediction" is only one element of a Total Flood Warning System. Other elements include interpretation (translating forecasts into impacts), message construction, communication, protective response, and review (ensuring continual improvement).

4 SUMMARY

The chapters in this book describe the state of practice in operational flood forecasting in many countries. They cover the various elements of the systems, such as data collection and modeling approaches, and provide some history that explains why systems have evolved to the way they are today. The systems have been shaped by their environments (natural, societal, and institutional). Each system has had to strike a complex balance — keeping up with the latest science while providing a solid operational service using available infrastructure. User requirements are also a "moving target," with increasingly sophisticated and specialized needs. There are also emerging opportunities (such as technologies) and threats (such as climate change).

When the authors describe the future directions of their systems, several themes emerge. Forecasting systems are growing in size (with the emergence of trans-national and global operational models), but also giving ever finer detail in space, particularly with the availability of high-resolution NWPs. Agencies are increasingly attempting to quantify accurately the certainty of their forecasts, but many struggle with communicating this technical and unfamiliar information to users. In an increasingly interconnected world, there are signs of international communities of forecasters sharing experiences and ideas, and adopting common tools and platforms, leveraging the benefits of working together towards a common goal.

We hope that the chapters contained within this book encourage others to identify and catalog flood forecasting efforts more comprehensively across the globe. The hope is that methods and procedures from the regional and national systems will be shared for the benefit of others as learning exercises. It is through sharing of ideas and experience that we all benefit. To this end, if you wish to make a contribution to a subsequent edition of this book, the editors welcome your contact.

REFERENCES

Adams, T., Chen, X., Davis, R., Heim, J., & Young, S. (2010). The Ohio River Community HEC-RAS Model. In: *American Soc. of Civil Engr., EWRI World Congress meeting in Providence, RI, May 17–20.*

Adams, T., Chen, X., Heim, J. E., Davis, R., & Young, J. S. (2012). NWS/OHRFC operational experience with the Ohio River Community HEC-RAS Model. In: *Amer. Met. Soc. annual meeting, New Orleans, LA, Jan. 2012.*

Adhikari, P., Hong, Y., Douglas, K. R., Kirschbaum, D., Gourley, J. J., Adler, R. F., et al. (2010). A digitized global flood inventory (1998–2008): Compilation and preliminary results. *Natural Hazards*, *55*, 405–422. http://dx.doi.org/10.1007/s11069-010-9537-2.

American Meteorological Society. (2000). Policy statement: Prediction and mitigation of flash floods. *Bulletin of the American Meteorological Society*, *81*, 1338–1340.

Arheimer, B., Lindström, G., & Olsson, J. (2011). A systematic review of sensitivities in the Swedish flood-forecasting system. *Atmospheric Research*, *100*, 275–284.

Ashley, S. T., & Ashley, W. S. (2009). Flood fatalities in the United States. *Journal of Applied Meteorology and Climatology*, *47*(3), 805–818.

Barnes, L., Schultz, D., Gruntfest, E., Benight, C., & Hayden, M. (2007). False alarms & close calls: A conceptual model of warning accuracy. *Weather and Forecasting*, *22*, 1140–1147.

Benight, C., Gruntfest, E., Hayden, M., & Barnes, L. (2007). Trauma & short fuse weather perceptions. *Environmental Hazards*, 7, 220–226.

Bhatt, G., Kumar, M., & Duffy, C. J. (2014). A tightly coupled GIS and distributed hydrologic modeling framework. *Environmental Modelling & Software*, *62*, 1–15.

Brakenridge, G. R., Kettner, A. J., Syvitski, J., Policelli, F., De Groeve, T., & Nghiem, S. (2011). Predicting and managing the effects of extreme floods using orbital remote sensing. *In: Extended abstract and talk: IEEE international geoscience and remote sensing symposium, Vancouver, Canada, August 1–5, 2011.*

Carroll, T., Cline, D., Fall, G., Nilsson, A., Li, L., & Rost, A. (2001). In: *NOHRSC operations and the simulation of snow cover properties for the coterminous U.S. proceedings of the 69th annual western snow conference; Sun Valley, Idaho; 2001 April 16–19.* (pp. 1–10).

Carroll, T. R., Cline, D. W., & Li, L. (2000). Applications of remotely sensed data at the National Operational Hydrologic Remote Sensing Center. In: *Presented at the IAHS, Remote Sensing and Hydrology 2000; Santa Fe, New Mexico; 2000 April, 2–7.*

Carter, N. H., Viña, A., Hull, V., McConnell, W. J., Axinn, W., Ghimire, D., et al. (2014). Coupled human and natural systems approach to wildlife research and conservation. *Ecology and Society*, *19*(3), 43. http://dx.doi.org/10.5751/ES-06881-190343.

Clark, R. A., Gourley, J. J., Flamig, Z. L., Yang, H., & Clark, E. (2014). CONUS-wide evaluation of National Weather Service flash flood guidance products. *Weather and Forecasting*, *29*, 377–392.

Clow, D. W., Nanus, L., Verdin, K. L., & Schmidt, J. (2012). Evaluation of SNODAS snow depth and snow water equivalent estimates for the Colorado Rocky Mountains, USA. *Hydrological Processes*, *26*, 2583–2591. http://dx.doi.org/10.1002/hyp.9385.

Cohen, S., Brakenridge, G. R., Kettner, A. J., Syvitski, J. P. M., Fekete, B. M., & De Groeve, T. (2012). Calibration of orbital microwave measurements of river discharge using a global hydrology model. In: *AGU Chapman conference, "Remote sensing of the terrestrial water cycle", Kona, Hawaii, USA, 19–22 February 2012.*

Danhelka, J. (2004). Uncertainty of hydrological forecasting due to inputting precipitation forecast and possible solution using probabilistic approach in The Czech Republic. In: *ACTIF 2nd workshop, Delft, 22–23 November.*

Demuth, J., Gruntfest, E., Drobot, S., Morss, R., & Lazo, J. (2007). Weather and Society ⋆ Integrated Studies (WAS⋆IS): Building a community for integrating meteorology & social science. *Bulletin of the American Meteorological Society*, *88*(11), 1729–1737.

Devia, G. K., Ganasri, B. P., & Dwarakish, G. S. (2015). A review on hydrological models. *Aquatic Procedia*, *4*, 1001–1007.

Diomede, T., Nerozzi, F., Paccagnella, T., & Todini, E. (2008). The use of meteorological analogues to account for LAM QPF uncertainty. *Hydrology and Earth System Sciences, 12*, 141–157. www.hydrol-earth-syst-sci.net/12/141/2008/.

Downton, M., Morss, R., Gruntfest, E., Wilhelmi, O., & Higgins, M. (2005). Interactions between scientific uncertainty & flood management decisions: Two case studies in Colorado. *Environmental Hazards, 6*, 134–146.

Drobot, S., Benight, C., & Gruntfest, E. (2007). Risk factors for driving into flooded roads. *Environmental Hazards*, 7, 117–134.

Ebert, B., Damrath, U., Wergen, W., & Baldwin, M. E. (2003). The WGNE assessment of short-term quantitative precipitation forecasts (QPFs) from operational numerical weather prediction models. *Bulletin of the American Meteorological Society, 84*, 481–492.

Emerton, R. E., Stephens, E. M., Pappenberger, F., Pagano, T. C., Weerts, A. H., Wood, A. W., et al. (2016). Continental and global scale flood forecasting systems. *WIREs Water*. http://dx.doi.org/10.1002/wat2.1137.

Gaume, E., Bernardara, P., Newinger, O., Barbuc, M., Bateman, A., Blaškovičová, L., et al. (2009). A compilation of data on European flash floods. *Journal of Hydrology, 367*(1–2), 70–78.

Gourley, J. J., Flamig, Z. L., Hong, Y., & Howard, K. W. (2014). Evaluation of past, present, and future tools for radar-based flash flood prediction. *Hydrological Sciences Journal, 59*, 1377–1389. http://dx.doi.org/10.1080/02626667.2014.919391.

Gourley, J. J., Yang, H., Flamig, Z. L., Arthur, A., Clark, R., Calianno, M., et al. (2013). A unified flash flood database across the United States. *Bulletin of the American Meteorological Society, 94*, 799–805.

Gruntfest, E. (2009). Editorial. *Journal of Flood Risk Management, 2*(2), 83–84.

Gruntfest, E., Ruin, I., & League, C. (2009). Learning from flash floods: Driver behavior in high-water conditions in Colorado water. *Newsletter of the Water Center of Colorado State University, 26*(4), 8–9.

Gruntfest, E., Showalter, P., & Ruin, I. (2008). Flash flood research — Past, present and future. *Natural Hazards Observer. 33*, 11–13.

Hayden, M., Drobot, S., Gruntfest, E., Benight, C., Radil, S., & Barnes, L. (2007). Information sources for flash flood warnings in Denver, CO and Austin, TX. *Environmental Hazards*, 7, 211–219.

Hirabayashi, Y., Mahendran, R., Koirala, S., Konoshima, L., Yamazaki, D., Watanabe, S., et al. (2013). Global flood risk under climate change. *Nature Climate Change, 3*(9), 816–821. http://dx.doi.org/10.1038/nclimate1911.

Huffman, G. J., Adler, R. F., Bolvin, D. T., & Nelkin, E. J. (2010). The TRMM multi-satellite precipitation analysis (TMPA). In F. Hossain & M. Gebremichael (Eds.), *Satellite applications for surface hydrology*. (pp. 3–22). Berlin: Springer Verlag. ISBN: 978-90-481-2914-0 (Chapter 1).

Jordan, R. (1990). *User's guide for USACRREL one-dimensional snow temperature model (SNTHERM.89)*. Hanover, New Hampshire: U.S. Army Cold Regions Research and Engineering Laboratory.

Kampf, S. K., & Burges, S. J. (2007). A framework for classifying and comparing distributed hillslope and catchment hydrologic models. *Water Resources Research, 43*, W05423. http://dx.doi.org/10.1029/2006WR005370.

Khan, S. I., Hong, Y., Vergara, H. J., Gourley, J. J., Brakenridge, G. R., De Groeve, T., et al. (2012). Microwave satellite data for hydrologic modeling in ungauged basins. *IEEE Geoscience and Remote Sensing Letters, 9*(4), 663–667.

Khan, S. I., Hong, Y., Wang, J., Yilmaz, K. K., Gourley, J. J., Adler, R. F., et al. (2011). Satellite remote sensing and hydrological modeling for flood inundation mapping in Lake Victoria Basin: Implications for hydrologic prediction in ungauged basins. *IEEE Transactions on Geoscience and Remote Sensing, 49*, 85–95. http://dx.doi.org/10.1109/TGRS.2010.2057513.

Kirk, G. S., Raven, J. E., & Schofield, M. (2013). *The presocratic philosophers: A critical history with a selection of texts* (2nd ed.). Cambridge: Cambridge University Press. 475 pp.

Kuhlman, K., Gruntfest, E., Stumpf, G., & Scharfenberg, K. (2009). In: *Beyond storm based warnings: An advanced WAS * IS workshop to study communication of probabilistic hazardous weather information American Meteorological Society.*

Liang, X., Lettenmaier, D. P., Wood, E. F., & Burges, S. J. (1994). A simple hydrologically based model of land surface water and energy fluxes for GSMs. *Journal of Geophysical Research, 99*(D7), 14415–14428.

Liu, J., Dietz, T., Carpenter, S. R., Alberti, M., Folke, C., Moran, E., et al. 2007a. Complexity of coupled human and natural systems. *Science, 14*, 1513–1516.

Liu, J., Dietz, T., Carpenter, S. R., Folke, C., Alberti, M., Redman, C. L., et al. 2007b. Coupled human and natural systems. *Ambio, 36*, 639–649.

Morss, R., Wilhelmi, O., Gruntfest, E., & Downton, M. (2005). Flood risk uncertainty & scientific information for decision-making: Lessons from an interdisciplinary project. *Bulletin of the American Meteorological Society, 86*(11), 1593–1601.

Ntelekos, A. A., Georgakakos, K. P., & Krajewski, W. F. (2009). On the uncertainties of flash flood guidance: Toward probabilistic forecasting of flash floods. *Journal of Hydrometeorology*, 7(5), 896–915.

Nurmi, P., Orfila, B., Roulin, E., Schroter, K., Seed, A., Szturc, J., et al. (2010). Propagation of uncertainty from observing systems and NWP into hydrological models: COST-731 Working Group 2. *Atmospheric Science Letters, 11*, 83–91.

Office of Hydrologic Development. (2009). *Integrated Water Resources Science and Services (IWRSS): An integrated and adaptive roadmap for operational implementation.* Silver Spring, MD: NOAA/U.S. National Weather Service. http://www.nws.noaa.gov/oh/nwc/IWRSS_ROADMAP_FINAL.pdf.

Overgaard, J., Rosbjerg, D., & Butts, M. B. (2006). Land-surface modelling in hydrological perspective — A review. *Biogeosciences, 3*, 229–241. www.biogeosciences.net/3/229/2006/.

Pagano, T. C. (2014). Evaluation of Mekong River commission operational flood forecasts, 2000–2012. *Hydrology and Earth System Sciences, 18*, 2645–2656. http://dx.doi.org/10.5194/hess-18-2645-2014.

Pagano, T. C., et al. (2014). Challenges of operational river forecasting. *Journal of Hydrometeorology, 15*, 1692–1707.

Pechlivanidis, I. G., Jackson, B. M., Mcintyre, N. R., & Wheater, H. S. (2011). Catchment scale hydrological modelling: A review of model types, calibration approaches and uncertainty analysis methods in the context of recent developments in technology and applications. *Global NEST Journal, 13*(3), 193–214.

Pomeroy, J. W., Gray, D. M., & Landine, P. G. (1993). The prairie blowing snow model: Characteristics, validation, operation. *Journal of Hydrology, 144*, 165–192.

Reed, S., Koren, V., Smith, M., Zhang, Z., Moreda, F., Seo, D.-J., et al. (2004). Overall distributed model intercomparison results. *Journal of Hydrology, 298*(1–4), 27–60.

Ruin, I., Creutin, J. D., Anquetin, S., Gruntfest, E., & Lutoff, C. (2009). Human vulnerability to flash floods: Addressing physical exposure and behavioral questions. In Samuels, et al. (Eds.), *Flood risk management: Research and practice*. London: Taylor & Francis Group.

Ruin, I., Lutoff, C., Creton-Cazanave, L., Anquetin, S., Borga, M., Chardonnel, S., et al. (2012). Toward a space-time framework for integrated water and society studies. *Bulletin of the American Meteorological Society, 93*, ES89–ES91. http://dx.doi.org/10.1175/BAMS-D-11-00226.1.

Smith, M. B., Georgakakos, K., & Liang, X. (2004a). Preface, The distributed model intercomparison project (DMIP). *Journal of Hydrology, 298*, 1–4.

Smith, M. B., & Gupta, H. V. (2012). The distributed model intercomparison project (DMIP) — Phase 2 experiments in the Oklahoma Region, USA. *Journal of Hydrology, 418–419*, 1–2.

Smith, M. B., Koren, V., Zhang, Z., Zhang, Y., Reed, S. M., Cui, Z., et al. (2012). Results of the DMIP 2 Oklahoma experiments. *Journal of Hydrology, 418–419*, 17–48.

Smith, M. B., Seo, D.-J., Koren, V., Reed, S. M., Zhang, Z., Duan, Q., et al. (2004b). The distributed model intercomparison project (DMIP): Motivation and experiment design. *Journal of Hydrology, 298*, 1–4.

Song, X., Zhang, J., Zhan, C., Xuan, Y., Ye, M., & Xu, C. (2015). Global sensitivity analysis in hydrological modeling: Review of concepts, methods, theoretical framework, and applications. *Journal of Hydrology, 523*, 739–757.

Spies, R. R., Franz, K. J., Hogue, T. S., & Bowman, A. L. (2015). Distributed hydrologic modeling using satellite-derived potential evapotranspiration. *Journal of Hydrometeorology, 16*, 129–146. http://dx.doi.org/10.1175/JHM-D-14-0047.1.

Sigma — Preliminary estimates for H1 2013a: Catastrophes cost global insurance industry more than USD 20 billion. http://www.swissre.com/media/news_releases/nr_20130821_sigma_natcat_estimates_H1_2013.html, 21 August 2013, Zurich.

Sigma World Insurance Database, http://www.sigma-explorer.com/, Swiss Re Economic Research & Consulting.

Tarboton, D. G., & Luce, C. H. (1996). *Utah energy balance snow accumulation and melt model (UEB)*. Utah Water Research Laboratory, Utah University and USDA Forest Service, Intermountain Research Station. 41 p.

World Meteorological Organization. (2011). Manual on flood forecasting and warning. Hydrology and water resources programme WMO-No. 1072 Geneva, Switzerland: World Meteorological Organization. 142 pp.

Wu, H., Adler, R. F., Hong, Y., Tian, Y., & Policelli, F. (2012a). Evaluation of global flood detection using satellite-based rainfall and a hydrologic model. *Journal of Hydrometeorology, 13*, 1268–1284.

Wu, H., Adler, R. F., Tian, Y., Huffman, G. J., Li, H., & Wang, J. (2014). Real-time global flood estimation using satellite-based precipitation and a coupled land surface and routing model. *Water Resources Research, 50*, 2693–2717. http://dx.doi.org/10.1002/2013WR014710.

Wu, H., Kimball, J. S., Li, H., Huang, M., Leung, L. R., & Adler, R. F. 2012b. A new global river network database for macroscale hydrologic modeling. *Water Resources Research, 48*, W09701. http://dx.doi.org/10.1029/2012WR012313.

Wu, H., Kimball, J. S., Mantua, N., & Stanford, J. (2011). Automated upscaling of river networks for macroscale hydrological modeling. *Water Resources Research*, 47, W03517. http://dx.doi.org/10.1029/2009WR008871.

Xuan, Y., Cluckie, I. D., & Wang, Y. (2009). Uncertainty analysis of hydrological ensemble forecasts in a distributed model utilising short-range rainfall prediction. *Hydrology and Earth System Sciences, 13*, 293–303.

PART 1

National and Regional Flood Forecasting Systems

CHAPTER 1

Australian Bureau of Meteorology Flood Forecasting and Warning

T.C. Pagano*, J.F. Elliott†, B.G. Anderson*, J.K. Perkins‡
*Australian Bureau of Meteorology, Melbourne, Australia
†Australian Bureau of Meteorology, Melbourne, Australia (Retired)
‡Australian Bureau of Meteorology, Brisbane, QLD, Australia

1 INTRODUCTION

Flood warning is one of the key measures used to mitigate the damaging impact of floods in Australia. Viewed as a total system, flood warning involves agencies at all levels of government working in partnership, with the Bureau playing a lead role as the national flood forecasting agency. In this chapter, the nature of the flood problem in Australia is briefly described in terms of the key demographic and climate influences, followed by a history of the development of flood forecasting in Australia. The organization of flood forecasting within the Bureau and the nature of the Bureau's role in the total flood warning system is provided, followed by a description of the current Bureau flood forecasting system and operations. This includes details of current hydrologic modeling tools and more recent developments, including the improved utilization of weather forecasting inputs. Finally, important operational challenges faced by flood forecasters are discussed, concluding with an outline of plans for the future direction of the service.

2 DEMOGRAPHICS, CLIMATE, AND FLOODS OF AUSTRALIA

Australia is a very sparsely populated continent, with a national density of around three persons per square kilometer, the third lowest in the world. Around two thirds of the overall population of nearly 24 million[1] live in the federal and state/territory capital cities (Figs. 1 and 2), almost all of which are close to coastlines. Overall population growth varies between 1% and 2% annually, although coastal populations are growing more rapidly, with

[1]http://en.wikipedia.org/wiki/Demographics_of_Australia, Accessed 01.02.15.

Flood Forecasting
http://dx.doi.org/10.1016/B978-0-12-801884-2.00001-3

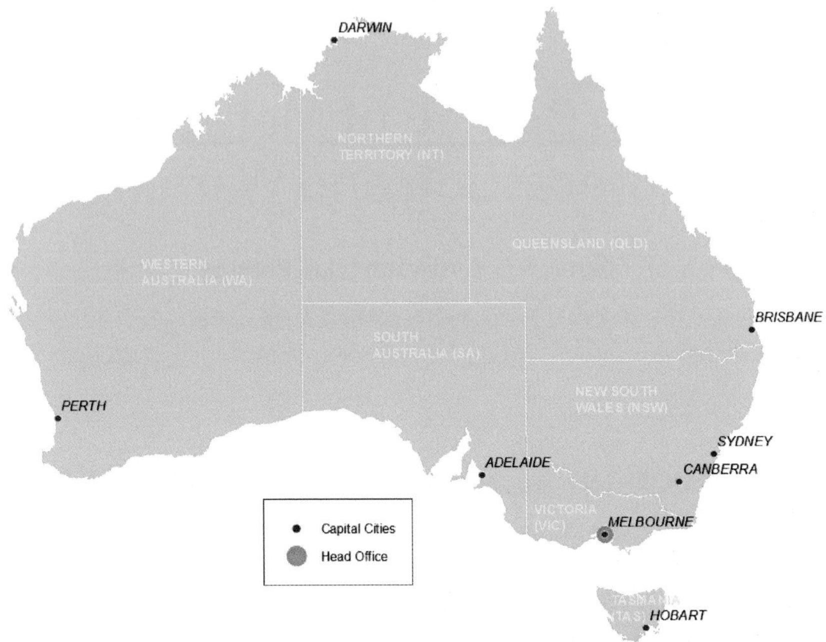

Fig. 1 Federal capital (Canberra) and capital cities of Australian states/territories (from Bureau of Meteorology, 2014). The Bureau of Meteorology has Regional Offices in every capital city and a Head Office in Melbourne.

the population of some regions of coastal Queensland increasing at 4.7% annually (Australian Bureau of Statistics, 2012).

In addition to the capital cities, there are significant population centers in the inland regional areas that were initially settled beside major rivers, as well as many smaller remote communities, including those of indigenous populations. An important contribution to population growth has been immigration. Australia is a multicultural country and for many residents, English is a second language. This is an important factor in emergency management planning, especially in the design of warning information.

Australia is the driest inhabited continent, with extreme variability in both annual rainfall (Fig. 3) and streamflow.[2] It features a wide range of climatic zones, from the tropical regions of the north, through the arid expanses of the interior, to the temperate regions of the south. Elevation also has an influence on rainfall, with the Great Dividing Range, the Australian Alps, and western Tasmania receiving higher rainfall totals.

[2]The following summary of Australian climate is based on the Australian Bureau of Statistics (2012).

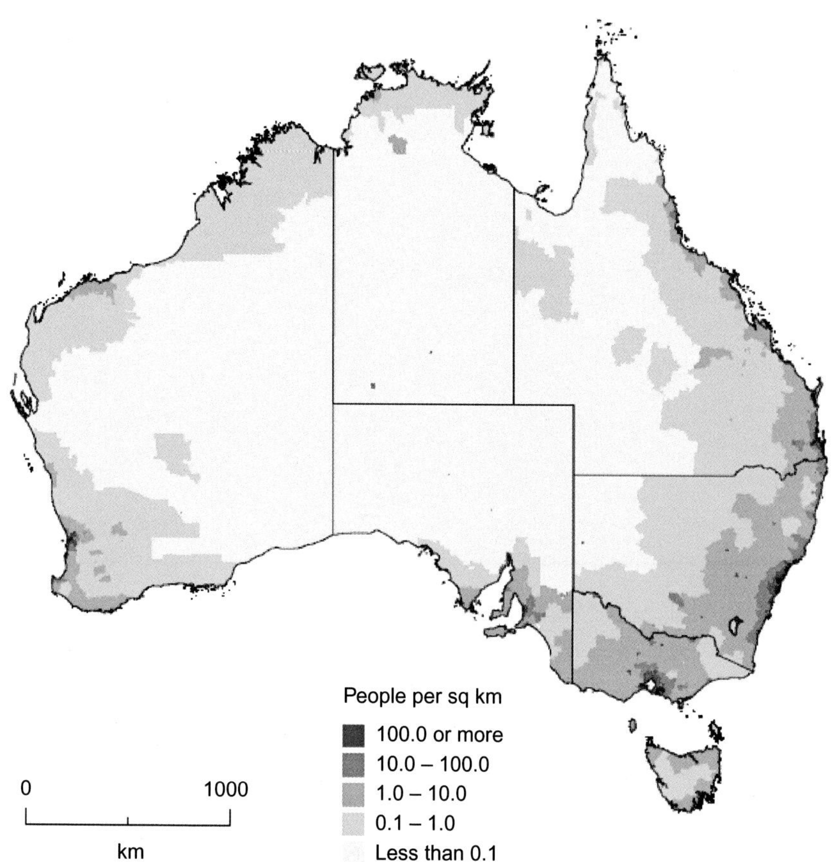

Fig. 2 Population density of Australia in Jun. 2010 (from Australian Bureau of Statistics, 2014). Note the relatively large areas of low population density. Regional population growth, Australia (3218.0).

There can be large seasonal fluctuations in both rainfall and temperature in parts of the country. In northern Australia, temperatures are warm throughout the year, with a wet season from approximately Nov. through Apr. — when almost all the rainfall occurs — and a dry season from May to Oct. Further south, temperature is more important in defining seasonal differences and rainfall is more evenly distributed throughout the year, reaching a marked winter peak in the south-west and along parts of the southern fringe.

The major driver of interannual climate variability, particularly in eastern Australia, is the El Niño-Southern Oscillation (ENSO) phenomenon. El Niño is an anomalous warming of the central and eastern tropical Pacific

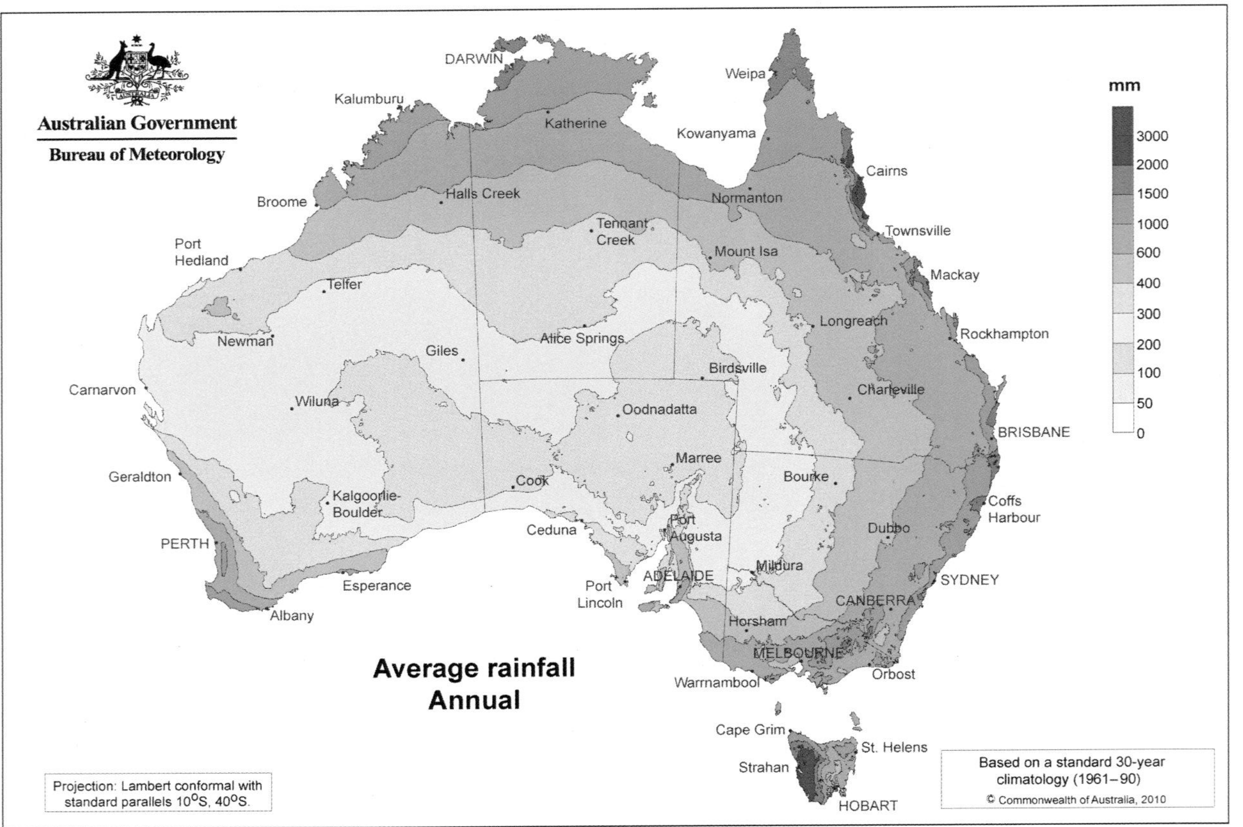

Fig. 3 Annual precipitation of Australia 1961–90. Bureau of Meteorology. (2000). *Climatic atlas of Australia — Rainfall* (p. 25). Melbourne, VIC: Bureau of Meteorology.

Ocean, while La Niña, the reverse phase of the system, is an anomalous cooling. The Southern Oscillation is the atmospheric counterpart to El Niño's oceanic changes. El Niño events are generally associated with a reduction in winter and spring rainfall across much of eastern, northern, and southern Australia. This can lead to widespread and severe drought, particularly in eastern Australia. Conversely, La Niña events are generally associated with wetter-than-normal conditions and have contributed to many of Australia's most notable floods (eg, 2010-12 in the eastern half of the country).

Tropical cyclones bring heavy rain as well as strong winds and are the cause of most of Australia's highest-recorded daily rainfalls. Although tropical cyclones rapidly lose their intensity on moving over land, rainfall associated with former cyclones often persists well after the destructive winds have eased and can occasionally bring heavy rains deep into the inland, causing widespread flooding. Parts of inland Western Australia receive 30–40% of their average annual rainfall from these tropical depressions, with certain places receiving their average annual rainfall within a 1- or 2-day period as a tropical cyclone (or ex-cyclone) passes by. Intense low pressure systems can also form outside the tropics, most commonly off the east coast, where they are known as "east coast lows." These systems can bring very strong winds and heavy rain, particularly where they direct moist easterly winds on their southern flank onto the coastal ranges of southern Queensland, New South Wales, eastern Victoria, and northeastern Tasmania.

Floods in Australia (Fig. 4) are predominantly caused by heavy rainfall, although extreme tides, storm surge, tsunami, snow melt, levee failure, groundwater surcharge, or dam break can also cause flooding (Middlemann, 2007). More recently, coastal flooding as a result of sea level rise precipitated by climate change is increasingly incorporated in planning and land management strategies.

Floods as a result of heavy rainfall can be broadly categorized as either flash floods or riverine floods. Flash floods can occur almost anywhere in Australia, and result from a relatively short, intense burst of rainfall, for example, during severe thunderstorms, tropical cyclones, and other high rainfall weather events. During these events, the drainage system may be unable to cope with the downpour and flow frequently occurs outside defined water channels. Areas with low-capacity drainage systems, whether natural, or artificial, are particularly vulnerable to flash flooding. Flash floods frequently occur over small areas, often embedded within regions of more widespread riverine flooding. Although flash floods are generally localized, they pose a significant threat to human life, because of the high flow velocities and rapid

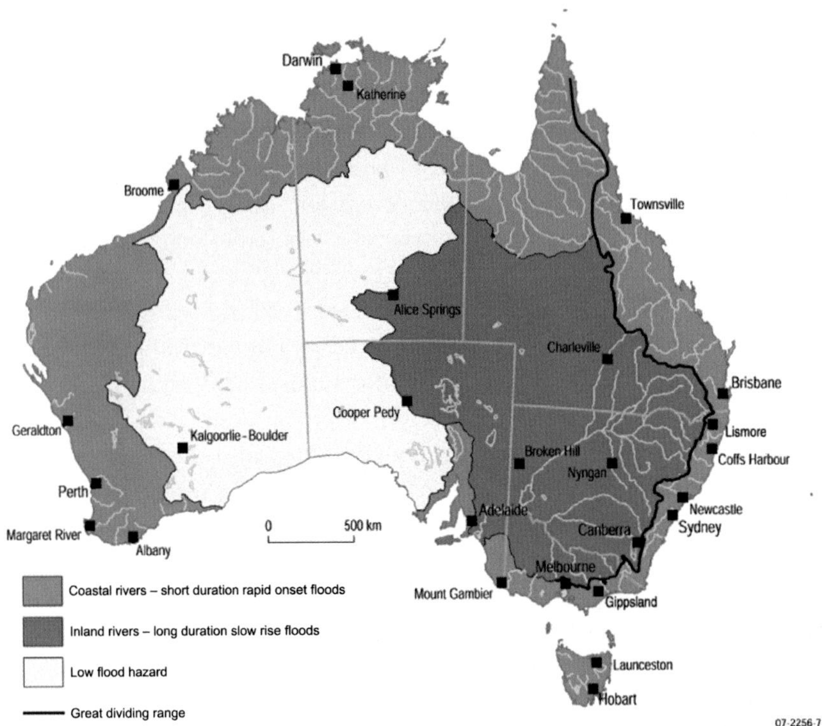

Fig. 4 Australian flood hazard map. From Geoscience, A. (2015). *Flood potential in Australia map*. http://www.ga.gov.au/metadata-gateway/metadata/record/69538/.

onset. Flash floods are normally defined as those floods that occur within 6 h of a rainfall event. Effective warning is extremely difficult because of the limited time available — often only 2 h or less — and the limited knowledge of the specific nature of the hazard in a particular area.

Riverine floods occur following heavy rainfall when watercourses do not have the capacity to convey the excess water. They occur in relatively low-lying areas adjacent to streams and rivers. In the flat inland regions of Australia, floods may spread thousands of square kilometers and last several weeks, with warning lead times of several days, sometimes weeks for the longer rivers. In the mountain and coastal regions of Australia, flooding may last only several days, but occurs more quickly with higher depths and faster flow velocities, causing significant inundation of towns and cities. Effective flood warnings can normally be provided in these flooding situations, with lead times typically ranging from 6 to 12 h.

The most recent two decades have been characterized by extreme hydrologic variability. The remainder of this section identifies notable recent disasters, and Section 3 discusses some of the impacts of these events on flood warning practices and institutions.

From 1996 to 2009, Australia experienced one of the worst droughts in its modern history (David Jones, personal communication, Mar. 4, 2015), which resulted in a drop in the numbers of flood warnings issued by the Bureau.

The drought was followed by an exceptional series of extreme weather events, which included severe and repeated flooding across many states. This variability of flooding can be seen in Fig. 5, which compares the number of flood warnings issued during part of the drought period (2003–09) with the number issued during the severe flood period (2010–12).

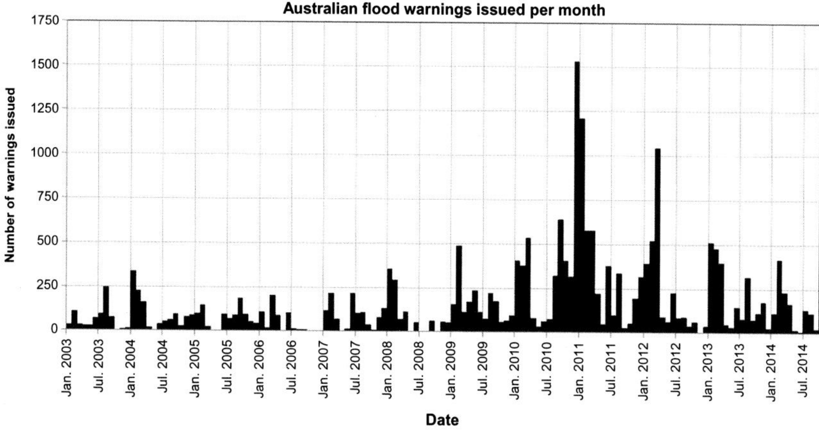

Fig. 5 Number of flood warnings issued nationally by month 2003–14. The early years were dominated by drought, whereas major floods occurred more recently. The Bureau also issues flood watches, which are not included in this graph but display a similar trend.

Prolonged and intensive rainfall over large areas of Queensland, coupled with already saturated catchments, led to significant flooding in Queensland in Dec. 2010 and Jan. 2011. More than 78% of Queensland (an area bigger than France and Germany combined) was declared a disaster zone, with over 2.5 million people affected. Some 29,000 homes and businesses suffered some form of inundation. There were 35 confirmed deaths from the flooding and $2.38 AUD billion in damages (Carbone & Hanson, 2012).

From Sep. 2010 to Feb. 2011, Victoria experienced some of the worst floods in the state's history. Several communities experienced flooding two or three times during the period. Around one-third of Victoria was affected, including some 70 local government areas. Initial estimates put the gross total cost of the floods at nearly $1.3 AUD billion.

3 HISTORY OF FLOOD FORECASTING AT THE BUREAU OF METEOROLOGY

In Australia, flood forecasting, and warning services are provided at a national level by the Bureau of Meteorology, working in close cooperation with agencies at the state and local government levels. The nature of this arrangement has evolved over the past century or so in accordance with the responsibilities of the various levels of government, as set out in the Constitution of Australia, and agreements between the states and the Commonwealth.

Prior to Federation (1901), the practice of meteorology in Australia was state-based. Under the Constitution, the power to make laws with respect to meteorological observations resides with the Federal Parliament. It was agreed by the states and the Commonwealth in 1906 that there should be a single Federal Meteorological Department responsible for both meteorological science and services to meet the needs of the Commonwealth and the states.

In pursuance of this intention, the Meteorology Act of 1906 established the office of the Commonwealth Meteorologist with responsibility for, among other things, "the display of ... flood signals." Early flood warning activity included the distribution of flood signals in the form of river height bulletins by telegraph, newspapers, and eventually radio to the masters of steamers plying the larger rivers, along with the radioing or telephoning of flood heights and general predictions to people in communities in the path of flood waters (Keys, 1997). Flags were also used, particularly to indicate the state of tides and their influence on flooding.

The Bureau of Meteorology formally commenced operation on Jan. 1, 1908, with a Head Office in Melbourne and State-based Divisional (now Regional) offices responsible for the delivery of services. Responsibility for water resource measurement and management, as well as emergency management, however, was a state responsibility and it proved effective for the Bureau's state-based offices to work closely with their state water and emergency management counterparts in delivering flood services. In 1908, some flood warning gauges operated by New South Wales and Queensland governments were taken over by the Bureau in conjunction with the consolidation of the state Meteorological Services. Also in that year, the

Bureau commenced the issue of generalized flood warnings, as needed, with daily weather forecasts. The operation of this service involved the Bureau in river gauging activities including working in cooperation with other agencies for the measurement of river levels. In 1936, the original functions of the Commonwealth Meteorologist in relation to floods were extended by including the establishment and maintenance of river gauges and the arrangement of communications for the distribution of warnings.

In the period from 1908 to the early 1950s, the Bureau, by general agreement, issued warnings of likely flooding, collected and disseminated reports of river heights from a network of flood gauges, and assumed financial responsibility for erecting and maintaining many of these gauges, in addition to paying flood height observers for their services. The Meteorology Act (from 1906) was repealed in 1955 and a new Act (The Meteorology Act 1955[3]), included in the functions of the Bureau as set out in s6(1)(c): "the issue of warnings of ... weather conditions likely to endanger life or property including weather conditions likely to give rise to floods ...".

In Apr. 1957, after a series of highly damaging floods in New South Wales, the Government instructed the Director of Meteorology to establish a Hydrometeorology Service, which included among its functions "systematic flood forecasting." The establishment of this new service brought about a major change in the level of flood warning service by following a more objective and systematic approach to flood forecasting, and taking advantage of new advances in hydrological forecasting and data collection technologies. Whereas the flood warning service had previously only involved very generalized statements about the extent of future flooding, this new service would support the preparation of quantitative forecasts of future river levels, thereby facilitating a much more detailed and effective response by the emergency management agencies and those at risk from flooding.

The first quantitative flood forecasting service established under this new arrangement was in the Macleay River valley in New South Wales in 1962. Specialist engineer hydrologists were subsequently engaged in Bureau state-based (regional) offices in Queensland, New South Wales, and Victoria, as well as its Head Office, and new flood forecasting systems were gradually established on the priority river basins. The development of these new systems was undertaken jointly between the Head Office and the respective Regional Offices, with the Head Office generally providing support through technique application and development, and the Regional

[3]http://www.comlaw.gov.au/Details/C2008C00066, Accessed 01.02.15.

Office supplying the essential data and field-based inputs. These new hydrologically based services were limited to the eastern states of Queensland, New South Wales, Victoria, and Tasmania, with the remaining states/territories (South Australia, Western Australia, and the Northern Territory) continuing to provide more generalized flood warnings as part of normal weather services.

The Bureau introduced flood forecasting and warning systems for many communities in the eastern states; however, progress stalled in the 1980s during a period of uncertainty surrounding the allocation of costs and responsibilities among the different levels of government for the provision of flood forecasting and warning services. Eventually, in Sep. 1987, the Bureau was provided with additional resources to upgrade its activities in flood warning and to extend these services to South Australia, Western Australia, and the Northern Territory. This included the provision of specialist hydrologists in the relevant Regional Offices and some supplementation of specialist staffing in the other offices.

In conjunction with this decision, the Commonwealth agreed to an arrangement for sharing responsibility for flood warning between the three levels of government. These arrangements involved the Bureau remaining as the lead agency with primary responsibility for forecasting system operation and the dissemination of flood forecasts and warnings, but with state and local government to share the costs of the data collection infrastructure and to ensure the forecast and warning information was used effectively at the local level as part of emergency management responsibilities. The system currently operates under the overall guidance of state/territory based Flood Warning Consultative Committees that advise on service priorities and help facilitate the cooperative arrangements that have been established between the three levels of government.

Near the most intense period of the 1995–2009 drought, the government's Water Act of 2007 gave the Bureau a wide range of responsibilities for the provision of various water information services. Among these, the Bureau was required to provide a range of new water availability forecasting services. One such service was the routine provision of 7 day forecasts of streamflow (SDF) for a range of water management purposes. In contrast to the current flood forecasting service, where hydrologic forecasts are only made during periods of flooding, the SDF forecasts were to be made on a continuous basis (eg, daily).

The Queensland floods of 2010–11 and the scale of the associated disaster led to the establishment of a Commission of Inquiry (COI, Queensland

Floods Commission of Inquiry, 2012). As part of the inquiry into water management, community vulnerability, and disaster response, there was an examination of the Bureau's forecasts and procedures for predictions of both rainfall and river heights (Bureau of Meteorology, 2011). Similarly, in the wake of the 2010–11 Victorian flooding, the Victorian Government announced a comprehensive review of flood warnings and emergency response efforts (Comrie, 2011). Both this review and the COI generated a wide range of recommendations, including many impacting the operation and delivery of flood warning services and related organizational arrangements for their respective states.

As a further response to this series of extreme weather events, the Australian Government announced a review of the Bureau of Meteorology's capacity to respond to future extreme weather and natural disaster events, and to provide seasonal forecasting services (generally referred to as the Munro Review, Munro, 2011). As part of the Australian Government response to the Munro Review findings, the Government provided the Bureau with funding of $63.3 million for additional frontline flood and weather forecasters, the establishment of a national desk for extreme weather, and other hazards-related investments over 2012–17 (Bureau of Meteorology, 2013). The Bureau is responding to recommendations from these reviews with a range of initiatives to improve the resilience and consistency of flood forecasting and warning services (Wilson, Anderson, Robinson, Comeadow, & Dale, 2014). For example, this includes the adoption of a new forecast production environment called the Hydrologic Forecasting System (HyFS).

4 CURRENT CHARACTERISTICS OF THE BUREAU AND ITS SERVICES

The Bureau of Meteorology is organized around a Head Office in Melbourne and Regional Offices in each state and the Northern Territory that deliver most of the public services. The overall organizational structure and the place of Flood Forecasting within the Hazards, Warnings, and Forecasts division is shown in Fig. 6. Flood forecasting and warning services are delivered through Flood Warning subsections in each Regional Office, with surge capacity provided through the National Operations group in Head Office.

The Bureau provides these services within the context of what is termed the total flood warning system. The existence of the total flood warning system recognizes that in order for flood warning to achieve the goal of helping

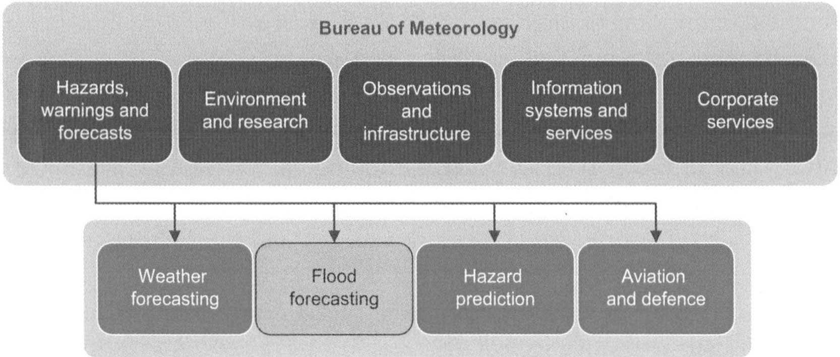

Fig. 6 Bureau of Meteorology top-level management structure. Bureau of Meteorology. (2014). *Flood forecasting and warning services — National directive* (p. 60). Melbourne, VIC: Bureau of Meteorology.

flood management agencies and individuals in flood-prone communities to understand the threat posed by developing floods so that they can take action to mitigate their effects, a set of connected activities (or elements) needs to operate together. The total flood warning system provides the overall context to guide how the Bureau and state/territory/local government agencies work together to achieve the common goal of effective flood warning.

At its simplest, an effective flood warning system has six components (from Australian Government, 2009, Fig. 7):

Monitoring and prediction: Detecting environmental conditions that lead to flooding, and predicting river levels during the flood.

Interpretation: Identifying the impacts of the predicted flood levels on communities at risk.

Message construction: Devising the content of the message that will warn people of impending flooding.

Communication: Disseminating warning information in a timely fashion to people and organizations likely to be affected by the flood.

Protective response: Generating appropriate and timely actions and behaviors from the agencies involved and from the threatened community.

Review: Examining the various aspects of the system with a view to improving its performance.

For a flood warning system to work effectively, these components must all be present and integrated, rather than operating in isolation from each other. To guide the design and operation of this system, a national manual of best practice in flood warning has been developed (Australian Government, 2009), supported by manuals for other aspects of flood management

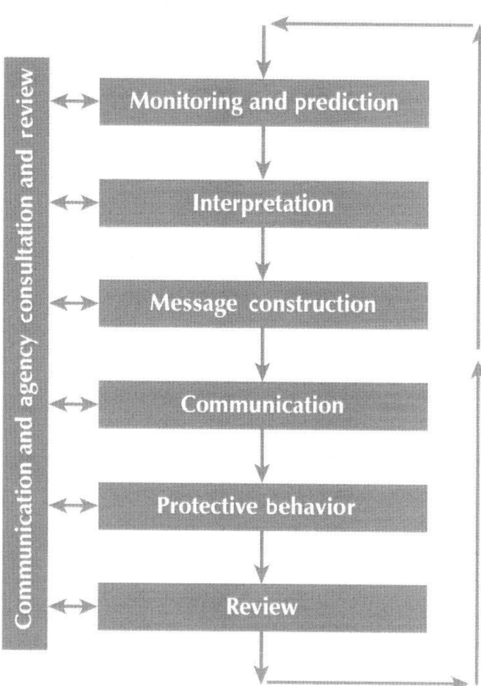

Fig. 7 The total flood warning system. Australian Government. (2009). Flood warning. In *Australian emergency management manual series* (Vol. 21, p. 79). Barton, ACT: Australian Government.

including flood preparedness, flood response operations, and a more general manual on best practice on flood risk management.

Currently the Bureau provides flood forecasting and warning services in approximately 150 river basins, and this involves modeling at over 2000 stream gauges and communities of interest. To support this service, the Bureau employs 48 specialist hydrologists and technical officers in the Flood Forecasting and Warning Branch, with 10 hydrographic staff in the Observations and Infrastructure Division who manage the flood warning network (maintaining both Bureau-owned and stations under contract from other agencies). The primary frontline forecasting is undertaken for each state/territory by staff in the Flood Warning subsection in the relevant Regional Office; however, arrangements are in place to utilize staff either from the National Operations group in Head Office or staff from other Regional Offices not currently experiencing flooding to support during peak workloads (described earlier as a "flying squad"). The details of these arrangements are set out in a National Directive which describes specific roles for staff during flood operations as well as the operating protocols that guide the involvement of staff from other offices.

All staff involved in operational flood forecasting and warning are subject to a structured process of competency assessment and development to ensure they are equipped with the necessary skills and capacities to perform their roles. This includes training on topics ranging from scientific modeling to handling customer inquiries. A key component is local knowledge about catchment hydrology, flood behavior and impacts, stakeholders and stakeholder needs, and the type of weather patterns and rainfall producing weather systems that cause floods in a particular area.

The principal flood warning products issued by the Bureau include flood warnings and flood watches. These are each used at different stages in the formation and evolution of a flood event according to a protocol established in consultation with the key service clients such as the emergency management agencies and local governments in order to meet the requirements of their flood response plans. This relationship between the Bureau and service clients is set out in detail in Service Level Specifications. A brief description of each product follows (examples are given in Boxes 1 and 2):

Flood watch: The primary purpose of a flood watch is to provide early advice that a significant risk of flooding is expected to affect specific

Box 1 Example Flood Watch Text for a Catchment in Tasmania (Excerpt)

Note: This Flood Watch provides early advice of the significant risk of flooding.

Flood watch for all northern and eastern river basins

Issued at 4:02 pm EST Sunday Aug. 11, 2013.

Flood watch number: 4.

In the 24 h to 9:00 am Sunday rainfall totals of less than 20 mm have been recorded in the Northern catchments. Since 9:00 am, rainfall totals of up to 12 mm have been recorded with an isolated total of 33 mm.

Heavy rainfalls are expected for the next 24 h ranging from 20 to 40 mm over the northern river basins and totals of 25–50 mm over the Huon River catchment.

The recorded rain has caused some stream rises. The forecast rainfall will cause stream and river rises with the potential for minor flooding to develop on Monday in the Derwent catchment and along streams and rivers in the northern river basins.

Catchments that may be affected include: North Esk, South Esk, Meander, Macquarie, Derwent, Jordan, and Coal Rivers.

Box 1 Example Flood Watch Text for a Catchment in Tasmania (Excerpt)—Cont'd

The Bureau of Meteorology will continue to monitor the situation and will issue catchment specific warnings if and when required. See www.bom.gov.au/tas/warnings to view all of the Bureau's current warning products.

FloodSafe advice is available at www.ses.tas.gov.au

Road closure information is available at www.police.tas.gov.au

For emergency assistance, call the SES on telephone number 132,500.

For life-threatening situations, call 000 immediately.

Next issue

The next Flood Watch is due to be issued by 10:00 am EST Monday Aug. 12, 2013.

Box 2 Example Flood Warning Text (Excerpt)

Initial minor flood warning for the Derwent River Basin

Issued at 10:46 am EDT on Thursday Jun. 9, 2011.

Flood warning number: 1.

Minor flooding is developing in the Clyde River and river tributaries in the lower River Derwent basin following moderate to locally heavy rainfall about the basin overnight Tuesday and Wednesday.

Strong river rises are developing in other areas but are not expected to exceed the minor flood level.

Clyde River

Minor flooding is expected in the Clyde River.

The Clyde River at Bothwell is expected to peak around 2.0 m during Thursday evening.

The Clyde River at Hamilton is expected to peak between 2.4 and 2.7 m during Friday evening.

Flood safety advice

FloodSafe advice is available at www.ses.tas.gov.au

Road closure information is available at www.police.tas.gov.au

Next issue

The next warning will be issued by 4.30 pm EST Thursday Jun. 6, 2011.

Latest river heights

Clyde River at Bothwell	1.83 m steady	10:00 am Thursday Jun. 9, 2011
Derwent River at Macquarie Plains	1.75 m rising	9:30 am Thursday Jun. 9, 2011

communities and to alert the relevant emergency service organizations. A flood watch can reference localized and/or flash flooding. The flood watch is a key product for emergency management agencies and other stakeholders. It focuses attention on the watch area and triggers any local surveillance activity. It enables emergency response agencies and other organizations to plan for the possibility of flooding in the context of their day-to-day operations.

Flood warning: the purpose of a flood warning is to warn that flooding is occurring or is expected to occur in a particular region, catchment, or location. For qualitative warnings this also includes expected severity levels. Flood warnings also provide specific river height information to flood management authorities; they are prepared and issued to meet forecast requirements and service levels as specified in the Service Level Specification for the region, and the Catchment Guide for the relevant catchment. These warnings are issued to the key emergency response agency and other stakeholders as specified, as well as the general public through the Bureau website.

These products are disseminated through the Internet as well as direct to some clients by email. In addition to these products, the service also includes the dissemination of an extensive set of data on present rainfall and river-level conditions to assist in preparation and interpretation of the warning products. Particular states have also experimented with special products for particular stakeholders, such as automated alerts when a rain gauge or stream gauge observes a value above a predetermined threshold.

Site-specific warnings of flash flooding are not provided as part of the flood warning service. Warnings of this type of flooding are covered by more generalized severe weather warnings that advise of the likelihood of intense rainfall with the potential to cause flash flooding. This service is supported by available weather radar information and provided by the Bureau as part of its severe weather service.

5 OPERATIONAL FORECASTING AND SYSTEMS

The Bureau has developed various systems to support the production and dissemination of operational forecasts (Fig. 8). The general aspects of the Bureau's systems are similar to those described elsewhere (eg, in other chapters of this book as well as Sene, 2008, 2010; World Meteorological Organization, 2011). The following sections describe some of the technical aspects specific to Australia.

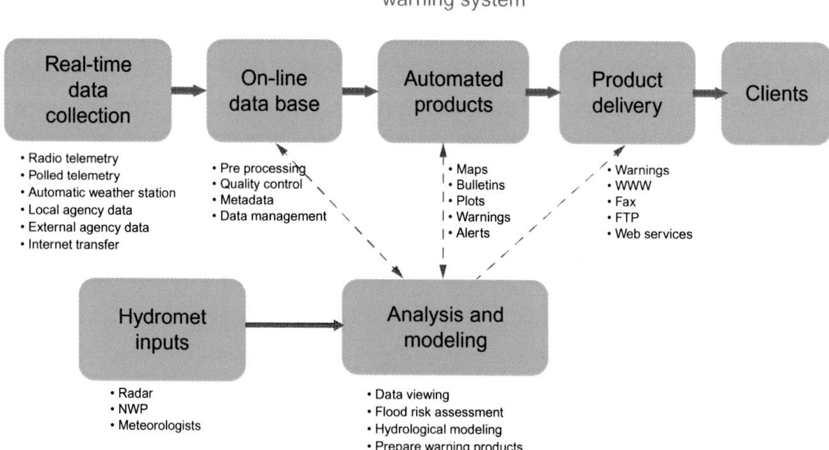

Fig. 8 Schematic of the Bureau of Meteorology's processes and products involved with the generation of flood forecasts and warnings.

5.1 Data Systems and Quality Control

The hydroclimatic variables of primary interest to the Bureau's flood forecasters are rainfall and river height. When necessary, forecasters also use information about reservoir storages and releases, as well as tide levels. The models used for the SDF service also use potential evapotranspiration data as forcings, which are derived from temperature measurements. The floods of very few regions in Australia are substantially affected by snowfall or snowmelt, and therefore measurements such as snow water equivalent are generally not available or used in the forecasting system.

In its role as the national weather agency, the Bureau has responsibility under the Meteorology Act (1955) for weather and climate monitoring and thus operates and maintains the major portion of the nation's rainfall (and other climate/weather) monitoring networks. Although the Bureau has established some river monitoring networks for the purpose of flood warning, data from other agency streamflow measurement networks at state, regional, and local levels are a necessary input to the national flood forecasting and warning system operation. State and territory governments are responsible for the assessment and management of water resources in their jurisdiction and undertake the majority of streamflow measurement and monitoring of the nation's rivers for this purpose. In some cases, to support the water resources activities, state and territory government agencies also undertake some rainfall monitoring. State and territory government agencies, and other agencies, supply the Bureau with their water data (primarily water level and rainfall data) for flood warning purposes or give the Bureau direct access to their automated stations.

In addition to the water information described earlier, in many areas of Australia, the Bureau and local governments cooperatively develop and operate rainfall and water level monitoring networks to support the provision of flood warning services. These systems have been established for both riverine flood warning situations involving relatively large catchments, and for specific flash flood warning systems operated by a local agency. Figs. 9 and 10 show the spatial coverage and ownership of rainfall and river-level gauges across Australia.

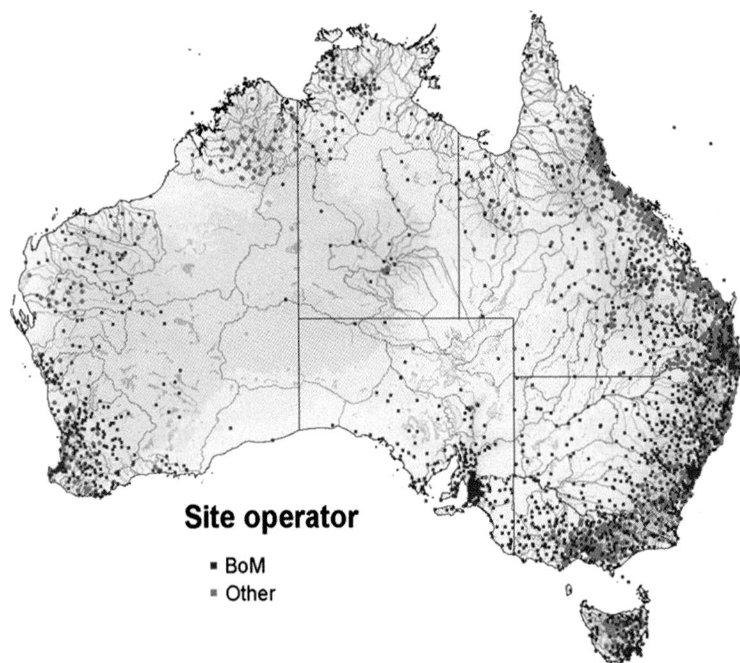

Fig. 9 Location of Australian Flood Warning Network rain gauges as of May 2013. *Red dots* are stations operated by the Bureau of Meteorology and *blue dots* are stations operated by partners. Approximately half of the >4100 rain gauges are operated by the Bureau.

The Bureau has operated a network of weather radars since 1948, when 15 naval radars were used to measure winds in the upper atmosphere, with the first radar-rainfall estimates occurring in Melbourne in 1953 (Seed, Siriwardena, Sun, Jordan, & Elliott, 2002). As of 2014, the Bureau has 69 radars, many of which are used to provide real-time estimates of rainfall. The

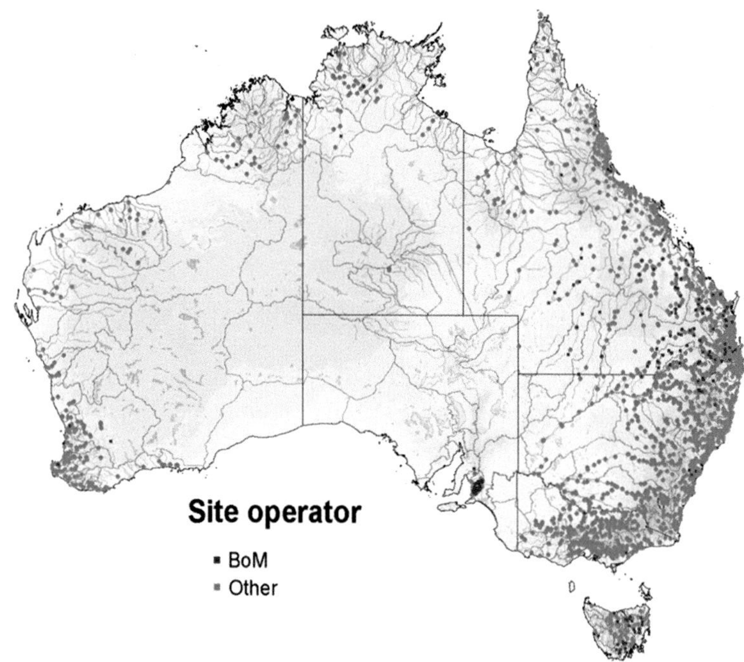

Fig. 10 Location of Australian Flood Warning Network river gauges as of May 2013. Only 5% of the approximately 2600 river gauges are operated by the Bureau.

Bureau also propagates and develops simulated storms 90 min into the future at 6 min resolution using the Short-Term Ensemble Prediction System (STEPS, Bowler, Pierce, & Seed, 2006). It is also running experimental systems to extend the STEPS predictions to 6 h ahead by blending it with numerical weather predictions (NWPs). Radar measurements are used extensively by meteorologists for nowcasting of heavy rainfall and severe weather, but procedures for using them quantitatively in flood forecasting are still being developed.

While flood warnings are based on river heights, hydrologic models simulate river discharge. Continuous discharge data are obtained by converting river heights using rating tables derived from historical manual gaugings. Rating tables are maintained by the data collection agency and are provided directly to the Bureau whenever they are updated and/or are available publicly on the Internet. At coastal locations, forecasters use dependent rating tables to adjust the river height according to the astronomical tides. Such dependent rating tables can also be used near the junctions of rivers where there are backwater effects (ie, the depth of a flood on a tributary can affect the river height upstream on the main stem and vice versa).

5.2 Modeling for Flood Forecasting

Aside from direct interpretation of the available data (eg, a particular river is nearly at moderate flood level), the standard tool of the forecaster is the hydrological model. The Bureau uses a variety of statistical and hydrologic simulation models to predict the timing and magnitude of future floods, as described in the following sections.

5.2.1 Statistical Models

The Bureau has a long history of using simple statistical tools that rely on the relatively high predictability of existing floodwaves traveling downstream. Typically, one axis contains the desired predictand (commonly the peak river height at a gauge of interest) and the other axis may contain a variety of predictors, most commonly peak river height at an upstream gauge, but also possibly rain gauge precipitation event totals (Fig. 11). Many (typically 10–30) historical events are then plotted, and a line of best fit derived statistically (by regression) or using forecaster expertise. Individual events may be labeled by their date or other relevant information. Outliers may be annotated. Although the correlation is not as strong, sometimes peak flow upstream is plotted against travel time between gauges (sometimes — but not always — large floods travel downstream faster than small floods).

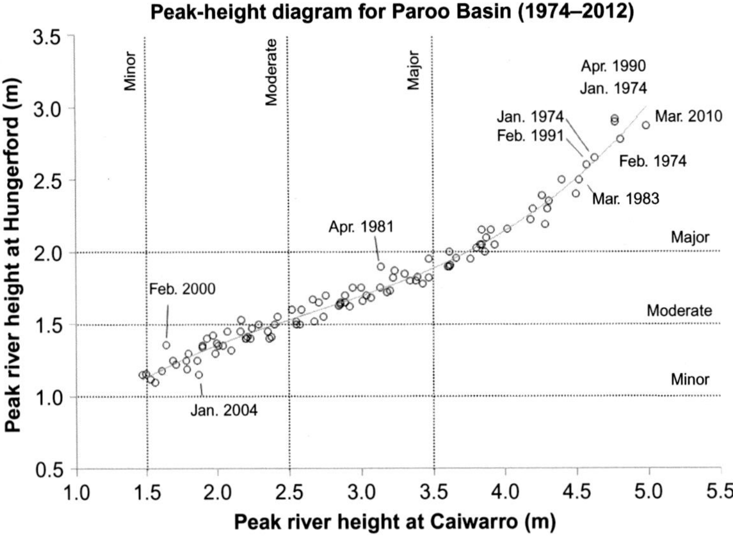

Fig. 11 Example of peak-height diagram. The axes are peak river heights for two locations in the Paroo Basin. The horizontal and vertical lines are flood classes (minor, moderate, and major), which give an indication of the severity of flood impacts.

An advantage of these sorts of tools is that they are relatively simple and easy to understand. Additionally, they can be printed and kept in binders as a handy reference for forecasters and in case of system/power failures. They also facilitate the forecaster comparing the current situation to past years and identifying possible analogues, which can be communicated to users during briefings and in warnings (eg, this is the largest flood since 1974). A drawback is that it requires the existence of upstream gauges and the forecast lead time is limited because the event must already be underway. These tools are of limited use in situations involving significant local area rainfall.

5.2.2 Hydrologic Simulation Models

In addition to statistical forecasting tools, flood forecasters also use a variety of dynamical simulation models. The Bureau uses spatially semidistributed rainfall-runoff models to simulate and predict streamflow. Here, elevation is used to delineate catchment boundaries above a given gauge (Fig. 12). The catchment is further subdivided into 5–25 relatively homogeneous model units (subareas) above each gauge. The number of subareas per river gauge reflects the density of the in situ raingauge network, with each subarea typically covering 30–50 km^2. Each subarea is connected by routing links to represent

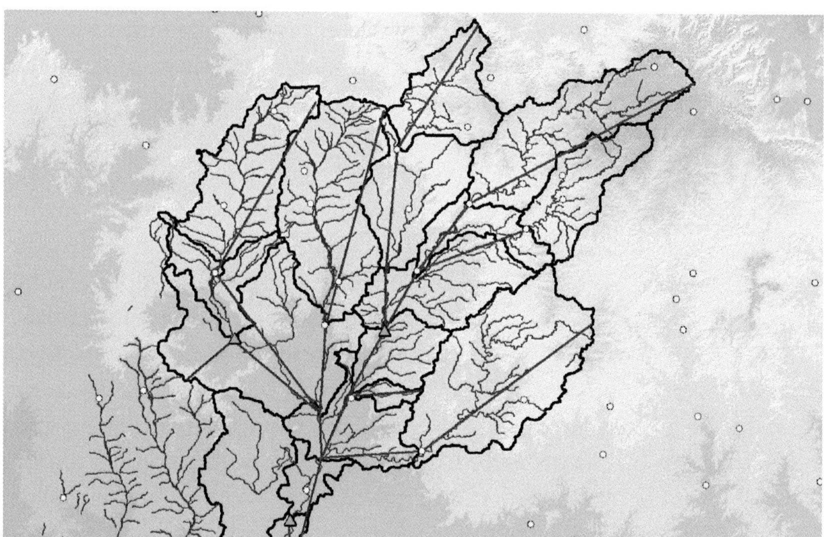

Fig. 12 Spatial representation of the Warrego River in Western Queensland. Streamlines are shown in *blue* and elevation is the colored background. Raingauges and streamgauges are shown as *yellow* and *cyan* icons, respectively. Flood model subareas are *black* outlines and routing links are shown in *red*.

the river network; the average link is 10-km long. Catchment delineation is currently done using a GIS package CatchmentSIM (Ryan, 2004).

Rainfall depth is calculated at the centroid of the subarea using an inverse distance weighted average of the total rainfall observed at the four nearest reporting rainfall stations. For the event-based models, the total rainfall for the duration of the event is spatially interpolated using a mix of pluviometric (eg, automated reporting hourly or at finer time intervals) and daily (eg, manually read) rainfall stations. This event total for each subarea centroid is then temporally disaggregated to the model timestep (typically hourly) using the relative pattern of the nearest reporting pluviometric station.

Bureau hydrologists primarily use event-based models, while continuous models have been introduced more recently as part of the 7-Day Streamflow Forecasting service. Continuous models explicitly account for the inter-storm variability of soil moisture by estimating losses from evapotranspiration and drainage. In contrast, event-based models require an externally supplied estimate of the initial loss of rainfall to soil moisture. This is often a subjective estimate based on forecaster interpretation of recent rainfall deficits. While this simplification may affect forecast accuracy, event-based models do not require serially complete precipitation forcings during the inter-storm period that continuous models do. The calibration, maintenance, and operation of continuous models require more data and effort than event-based models.

A drawback of event-based loss models, however, is that initial losses are highly uncertain before rising waters appear in the headwaters of the catchment. After the start of the flood, the losses can be estimated, for example, by matching observed to simulated flows through adjustment of the loss parameters. Forecast lead time is then limited to the travel time of the flood downstream. The historical reliance on event-based models in Australia was partly driven by a lack of data, but also because of the ephemeral nature of most catchments. The demand for longer lead times and more accurate forecasts has driven the Bureau's recent investment in continuous models.

The primary event-based model for Bureau flood forecasters is the Unified River Basin Simulator (URBS, Carroll, 2007), which was introduced to the agency in the early 1990s. The most common operational implementation of URBS involves the estimation of local runoff through an initial loss of rainfall to soil moisture deficits, followed by ongoing losses. Gross rainfall is interpolated to the subarea centroid using nearby rain gauge measurements. No runoff is produced until the initial loss is satisfied, otherwise a continuing/proportional loss is further removed and the remainder is effective (net) rainfall. The effective rainfall becomes runoff and can then be conveyed across the landscape using 1-parameter nonlinear reservoir catchment routing and/or 2-parameter Muskingum channel routing. URBS also has the ability to model reservoirs (the volume before spill or by an equation relating storage to discharge), baseflow, losses, and variable infiltration/recovering loss, and these features are used as needed.

Initial estimates and reasonable ranges of model parameters are determined by calibration of historical flood events. Typically, model setup has been done with manual calibration based on hydrologist experience and expertise, although the agency has recently begun to use Shuffle Complex Evolution to optimize parameters (Duan, Sorooshian, & Gupta, 1994). In real-time, forecasters modify the loss and river routing parameters to improve the simulated fit to the observed hydrograph.

The Bureau has recently adopted the *Génie Rural a 4 Parametres* (GR4J) continuous model for its Short-Term Streamflow Forecasting service. GR4J was developed by the *Institut national de Recherche en Sciences et Technologies pour l'Environnement et l'Agriculture* (IRSTEA) in France using a "top down" approach, testing 235 model structures and combining parts of 19 well-known models (Perrin, Michel, & Andreassian, 2001, 2003). IRSTEA tested the models using data from hundreds to thousands of catchments across four continents including Australia. The best-performing model with the least number of parameters became GR4J. This is in contrast with "bottom up" approaches, where equations of individual physical processes are pieced together into a complete model. GR4H is the hourly timestep variant of GR4J.

The Bureau of Meteorology selected the GR models based on the results of model comparisons done by several independent groups in the Commonwealth Scientific and Industrial Research Organisation (CSIRO), using data from hundreds of Australian catchments (Pagano, Hapuarachchi, & Wang, 2009). When calibrating the model parameters to get a good fit between simulated and observed runoff, GR4J performed about as well as many other models. However, when testing the model on an independent dataset, GR4J performed significantly and consistently better than the other models.

The GR models use rainfall and potential evapotranspiration as inputs. GR4J and GR4H contain two soil water stores and two unit hydrographs. The models' processes include evaporation, percolation, and both fast and slow pathways for runoff generation. The model was designed as a spatially lumped model, but the Bureau applies it in a semidistributed manner using Muskingum channel routing, similar to the URBS models.

The model has four parameters: two controlling the capacities of the soil water stores, one controlling the time length of the unit hydrographs, and one groundwater exchange coefficient. When this last parameter has a value other than zero, water is artificially added or subtracted from the water balance. Other modelers have similar corrections (eg, apply a multiplier to rainfall, potential evaporation, or streamflow), but tests have shown that the GR model's process is the most effective (Le Moine, Andreassian, Perrin, & Michel, 2007) and in some karstic catchments has a physical interpretation (Le Moine, Andreassian, & Mathevet, 2008). Parameters are automatically calibrated using Shuffled Complex Evolution optimization. These parameters are fixed for real-time operations.

5.2.2.1 Reservoir Modeling

URBS contains routines for simulating the behavior of reservoirs and other controlled structures. Most commonly, the forecaster specifies the volume remaining before spill at the start of the event. While the reservoir continues to fill, no streamflow passes downstream. Once the reservoir is full, some or all of the inflow becomes outflow. There are also options to specify a lookup-table or equation relating inflow or storage to outflow. If the past or future reservoir releases are known (eg, provided by the operator), they can replace the simulated outflow and are routed downstream.

5.2.2.2 Weather Predictions (QPF)

Before significant rain has fallen, unknown future rainfall is easily the greatest uncertainty in most Bureau river forecasts. Therefore, hydrologists have a keen interest in weather forecasts and track them at least daily during intensive operations. The Bureau uses quantitative precipitation forecasts (QPF) from models and systems, but also engages meteorologists to obtain guidance specifically tailored to the geographic area and forecast time which the hydrologist identifies will affect the flood forecast. Hydrologists may also force the models with fixed "what if" scenarios (eg, 50 mm, 100 mm, twice or half the rainfall forecast). Occasionally, zero future rainfall is used to estimate the least possible flooding expected.

Historically, meteorologists communicated rainfall forecasts to the hydrologists through a conversation and/or hand-drawn policy maps indicating, for example, that a particular area might receive "40–70 mm, with localized 100 mm totals from Sunday night to Saturday morning." These forecasts would be consistent with the official forecasts communicated elsewhere throughout the Bureau. Such scenarios were forced into the hydrologic models, specifying a storm total and duration.

With the adoption of HyFS (described in Section 5.3), hydrologists now have direct access to the same NWP model outputs used to generate the weather forecasts. These come from the Bureau of Meteorology's ACCESS (Australian Community Climate and Earth-System Simulator) system, which consists of the following configurations and domains (Fig. 13, Table 1).

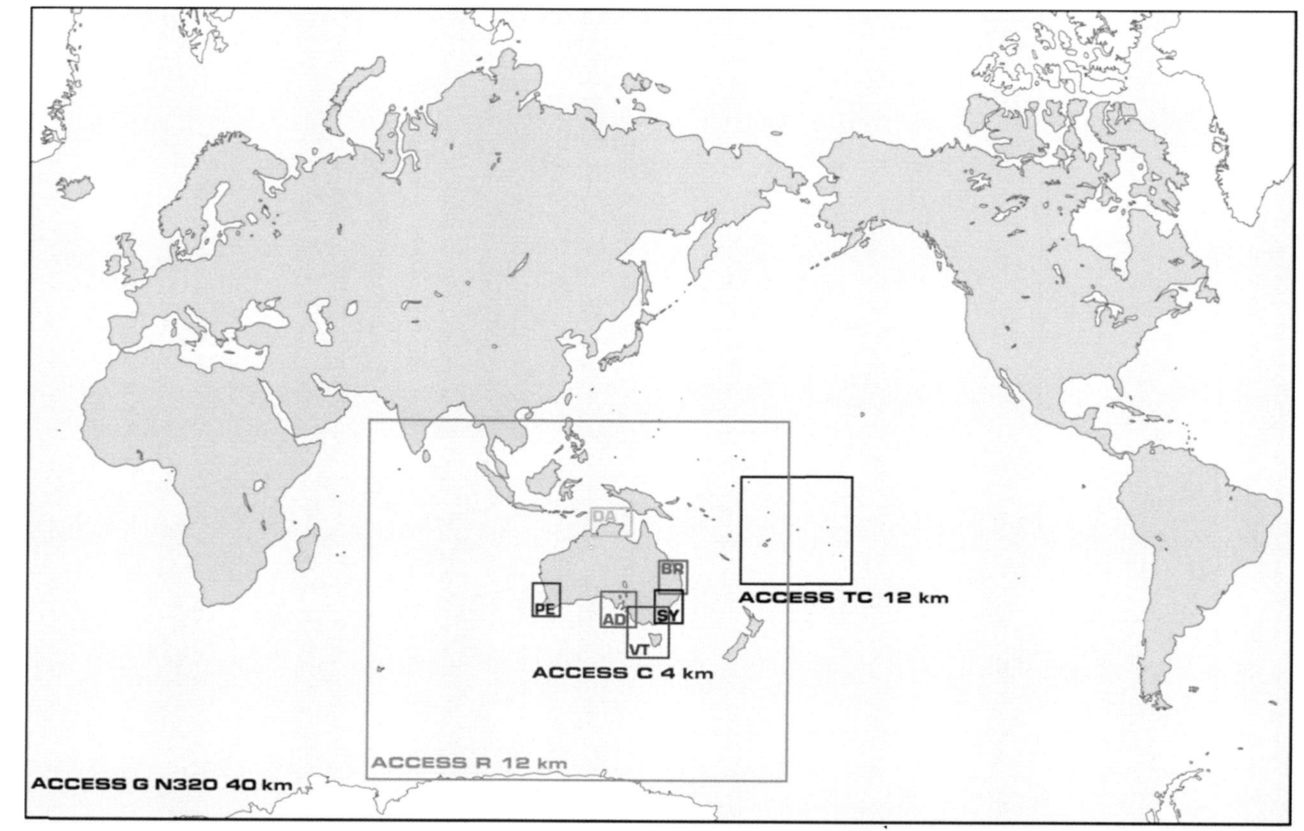

Fig. 13 Map of spatial domains of ACCESS NWP models. From http://web.bom.gov.au/nmoc/access/images/aps1-domains.png. Accessed 03.02.15. DA, Darwin; PE, Perth; AD, Adelaide; VT, Victoria-Tasmania; BR, Brisbane; SY, Sydney.

Table 1 Domains, resolutions, lead times, and release times of the ACCESS NWP model outputs as of Mar. 2014

Model	Domain	Resolution	Lead time	Release time (UTC)
ACCESS-G	Global	~40 km	10 days	00, 12
ACCESS-R	Regional	0.11 degree (~12 km)	3 days	00, 12
ACCESS-C[a]	City	0.036 degree (~4 km)	1.5 days	00, 06, 12, 18
ACCESS-TC	Relocatable	0.11 degree (~12 km)		00, 12

[a] ACCESS-C consists of a set of models covering domains around Sydney, Darwin, Melbourne, Brisbane, Perth, and Adelaide.

Currently, the SDF service drives the continuous hydrologic model with deterministic ACCESS-G output. There are, however, known limitations to the approach — such as the systematic biases of the weather model outputs — and the Bureau is developing seamless rainfall forecast systems that join forecasts from multiple lead times and resolutions as well as removing the biases.

In addition, the Bureau is creating "NexGen" products, meteorologist-edited probabilistic gridded forecasts for a range of elements including temperature, wind, humidity, and rainfall. This technology is based on the Graphical Forecast Editor system developed by the US National Weather Service. The suite includes 3-hourly mean, daily 50%, 25%, and 10% probability of exceedance precipitation. The probabilistic grids are not entirely suitable for direct input into the hydrologic model because they do not have the space-time characteristics of realistic rainfall. The advantage of using the NexGen products, however, is that they are consistent with the official forecasts of the Bureau and are enhanced by meteorologist expertise. The Bureau is developing policies and tools for the best use of this information.

5.2.2.3 Data Assimilation

A significant element of hydrologic forecasting is keeping model simulations consistent with recent observations of nonforcing variables (eg, soil moisture, streamflow). Differences between simulations and observations can persist over time, and adjustment of the simulations to compensate for errors can lead to more skilful predictions.

Bureau flood forecasters perform subjective data assimilation by tuning the URBS model parameters to achieve a better fit between the simulated and observed hydrographs. Additionally, at the hydrologist's discretion (depending mostly on the quality of the data), URBS' simulated streamflow can be replaced with observed streamflow and further routed down the river network in a process called "matching" (which in other contexts has been called "flow insertion," Cole, Robson, Howard, Bell, & Moore, 2011). This is particularly useful when rainfall has ceased and an observed floodwave is traveling downstream. The SDF service has the same matching/flow insertion options as the URBS models, except that SDF also uses linear error correction (Pagano, Wang, Hapuarachchi, & Robertson, 2011) to statistically shift the shorter lead-time forecasts based on the difference between simulated and observed at the start of the forecast period.

5.3 Production Environment: HYMODEL and HyFS

The Bureau currently operates the URBS model in a framework known as HYMODEL, which it developed in the early 2000s. It has a web-browser based interface that can control workflows for nearly all aspects of forecast production, including data preparation, visualization, data editing, model running, parameter adjustment, calibration, and report generation. While this system was effective for many years, it lacked extensibility and was difficult to connect to new data and forecast streams.

In 2015, the Bureau will complete the commissioning of a new forecast production environment the HyFS. It will use the Delft Flood Early Warning System (FEWS, Werner, Cranston, Harrison, Whitfield, & Schellekens, 2009; Werner et al., 2013) forecasting framework that has also been adopted by the US National Weather Service, the UK Environment Agency, and many other national hydrological forecasting agencies.

HyFS will be a 24-h per day, 7-day per week "category 1" supported system, meaning that its failure has immediate and serious impacts on essential Bureau operations and delivery of essential services, including external organizations. The Bureau has three levels of system support, and category 1 ensures the highest level of reliability. HyFS is also designed to be supported during full disaster recovery situations, as well as remote support for local forecasting operations. All software and metadata maintenance and development activities are done using structured version control and change management procedures.

HyFS is also supported by a significantly improved data collection system (Hydrologic Data Systems, HyDS, Fig. 14), which includes real-time data for more than 6000 rainfall and river sensors collected by the Bureau and the cooperative agencies that support flood warning services in Australia. HyFS also provides an advanced user interface that enables forecasters to

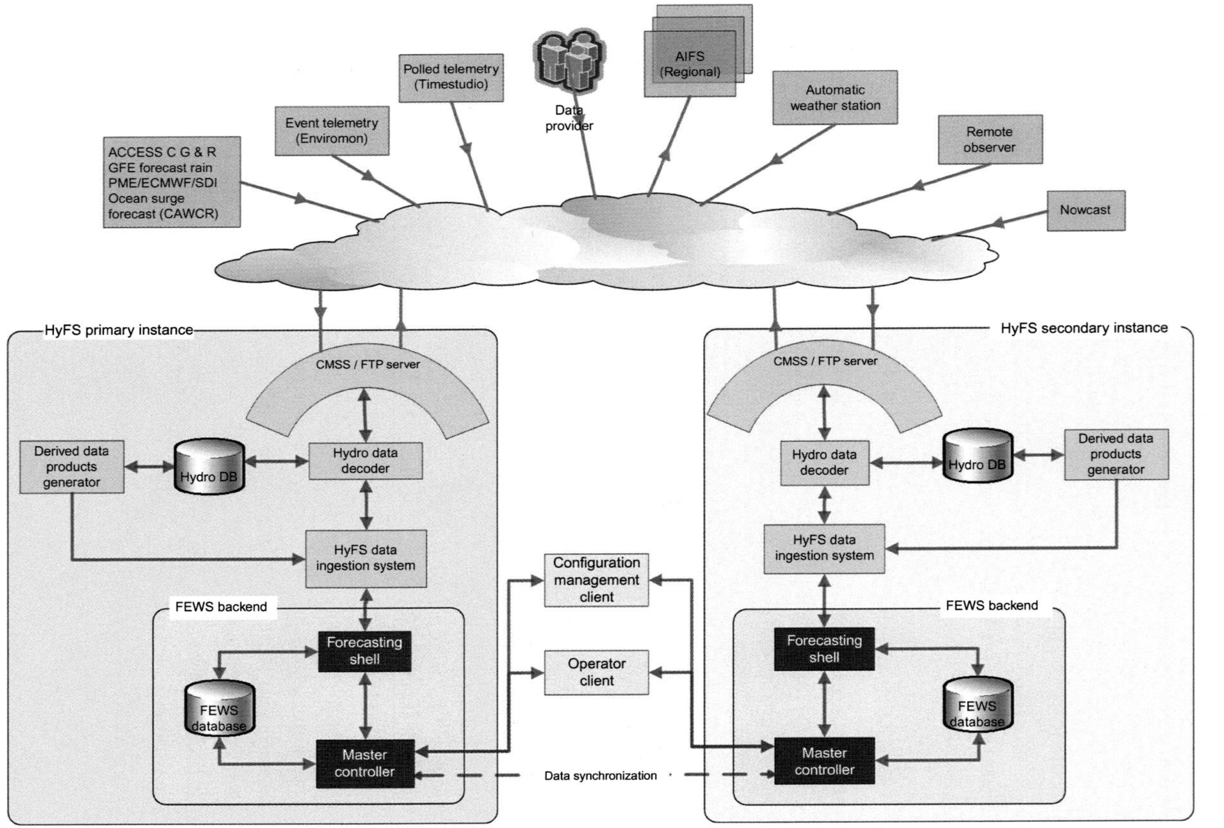

Fig. 14 Data management infrastructure to support HyFS for flood forecasting operations. Data enters the system from myriad sources (top), and is ingested into redundant internal data management systems (bottom, left, and right). The systems are synchronized for easy failover management.

visualize and quality control observations, but also provides automated data quality control tools.

In addition to providing the same functionality as HYMODEL, HyFS will facilitate the better use of advanced rainfall forecasting products, including the ability to work with ensemble forecasts, ingesting NWP model outputs and radar-rainfall quantitative precipitation estimates and nowcasts (Fig. 15).

The use of forecasts from the Bureau's ocean model, astronomical tide predictions from the National Tide Centre, and planned storm surge model will also be supported (Wilson et al., 2014). Flood forecasts and SDF will be produced in the same environment. This will facilitate the maintenance

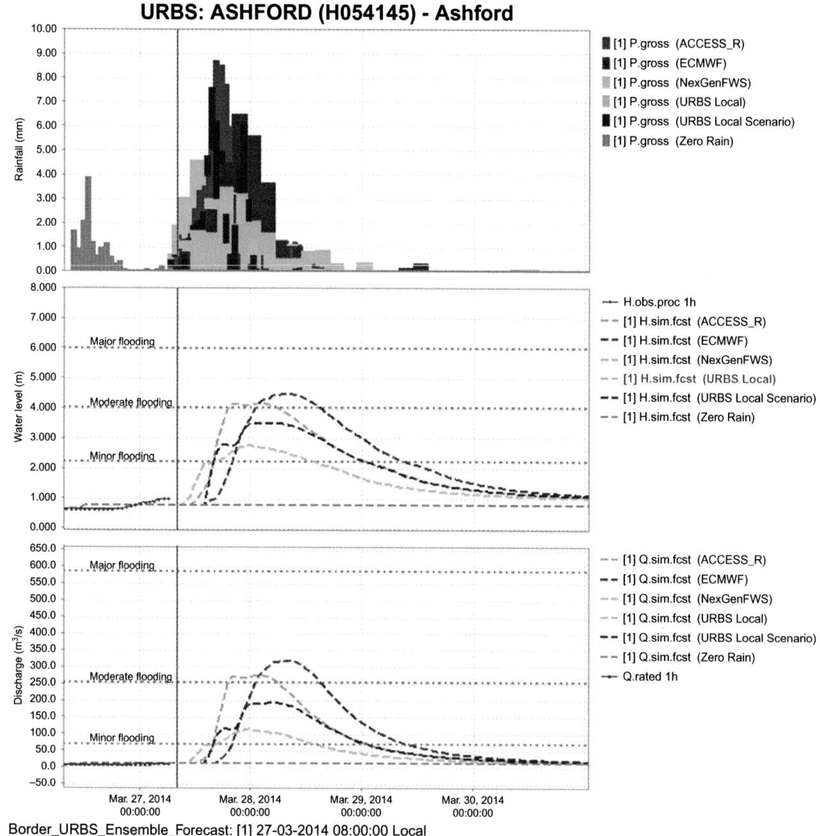

Fig. 15 Ensemble rainfall forecast (top) being used to drive an ensemble flood forecast for water level (middle) and discharge (bottom) for a location in northeastern New South Wales in Mar. 2014. Here, various sources of rainfall forecast guidance indicate various severities of flooding, highlighting the forecast uncertainty. Observed data are shown to the left of the vertical *blue* line.

and operations of the SDF products, and also enable the use of continuous models for flood forecasting.

One of the most appealing aspects of HyFS is its extensibility and customizability, such as the relative ease in which new data streams or models can be integrated into the production environment. HyFS provides the Bureau with a system that can evolve to meet future service demands and implement new advances in hydrological science. It is specifically designed to facilitate the transfer to research to the operational forecaster's desk.

Finally, among other features, FEWS has built-in scenario simulators (De Kleermaeker & Arentz, 2012) for training and the assessment of forecaster competency. This aids in the use of HyFS as an expert forecasting system.

5.4 Role of Expertise

Bureau hydrologists are actively engaged in the process of flood forecasting. Although data collection and transmission is largely automated, much of the hydrologists' work involves retrieving, cleaning, and infilling time-series data. Hydrologists often visually inspect data not only to assess its quality but also to recognize if any immediate action is necessary (eg, a reservoir has just filled and so flood warning is necessary for those downstream). Hydrologists often talk with meteorologists about the expected rainfall and use their judgment to determine what plausible scenarios should drive the rainfall-runoff models. Hydrologists use expertise to manually calibrate model parameters on historical events during model setup, but also tune parameters as events are unfolding. Further, hydrologists are involved in many aspects of product generation, dissemination, and communication, often directly engaging key stakeholders by telephone and/or email.

In contrast, the SDF service is primarily automated, similar to the Bureau's seasonal streamflow forecasting service (Plummer et al., 2009) where the direct model output is itself the official forecast. Duty officers monitor the status of the model runs but largely do not intervene in the forcing data, parameters, or states. This system has since gone officially live and is no longer in a trial period, with products delivered to a webpage accessible only by registered users. During this trial, the Bureau is using hydrologists to monitor the quality of the SDF products daily by comparing the simulations to recent observations.

6 OPERATIONAL CHALLENGES

The Bureau of Meteorology faces many of the same challenges as flood forecasting agencies overseas, such as issues related to data, modeling, warning/communication, and institutional factors (Pagano et al., 2014). However, some of these issues are exacerbated by the hydroclimatic and demographic characteristics of Australia.

6.1 Data and Services

Australia's hydroclimatic networks largely cover populated areas. Accordingly, there is extreme variability in gauge density. For example, the Northern Territory as a whole averages 1 gauge per 13,360 km^2. While this is well below WMO recommended standards, most of this area does not have flowing rivers. By comparison, metropolitan areas have quite dense monitoring networks. As water resource monitoring is largely the responsibility of agencies other than the Bureau, the networks have been developed mainly to satisfy purposes other than flood forecasting, such as water supply planning and management. As a consequence, these networks do not necessarily to reflect the data density required for flood forecasting models. For example, there are often few measurements in headwater regions.

Where the population density is very low, the Bureau may offer only generalized nonquantitative flood warning services. In contrast, the highest-priority and best-monitored catchments have full quantitative prediction services, aligned with either higher population densities or a more significant flood risk profile for the area. Because the level of service and staff varies according to the affected population and flood threat (eg, most of the hydrologists reside in Queensland, New South Wales, and Victoria), providing a service in smaller population regions, especially semi-arid regions with high variability, can be challenging.

For example, during the drought in 2006–08, South Australia issued no flood warnings, whereas during this same relatively dry period, Queensland issued more than a thousand. Long stretches between major floods also presents challenges in terms of staff gaining experience, retaining skills, remembering how systems operate, and maintaining engagement with stakeholders. The Bureau is addressing this by using a system of cluster regions and developing a centralized National Operations Unit that can provide remote support (eg, forecasters in Melbourne can monitor and generate products for the Northern Territory) or surge support, where forecasters from several states can fly into a flood-affected region. Such an approach requires increased standardization of systems and training.

Monitoring networks in Australia are owned and maintained by myriad agencies, and data collection and transmission methods are similarly diverse. This creates a data management burden for the Bureau and makes some data practically unavailable. Many of the sites providing river-level data may not be fully rated, and the repair of faults in communication hardware do not attract the urgency demanded of critical real-time operations. A range of data reporting protocols are also used, including event-based reporting, regular timed observations from polled systems, and discontinuous time series. A reliance on other agency polling systems to gather and transfer some data can introduce delays and reduce control over the data collection process.

Quality control and processing of data from such a diverse range of sources presents challenges. Only recently did over 200 of these water agencies move towards adopting a common format (Walker, Taylor, Cox, & Sheahan, 2009) for transferring hydrological data to the Bureau. As a result of this and other data management developments, the Bureau began serving river height and discharge data for the period of record of thousands of gauges freely on the Internet in 2014. However, a lack of easily accessible and quality controlled subdaily rainfall and discharge data for long periods of record has hindered research improving Australian flood forecasting systems (Pagano et al., 2009).

6.2 Modeling and Forecasting

As described earlier, to date, hydrological modeling has been event-based, using fairly simple models to estimate runoff volumes from rainfall. Recent improvements have seen the gradual introduction of continuous models. One advantage of the current event-based models is that they are simple to "adjust" during a flood event to better match the modeled and observed river behavior, and forecasters became quite skilled at this. Continuous (and perhaps more automated) models are more complex and the sort of adjustment that is needed to keep the modeled river levels accurately following the observed levels is not so obvious. While there are techniques for "updating" the continuous models, these are not necessarily intuitive to the forecaster and so a separation between the model and the forecaster can develop, denying the forecasting process the intuitive understanding of catchment (and model) behavior that a forecaster may be able to bring. The challenge is how best to integrate the expertise of the hydrologist in the forecast process, while streamlining his/her workflow as much as possible.

Finally, with the adoption of HyFS, hydrologists will have a much greater technical ability to directly input NWP-based rainfall forecasts into the hydrologic models, but the policy and training around such procedures are still under consideration. Methods are still in development for driving flood models using NexGen gridded official precipitation forecasts.

6.3 Warning and Communication

The Bureau role in the total flood warning system has predominantly been to provide forecasts of future river levels at key locations in a river system. Other agencies take these forecasts and convert them into flood impacts in the area surrounding the forecast location and activate flood response plans. Increasingly, agencies responsible for the interpretation of the Bureau's flood forecast are utilizing hydraulic modeling to generate flood maps, both in advance of the flood event and in real-time during the event. In some countries, the forecasting agency operates the hydraulic models and provides the real-time flood map as an output product rather than just the single forecast height. A properly prepared flood map is much more useful to the total flood warning process, and the challenge for Bureau forecasters is either to build the capacity to generate real-time flood maps or to develop the necessary links with the agencies currently engaged in flood mapping, to build a seamless service. With the growing number of agencies engaged in flood mapping, a related challenge is to manage the consistency and quality of the various products.

Effectively utilizing social media is a current operational challenge for forecasting agencies. This includes both the delivery of information as well as effectively harnessing the capacity of individuals in the flood-affected areas to provide "on the ground" information that may, for example, supplement modeling efforts. The benefit of geo-targeted warnings is also evident, although the associated technologies and policies are still in development.

Despite significant technological improvements in the various elements of the total flood warning system, the overall effectiveness of flood warning in mitigating flood damages may not always improve at a similar rate. While this is not a specific challenge for a forecasting agency, as a partner in the total system the Bureau needs to continue to work closely with the other agencies to ensure it is contributing as effectively as possible to the ultimate goal of the system. In particular, improvements in areas around effective risk communication and assisting those in the floodplain to recognize their level of risk and plan to take effective action have proven difficult in the past. Improvements in these areas are likely to be more effective if agencies follow a common strategy consistent with their respective roles within the total flood warning system.

6.4 Institutional Factors

Because of the way responsibilities are shared among the different levels of government in Australia, effective flood forecasting and warning has to involve a partnership between all three levels of government (Bureau of Meteorology, 2014). This applies to activities in emergency management as well as flood risk management and mitigation. While the individual agencies involved aim to work closely together, they can each be clients of different funding arrangements that may not always have the same goal when building new and improved flood warning systems. The data collection system can be seen as an end in itself, and the need for related investment in maintenance and in other elements of a total warning system is often neglected. The challenge for the forecasting agency is to seek to influence such projects such that the overall goal is prioritized, full operational costs are recognized, and the longer-term sustainability of the system is properly considered.

6.5 Understanding User Needs

Establishing a clear specification of the forecasting objective should be the first stage of the development and operation of a forecasting system. This requires a dialogue between the forecasting agency and the stakeholder (eg, emergency response group(s)). The latter need to be clear in their understanding of the flood hazard to which their constituent community is exposed and how advance information about the extent and nature of flooding can be utilized to make that community safer. This requires thinking through what that community can reasonably do to protect itself (evacuate, lift belongings, erect sandbag barriers, etc.) and the logistics/details of how they might undertake those tasks. This enables critical forecast locations and flood levels for each location to be specified and meaningful warning lead times to be established so these protective tasks can be planned and carried out effectively. These are most appropriately tasks for the emergency agencies. However, this information is often not available, which requires the forecasting agency to work with an incomplete understanding of the true needs of the forecasting system in any particular event. An example of where this challenge has been well met is in New South Wales, where the State Emergency Service has prepared a State Flood Plan in which the detailed forecasting requirements for each river valley have been agreed with the Bureau. However in other parts of the country this sort of information is only just starting to be collated systematically, facilitated by Service Level Agreements between the Bureau and its key clients.

7 FUTURE DIRECTIONS

A variety of recent improvements in Australia's flood warning system have focused on aspects such as staffing and operational arrangements, centralized systems, and improved modeling and products (described further by Wilson et al., 2014).

Between 2013 and 2017, the Bureau will increase the number of frontline flood forecasters by 23. Recent increases resulted in the expansion of a National Operations Unit to provide centralized support for regional forecasters. The Bureau is also developing and implementing a competency-based training program which involves a variety of training modules and assessments. Evidence of experience and training are required to qualify for different roles within the operational environment, such as Lead Flood Warning Duty Officer or Flood Data and Systems Duty Officer.

Increasingly, the Bureau is centralizing and standardizing its data management, forecasting environment and product generation infrastructure. There has been a substantial investment in the next generation flood forecasting system (HyFS), and developments of HyDS, improved archival of data and model runs, and an increase in verification activities.

Previous sections have mentioned the increased use of NWPs in flood forecasting. The Bureau is developing a suite of "seamless" rainfall forecasts (Cooper & Seed, 2013), for time scales ranging from 10 min to 10 days, with spatial resolutions from 1 to 100 km. The NWPs from several systems will be blended and calibrated (de-biased) using the STEPS (Bowler et al., 2006) and their uncertainty quantified through a set of 50 physically realistic ensembles. In 2014, the Bureau also engaged in a Forecast Demonstration Project where NWP models were updated hourly and run at higher than normal resolution (ie, 1.5 km). As part of this campaign, hydrologists modified the SDF models (ie, GR4H) to provide 1-, 3-, and 6-h duration Flash Flood Guidance using the method of Georgakakos (2006). The Bureau is viewing options for using advanced rainfall forecast products in operational flood forecasting, generating ensemble flood and SDF products, and increasing the automation of flood forecasting guidance.

The Bureau has a substantial initiative to standardize the language and structure of its flood warnings, and has plans to introduce spatial displays of flood warnings, such as warnings that are relevant to particular river reaches. The Bureau is increasing the navigability and interactivity of its flood information website, with an emphasis on mobile device users. As of Dec. 2014, the Bureau had the most popular Australian Government Facebook page, and in Feb. 2015 it joined Twitter. The Bureau is continuing to develop services for social media channels.

An important outcome of the reviews following the severe flooding of the past 4–5 years has been the need to clarify and communicate the working arrangements and responsibilities among the various stakeholders in the total flood warning system. This work included the preparation of documentation describing the national arrangements as well as reviewing and standardizing the service levels between the Bureau and client agencies in each jurisdiction. This work is currently progressed through a number of taskforce groups that have been established under the auspices of the Australian New Zealand Emergency Management Committee (ANZEMC). When complete, this information will be publicly available through the Bureau website.

8 SUMMARY

The provision of warning information to those at risk from flooding has been present since the early 1900s. Initially based on meteorological inputs only, the service soon matured to incorporate scientific hydrology and has grown to be a well-established forecasting service. This chapter has traced the development of the forecasting service and described the important climatological and demographic features of Australia that have influenced this development and led to the current forecasting and warning system. Characteristics of this system particular to Australia have been described along with the key challenges currently faced by the Bureau as the lead national flood forecasting agency. Severe flooding in the last few years identified areas of weakness in the system and gave impetus to a program of upgrades and improvement. Recent achievements have been described, along with plans to further improve the Bureau's national operations and systems. Each country works to develop their national flood forecasting and warning service as a balance between the nature of the flood problem, the prevailing organizational framework of relevant agencies, and available technological opportunities and capabilities. The system developed to date in Australia has evolved with the aim of achieving this balance, but may also include lessons and examples that are helpful to others.

ACKNOWLEDGMENTS

Thanks are given to Soori Sooriyakumaran and Dasarath Jayasuriya for their reviews of this chapter. We are also appreciative of Tanya Smith's editorial improvements.

REFERENCES

Australian Bureau of Statistics. (2012). *Year book Australia*. Canberra: Australian Bureau of Statistics.

Australian Bureau of Statistics. (2014). *Regional population growth, Australia, 2012–13*. Canberra: Australian Bureau of Statistics.

Australian Government. (2009). Flood warning. *Australian emergency management manual series: Vol. 21*. Barton, ACT: Australian Government. p. 79.

Bowler, N. E., Pierce, C. E., & Seed, A. W. (2006). STEPS: A probabilistic precipitation forecasting scheme which merges an extrapolation nowcast with downscaled NWP. *Quarterly Journal of the Royal Meteorological Society, 132*(620), 2127–2156.

Bureau of Meteorology. (2011). *Report to Queensland Floods Commission of Inquiry: Provided in response to a request for information from the Queensland Floods Commission of Inquiry received by the Bureau of Meteorology on 4 March 2011*. Melbourne, VIC: Bureau of Meteorology. p. 492.

Bureau of Meteorology. (2013). *Australian Government Response to the review of the Bureau of Meteorology's capacity to respond to future extreme weather and national disaster events and to provide seasonal forecasting services*. Melbourne, VIC: Bureau of Meteorology. p. 40.

Bureau of Meteorology. (2014). *Flood forecasting and warning services — National directive*. Melbourne, VIC: Bureau of Meteorology. p. 60.

Carbone, D., & Hanson, J. (2012). Floods: 10 of the deadliest in Australian history. *Australian Geographic*.

Carroll, D. G. (2007). *Unified River Basin Simulator (URBS): A rainfall runoff routing model for flood forecasting and design (v 4.30)*. Mount Gravatt, QLD: Don Carroll Project Management.

Cole, S., Robson, A., Howard, P., Bell, V., & Moore, B. (2011). Probabilistic flood forecasting for small catchments using the G2G model. In *Paper presented at the British Hydrological Society/Institute of Civil Engineers South West Region Workshop on "Flood forecasting in small catchments," University of Plymouth, June 17, 2011*.

Comrie, N. (2011). *Review of the 2010–11 flood warnings & response*. Melbourne, VIC: Victorian Government. p. 236.

Cooper, S., & Seed, A. (2013). Seamless rainfall forecasts for 0–10 days. In *Paper presented at the observing, estimating and forecasting rainfall: From science to applications, Melbourne, Victoria, Australia, October 21–23, 2013*.

De Kleermaeker, S., & Arentz, L. (2012). Serious gaming in training for crisis response. In *Paper presented at the 9th international ISCRAM conference, Vancouver, Canada, April 2012*.

Duan, Q., Sorooshian, S., & Gupta, V. K. (1994). Optimal use of the SCE-UA global optimization method for calibrating watershed models. *Journal of Hydrology, 158*(3–4), 265–284.

Georgakakos, K. P. (2006). Analytical results for operational flash flood guidance. *Journal of Hydrology, 317*(1), 81–103.

Keys, C. (1997). The total flood warning system: Concept and practice. In J. W. Handmer (Ed.), *Flood warning issues and practice in total system design* (pp. 13–22). London: Flood Hazard Research Centre, Middlesex University.

Le Moine, N., Andreassian, V., & Mathevet, T. (2008). Confronting surface and groundwater balances on the La Rochefoucauld-Touvre karstic system (Charente, France). *Water Resources Research, 44*(3). http://dx.doi.org/10.1029/2007wr005984.

Le Moine, N., Andreassian, V., Perrin, C., & Michel, C. (2007). How can rainfall-runoff models handle intercatchment groundwater flows? Theoretical study based on 1040 French catchments. *Water Resources Research, 43*(6). http://dx.doi.org/10.1029/2006wr005608.

Middlemann, M. H. (2007). *Natural hazards in Australia: Identifying risk analysis requirements*. Canberra: Geoscience Australia.

Munro, C. (2011). *Review of the Bureau of Meteorology's capacity to respond to future extreme weather and natural disaster events and to provide seasonal forecasting services*. Canberra, ACT: Department of Sustainability, Environment, Water, Population and Communities. p. 106.

Pagano, T. C., Hapuarachchi, H. A. P., & Wang, Q. J. (2009). *Development and testing of a multi-model rainfall-runoff streamflow forecasting application*. Water for a healthy country flagship report, Melbourne, VIC: Commonwealth Scientific and Industrial Research Organisation. p. 40.

Pagano, T. C., Wang, Q. J., Hapuarachchi, P., & Robertson, D. (2011). A dual-pass error-correction technique for forecasting streamflow. *Journal of Hydrology, 405*(3–4), 367–381. http://dx.doi.org/10.1016/j.jhydrol.2011.05.036.

Pagano, T. C., Wood, A. W., Ramos, M.-H., Cloke, H. L., Pappenberger, F., Clark, M. P., et al. (2014). Challenges of operational river forecasting. *Journal of Hydrometeorology, 15*(4), 1692–1707. http://dx.doi.org/10.1175/jhm-d-13-0188.1.

Perrin, C., Michel, C., & Andreassian, V. (2001). Does a large number of parameters enhance model performance? Comparative assessment of common catchment model structures on 429 catchments. *Journal of Hydrology, 242*(3–4), 275–301.

Perrin, C., Michel, C., & Andreassian, V. (2003). Improvement of a parsimonious model for streamflow simulation. *Journal of Hydrology, 279*(1–4), 275–289. http://dx.doi.org/10.1016/s0022-1694(03)00225-7.

Plummer, N., Tuteja, N., Wang, Q., Wang, E., Robertson, D., Zhou, S., et al. (2009). A seasonal water availability prediction service: Opportunities and challenges. In *Paper presented at the 18th World IMACS/MODSIM Congress*.

Queensland Floods Commission of Inquiry. (2012). *Queensland Floods Commission of Inquiry: Final report*. Brisbane, QLD: Queensland Floods Commission of Inquiry. p. 659.

Ryan, C. (2004). *CatchmentSIM: A terrain processing and hydrologic analysis model*. Cooperative Research Centre Toolkit version 1.26.

Seed, A., Siriwardena, L., Sun, X., Jordan, P., & Elliott, J. (2002). *On the calibration of Australian weather radars*. Technical report, Melbourne, VIC: Cooperative Research Centre for Catchment Hydrology. p. 41.

Sene, K. (2008). *Flood warning, forecasting and emergency response*. New York: Springer.

Sene, K. (2010). *Hydrometeorology: Forecasting and applications*. New York: Springer.

Walker, G., Taylor, P., Cox, S., & Sheahan, P. (2009). Water Data Transfer Format (WDTF): Guiding principles, technical challenges and the future. In *Paper presented at the 18th World IMACS Congress and MODSIM09 International Congress on Modelling and Simulation, Modelling and Simulation Society of Australia and New Zealand and International Association for Mathematics and Computers, in Simulation, Cairns, Queensland, Australia*.

Werner, M., Cranston, M., Harrison, T., Whitfield, D., & Schellekens, J. (2009). Recent developments in operational flood forecasting in England, Wales and Scotland. *Meteorological Applications, 16*(1), 13–22. http://dx.doi.org/10.1002/met.124.

Werner, M., Schellekens, J., Gijsbers, P., van Dijk, M., van den Akker, O., & Heynert, K. (2013). The Delft-FEWS flow forecasting system. *Environmental Modelling & Software, 40*, 65–77.

Wilson, D., Anderson, B., Robinson, J., Comeadow, S., & Dale, E. (2014). The Bureau of Meteorology's recent improvements in Australia's flood warning system. In *Paper presented at the hydrology and water resources symposium 2014, Barton, ACT, Australia*.

World Meteorological Organization. (2011). *Manual on flood forecasting and warning: WMO-No.1072*. Geneva, Switzerland: Hydrology and Water Resources Programme. p. 142.

CHAPTER 2

Hydrological Forecasting Practices in Brazil

F.M. Fan, R.C.D. Paiva, W. Collischonn
Federal University of Rio Grande do Sul (UFRGS), Porto Alegre, Brazil

1 INTRODUCTION

This chapter presents the current hydrological and flood forecasting practices in Brazil, including the main forecast applications, the different kinds of techniques that are currently being employed and the institutions involved in forecast generation. A brief overview of Brazil is provided in Section 2. In Section 3, a general discussion about the Brazilian practices on hydrological short- and medium-range forecasting is presented. Section 4 presents detailed examples of some hydrological forecasting systems that are operational or in a research/preoperational phase. Finally, in the conclusion section, some suggestions are given about how the forecasting practices in Brazil can be understood nowadays, and what the perspectives are for the future.

2 GEOGRAPHY, CLIMATE, AND FLOODS IN BRAZIL

Brazil is the largest country in South America, covering 8,515,767 km^2, which corresponds to 47% of the continent (Fig. 1). Two of the world's major river basins are located in Brazil, namely the Amazon and La Plata, and a great part of Brazilian rivers and water resources are shared among Brazil and several South American neighboring countries.

The Brazilian territory covers a vast area with diverse climate, vegetation, topographic, and hydrological characteristics. A tropical climate can be found in the Amazon region in the north, with high precipitation rates, dense forest vegetation, and large rivers with massive wetlands that are seasonally flooded. The climate shift to arid conditions at northeastern regions, where intermittent rivers can be found. Central Brazil is the home of the Cerrado biome, which is covered by agricultural and pasture lands and a savannah-type natural vegetation. The climate is seasonally driven and precipitation occurs mostly on the south hemisphere in the summer. The southern regions present a temperate climate with intense precipitation occurring year round.

Flood Forecasting
http://dx.doi.org/10.1016/B978-0-12-801884-2.00002-5

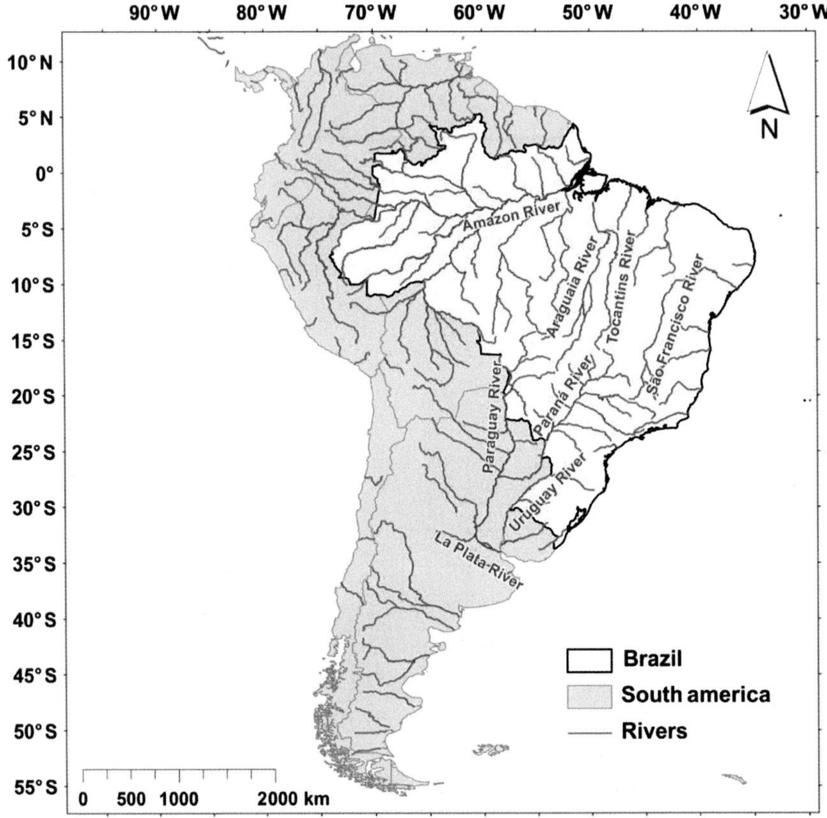

Fig. 1 Brazil location within South America and main rivers.

Although part of the international attention on Brazil's water resources focuses on the Amazon basin, most of the country's population is located in the south-southeast region and along the coast, as can be seen by the distribution of municipalities (Fig. 2). Important economical and societal impacts of flood hazards are frequent, some related to flooding of the large Brazilian rivers and several others to smaller rivers. Since the 1970s, more than 19 million Brazilians have been affected by floods, according to Guha-Sapir, Below, and Hoyois (2015). Another important issue in Brazil is the use of water resources for energy production. A large amount of the major rivers is regulated by reservoir of hydropower dams (Fig. 2), and hydrological information is important not only for the optimization of energy generation but also for operations targeting the mitigation of floods.

This brief description of Brazil highlights the important role that flood forecast information may have, but also some challenging features for the

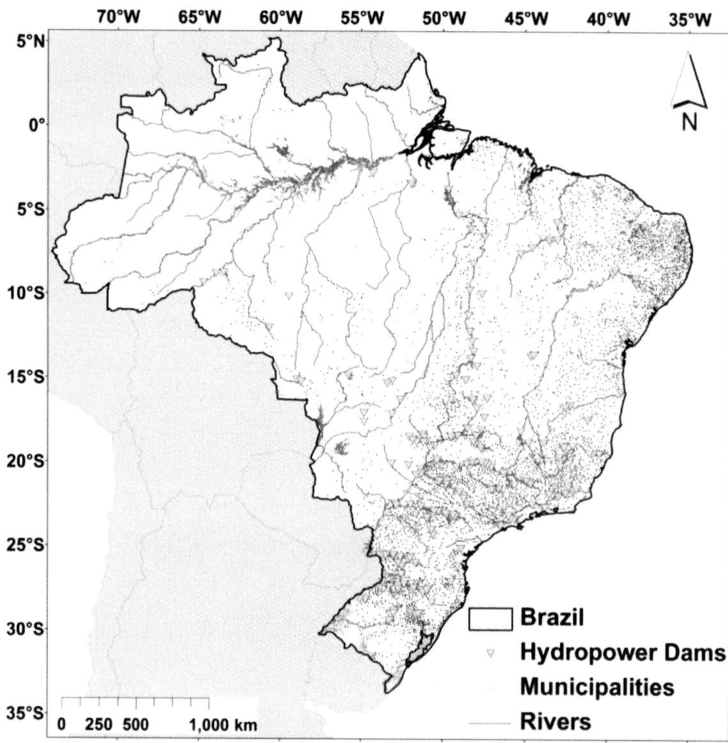

Fig. 2 Brazilian municipalities, main hydropower dams, and rivers.

development of hydrological forecast systems, including: (i) the continental scale of the region; (ii) the trans-boundary nature of some basins; (iii) the diversity of climates and hydrological characteristics; (iv) flood hazards occurring at multiple spatiotemporal scales and the need for forecast information for both larger and smaller rivers; and (v) the extensive river flow operation by hydropower plants.

3 GENERAL OVERVIEW OF HYDROLOGICAL FORECASTS IN BRAZIL

Short- and medium-range flow forecasts are mainly used in the Brazilian context for two purposes: (i) the scheduling of hydropower reservoirs operation; and (ii) flood forecasting in vulnerable locations. There is a greater emphasis on the first one. As a consequence, it is not possible to describe Brazilian forecasting practices without talking about hydropower reservoirs and power generation, as some reservoirs operate to mitigate flood impacts on upstream and downstream cities (Araujo et al., 2014; Fan, Collischonn, Meller, & Botelho, 2014).

The National System Operator (*Operador Nacional do Sistema*, or ONS) is responsible for the coordination of power generation facilities and electricity transmission in the National Interconnected System (*Sistema Interconectado Nacional*, or SIN), under the supervision and regulation of the National Electric Energy Agency (*Agência Nacional de Energia Elétrica* or ANEEL).

Under normal flow conditions, ONS uses forecasts of average daily inflow with lead times up to 14 days to schedule the hydropower generation in the system. These forecasts are generated by ONS or by the dam operation agents (responsible for the plants) themselves (Costa, Raupp, Damazio, Oliveira, & Guilhon, 2014; Oliveira, Guilhon, Costa, Raupp, & Damazio, 2014; ONS, 2011, 2012; Zambon et al., 2012; Zambon, Barros, Gimenes, Bozzini, & Yeh, 2014; Zambon, Barros, & Yeh, 2014).

Methods used to produce such forecasts are diverse. Many plants still use forecasting models with a poor physical basis, such as PREVIVAZH, which is a model based on weekly forecast desegregation from the inferred trend of recent past flows and daily synthetic series of natural flows (Guilhon, Rocha, & Moreira, 2007; ONS, 2011, 2012).

Aiming at a better forecast performance, ONS recently organized a study to assess different alternatives for predicting daily inflow to the hydroelectric power plants (HPPs) that integrate the SIN (Guilhon et al., 2007). The studied HPP dams are located in the basins of the rivers Iguaçu, Paraná, and Paranaíba. The forecast methods ranged from process-based modeling using lumped or distributed conceptual models, through hybrid methodologies, stochastic models, artificial intelligence, and data mining techniques.

At the end of the work, the main conclusion was that most of the alternatives outperformed PREVIVAZ. Also, it is concluded that using information about future precipitation resulted in an improvement to the quality of forecasted streamflow (Guilhon et al., 2007).

After this work, a greater diversity of models for predicting inflow to Brazilian hydroelectric plants has been adopted. The use of information arising from weather forecasts obtained by meteorological models also increased. For example, in the São Francisco River basin, in the Três Marias HPP, a model based on neural networks presented by Gomes, Montenegro, and Valença (2010) is used, and there are recent efforts to use forecasts based on a physically based distributed MGB-IPH model (Fan, Collischonn, Meller, et al., 2014). The same model is being used at the Parnaíba River, São Simão HPP (Collischonn et al., 2007), and in the Upper Uruguay River basin at Campos Novos HPP and Barra Grande HPP (Fan et al., 2014c). For the Iguaçu River, a hydrological forecasting system named SISPSHI, presented by Araujo et al. (2014), is used, applying quantitative precipitation forecasts (QPFs) results to estimate river flow at 21 sites along the whole watershed, including Foz de Areia HPP and Segredo HPP.

When the inflows to SIN reservoirs are not at normal conditions, but in a state of attention, alert, or emergency (defined by the occupation of expected volumes and violation of hydraulic constraints), the operation of the reservoirs is usually not controlled by ONS, but by the hydropower companies following a few guidelines established by ONS and ANEEL (Costa et al., 2014; Oliveira et al., 2014; ONS, 2011, 2012; Zambon et al., 2012; Zambon, Barros, Gimenes, et al., 2014; Zambon, Barros, & Yeh, 2014).

For such cases, forecast information also has great value to support decision-making, especially during floods when the anticipated knowledge about an event provides valuable time for reservoir operation. Some reservoirs, such as the ones from Três Marias HPP, Estreito HPP, Foz de Areia HPP, Segredo HPP, and even Itaipu HPP, have operational restrictions to help reduce or avoid downstream and upstream floods, and ensure the dam's safety.

Aiming at a better operation during floods, some Brazilian hydropower companies are funding the development of custom hydrological forecasting systems. Such systems are designed to provide detailed forecasts for specific hydropower plants and at an hourly temporal resolution that would allow operational decision-making and mitigate any violation of operational restrictions.

Another use of forecasts that is growing in Brazil relates to human, economic, and sanitary impacts of floods. Flood inundation caused large negative impacts in the last decades, and according to the Emergency Events Database EM-DAT (CRED, Université Catholique de Louvain, Brussels, http://www.emdat.be), Brazil is among the 10 countries most affected by flooding in the world.

Great advances are expected for future years concerning the development of flood alerts systems, especially after the Brazilian federal government established a Center for Natural Disaster Monitoring and Alert (CEMADEN), led by the Ministry of Science and Technology (MCT). CEMADEN aims at developing, testing, and implementing monitoring and forecast systems for natural disasters in Brazil. However, as it is a recently created center, operational flood forecast systems have not been developed yet.

Most of the operational flood forecast systems in Brazil are led by the Geological Survey of Brazil (CPRM). It operates forecast systems in vulnerable areas at some rivers, such as Doce (MG), Caí (RS), Taquari (RS), Parnaíba (Piauí), Muriaé (RJ e MG); Negro (AM), Acre (AC), and Branco (RR). The forecasts are available at the CPRM website (http://www.cprm.gov.br) in a system called *Sistema de Alerta Contra Enchentes*. These forecasts are performed using data-based models, showing good performance for short lead times, usually less than 3 days.

An example of operational forecast system is the one at the river Itajaí (http://www.comiteitajai.org.br/alerta), which is held by the local basin committee in the state of Santa Catarina. This system also uses a data-based model with a short lead time (Cordero, Momo, & Severo, 2011; Pinheiro, 2003). Another example is given by Silva et al. (2014), which shows the MAVEN system (Monitor AVançado de ENchentes — *Advanced Floods Monitor*), an automated flood forecasting system based on the HEC-HMS/RAS (Hydrologic Engineering Centers - Hydrologic Modeling System/River Analysis System) models for subbasins in the Pernambuco state (Brazilian northeast).

Those early flood forecasting systems mentioned above are only some samples of works being developed in the country. More works not mentioned here are also known to be under development. Besides the short- and medium-range forecasting examples cited, it is important to note that there are also efforts being made on long-term seasonal forecasts (Collischonn & Tucci, 2005; Reis, Nascimento, & Martins, 2007). Operational systems of this type are widely used in the country to evaluate water availability aiming at agricultural and energetic issues. However, these systems are not the focus of this chapter.

4 EXAMPLES OF OPERATIONAL AND PREOPERATIONAL FORECASTING SYSTEMS

This section presents detailed descriptions of some selected operational and research or preoperational hydrological forecast systems at different regions in Brazil: the Doce River basin system in the southeast region, the Upper São Francisco River basin also in the southeast, the Upper Uruguay River basin system in the south, and the experimental Amazon River basin model in the north.

4.1 Doce River Flood Forecasting System Maintained by the Brazilian Geological Service (CPRM)

The Doce River basin is located in the southeastern region of Brazil (Fig. 3) and drains 83,400 km^2, 86% of which is inside Minas Gerais state, and the remaining 14% in Espírito Santo state. Near to the Doce at the city of Colatina (about 120,000 inhabitants), which is impacted by floods.

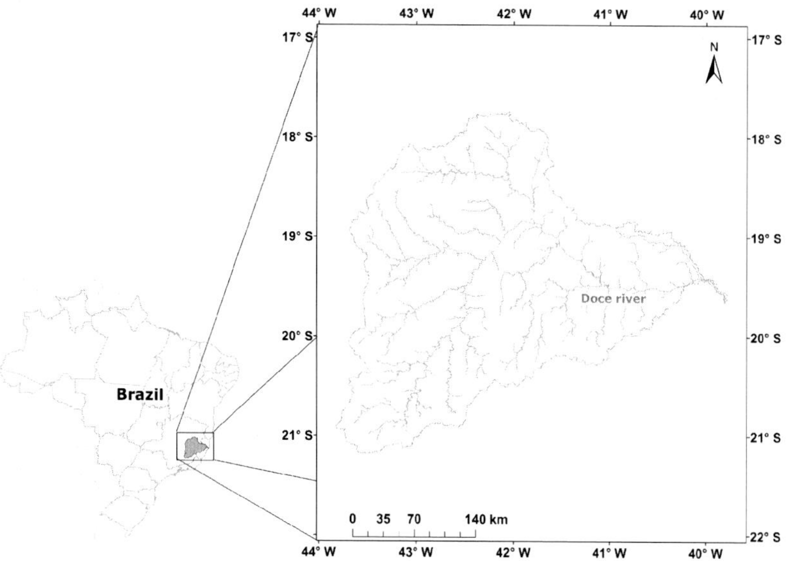

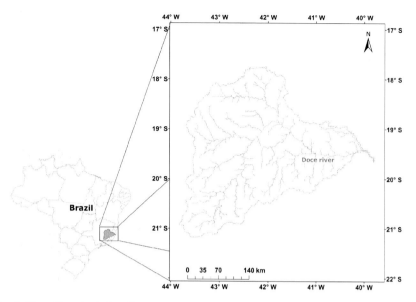

Fig. 3 Doce River basin location.

The Doce basin flood early warning system (FEWS) is operated 24h a day during the rainy season (Dec. to Mar.) by CPRM. Hydrometeorological data is collected from 45 stations throughout the basin. This data is processed by the CPRM team to generate weather and water-level forecasts that are produced and distributed to municipal authorities by email. Such data and forecasts are also disseminated through the CPRM web portal. The forecasting system team works 24h every day and includes hydrology engineers, meteorologists, field teams, and technicians.

The system is based on simple statistical linear models to forecast streamflow 12h ahead in downstream locations based on upstream gauging stations. One example of the system operation in 2013 is showed by Matos, Davis, Silva, Almeida, and Candido (2014). In this year the South Atlantic Convergence Zone caused high rainfall, above the historical average in the eastern portion of the state of Minas Gerais and central portion of the state of Espírito Santo. This system was the causative agent of the severe flooding that occurred in the town of Colatina and in other cities nearby. The warning system of the Doce River basin, which completed 17 years of operation in this rainy season of 2013/2014, was essential for the monitored cities to minimize the negative impacts of flood.

4.2 Upper Uruguay River Basin Hydropower Reservoir Inflow Forecasting System

The Upper Uruguay River forecasting system is an operational platform first presented by Fan et al. (2014c) and Fan et al. (2012). The systems has been operational since Jul. 2013 and provide daily forecasts for Barra Grande HPP and the Campos Novos HPP, located at the rivers Pelotas and Canoas, respectively (Fig. 4).

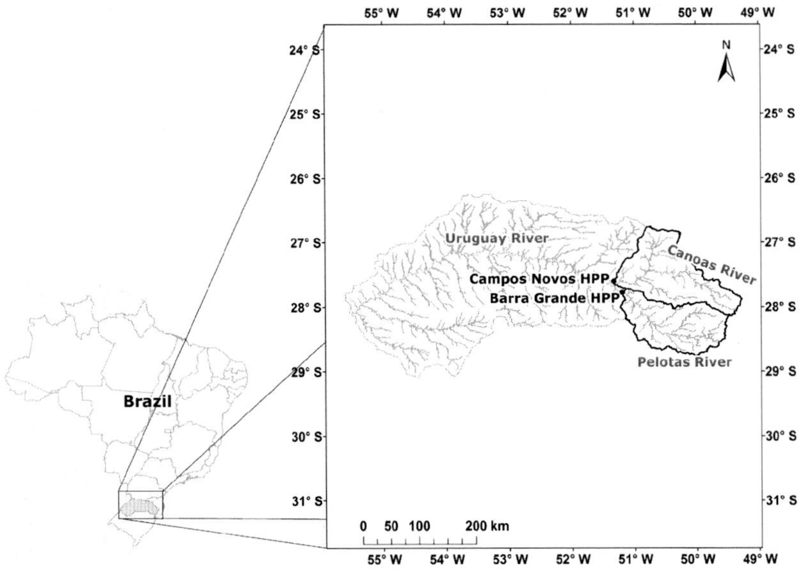

Fig. 4 Position of the Upper Uruguay River basin within South America. The main subbasins that the system is focused on are highlighted in the upstream area.

Early flood forecasts are especially important for this basin as it has a fast surface runoff component, as a consequence of topographical and geological local conditions. Also, high precipitation rates can occur any time during the year, causing a rapid increase in river discharge.

The system provides deterministic streamflow forecasts with 10 days lead time. It is on the process-based distributed hydrological model MGB-IPH (Collischonn et al., 2007) that uses input information from: (i) real-time rainfall; (ii) streamflow data from automatic in situ gauges; and (iii) QPFs from a Numerical Weather Prediction (NWP) model.

Recent past measured rainfall data recorded by the telemetric gauging network is fundamental for the system. This rainfall information goes through an inverse distance weighted interpolation process for serving as input information to spin up the hydrological model. Also, the streamflow telemetric data in known locations within the basin are used in the MGB-IPH hydrological data assimilation scheme (Fan, Meller, & Collischonn, 2015; Paz, Collischonn, Tucci, Clarke, & Allasia, 2007).

Information about future rainfall currently comes from the NWP model Eta-15km (Chou et al., 2007). The Eta-15km model is a version of the Eta model with spatial resolution of 15 km that provides forecasts over South America twice a day with lead time of 7 days. Eta forecasts are available online at the Brazilian meteorological center CPTEC (*Centro de Previsão de Tempo e Estudos Climáticos*) website. The total lead time of the issued hydrological forecasts by the system is 10 days (240 h). For the first 7 days, the CPTEC Eta 15km rainfall forecasts are used as an input, and for the last 3 days, precipitation is considered null.

The system operation itself is done automatically. Results are monitored daily by a team of researchers from the Large Scale Hydrology Group (www.ufrgs.br/lsh) and have been used not only for research purposes on operational hydrology, but mainly for the Barra Grande and Canoas reservoirs operation in the Upper Uruguay River basin. Results from the system are sent daily to ONS and to the dam's operators after quality verification. An example of a forecast issued by this system is shown in Fig. 5.

System performance was evaluated for its first year of operation by Fan, Pontes, et al. (2014), as presented in Fig. 6. Values of the Nash Sutcliffe coefficient were positive for lead times lower or equal than 220 h (around 9 days), indicating the model is informative until this lead time. The Persistence Coefficient has high values from early lead times. These high values show that using the last observed values as a prediction is not very convenient in comparison to using the forecasting system from the early lead times of the forecast. This is probably due to the river basin's fast response to rainfall. By using a forecasting system point of view, this result can be considered very positive, supporting the system's importance.

Visual analysis of forecasts usually shows that the predictions have good agreement with observed flows, predicting the possible occurrence of floods 5-7 days in advance, which is generally confirmed by the predictions issued with smaller lead times. This is consistent with the performance lead time analysis, where we showed that the best performance of the model is for lead times up to 2 days, and also an adequate performance can be obtained with lead times up to 210-220 h.

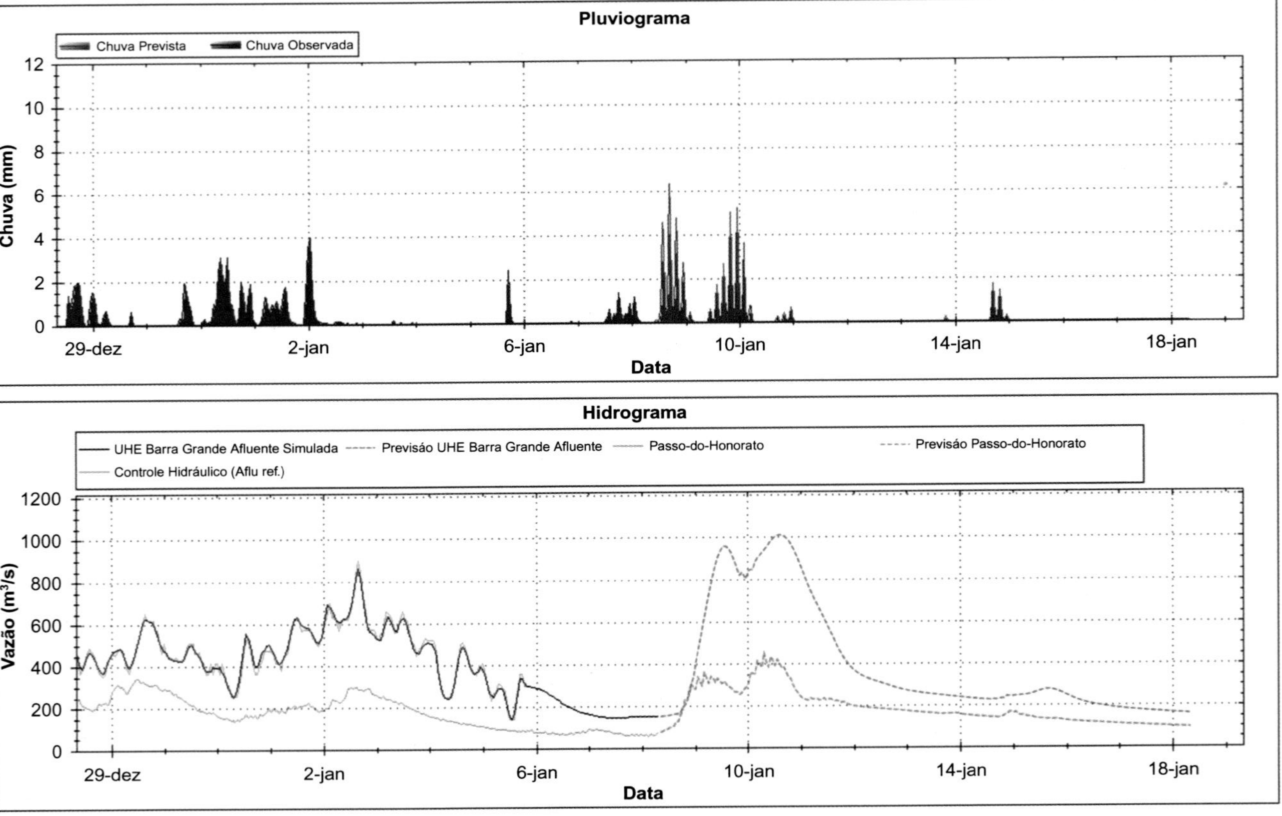

Fig. 5 Sample forecast issued by the Upper Uruguay System. *On the top*, the rainfall forecast. *On the bottom*, the streamflow forecast. *Straight lines* are observed values, and *dashed lines* are forecasted values.

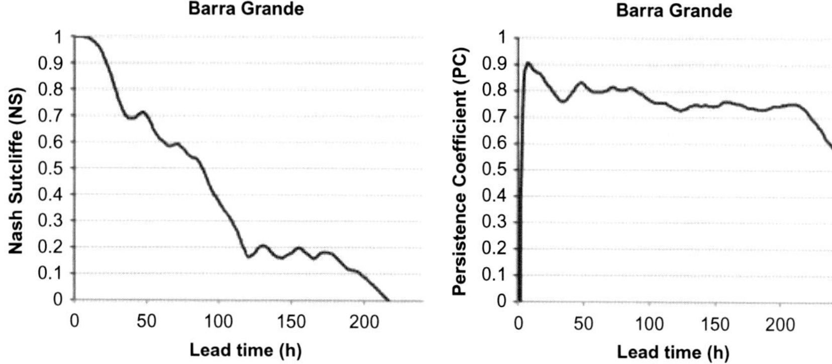

Fig. 6 Upper Uruguay first year of operation performance analysis. Data obtained from Fan, F. M., Pontes, P. R. M., Beltrame, L. F. S., Collischonn, W., & Buarque, D. C. (2014). Operational flood forecasting system to the Uruguay River basin using the hydrological model MGB-IPH. In *Proceedings of 6th international conference on flood management — ICFM6*. São Paulo: ABRH.

Despite the overall good results of the system, the flow forecasts are not free of errors. For example, some precipitation events were not captured by the existing gauge network, causing errors in simulated discharges. Flood forecasts in the Uruguay basin from this system could benefit from better rainfall telemetric network or radar precipitation estimates, improvements on the data assimilation methods, and the use of probabilistic precipitation forecasts.

4.3 São Francisco River Basin Ensemble Forecasting System

The São Francisco River forecasting system was developed to provide inflow information to the Três Marias HPP basin that is operated by CEMIG (Companhia Energética de Minas Gerais) and is located at the upper part of the São Francisco River basin, in southeastern Brazil. At the dam location, the drainage area of the São Francisco River is of approximately 50,000 km^2. However, due to possible impacts of reservoir operation downstream of the dam, the area of interest in this system extends to Pirapora city, located 120 km downstream of the dam, where the total drainage area consists of about 60,000 km^2 (see Fig. 7).

Reservoir operation targets multiple purposes, as hydropower generation and flood control the dam and flood mitigation at Pirapora city, upstream. However, floods in Pirapora are partly caused by the inflow of the Abaeté River, located downstream of the dam. This river drains 5224 km^2 of a region with high slopes and as a consequence, it presents rapid floods.

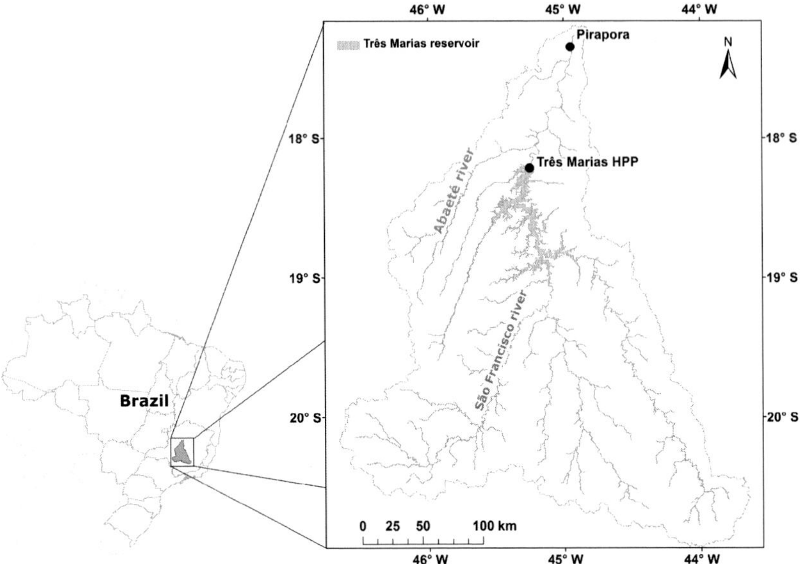

Fig. 7 São Francisco River basin location, up until the confluence with Das Velhas River.

When floods are verified or forecasted in the Abaeté River, the HPP operators release less water from the reservoir to account for the flood peaks expected from the river. Because of this, two main places are crucial for forecasting results within the basin: the Três Marias HPP Reservoir inflow, and the Abaeté River flow.

The forecast system is based on the large-scale distributed hydrological model MGB-IPH (Collischonn et al., 2007; Fan, Collischonn, Meller, et al., 2014) and runs at an hourly time step. It uses real-time information from a network of rainfall and streamflow gauges. It also uses an empirical data assimilation method to take into account real-time streamflow observations to update its state variables. After the initial developments (Fan, Collischonn, Buarque, Tucci, & Botelho, 2013), the system was upgraded to work with ensemble forecasts from the Global Ensemble Forecasting System (GEFS) v2 model (Fan, Collischonn, Meller, et al., 2014; Fan, Collischonn, Quiroz, et al., 2014; Fan, Pontes, et al., 2014). Then, in a second phase of developments, the system was coupled with the Deltares FEWSs (Werner et al., 2013), and with the Real Time Control Tools (RTC-Tools — Schwanenberg, Becker, & Xu, 2015). This coupled version of the model allows for the use of multiple QPF from different NWP models, including meteorological ensemble prediction systems (EPSs) and the application of

RTC-Tools reservoir operation optimization algorithms, including stochastic optimization. Details about this coupled version of the system are presented by Schwanenberg, Fan, et al. (2015).

Fig. 8 shows a sample ensemble forecast from the system. This forecast was issued using QPF from the CPTEC global EPS to force the hydrological model. The São Francisco forecasting system is currently used by CEMIG to support decisions related to the Três Marias reservoir operation, especially for flood control purposes.

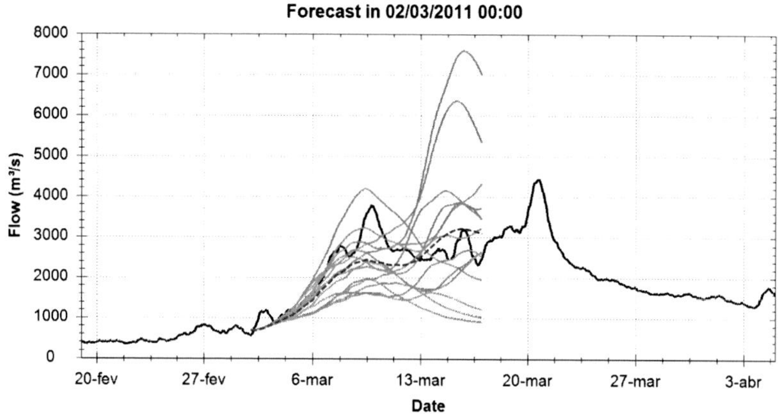

Fig. 8 Sample ensemble forecast from the São Francisco forecasting system.

4.4 Tocantins River Basin Forecasting System

The Tocantins River is an important Brazilian river for hydropower production, with several plants along the main stem and its tributaries (Fig. 9). The Tocantins basin is located in central-north Brazil, and covers 300,000 km^2 of drainage area upstream its confluence with the Araguaia River. Flooding occurrences along this river are very important for hydropower operation, as they affect several cities downstream and their inhabitants.

A hydrological forecasting system (Fan, Collischonn, Quiroz, et al., 2014) was developed in order to assist the decision-making of reservoir operation for flood control along the Tocantins River basin, taking advantage of the existence of real-time gauging networks. The forecasting system has been used operationally since mid-2012 and is based on the MGB-IPH large-scale distributed hydrological model (Collischonn et al., 2007).

The most important characteristics of the forecasting system are: (i) the use of real-time satellite precipitation data from Tropical Rainfall Measuring

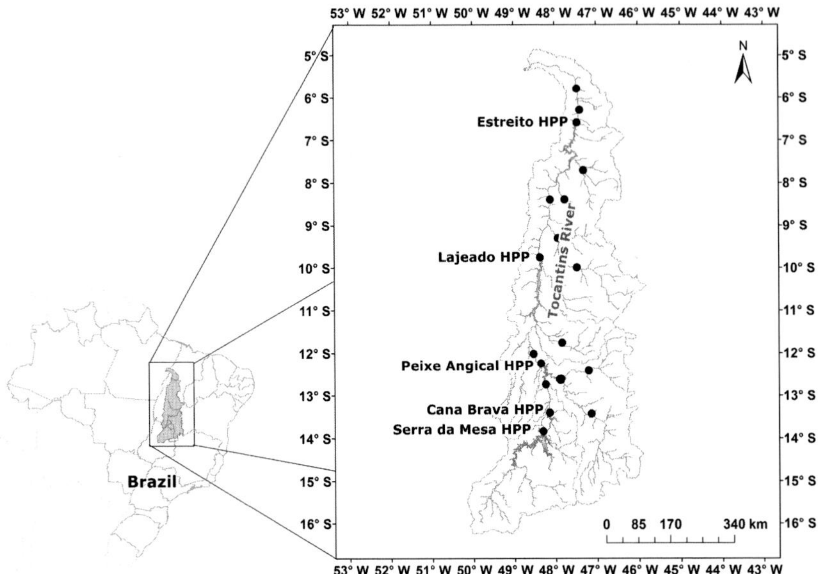

Fig. 9 Map of the Tocantins River basin location, showing hydropower plants on the main river axis, and points where streamflow and rainfall data are available in real-time at gauging stations or dams.

Mission (TRMM) merged with rain gauge information, to reduce the impact of rain gauge scarcity in the basin; (ii) the use of the numerical 7-day lead time deterministic rainfall forecast from the Brazilian Agency for Climate Prediction (CPTEC) and of the 16-day lead time precipitation ensemble forecasts provided by the GEFS, from the Second Generation of National Oceanic and Atmospheric Administration (NOAA) Global Ensemble Reforecast Data Set, as meteorological forecast inputs; and (iii) the use of a custom-designed graphical interface, coupled with an open-access Geographic Information System (GIS) platform, to allow for the forecasting system data management and operations.

The basic information necessary for conducting hydrological forecasts are the observed data of rainfall and streamflow up to the start time of the forecast. Observed streamflow data is important for updating the initial conditions of the hydrological model through data assimilation, and has a positive impact on the quality of forecasts during the early part of the forecasting lead time. This effect occurs because the quality of short-lead forecasts is dependent more on the existing water in the basin than on forecasted precipitation. The Tocantins river forecasting system uses observed hourly data of rainfall and streamflow at gauging stations and hydropower plants, as shown in Fig. 1. However, the rain gauge network in the basin is rather sparse, and irregularly distributed. The total number of gauging stations available is 16 (~1 gauge per 17,000 km^2), but some areas of the basin are not covered. Due to this poor gauge coverage, satellite-derived rainfall is used as complementary data, as described below.

Given the sparseness and irregularity of the existing rain gauge network, which greatly contributes to uncertainty in the streamflow forecasting, the possibility of using estimated rainfall from satellite measurements as alternative and complementary data sources is extremely attractive and may allow for a reduction of the uncertainty impacts. In this sense, rainfall products of the Multisatellite Precipitation Analysis from the TRMM (Huffman et al., 2007) are among the most used for this purposes, especially the TRMM-3B42RT product available in near real-time at a spatial resolution of 0.25 × 0.25 degrees, and at a temporal resolution of 3 h, favoring its use in operational applications such as streamflow forecasting.

In the Tocantins River basin forecasting system, a regular grid combination of satellite and gauge data called MergeHQ is applied in the Tocantins region. The methodology is based on the work of Rozante, Moreira, De Goncalves, and Vila (2010), but over an hourly time step.

The procedure of streamflow forecasting within the forecasting system requires information about the future rainfall. The forecasting system allows for the use of the following future rainfall data sources: QPFs from NWP models; null rainfall assumption; or manually estimated future rainfall based on a consensus forecast by a team of experienced meteorologists in the region. Two main sources of QPF were considered:

(i) deterministic forecasts from the Brazilian CPTEC Eta 15 km regional model;

(ii) ensemble forecasts from NOAAs Global Ensemble Forecasting System version 2.

The Eta-15 km model is a version of the Eta model with grid resolution of 15 km and results of predictions over the whole of South America. The ETA model is a regional NWP model operationally used by CPTEC-INPE to produce weather forecasts since the 1990s (Chou et al., 2007). Data from the model forecasts is available daily at the CPTEC Internet portal with a lead time of 7 days. The QPF has a temporal resolution of 3 h and is available for two times a day (for 00 UTC and at 12 UTC).

The second QPF source is the meteorological ensemble provided by the GEFS, from the Second Generation of NOAA Global Ensemble Reforecast

Data Set, recently available and maintained by the National Center for Environmental Prediction of the NOAA of the United States (Hamill et al., 2013). The data is available online at http://www.esrl.noaa.gov/psd/forecasts/reforecast2/download.html. The set of predictions is an ensemble of 11 members, extending up to 16 days of lead time, with new forecasts every day.

Regarding GIS integration, the Tocantins River forecasting system was developed coupled to a GIS called MapWindow GIS® (Ames, Michaelis, Anselmo, Chen, & Dunsford, 2008). This platform is open-source free software that contains a large number of basic GIS functions. A plugin was developed in Visual Basic .NET programming language containing the forecasting system. This coupling facilitates the data processing for model input, the system operation, and the assessments of results at multiple points over the basin (Fan & Collischonn, 2014).

Fig. 10 shows a sequence of ensemble streamflow forecasts for the Jan. 2012 flood, when inflow to the Estreito reservoir peaked at approximately 14,000 m^3/s. This peak occurred in Jan. 7, as can be seen by the blue hydrograph that shows estimated inflow to the reservoir. The ensemble streamflow forecasts were obtained by running the hydrological model using observed rainfall data up to the time of the start of the forecast, and using 11 different rainfall forecasts given by the 11 members of GEFS meteorological forecast ensemble.

Ensemble forecasts can also be assessed by showing exceedance diagrams (Fig. 11). Exceedance diagrams show boxes with numbers that represent the percentage of ensemble members that exceeds a defined threshold. It was adopted a threshold of 8000 m^3/s, a value that can be considered as a limit of alert for the Tocantins riverside region. Colors in the diagram relate to percentage of members above the threshold, from light gray (low percentage) to dark gray (high percentage). In the diagram shown in Fig. 11, one can see that 18% of the ensemble members predicted peaks above the threshold on Dec. 30, 8 days prior to its occurrence. The signal was persistent during the next 4 days, although the percentage was slightly reduced on Jan. 3. From Jan. 4 on, the forecasts consistently show 100% of the ensemble members above the threshold. The dark gray boxes in the last line show the days when the threshold was actually exceeded (from Jan. 7 to 19).

Results shown here strongly suggest that ensemble forecasts in the Tocantins River basin are useful to predict the occurrence of floods some days in advance. For the major flood analyzed, the signal of the predicted flow was observed 3 days ahead with a probability of occurrence greater than 55%; a good indicator for flood anticipations.

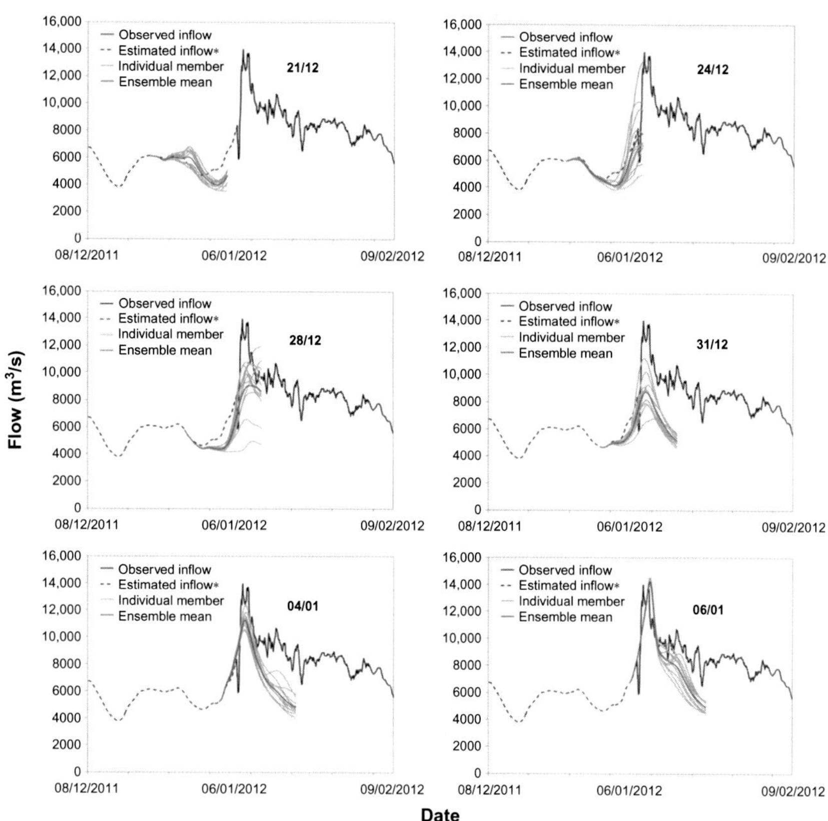

Fig. 10 Sequence of forecasts for the Jan. 2012 flood peak, highlighting the day/month in which the forecast was issued, at the UHE Estreito dam (* observed inflow was estimated by reservoir water budget from Jan. 06; prior to this data, observed inflow was estimated using the hydrological model simulation results).

Ensemble (GEFS v2)	29/12/2011	30/12/2011	31/12/2011	01/01/2012	02/01/2012	03/01/2012	04/01/2012	05/01/2012	06/01/2012	07/01/2012	08/01/2012	09/01/2012	10/01/2012	11/01/2012	12/01/2012	13/01/2012	14/01/2012	15/01/2012	16/01/2012	17/01/2012	18/01/2012	19/01/2012	20/01/2012	21/01/2012	22/01/2012	23/01/2012	24/01/2012
29/12/2011	0	0	0	0	0	0	0	0	0	0	9.1	27	27	18	18	9.1											
30/12/2011		0	0	0	0	0	0	0	0	18	45	64	55	45	18	18	0										
31/12/2011			0	0	0	0	0	0	0	18	64	73	73	36	9.1	0	0	0									
01/01/2012				0	0	0	0	0	0	27	73	91	91	73	27	9.1	9.1	0	0								
02/01/2012					0	0	0	0	0	36	91	100	100	100	82	45	18	9.1	9.1	9.1							
03/01/2012						0	0	0	0	0	36	64	64	36	0	0	0	0	0	0	0						
04/01/2012							0	0	0	100	100	100	100	100	73	9.1	0	0	0	0	0	0					
05/01/2012								0	0	100	100	100	100	100	100	64	9.1	0	0	0	0	0	0				
06/01/2012									0	100	100	100	100	100	100	91	73	73	55	18	0	0	0	0			
07/01/2012										100	100	100	100	100	100	100	100	100	100	100	73	36	0	0	0		
08/01/2012											100	100	100	100	100	100	100	100	100	100	91	55	18	0	0	0	
09/01/2012												100	100	100	100	100	100	100	100	100	100	73	64	55	55	55	36
Observed																											

Fig. 11 Threshold exceedance diagram for the flood in the 2011/12 rainy season, considering a threshold of 8000 m³/s. Numbers and shades indicate the percentages of ensemble members that exceeded the threshold.

The Tocantins system is distinctive for its large scale, covering an area of several hundred thousand square kilometers, and for its use of both deterministic and ensemble precipitation forecasts. The main limitations of the forecasting system are related to the low density of telemetric rain and river gauging stations, which we expected to be overcome with the use of satellite precipitation data. Another important limitation for results assessment is related to the uncertainty in the estimation of actual reservoir inflow, which is calculated by a reservoir water balance method that can introduce high frequency noise in the data. Nevertheless, the hydrological forecasting system gives some added value to the flood control in the Tocantins River basin, allowing for the anticipation of important flood events.

4.5 Experimental River Flow Forecasts in the Amazon

This section presents a discharge forecast system prototype for the Amazon River basin developed at IPH/UFRGS (Paiva, Collischonn, Bonnet, de Gonçalves, et al., 2013). The Amazon (Fig. 12) is known as the largest hydrological system of the world (~6 million km^2 of surface area), contributing ~15% of the total fresh water released into the oceans. It has large rivers with extensive seasonally flooded wetlands and complex hydraulics, besides high precipitation rates, with high spatial variability and contrasting rainfall

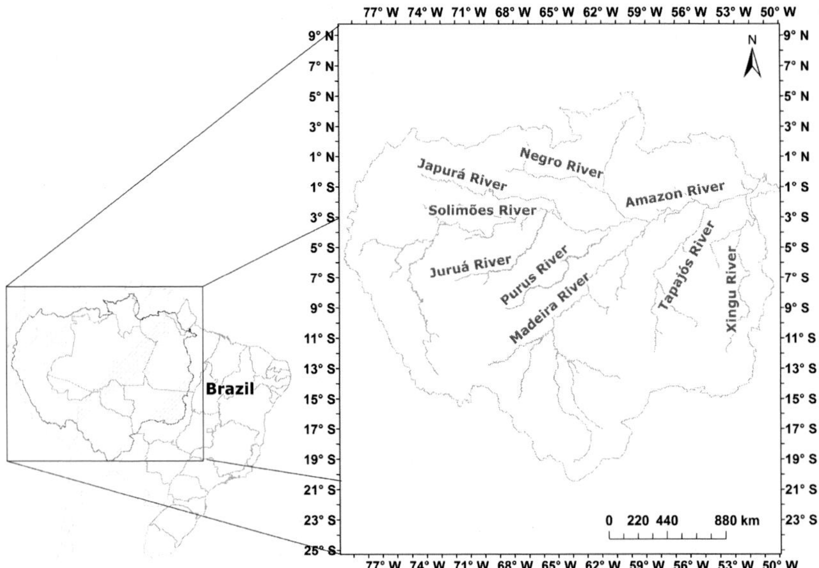

Fig. 12 Amazon River basin location.

regimes across the basin. Several extreme hydrological events have occurred recently in the Amazon (Marengo and Espinoza, 2015), for instance, the 2009, 2012, and 2014 floods, and the 2005 and 2010 droughts. These extreme events caused several impacts on the local population, since most settlements lie along rivers where susceptibility to floods is large. Also, the local population strongly depends on these rivers for transportation of people and goods, agriculture, and generation of hydroelectricity, among others. A system capable of predicting such extreme events in space and time is necessary, as each of these floods and droughts is particular, with different degrees of severity over different parts of the basin.

The forecast system is composed of: (i) a distributed hydrological model; (ii) a data assimilation system to estimate hydrological initial states (eg, soil moisture, surface, and groundwater storage); and (iii) ensemble precipitation forecasts based on historical data. The configuration of the system was chosen based on experiments (Paiva, Collischonn, Bonnet, & De Gonçalves, 2012) showing that the hydrological predictability in the Amazon is governed by hydrological initial states due to the size of the basin and physical features such as the large travel times of flood waves.

The process-based hydrological model called MGB-IPH (Collischonn et al., 2007) was used coupled with a river hydrodynamic module using a storage model for floodplains (Paiva, Collischonn, & Tucci, 2011). The model was forced using remotely sensed precipitation estimates from the TRMM 3B42 v6 product (Huffman et al., 2007), showing good performance when extensively validated against in situ discharge and stage measurements and also remotely sensed data, including radar altimetry-based water levels, gravimetric-based terrestrial water storage, and flood inundation extent, as described in Paiva, Collischonn, Bonnet, Buarque, et al. (2013).

Hindcast streamflow forecasts were generated using an ensemble streamflow prediction (ESP) approach (Day, 1985), as described in Paiva, Collischonn, Bonnet, de Gonçalves, et al. (2013). The model uses estimates of initial conditions derived from a data assimilation scheme and runs forced by an ensemble of observed meteorological data from past years. An estimate of initial conditions is computed during the spin-up period using the hydrological model driven satellite precipitation from TRMM Merge (Rozante et al., 2010) and updated using data assimilation of observations up to the time of forecast. The data assimilation scheme is based on the Ensemble Kalman Filter technique and it assimilated in situ discharge data from 12 gauges located at the Amazon and main tributaries to update model states before starting a forecast. Then, an ensemble forecast is obtained using observed precipitation data resampled from past years from TRMM 3B42 data. The ESP runs generated decadal forecasts up to 90 days' lead time for the 2 year period of 2004–05. Forecasts were evaluated only by deterministic means by averaging ensemble values into a single forecast using the skill score SS_{cli}, which compares the performance of the model forecasts (Q_{for}) with a control forecast (Q_{cli}) based on climatology $SS_{cli} = 1 - \sum_t \left(Q^t_{obs} - Q^t_{for}\right)^2 / \sum_t \left(Q^t_{obs} - Q^t_{cli}\right)^2$, where Q_{obs} are discharge observations. Positive SS_{cli} values show an improvement over a reference forecast based on discharge climatology.

The experiments show that these methods are able to provide relatively accurate streamflow forecasts in the Amazon basin. Fig. 13 shows analyses of hindcast forecasts at the Solimões/Amazon main stem. The model was able to forecast discharges with relatively high accuracy even for very large lead times (90 days) and forecasts are markedly better than simply using discharge climatology, as shown by positive values of SS_{cli} skill score. Even though the agreement between model values and observations decreases as a function of lead time, SS_{cli} remains high, showing that it would be possible to produce accurate forecasts at the Amazon main river for even larger lead times. Also, the model successfully predicted the severe 2005 drought at the Solimões/Amazon main stem. In this year, discharges dropped ~1 month earlier than normal (Fig. 8) and river levels fell to historically low levels, causing navigation to be suspended (Marengo et al., 2008). Even so, the model was able to predict these low flows ~90 days ahead. A similar analysis was performed for 109 gauging stations, as shown in Fig. 13. Forecasts for a smaller lead time (5–15 days) were relatively accurate, with positive SS_{cli} values at several gauges. However, the quality of the forecasts decreased as a function of lead time (eg, 90 days). It becomes very poor at smaller rivers but remains meaningful with positive SS_{cli} values mainly at gauging stations within large rivers such as the Solimões/Amazon main stem and some of its main tributaries.

These results demonstrate the potential for developing stream flow forecasts with large lead times in the large Brazilian river basins, such as the Amazon, founded on large-scale hydrological models based mostly on initial states gathered with proper data assimilation schemes, and using past climate with the ESP approach. Also, results point to the potentiality of providing hydrological forecasts at poorly monitored regions by using mostly remotely sensed information.

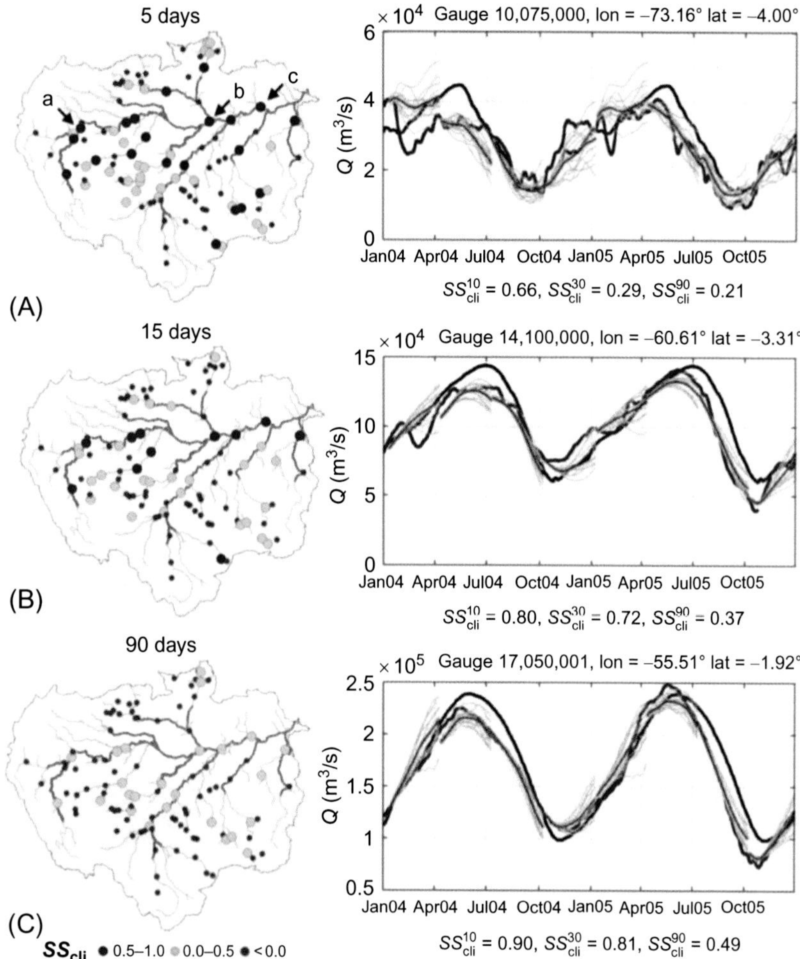

Fig. 13 Evaluation of streamflow forecasts over the Amazon basin. Spatial distribution of the skill score SS_{cli} for 5, 15, 90 days lead time. Observed *(blue)*, climatological *(black)* discharges, ensemble forecasts *(gray)* together with ensemble mean *(red)* at (A) Upper Solimões River at Tamishiyacu, (B) Solimões River at Manacapuru, and (C) Amazon River at Óbidos. Presented forecasts started each 10th Jan., Apr., Jul., and Oct.

5 CONCLUSIONS

Operational hydrologic forecasting in Brazil is still evolving and great opportunities exist. This overview of the hydrological forecasting practices in Brazil demonstrate the challenges in developing forecasting systems in Brazil, such as: (i) its continental scale; (ii) the trans-boundary nature of some basins; (iii) the diversity of climates and hydrological characteristics; (iv) flood hazards occurring at multiple spatiotemporal scales; and (v) the extensive river flow operation by hydropower plants.

There are several efforts to develop forecast systems in Brazil, ranging from simple statistical models to state-of-the-art methods such as distributed hydrological models coupled to ensemble meteorological forecasts. These initiatives are somehow decentralized, as current hydrological forecast systems are being developed for specific rivers at universities, regional hydrologic centers, and by the hydropower sector. Although there is knowhow on state-of-the-art hydrological forecasts in Brazil, it has not, as of 2015, been converted into a countrywide operational flood forecast system.

Some of the future challenges are the mapping of hydrological predictability and the development and evaluation of state-of-the-art techniques in different regions and scales, as well as the implementation of operational systems for the first time in most Brazilian rivers. For example, although ensemble forecasts have shown good performance, this technique is still not used operationally in Brazil. The development of science and technology in the operational flood forecast field is still evolving in Brazil, contrasting with the great benefits that such systems could bring to decision-making and mitigation of flood impacts. Brazil is highly dependent on its water resources for energy production and floods greatly affect a significant number of people in several regions of Brazil. Nevertheless, little is known about the performance of state-of-the-art techniques.

Based on the current panorama, it is possible to see that nowadays the development and operation of forecasting systems in Brazil is generally decentralized, with different responsible centers focusing on specific river basins. Even for the case of the systems maintained by CPRM, developments are from regional offices. Also, there is not a clear picture about the exchange of information between forecasting centers, for example, between operational forecasts generated to be used by hydropower companies and local early flood warning systems for the same river basin. This happens even though forecasting systems for hydropower reservoirs are, to some extent, intended to help flood warning and flood response. Also, currently in Brazil there are not countrywide or continental forecasting systems, such as the EFAS - European Flood Awareness System (Younis, Ramos, & Thielen, 2008) and the AFFS - African Flood Forecasting System (Thiemig, Bisselink, Pappenberger, & Thielen, 2014), although propositions for the development of such systems are being discussed. This kind of system could integrate forecasts from multiple centers (eg, the ones cited here), and run its own forecasting models that would complement information where forecasts already exist, and generate new information for other places.

We believe that the Brazilian scenario may change in the coming years, with the growth of regional and countrywide forecast systems through the development and transfer of technological knowledge and the growth of institutions such as CEMADEN.

REFERENCES

Ames, D. P., Michaelis, C., Anselmo, A., Chen, L., & Dunsford, H. (2008). MapWindow GIS. *Encyclopedia of GIS.* New York: Springer (pp. 633–634).

Araujo, A. N., Breda, A., Freitas, C., Leite, E. A., Gonçalves, J. E., Calvetti, L., et al. (2014). Hydrological and meteorological forecast combined systems for flood alerts and reservoir management: The Iguaçu River basin case. In *Proceedings of 6th international conference on flood management — ICFM6, São Paulo/SP.*

Chou, S. C., CataldI, M., Osorio, C., Guilhon, L. G., Gomes, J. L., & Bustamante, J. (2007). Análise das previsões de precipitação obtidas com a utilização do modelo Eta como insumo para modelos de previsão semanal de vazão natural. *Revista Brasileira de Recursos Hídricos, 12*(3), 5–12.

Collischonn, W., & Tucci, C. E. M. (2005). Previsão Sazonal de Vazão na Bacia do Rio Uruguai 1: Ajuste e Verificação do Modelo Hidrológico Distribuído. *Revista Brasileira de Recursos Hídricos, 10*(4), 43–59.

Collischonn, W., Tucci, C. E. M., Clarke, R. T., Chou, S. C., Guilhon, L. G., Cataldi, M., et al. (2007). Medium-range reservoir inflow predictions based on quantitative precipitation forecasts. *Journal of Hydrology, 344*, 112–122.

Cordero, A., Momo, M. R., & Severo, D. L. (2011). Previsão de cheia em tempo atual, com um Modelo Armax, para a cidade de Rio do Sul-SC. In *Anais do XIX Simpósio Brasileiro de Recursos Hídricos.*

Costa, F. S., Raupp, I. P., Damazio, J. M., Oliveira, P. D., & Guilhon, L. G. F. (2014). The methodologies for the flood control planning using hydropower reservoirs in Brazil. In *Proceedings of 6th international conference on flood management — ICFM6, 2014, São Paulo.* São Paulo: ABRH.

Day, G. N. (1985). Extended streamflow forecasting using NWSRFS. *Journal of Water Resources Planning and Management, 111*, 157–170.

Fan, F. M., & Collischonn, W. (2014). Integração do Modelo MGB-IPH com Sistema de Informação Geográfica. *Revista Brasileira de Recursos Hídricos, 19*, 243–254.

Fan, F. M., Collischonn, W., Buarque, D. C., Tucci, C. E. M., & Botelho, L. C. M. (2013). Desenvolvimento e Avaliação de um Sistema de Previsão Hidrológica Distribuída para a Região da Uhe Três Marias. In *Anais do XX Simpósio Brasileiro de Recursos Hídricos, Bento Gonçalves, RS.*

Fan, F. M., Collischonn, W., Meller, A., & Botelho, L. C. M. (2014a). Ensemble streamflow forecasting experiments in a tropical basin: The São Francisco River case study. *Journal of Hydrology, 519*, 2906–2919. http://dx.doi.org/10.1016/j.jhydr.

Fan, F. M., Collischonn, W., Quiroz, K., Sorribas, M. V., Buarque, D. C., & Siqueira, V. A. (2014b). Ensemble flood forecasting on the Tocantins River — Brazil. *Geophysical Research Abstracts, 16*, 1818.

Fan, F. M., Meller, A., & Collischonn, W. (2015). Incorporação de filtro numérico de separação de escoamento na assimilação de dados para previsão de vazões utilizando modelagem hidrológica. *Revista Brasileira de Recursos Hídricos, 20*(2), 472–483.

Fan, F. M., Pontes, P. R. M., Beltrame, L. F. S., Collischonn, W., & Buarque, D. C. (2014c). Operational flood forecasting system to the Uruguay River basin using the hydrological model MGB-IPH. In *Proceedings of 6th international conference on flood management — ICFM6*. São Paulo: ABRH.

Fan, F. M., Pontes, P. R. M., Collischonn, W., & Beltrame, L. F. S. (2012). Sistema de Previsão de Vazões para as Bacias dos Rios Taquari-Antas e Pelotas. In *Anais do XI Simpósio de Recursos Hídricos do Nordeste, João Pessoa, PB*.

Gomes, L. F. C., Momntenegro, S.M.G.L., & Valença, M. J. S. (2010). Modelo Baseado na Técnica de Redes Neurais para Previsão de Vazões na Bacia do Rio São Francisco. *Revista Brasileira de Recursos Hídricos, 15*(1), 5–15.

Guha-Sapir, D., Below, R., & Hoyois, Ph. (2015). EM-DAT: International Disaster Database, http://www.emdat.be, Université Catholique de Louvain, Brussels, Belgium.

Guilhon, L. G. F., Rocha, V. F., & Moreira, J. C. (2007). Comparação de métodos de previsão de vazões naturais afluentes a aproveitamentos hidroelétricos. *Revista Brasileira de Recursos Hídricos, 12*(3), 13–20.

Hamill, T. M., Bates, G. T., Whitaker, J. S., Murray, D. R., Fiorino, M., Galarneau, T. J., Jr., et al. (2013). NOAA's second-generation global medium-range ensemble reforecast data set. *Bulletin of the American Meteorological Society, 94*, 1553–1565. http://dx.doi.org/10.1175/BAMS-D-12-00014.1.

Huffman, G. J., Adler, R. F., Bolvin, D. T., Gu, G., Nelkin, E. J., Bowman, K. P., et al. (2007). The TRMM Multisatellite Precipitation Analysis (TMPA): Quasi-global, multiyear, combined-sensor precipitation estimates at fine scales. *Journal of Hydrometeorology, 8*(1), 3–55.

Marengo, J. A., & Espinoza, J. C. (2015). Review Article. Extreme Seasonal Droughts and Floods in Amazonia: Causes, Trends and Impacts. *International Journal of Climatology*. http://dx.doi.org/10.1002/joc.4420.

Marengo, J., Nobre, C., Tomasella, J., Oyama, M., de Oliveira, G., de Oliveira, R., et al. (2008). The drought in Amazonia in 2005. *Journal of Climate, 21*, 495–516.

Matos, A. J. S., Davis, E. G., Silva, A. J., Almeida, I. S., & Candido, M. O. (2014). Assessment of a real-time flood forecasting at the Doce River basin: Summer 2013 event. In *Proceedings of 6th international conference on flood management — ICFM6, São Paulo*.

Oliveira, P. D., Guilhon, L. G. F., Costa, F. S., Raupp, I. P., & Damazio, J. M. (2014). The operation of flood control in large hydroelectric power systems — The Brazilian experience. In *Proceedings of 6th international conference on flood management — ICFM6, São Paulo*.

ONS (2011). Operador Nacional do Sistema. Procedimentos de Rede Submódulo 9.5: Previsão de Vazões e Geração de Cenários de Afluências (Vol. 2, 9 pp.).

ONS (2012). Operador Nacional do Sistema. Diretrizes para as Regras de Operação de Controle de Cheias — Bacia do Rio São Francisco (Ciclo 2012–2013). ONS RE 3/166/2012 (158 pp.).

Paiva, R. C. D., Collischonn, W., Bonnet, M.-P., Buarque, D. C., Frappart, F., Calmant, S., et al. (2013a). Large scale hydrologic and hydrodynamic modelling of the Amazon River basin. *Water Resources Research, 49*, 1226–1243. http://dx.doi.org/10.1002/wrcr.20067.

Paiva, R. C. D., Collischonn, W., Bonnet, M. P., & De Gonçalves, L. G. G. (2012). On the sources of hydrological prediction uncertainty in the Amazon. *Hydrology and Earth System Sciences, 16*, 3127–3137. http://dx.doi.org/10.5194/hess-16-3127-2012.

Paiva, R. C. D., Collischonn, W., Bonnet, M.-P., de Gonçalves, L. G. G., Calmant, S., Getirana, A., et al. (2013b). Assimilating in situ and radar altimetry data into a large-scale hydrologic-hydrodynamic model for streamflow forecast in the Amazon. *Hydrology and Earth System Sciences, 17*, 2929–2946. http://dx.doi.org/10.5194/hess-17-2929-2013.

Paiva, R. C. D., Collischonn, W., & Tucci, C. E. M. (2011). Large scale hydrologic and hydrodynamic modeling using limited data and a GIS based approach. *Journal of Hydrology, 406*, 170–181, http://dx.doi.org/10.1016/j.jhydrol.2011.06.007.

Paz, A. R., Collischonn, W., Tucci, C., Clarke, R., & Allasia, D. (2007). *Data assimilation in a large-scale distributed hydrological model for medium range flow forecasts. Vol. 313.* Wallingford, UK: IAHS Press, IAHS Publication (pp. 471–478).

Pinheiro, A. (2003). Modelos de previsão de cheias. In F. E. Pinheiro (Ed.), *Enchentes na bacia do rio Itajaí: 20 anos de experiências.* Blumenau: Edifurb.

Reis, D. S., Jr., Nascimento, L. S. V., & Martins, E.S.P.R. (2007). Avaliação do ensemble da previsão climática de chuva no estado do ceará com base em modelos numéricos de clima. In *Anais do XVII Simpósio Brasileiro de Recursos Hídricos.*

Rozante, J. R., Moreira, D. S., De Goncalves, L. G. G., & Vila, D. A. (2010). Combining TRMM and surface observations of precipitation: Technique and validation over South America. *Weather and Forecasting, 25,* 885–894.

Schwanenberg, D., Becker, B. P. J., & Xu, M. (2015). The open real-time control (RTC)-tools software framework for modeling RTC in water resources systems. *Journal of Hydroinformatics, 17*(1), 130. http://dx.doi.org/10.2166/hydro.2014.046.

Schwanenberg, D., Fan, F. M., Naumann, S., Kuwajima, J. I., Montero, R. A., & Assis dos Reis, A. (2015). Short-term reservoir optimization for flood mitigation under meteorological and hydrological forecast uncertainty. *Water Resources Management, 29,* 1635–1651. http://dx.doi.org/10.1007/s11269-014-0899-1.

Silva, E. R., Oficialdegui, E., Cirilo, J. A., Ribeiro Neto, A., Dantas, C. E. O., & Santos, K. A. (2014). Monitor avançado de enchentes (MAVEN): A hydroclimatologic computational framework for early flood alert systems. In *Proceedings of 6th international conference on flood management — ICFM6, São Paulo.*

Thiemig, V., Bisselink, B., Pappenberger, F., & Thielen, J. (2014). A pan-African flood forecasting system. *Hydrology and Earth System Sciences Discussions, 11,* 5559–5597. http://dx.doi.org/10.5194/hessd-11-5559-2014.

Werner, M., Schellekens, J., Gijsbers, P., Van Dijk, M., Van Den Akker, O., & Heynert, K. (2013). The Delft-FEWS flow forecasting system. *Environmental Modelling & Software, 40,* 65–77. http://dx.doi.org/10.1016/j.envsoft.2012.07.010.

Younis, J., Ramos, M.-H., & Thielen, J. (2008). EFAS forecasts for the March–April 2006 flood in the Czech part of the Elbe river basin — A case study. *Atmospheric Science Letters, 9,* 88–94.

Zambon, R. C., Barros, M. T. L., Gimenes, M., Bozzini, P. L., & Yeh, W. W. G. (2014a). Flood control and energy production on the Brazilian hydrothermal system. In *Proceedings of 6th international conference on flood management — ICFM6, São Paulo.*

Zambon, R. C., Barros, M. T. L., Lopes, J. E. G., Barbosa, P. S. F., Francato, A. L., & Yeh, W. W. G. (2012). Optimization of large-scale hydrothermal system operation. *Journal of Water Resources Planning and Management, 138,* 135–143.

Zambon, R. C., Barros, M. T. L., & Yeh, W. W. G. (2014b). Brazilian hydrothermal system operation: Interconnected large system or isolated subsystems? In *World environmental and water resources congress, Portland* (p. 1926). Reston: American Society of Civil Engineers.

Fernando Mainardi Fan received a PhD in Water Resources and Environmental Sanitation, and an Msc in Water Resources and Environmental Sanitation. An environmental engineer, he graduated from the Federal University of Rio Grande do Sul (UFRGS), in southern Brazil. He is currently a temporary professor and fellow researcher at UFRGS Institute of Hydraulic Research (Instituto de Pesquisas Hidráulicas — IPH). His main research expertise is water resources studies, hydrological forecasting, and its applications.

Rodrigo C.D. Paiva is a civil engineer and received a PhD in Water Resources and Environmental Sanitation from the Federal University of Rio Grande do Sul (UFRGS), Brazil, and Université Toulouse III Paul Sabatier, France. He is currently a professor and researcher at UFRGS Institute of Hydraulic Research (IPH), Porto Alegre, Brazil. His areas of research interest are large-scale hydrology, hydrological processes of South American systems, hydrologic modeling, remote sensing, and its applications.

Walter Collischonn is a mechanical engineer and received a PhD in Water Resources and Environmental Sanitation from the Federal University of Rio Grande do Sul (UFRGS), Brazil. He is currently a professor of hydrology and researcher at UFRGS Institute of Hydraulic Research (IPH), Porto Alegre, Brazil. His main research activities are related to development of large-scale hydrology models, hydrological forecasts, GIS methods for water resources and South American hydrology.

CHAPTER 3

The Development and Recent Advances of Flood Forecasting Activities in China

Z. Liu
Hydrological Forecast Center, The Ministry of Water Resources of China, Beijing, China

1 INTRODUCTION

Flooding has been the most severe hazard in China from time immemorial, due to its special geographical, climatic, and socioeconomic conditions. Although a few floods in west China are subject to the mixed-type floods from snow and glacier melting as well as local storms, the majority of floods in China are caused by rainstorms in combination with the coastal storm surges. According to historical records, big flood disaster events averaged once every 2 years before 1949. Beginning from the 20th century, many flood fatalities have been recorded in China. In 1931, disastrous floods hit the Huaihe River basin and the Yangtze River basin, affected over 51 million people, and caused 400,000 deaths. The average annual death toll in recent floods indicates a decreasing trend from 8571 (1950s) to 1454 (2000s), as flood control systems have been improved significantly since 1950.

It is the exponentially increasing impact of flooding that has raised the profile of the practice of flood forecasting and warning. Since the late 1980s there has been a move away from the primacy of major structural measures for flood control towards a more integrated approach, of which flood forecasting and warning is a component. The Integrated Flood Management Concept encourages a shift from the traditional, fragmented, and localized approach, towards the use of the resources of a river basin as a whole, while at the same time providing protective measures against losses due to flooding (Arduino, Reggiani, & Todini, 2005; Linnerooth-Bayer & Amendola, 2003).

Due to global warming, social and environment changes, and climate systems, flooding is becoming more extreme, more widespread, and more frequent in China, increasing the risks and the costs of flood disasters

Flood Forecasting
http://dx.doi.org/10.1016/B978-0-12-801884-2.00003-7

(Wu, Lu, Liu, et al., 2013). This has given rise to increased emphasis on the improvement of operational flood forecasting and warning, and the enhancement and refinement of flood-risk (understood as a product of probability and consequences) management systems.

This chapter presents an overview of Chinese experiences in coping with floods and gives lessons from disastrous flood events. The chapter also describes the development of hydrological forecasting and prediction technology and methods that support flood management in China, focusing on recent advances in flood forecasting and warning. Furthermore, challenges and expectations in China's hydrological forecasting and prediction are also presented in the chapter.

2 EVOLUTION OF FLOOD CONTROL AND MANAGEMENT IN CHINA

For thousands of years, China has suffered greatly from flooding on large and small rivers. With the deepening understanding of the disaster risk change, as well as socioeconomic development, flood control strategies and measures in China have been continuously improved in the process of coping with floods over the years. In general, three phases are identified in the evolution of flood control and management in China, namely from flood defense, flood diversion, and storage, to ensure harmonious coexistence between humans and water.

Flood defenses

The watercourse regulation masterminded by DAYU, the most significant historic figure in China with merit in flood mitigation about 2000 years ago, marked the beginning of traditional approach to prevention and control of river flooding, based on the use of dykes and channel improvements. Dykes were a major means of flood prevention in China until the first half of the 20th century.

Flood diversion and storage

Introduction of damming technology to China in the 20th century started the phase of water impounding for flood prevention and control in the country. During this stage, measures such as building reservoirs, delimiting floodplains, and opening up floodways were used to temporarily hold and vent extra flood water volume when the floods exceeded the carrying capacity of river channels, adding to the original structural measures of levee reinforcement and maintenance, and the realignment of river courses.

Harmonious coexistence between humans and water

A significant breakthrough emerged on the adjustment of flood control strategy. A strong emphasis was given on the deployment of flood control development within the overall framework of integrated water resources management. System theory and risk management methods were also adopted in dealing with flooding issues, representing a shift from flood control paradigm to flood risk management.

Since the founding of the People's Republic of China in 1949, the government has attached high significance to the prevention and control of floods. Continuous efforts have been exerted to summarize experiences and lessons from major flood events and the role played by innovative work mechanisms, enhancement of structural development and improvement of nonstructural measures, and this has led to remarkable progress in capacity building for flood prevention and management. The accomplishments of China in this regard during the past 6 decades encompass the following three phases of flood control and management.

2.1 Start-up Phase: Before the Mid 1970s

The 1950s and 1960s saw increasing efforts to harness major rivers, control and prevent floods. In 1950, the State Flood Control Headquarters and the Yellow River Flood Control Headquarters were formally established, followed by the setting up of the Yangtze River Flood Control Headquarters and some key provincial flood control headquarters in succession. During this period, based on the principles of attaching equal importance to enhancing floodwater storage and discharge capacity, with the emphasis on the latter, the comprehensive management planning for the Huaihe River, the Yellow River, the Yangtze river, and the Haihe River basin have been successively carried out, which laid a solid foundation for harnessing rivers afterwards. Moreover, an engineering system for flood control has been preliminarily constructed, including the construction of trunk embankments for protecting major rivers, building numbers of large and medium-sized reservoirs, completing flood storage and detention areas on major rivers and progressively training major rivers.

2.2 Development Phase: From the Mid 1970s to 1998

In August of 1975, the Great Flood of the Huaihe River, caused by the heavy rainstorm and dam-failures, devastated Henan province, killing 26,000 people. A subsequent profound reflection on the "75.8" flood led to the transition towards a broader, comprehensive flood management approach where the emphasis shifted from engineering measures to a broad portfolio of structural and nonstructural measures for managing flood risk. From this time onwards, the national specification on design flood estimation was revised, and the standard of flood control projects has been significantly enhanced. Besides, investment on the water conservancy has been progressively increased, with the emphasis on dyke reinforcement and renovation on major rivers, danger control and reinforcement of dangerous reservoirs by stages and in groups, and construction of flood storage and detention areas with a view to guaranteeing the safety of people living there. By 1998, continuous efforts had made to improve the nonstructural measures including development of early warning and forecasting systems, formulating flood disaster preparation plans, improvement of the legal framework, establishing the administrative leader responsibility system, strengthening the organizational and institutional system, and setting up flood relief rescue teams.

2.3 Improvement Phase: From 1998 to the Present Day

In 1998, the most devastating floods occurred in some major rivers such as the Yangtze, Songhuajiang, Pearl, and Min Rivers, which left over 4000 people dead and caused material damage of 30 billion US$. In the aftermath of the Great Flood of 1998, the central government launched the strategic plan encompassing post-disaster reconstruction, renovation of rivers and lakes, start construction of water conservancy, and developed the water governance policy, namely overall planning and comprehensive treatment, promoting what is beneficial and abolishing what is harmful, attaching equal importance to increasing income and reducing expenditure, and putting equal emphasis on flood-control and anti-drought. Thus, flood control work entered a new period of development.

The flood risk concept was introduced and gradually applied in flood management in China. In 2003, the China Ministry of Water Resources proposed to redirect flood prevention from flood control to flood management, which aims to enhance understanding about interconnecting systemic issues and risk awareness. The flood prevention efforts thus shifted from attempts to eliminate floods forward to building capacity to endure floods with certain degrees of risk. In flooding areas, community management was introduced to regulate human activities. Appropriate and feasible flood control standards were established together with flood prevention schemes and flood regulation plans, where a variety of measures were taken to ensure safety under the established flood control standard and minimize loss caused by exceeding standard floods. Meanwhile, efforts were also taken to utilize storm water as a complementary resource for water supply.

3 DEVELOPMENT OF OPERATIONAL HYDROLOGICAL FORECASTING AND PREDICTION IN CHINA

In response to growing demands and development of forecasting theory as well as computer science, China has built up its hydrological forecasting and prediction technology and capacity.

3.1 Hydrological Information and Forecasting Administration

Hydrological administration in China is implemented at central and local levels. The Bureau of Hydrology of the Ministry of Water Resources (MWR-BoH), is the body responsible for central management, while there are bureaus of hydrology or bureaus of hydrology and water resources survey in the 7 river basin authorities under the Ministry and 31 provinces (autonomous region and municipalities) carry out local hydrological management.

The MWR-BoH is a nonprofit agency with administrative functions and responsibilities for hydrological industry administration all over China, which organizes and provides guidance for water-related activities. These include monitoring and analysis of surface and ground water quantity and quality, water resources evaluation, collecting, processing and predicting of hydrological and rainfall data in key areas for flood control and major reservoirs, ensuring communication and informationazation in the water sector, and guaranteeing safety of communication and the computer network of Office of State Flood Control and Drought Relief Headquarters and the Ministry of Water Resources.

The Hydrological Forecast Center of the Ministry of Water Resources is a nonprofit agency with administrative functions and responsibilities for hydrological information and forecast, and hydrometeorological administration all over China. The center organizes and provides guidance for hydrological information and forecast-related activities which include guiding technical standards, supervising its implementation, managing hydrological information for central flood-drought-reporting stations, organizing to provide all sorts of information for national flood control and drought relief, issuing real-time hydrological information and forecast, and so on.

3.2 Hydrological Monitoring and Data Transmission

3.2.1 Hydrological Monitoring

Hydrological monitoring stations in China are divided into two types: state level basic hydrometric stations and special stations (Zhang and Liu, 2006). By the end of 2013, there were 86,554 hydrometric stations maintained by the water sector in China, including 4195 hydrological stations, 9330 water level stations, 43,028 rainfall stations (see Fig. 1), 1912 soil moisture stations, and 16,407 groundwater monitoring stations, among which were 24,518 flood-drought-reporting stations and 1300 hydrological forecasting stations.

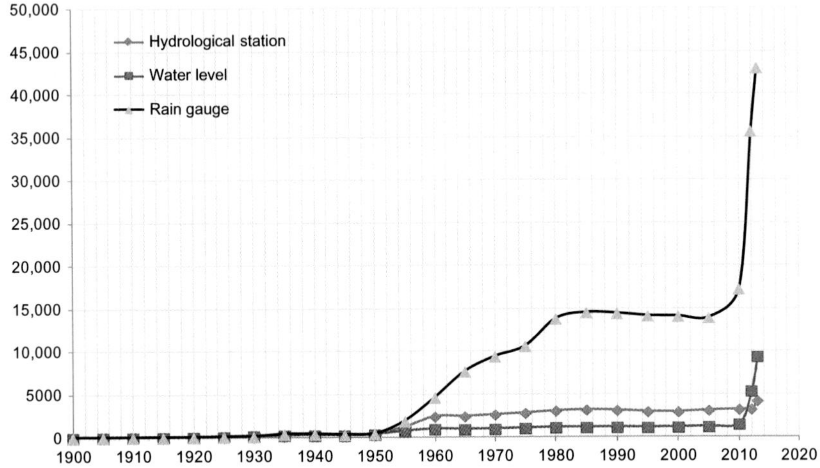

Fig. 1 Development of hydrometric station network in China.

At present, hydrological observation items include (Fig. 2), evaporation (Fig. 3), discharge (Figs. 4–8), stage, sediment (Fig. 9), water quality, groundwater, water temperature, ice slush, etc. The types of hydrological stations in China include perennial stationary gaging, stationary gaging during flood season, full-year touring gaging, entrust gaging, etc. The discharge measurement is done mainly by using cableway and hydrometric boat. Current-meter method is commonly used for discharge measurement while the float method, discharge calculation by hydraulic method, and flow measurement by hydraulic structure are used in the case that ordinary hydrometry facilities are destroyed due to major floods. Presently, some new techniques and methods such as the Moving-boat method by using Acoustic Doppler Current Profiler are being applied to flow measurement in China (Figs. 2–9).

3.2.2 Data Transmission

In flood control and drought relief, it is necessary to collect rainfall and weather information, including the observed data in more than 2000 weather stations, satellite nephograms, weather radars reports, live weather reports from ground and high altitude, all types of weather forecasts faxes, and observed rainfall data by the hydrological units and departments.

Fig. 2 Rainfall observation site.

Fig. 3 Evaporation observation site.

Fig. 4 Hydrometric boat.

Fig. 5 Measurement from bridge.

Fig. 6 Flow measurement is made with electric wave current-meter.

Fig. 7 Flow measurement is made with Doppler current-meter.

Fig. 8 Flood peak measuring.

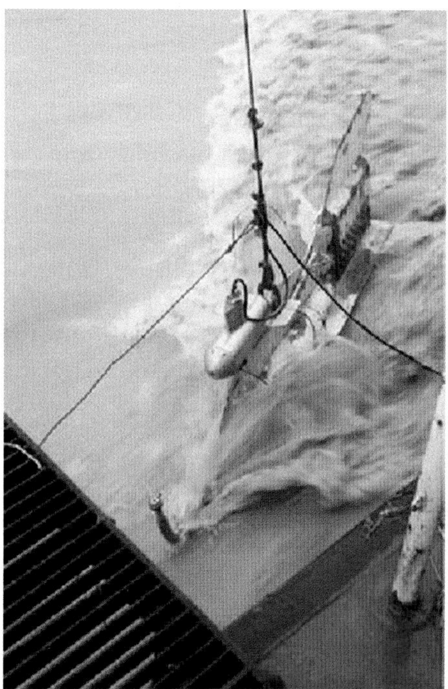

Fig. 9 Online sediment measurement is realized by a vibrational sediment meter.

Relevant information from China Meteorological Administration (CMA) is transmitted through special communication lines to the Bureau of Hydrology (BOH), MWR, and other hydrological departments at different levels.

Real-time hydrological information is transmitted by means of GSM, GPRS and telephone, satellite, and transmitter-receiver, then transferred to the BOH, MWR through Wide Area Network (WAN). The information is automatically received, processed, and stored by computer system.

Hydrological information transmission technology in China has been improved. By 1995, the traditional telegraph transmission way had been replaced by public packet switched data network (CHINAPAC X.25) to achieve a countrywide real-time hydrological information transmission by WAN. By 2005, with an established broadband network of 2 Mbps, using hydrological information code way, the real-time hydrological information was being transmitted rapidly. By 2011, the Hydrological Information Transmission System had replaced the hydrological information code way to realize a comprehensive and efficient transmission for hydrological information. At present, about 86% of the hydrological stations measure and report information in an automatic way; about 98% of the hydrological information from the hydrological stations can be transmitted to BOH, MWR, within 30 min.

3.3 Hydrological Forecasting Method and Model

The constraints of hydrological data availability and computer science in the past limited China's hydrological forecasting methods to those based on experience. In the 1960s, hydrologic models started to be developed and applied following the introduction of mathematical simulation into the field of hydrology. China focused on bringing in suitable hydrologic models and developing those to be applicable to local hydrological conditions. Currently there are basically two types of hydrological forecasting methods: the empirical hydrological forecasting method and that using conceptual hydrologic models.

The empirical hydrological forecasting method comes from the experiences that hydrological practitioners gathered over a long time, and has proven to be very effective in operational hydrological forecasting. At present, the 7 major river basin authorities in China have compiled complete empirical hydrological forecasting schemes, and the 600 plus hydrological forecasting stations in China have nearly 1000 schemes. In general, the empirical hydrological forecasting schemes apply such methods as antecedent precipitation index (API), corresponding stage (discharge) method, resultant discharge method, stage (discharge) fluctuating rate method, multi-factors combined axes correlation method, rainfall runoff correlation method, and so on.

China has extensive territories, with varied underlying surface conditions and different climates such as humid, dry, semi-humid, and semi-dry climates. For decades, Chinese hydrologists have made efforts to learn from international hydrologic models and succeeded in developing a series of river basin hydrologic models applicable to local conditions. The Xin'anjiang model proposed by Prof. Zhao Renjun in 1973 is the representative model that was developed. The main feature of the model is the concept of runoff formation on repletion of storage, which means that runoff is not produced until the soil moisture content of the aeration zone reaches the field capacity, and thereafter runoff equals the rain fall excess without further loss. The model can be used for real-time flood forecasting, investigations for gaining understanding of hydrological processes and study of the impacts of land use change on runoff, etc. Nowadays, hydrologic models used in operational hydrological forecasting in China comprise three types: those developed by Chinese professionals, international models, and adapted international models. Locally developed models include examples of the Xin'anjiang model (Zhao et al., 1980; Zhao, 1992; Zhao & Liu, 1995), the double excess runoff yield model, the Hebei storm flood model, the Jiangwan Bay runoff model, and the double attenuation curve model. Tank, Sacramento, NAM, and SMAR are major international models that are used, while the continuous API model and SCLS model were also introduced and modified to adapt to local conditions (WMO, 2011).

In recent years, environmental changes have brought challenges to river basin hydrological forecasting. Distributed hydrological models based on underlying terrain geographic information were developed and started to be used in flood forecasting and warning for flash floods as well as medium-small river flooding.

3.4 Hydrological Forecasting System

Computer science and network technology are indispensable for the development and application of Chinese hydrological forecasting systems. In the early 1980s, China developed the real-time flood forecasting system run on the VAX machines and on a single computer. As the consequence of the limited database management and software development technology at that time, data were mostly managed in a file system, and online operational forecasting was at a very low level with very simple functions, mainly programming existing forecasting schemes and models. From 1990 onwards, hydrologists in China learned from peers in developed countries with more mature and advanced technologies, and achieved progress in

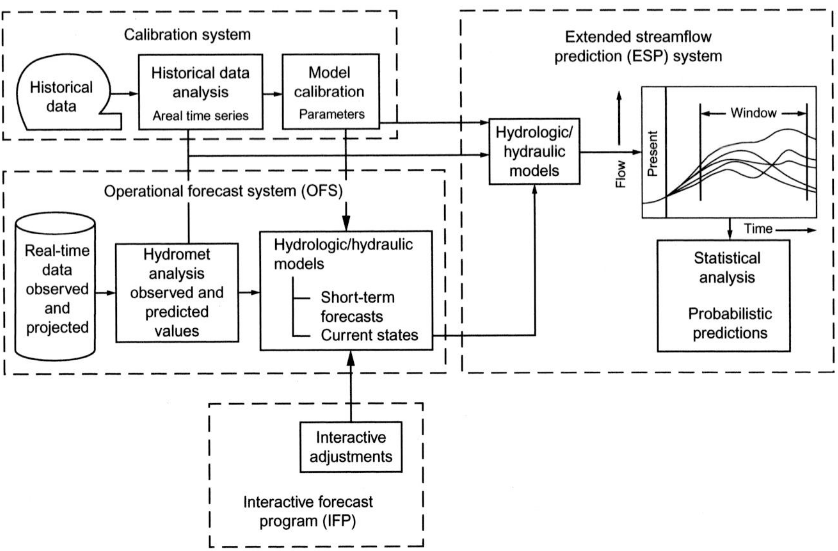

Fig. 10 Overall structure chart of China National Flood Forecasting System.

developing forecasting systems with functions of database management, model parameter calibration and optimized calculation, operational forecasting, real-time forecast updating, and visualized data and forecasting results. After the 1998 Great Flood, the Chinese hydrological sector made efforts to enhance the hydrological monitoring capacity and develop the universal flood forecasting system software platform: China National Flood Forecasting System (CNFFS) (Zhang & Liu, 2007). The CNFFS consists of three major components in a calibration system, operational forecast system, and extended stream flood prediction system, which are shown in Fig. 10. The CNFFS contains common forecasting models and a method library allowing prompt building of forecasting schemes. Many different flood forecasting models are employed in the system, including the Xin'anjiang model, and other models such as API, Sacramento, Tank, SMAR, and the synthetic constrained linear system (SCLS), as listed in Table 1.

Most large reservoirs and key medium-sized reservoirs built for flood prevention purpose have established telemetry systems, flood forecasting, and regulation systems. The CNFFS has been successfully used in 25 national and provincial hydrology departments in support of flood management at different levels in the country (Figs. 11-20).

Table 1 Flood forecasting models employed in the CNFFS

No.	Name
1.	Xin'anjiang model
2.	API model
3.	Jiangwan runoff model
4.	Hebei storm flood model
5.	Shanbei model
6.	Xin'anjiang model for semi-arid area
7.	Liaoning model
8.	Double attenuation curve model
9.	Double excess runoff yield model
10.	SMAR model
11.	NAM model
12.	Tank model
13.	Sacramento model
14.	SCLS model
15.	Index recession method
16.	Recession curve method
17.	Unit hydrograph method

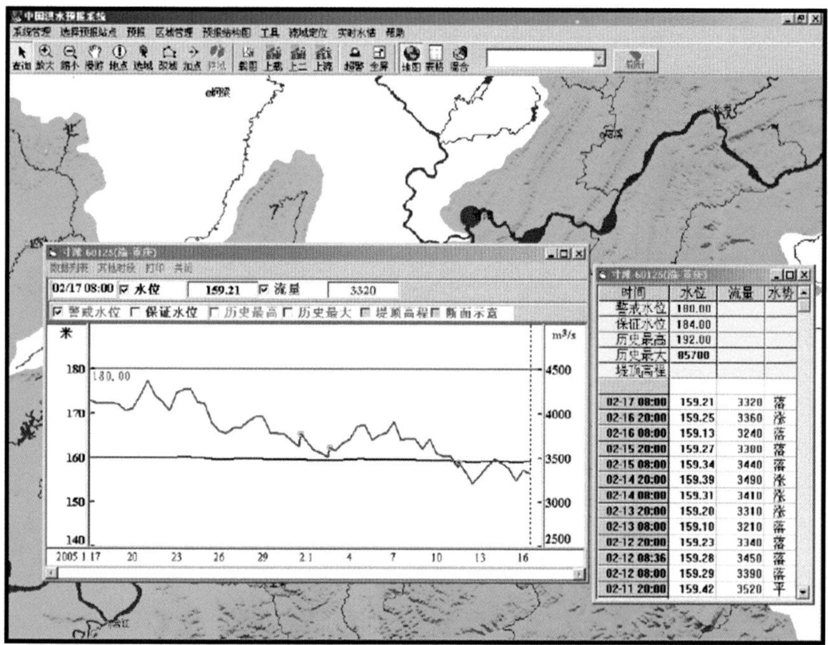

Fig. 11 GIS functions in CNFFS.

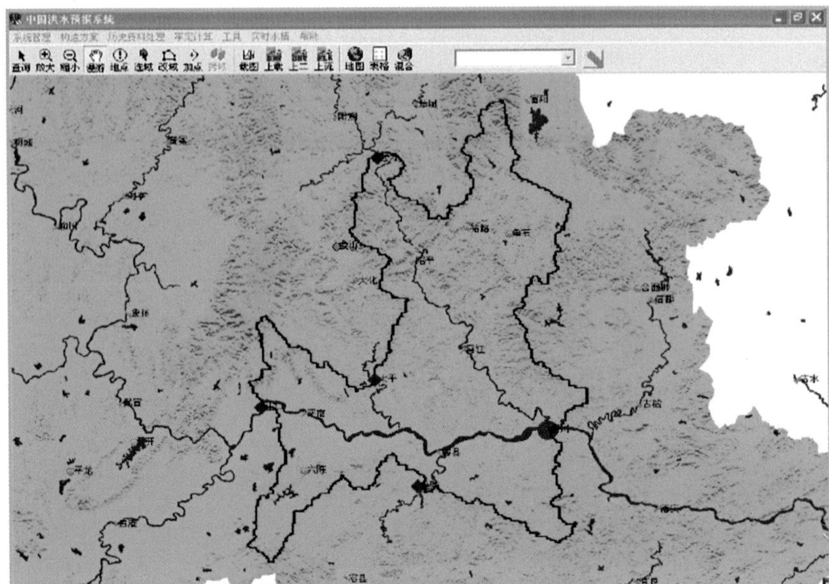

Fig. 12 Interface of automatic generation of catchment boundary in CNFFS.

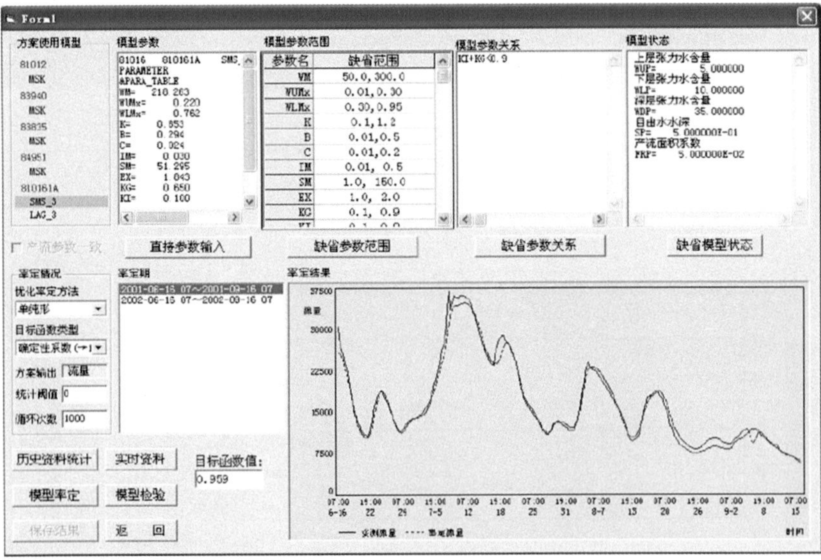

Fig. 13 CNFFS interface of model parameter calibrating.

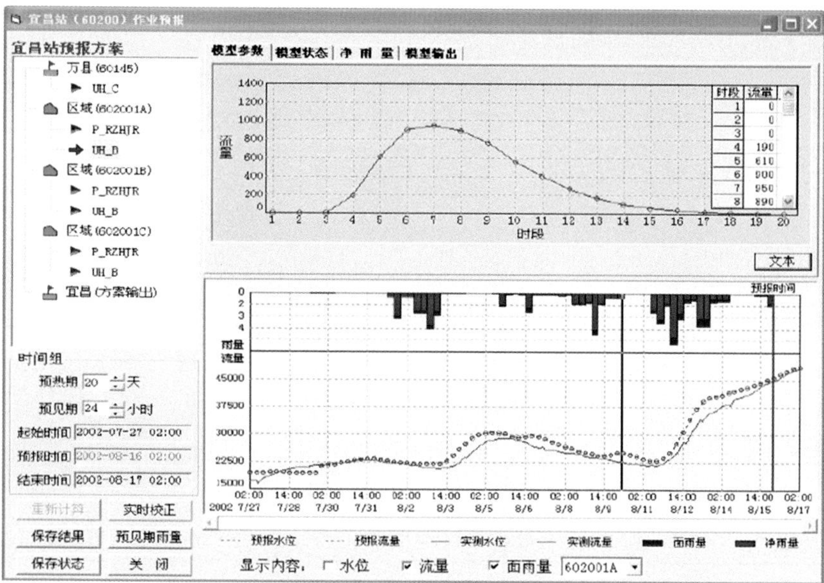

Fig. 14 CNFFS interface of real-time operational forecasting.

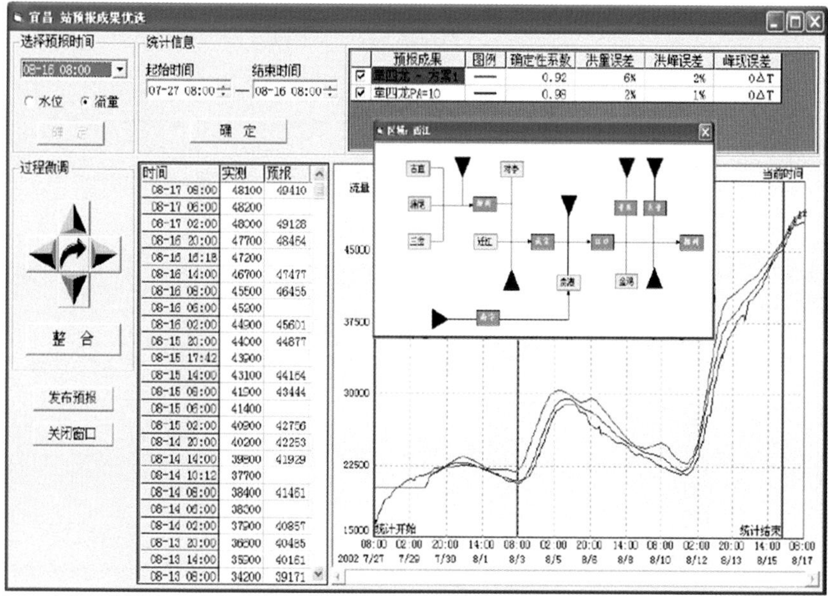

Fig. 15 CNFFS interface of forecasted result optimizing.

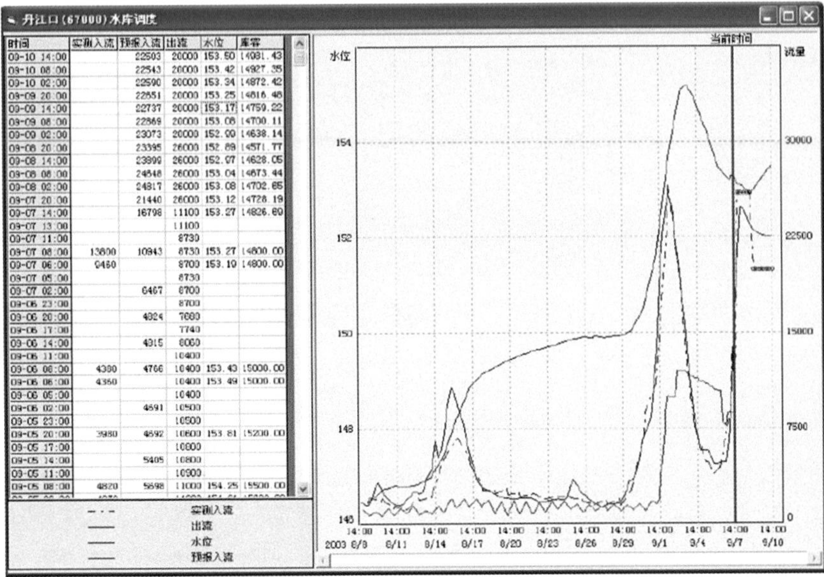

Fig. 16 CNFFS interface of reservoir forecasting and operation.

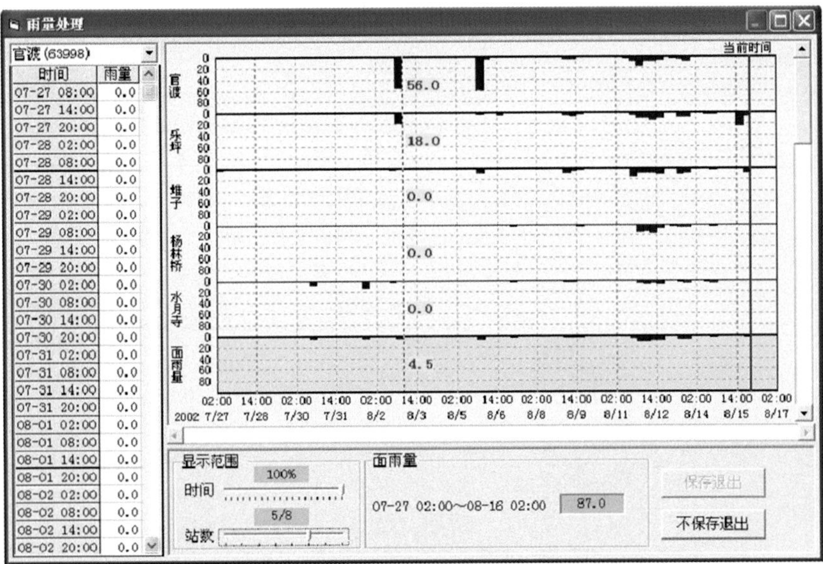

Fig. 17 Man-machine alternative interface of rainfall amending in CNFFS.

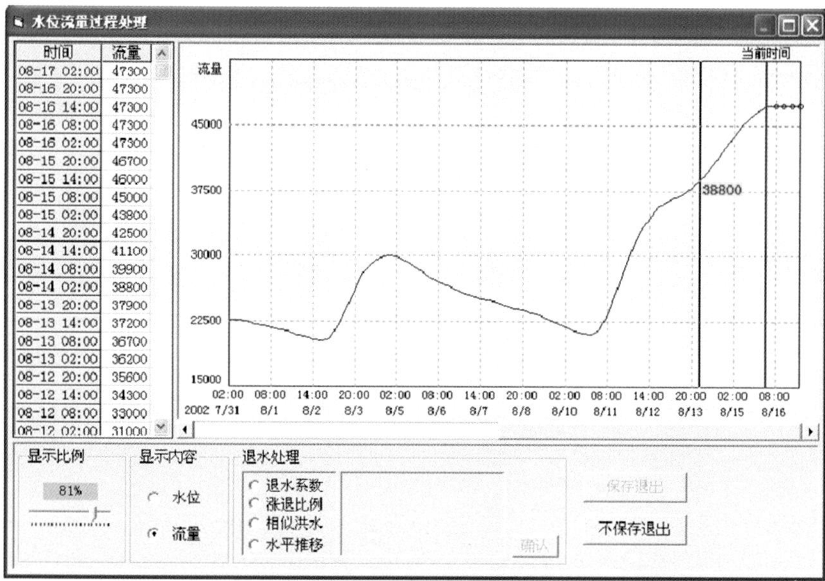

Fig. 18 Man–machine alternative interface of discharge amending in CNFFS.

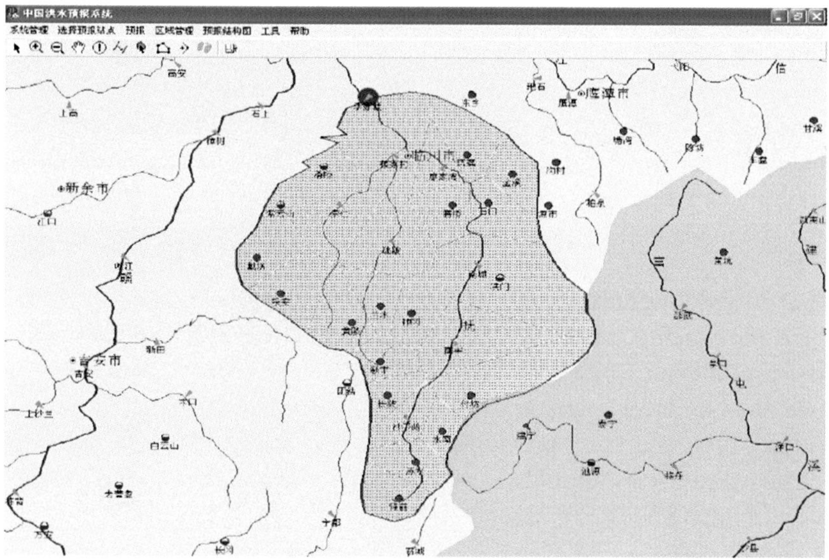

Fig. 19 CNFFS interface of Thiessen polygons creating.

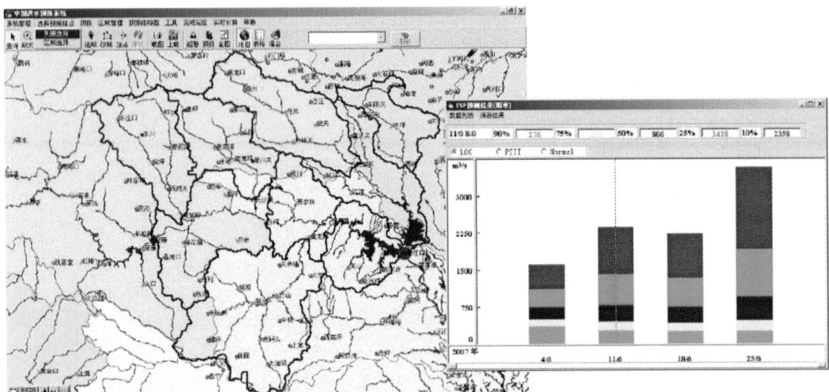

Fig. 20 Ensemble streamflow prediction component in CNFFS.

3.5 Application of New Technology in Hydrological Forecasting and Prediction

The scientific and technological development in recent years has enabled China to apply new and advanced technology to the hydrological forecasting, which improves the forecasting software and operational forecasting accuracy. The following are three representative examples.

3.5.1 Rainfall and Flood Joint Forecast

In real-time operational forecasting, the rainfall and flood joint forecast technology applies satellite and radar images for rainfall quantitative analysis and estimation. In the first step, the meteorological department provides data needed by the hydrological department for rainfall forecasts and prediction in terms of location, quantity, and time. The next step is the flood forecast. The results of the two steps are then combined to improve the lead time and prediction accuracy of hydrological forecasting and prediction.

3.5.2 Interactive Forecast Program

Since introducing the NWSRFS from NOAA of USA and Interactive Forecast Program (IFP), an interactive river forecast proto-system has been established for river basins in China. IFP combines the operational forecast system (OFS) and GUI, providing hydrological forecasters with information required for assessing data or simulating results, and verifying interactively to improve forecast accuracy.

3.5.3 Application of DEM, RS, and GIS Technologies

More recently, deterministic spatially distributed hydrologic models have been gradually used in China. Based upon equations representing the storage and the movement of water in the soil and on the surface, distributed models are potentially easier to calibrate by relating the values for their physically meaningful parameters to additional information provided in the form of Digital Elevation Maps, Soil Maps, and Land Use Maps. These models account for the spatial distribution of rainfall (now available at pixels of between 1×1 km and 10×10 km from RADAR images or in terms of QPF from NWP models) and are better equipped to address the problem of extrapolating the parameter values to the ungaged sub-catchments, a problem that has not yet found a proper solution. All these potential properties, combined with the increased efficiency of computers, have given rise to a number of fully distributed rainfall-runoff models, some of which, for instance, the TOPKAPI model (Liu, Martina, & Todini, 2005; Liu, Tan, Tao, & Xie, 2008), have been incorporated into real-time flood forecasting systems.

4 CONCLUSIONS AND OUTLOOK

Flood disasters in the past decades have become more frequent and devastating in China. The Chinese government gives high priority to flood management through both structural measures and nonstructural measures with the aim of enhancing flood prediction precision, and to assist in decision-making. However, for the larger flood events, nonstructural measures, such as the hydrological monitoring and forecasting, play very important roles.

Compared to developed countries with more advanced technology in hydrological services, China still lags behind in terms of new technology use and information service, but is rapidly catching up. China's hydrological forecasting services are expected to further develop in two aspects: firstly, the development of large-scale flood forecast and early warning systems for more regions can help flooding prevention in mountainous areas and in small-to-medium-sized rivers, and secondly, the development of improved medium-to-long-term streamflow forecasts can provide data for integrated water resources management.

Improved forecasts and predictions call for more collaboration between hydrologists and meteorologists and increased utilization of integrated data. Uncertainty must now be accounted for in decision-making at all levels of flood risk management, and this is important for communicating uncertainty in predictions to government and the public.

It is anticipated that full use will be made of new technologies, tools, and products to improve outputs of models, predictions and forecasting. Among the cutting-edge technologies and tools available for improved hydrological modeling and forecasting are the use of radar and satellite-based rainfall estimation, quantitative precipitation forecasts, and numerical weather prediction. Combining such technologies and tools with the significant advances in spatial modeling technologies such as digital elevation models and GIS tools have great potential to improve significantly the accuracy of the outputs of hydrological models. Such improved modeling techniques will deliver high-quality products, advice, and decision support systems such as flood forecasts and warnings, flood plain and inundation maps, and flow forecasts for reservoir operation, among others.

REFERENCES

Arduino, G., Reggiani, P., & Todini, E. (Eds.), (2005). Special issue: Advances in flood forecasting. *Hydrology and Earth System Sciences, 9*(4), 280–284.

Linnerooth-Bayer, J., & Amendola, A. (2003). Introduction to special issue on flood risks in Europe. *Risk Analysis, 23*(3), 537–543.

Liu, Z., Martina, M., & Todini, E. (2005). Flood forecasting using a fully distributed model: Application to the upper Xixian catchment. *Hydrology and Earth System Sciences, 9*(4), 347–365.

Liu, Z., Tan, B., Tao, X., & Xie, Z. (2008). Application of a distributed hydrologic model to the catchments of different characters for flood forecasting. *ASCE Journal of Hydrologic Engineering, 13*(5), 378–384.

WMO. (2011). *Manual on flood forecasting and warning.* WMO Publication. No. 1072, 48–49, IAHS Press, Wallingford, UK.

Wu, Z., Lu, G., Liu, Z., et al. (2013). Trends of extreme flood events in the pearl river basin during 1951–2010. *Advances in Climate Change Research, 4*(2), 110–116.

Zhang, J., & Liu, Z. (2006). Hydrological monitoring and flood management in China. In I. Tchiguirinskaia, K. Thein, & P. Hubert (Eds.), *305. Frontiers in flood research* (pp. 93–102): IAHS Publications, IAHS Press, Wallingford, UK.

Zhang, J., & Liu, Z. (2007). Hydro-meteorological monitoring and operational hydrosystems for flood management in China. In *311. Proceedings of the international symposium on methodology in hydrology held in Nanjing, China.* (pp. 3–9): IAHS Publications, Wallingford, UK.

Zhao, R. J. (1992). The Xinanjiang model applied in China. *Journal of Hydrology, 135*, 371–381.

Zhao, R. J., & Liu, X. R. (1995). The Xinanjiang model. In V. P. Singh (Ed.), *Computer models of watershed hydrology.* Colorado: Water Resources Publications.

Zhao, R. J., Zhang, Y. L., Fang, L. R., LIU, X. L., & Zhang, Q. S. (1980). The Xinanjiang model. *Hydrological Forecasting Proceedings, Oxford Symposium, IAHS, 129*, 351–356.

CHAPTER 4

A Regional Perceptive of Flood Forecasting and Disaster Management Systems for the Congo River Basin

R.M. Tshimanga*,†, J.M. Tshitenge*,†, P. Kabuya*,†, D. Alsdorf‡, G. Mahe§, G. Kibukusa¶, V. Lukanda*

*University of Kinshasa, Kinshasa, Democratic Republic of Congo
†CB-HYDRONET, Kinshasa, Democratic Republic of Congo
‡The Ohio State University, Columbus, OH, United States
§Laboratoire HydroScience, IRD, Montpellier, France
¶Ministère des Affaires Sociales, Actions Humanitaires et Solidarité Nationale, Kinshasa, Democratic Republic of Congo

1 INTRODUCTION

The Congo River Basin is geographically located in Africa within 9° N, 12° E to 13.30° S, 34° E and encompasses nine political boundaries: Angola, Burundi, Central African Republic, Democratic Republic of Congo, Cameroon, Republic of Congo, Rwanda, Tanzania, and Zambia (Fig. 1). The region of the Congo River Basin covers a wide range of climatic conditions and is prone to recurrent extreme events and the associated disasters, including floods, droughts, cyclones, and changes in the patterns of rainfall and temperatures. Associated with this challenge are rapid population growth, uncontrolled urbanization, deforestation, and land degradation, which represent a serious challenge to sustainable development. Climate change may further exacerbate these problems with implications on agriculture, biodiversity, hydropower, and health (eg, water-related diseases).

Flood disasters have always been reported in the Congo River Basin with some times significant damages to socio-economic lives and to the environment. While these impacts are real in the basin, efforts to mitigate them or to enable the capacity of people to minimize their costs remain very limited. This is a result of a lack of understanding of processes governing the dynamics of the extreme events, a lack of adequate data and information necessary to make sound predictions, and insufficient infrastructures, compounded by the lack of a clear policy towards flood management systems.

Flood Forecasting
http://dx.doi.org/10.1016/B978-0-12-801884-2.00004-9

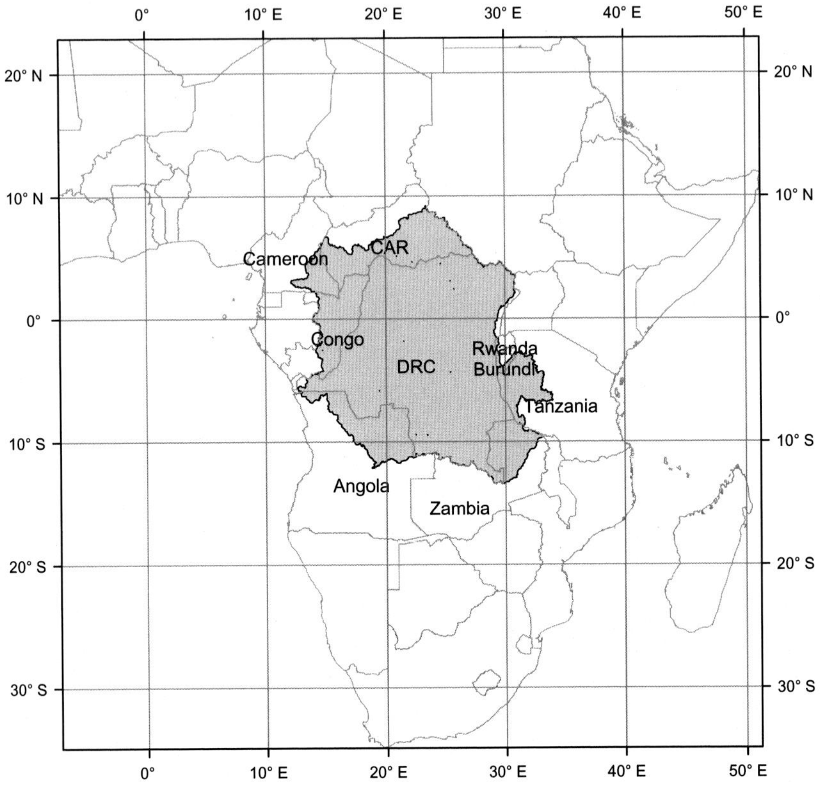

Fig. 1 Map of the Congo River Basin showing political boundaries.

The lack of information on the current state of climate at different scales in this large river basin, and how the frequency, intensity, duration and extent of flood events will change in the future based on changes of environmental conditions (eg, climate-land use changes) represents a bottleneck to sustainable planning. A critical lack of skills, infrastructures, and human and financial resources as well as institutional challenges undermines the ability to implement an effective flood management system for the benefit of climate change resilience and socio-economic development.

The present chapter provides an exploratory assessment of the adequacy and effectiveness of flood forecasting systems in the Congo River Basin with the aim to identify challenges that could be addressed towards an effective flood and disaster management. Identification of the basin-wide characteristics driving the momentum of floods, identification of sensitive areas and vulnerability mapping, the adequacy and performance of flood monitoring systems, existence of supportive technologies and flood management strategies including structural and nonstructural, constitute the main topics that will be explored in this chapter.

2 PHYSIOGRAPHIC SETTING AND PHYSICAL CHARACTERISTICS OF THE CONGO RIVER BASIN

The geomorphologic classification of the central African land surface (Burke & Gunnell, 2008) includes the presence of topographic highs, also called "swells" (Kadima, Delvaux, Sebagenzi, Tack, & Kabeya, 2011) or "rises" (Runge, 2008), which surround the central part of the Congo River Basin. The Atlantic rise encompasses the streams of the western right bank of the Congo River. The main rivers generated in this area are known as the Sangha, Mossaka, Alima, Nkeni, and Lefini. Further northeast of the Congo River Basin, there is the Asante rise which encompasses the streams that drain the basin starting from its drainage divides with the Chari and Nile Basins to the main trunk of the Congo River. This drainage unit is known as the Oubangui Basin, the name of the main stream that connects all the upstream tributaries to the main trunk of the Congo River. The eastern part of the Congo River Basin is flanked by the Mitumba Mountains, which mark a clear drainage divide between the Congo and the Nile Basins. The main streams generated from these highlands are the Aruwimi, Tshopo, and Lindi rivers, which are connected to the Congo river at Kisangani, where it is actually known as the Congo River. The southern rim of the Congo River Basin is flanked by the Lunda rise, which is shaped by the Angolan highlands in the southwest and the Shaba, or North Zambian, swell in the southeast. The main rivers of the Southern Congo Basin, notably the Kasai and Lualaba rivers, rise in the highlands of the Lunda Rise and the Shaba Swell, respectively. The river generated from these physiographic features runs over 4375 km before pouring its average flow of over 41,000 $m^3 s^{-1}$ into the Atlantic Ocean, thus draining an area of about $3.7 \times 10^6 km^2$. From the plateaus of Katanga, the river first flows north, then west and south, crossing the equator twice in a great arc as it traverses a vast swampy basin over 4375 km from east to west and up to 850 km from north to south. In its middle course, the Congo River varies in width between 3 and 15 km and loses only 115 m in elevation over a river distance of 1740 km between Boyoma Falls (0° 29′N/25° 12′E) and Malebo Pool (4° 11′S/15° 35′E) (Hughes & Hughes, 1987). During the high water periods, vast areas of land adjacent to rivers in the central basin are flooded. Fig. 2 shows the physical layout of the main rivers of the Congo Basin.

The main part of the basin has low slopes, but many of the headwaters have steeper topography (Runge, 2008), from which flow the four main tributaries that meet in the central basin and constitute the main stream of the Congo River. Vegetation varies from open savannah grassland and woodland in the upland areas to tropical rainforest in the central basin. The central part

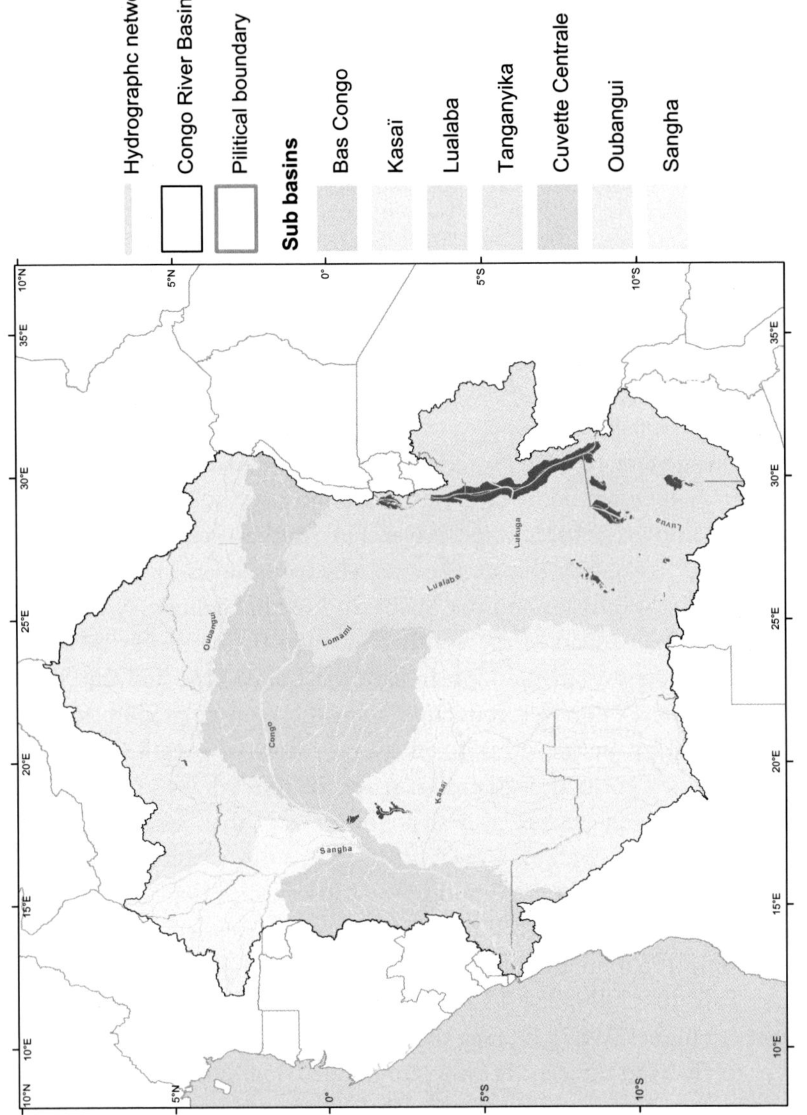

Fig. 2 Physical layout of the Congo Basin showing the main rivers and their sources.

of the basin is covered by unconsolidated Cenozoic sediments, whereas the primary catchments that feed into the central basin have deeply weathered Mesozoic and Precambrian rocks (Runge, 2008). Some studies conducted in the basin show evidence of a markedly unstable water level during the second half of the last century, induced by rainfall variations, the influence of which was considerably modified by the soil type and the geology of the terrain (Laraque, Mahé, Orange, & Marieu, 2001). A so-called "see-saw" phenomenon (Eltahir, Loux, Yamana, & Bomblies, 2004) suggests the interaction of hydrological processes between the Congo and Amazon Basins. This phenomenon was inferred from an anti-correlation in runoff anomalies between the two basins by using satellite rainfall and river flow data, suggesting that floods over the Amazon Basin tend to coincide with drought over the Congo River Basin and vice versa.

The main tributaries cross areas of various heights, slopes, soils, and geologies before discharging their flows from the basin into the Atlantic Ocean. Fig. 3 shows the main relief regions that characterize the Congo River

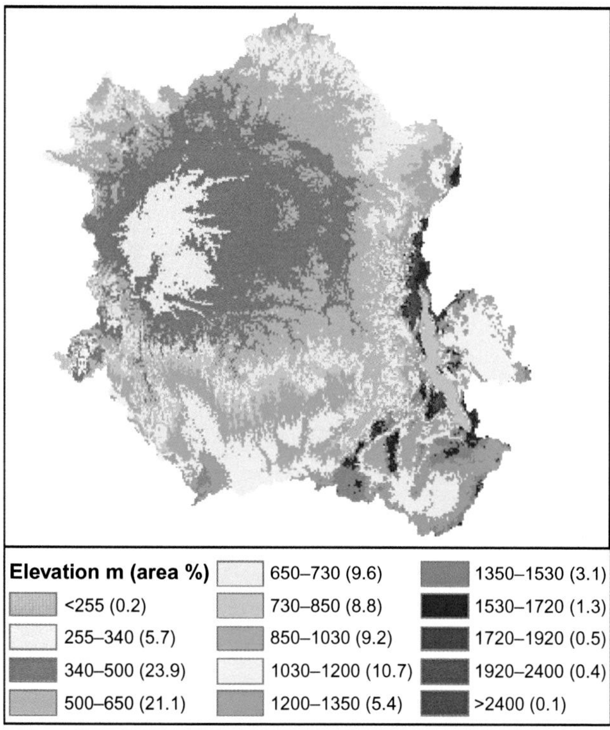

Fig. 3 Physical layout of dominant land elevation areas showing the percentage of areas occupied by each of the predefined elevation classes for the Congo Basin.

Basin. These regions are set in the form of concentric layers for which the surface area decreases with the increasing elevation. Based on this classification, it can be seen that the terrain elevation of most of the basin is between 340 and 650 m, representing more than 40% of the basin area. A slope map derived from a combination of various slope classes shows that about 51% of the basin area is comprised between 0.25% and 2%, a class of flat wet, and about 10% comprised between 0% and 0.25%, a class of flood plain (Tshimanga, 2012).

3 FLOOD-BEARING PROCESSES IN THE CONGO RIVER BASIN

Central to flood management is the ability to predict hydrological responses under different spatial and temporal conditions, including stationarity and non-stationarity. A description of the dynamics of the atmosphere-land surface and sub-surface processes is essential for successful prediction. Climate, land cover and land use, and the conditions of change are essential for understanding these dynamics.

3.1 Hydro Climate Processes

Climate is undeniably the main controlling factor of the basin hydrology and much of the observed variability in streamflow is related to the variability in climate. Despite the predicted impacts of climate variability and changes on water resources, it is important not only to quantify the impacts but also to understand the processes that drive the momentum of change in climate variables at local and regional scales. Depending on region and timescale, it has been observed that past hydrological changes in Africa have been linked to various climatic processes (Schefuss, Schouten, & Schneider, 2005).

The Congo River Basin represents a climatic transition zone between Northern and Southern Africa, and Eastern and Western Africa (Balas, Nicholson, & Klotter, 2007), thus making the climate variability remarkably complex. Fig. 4 shows the distribution of the mean annual rainfall over the Congo Basin, computed using the Climate Research Unit (CRU) gridded dataset. Many studies have shown that the climatology over tropical Africa in general, and particularly the Congo Basin, is influenced by many factors, which depend on atmospheric-ocean interactions and the monsoonal processes (Balas et al., 2007; Farnsworth, White, Williams, Black, & Kniveton, 2011). The inter tropical convergence zone (ITCZ), sea surface temperatures

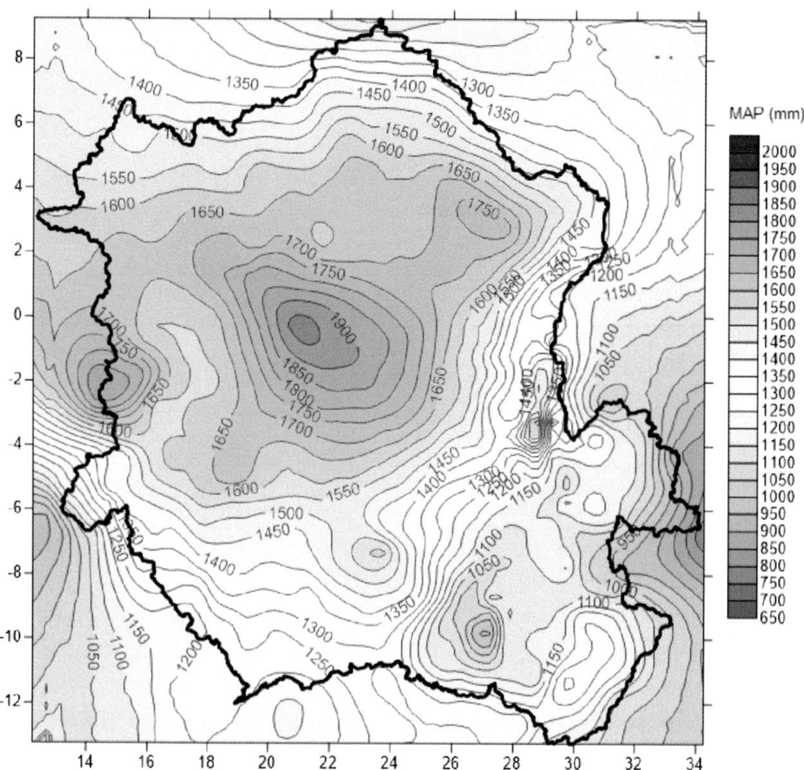

Fig. 4 Distribution of the mean annual rainfall over the Congo Basin, computed using the Climate Research Unit gridded dataset.

(SSTs), atmospheric jets (Central African jets), and meso-scale convective systems are the main drivers that modulate climate variability over the region of the Congo River Basin, with an indication of the role of the tropical easterly jet (TEJ), the westerly African jet, African easterly jet, the southern African easterly jet, the Atlantic/Indian Ocean SST anomalies, and the impact of El Nino Southern Oscillation (ENSO) upon the mechanisms governing wet and dry years (Farnsworth et al., 2011; Poccard, Janicot, & Camberlin, 2000).

The concept of the ITCZ over Africa refers to a band that follows the sun, migrating to the northern hemisphere during the boreal summer and to the southern hemisphere during the austral summer (Nicholson, 2009), thus having a direct influence on the rainfall variability through perturbation of the strength and position of the rainfall belt (Farnsworth et al., 2011). The rainfall belt is therefore defined as the variability in rainfall caused by intensity and position. Convergence and uplift occur during the seasonal ITCZ movement towards the equator, but strong precipitation occurs only where the moist layer is sufficiently thick to support deep clouds and convection.

The migration has been shown to be related to a bimodal pattern of rainfall over the Congo River Basin with both dry and wet seasons occurring in different regions of the basin and at the same time (Hughes & Hughes, 1987; Mahé, 1993). Many studies carried out have concluded that there is a strong correlation between the ITCZ and the occurrence of rainfall or rainfall variability over the tropical equatorial region of Africa (Farnsworth et al., 2011; Mahé, 1993). A recent study by Nicholson (2009) showed that the tropical rain belt is produced by a large core of ascent lying between the African easterly jet and the TEJ, and not necessarily represented by the ITCZ. The rainfall is therefore distributed through the southern track of the African easterly waves, which correspond to the African easterly jet and the TEJ. However, this emerging theory contradicts previously suggested theories about the ITCZ and its relation to the tropical rain belt as well as its implication for interannual and multi decadal variability over the region of West Africa including the Congo River Basin. Nicholson (2009) points to the outdated 1950s concept that has perpetuated the current widely prevailing but erroneous views of the ITCZ, which has an implication for seasonal forecasting.

Besides the role of the ITCZ in rainfall variability over the Congo Basin, SST anomalies also have a direct effect on the regional circulation, which alternatively induces seasons of wetter and drier conditions as well as warm and cold anomalies (Farnsworth et al., 2011). Atmospheric moisture transport onto the Central African region is modulated through important changes in the SST patterns (Schefuss et al., 2005). Much of the rainfall variability over the Congo Basin could be explained through the Atlantic and Indian Ocean's SST anomalies such as the Atlantic Nino, the inter-hemispheric mode, and the ENSO (Balas et al., 2007; Farnsworth et al., 2011; Paeth & Friederichs, 2004). During the boreal summer, greater rainfall over the Congo Basin is attributed to equatorial warming (Atlantic Nino) that creates warm SST anomalies displacing convection southeastwards. During this period, the northernmost part of the basin, mainly composed of the Oubangui and Sangha subbasins, experiences dry conditions. The reverse is observed when greater upwelling in the equatorial Atlantic creates cold SST anomalies. Mahé (1993) reports a decrease of the SSTs along the equator from Jun. to Oct., and also during Jan. and Feb. The decrease is caused by equatorial upwelling, which in turn is characterized by a direct action of the wind on surface waters, a divergent effect of the equatorial circulation that is influenced by the Coriolis forces, and the oceanic wave arising in the equatorial Atlantic. In the inter-hemispheric mode, there is a change in the SST warm or cold anomalies due to changes in the above mentioned dipole, which contributes to shifting the predominant source regions of atmospheric moisture of the Atlantic SST (Balas et al., 2007; Farnsworth et al., 2011). SST is less important during the boreal winter when its influence is mainly concentrated within the tropical band south of 10° N — that is, when the ITCZ is located south of the equator. The contrast is obvious during the boreal summer where the SST induced fraction of total rainfall variance amounts to at least 10% over the entire continent of Africa north of 10° S (Paeth & Friederichs, 2004).

The so-called atmospheric jets or Central African jets (Farnsworth et al., 2011) are an important component of atmospheric circulation that play a significant role in rainfall processes and the position of the rainfall belt over the Congo Basin through impacts upon the African easterly wave production and modulation as well as the impact of vertical sheer upon deep convection (Farnsworth et al., 2011; Nicholson, 2009). The transport of energy over the Congo River Basin is related to the mesoscale convective systems such as cloud and thunderstorms that ensure vertical motion in the transport processes (Farnsworth et al., 2011).

In general terms, the central part of the basin is characterized by high rainfall associated with high temperature. Away from the central basin, there is a decrease in the mean annual rainfall. This decreasing trend is accentuated in the southeastern part of the basin as well as in the extreme north and the lower parts of the basin. These areas are also characterized by high evapotranspiration and low temperature compared to the central basin. This spatial variability is accentuated by the variability in time. Fig. 5 shows the variability in monthly rainfall for the basin. The spatio-temporal variability is partly driven by the seasonal migration of the rainfall belt across the basin. The rainfall belt first appears in the northern part of the basin in a scattered form in May, but constitutes a consistent mass by Jun. From Jun., the mass of the rainfall belt starts its migration towards the south until Mar. A recession of the rainfall belt appears first in a small portion of the northern Oubangui in Oct. and it is only during Dec. that this recession is complete over the northern part of the basin, while remaining only concentrated in the southern part where its strength diminishes gradually from Jan. to Mar. The recession of the rainfall belt over the Congo River Basin is complete in Apr. where there is low rainfall over the basin.

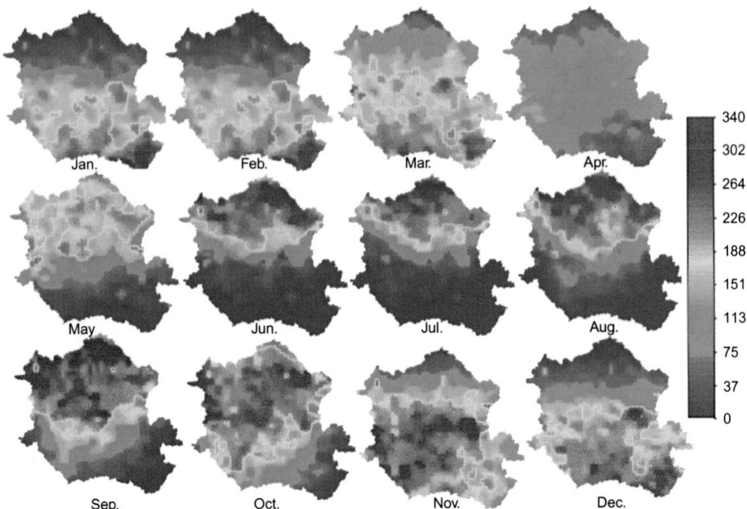

Fig. 5 Monthly distribution of the long-term average rainfall over the Congo River Basin.

The seasonal cycle in the basin is characterized by a bimodal pattern of the rainfall distribution with maximum rainfall values in Mar., Apr., Oct., and Nov. (Beighley et al., 2011; Juarez, Robin, Li, Fu, & Fernandes, 2009). The rainy season in the north coincides with the dry season in the south and vice versa; so heavy rain in the north tends to compensate for light rain in the south, thus maintaining downstream river flow stability throughout the year. Nevertheless, levels in the watercourses of the flat central basin normally exhibit two maxima and two minima each year.

Fig. 6 shows the daily trend in streamflow (1903–2012) for the main gaging site of the Congo Basin located at Kinshasa. The location of this gage ensures a streamflow record for about 98% of the basin drainage area. The

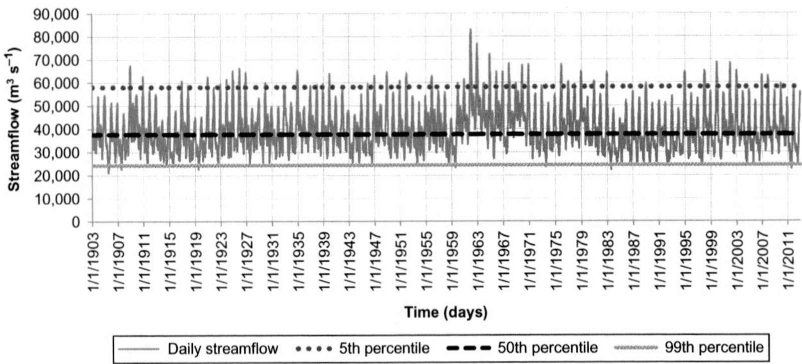

Fig. 6 Daily trend of streamflow (1903–2012) for the main gaging site of the Congo Basin located at Kinshasa.

daily hydrograph is plotted against 99th, 50th, and 5th percentiles of time that flow is equaled or exceeded for high, medium, and low flows, respectively. The annual maximum, mean, and minimum flows (extremes high and low flow) are presented in Fig. 7, and their seasonal distribution in Fig. 8.

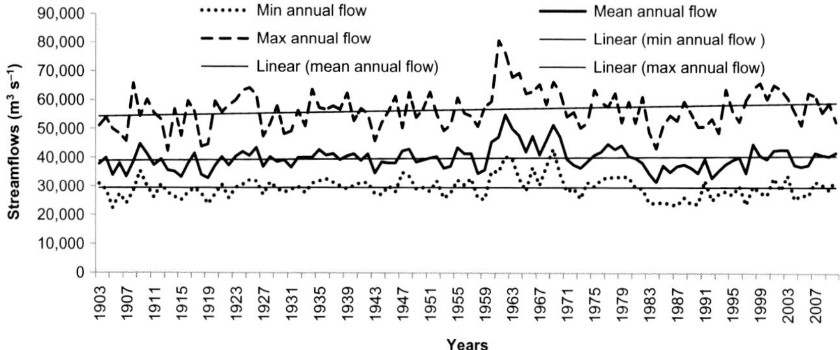

Fig. 7 Annual maximum, mean, and minimum flows (extremes high and low flow) for the main gaging site of the Congo Basin located at Kinshasa.

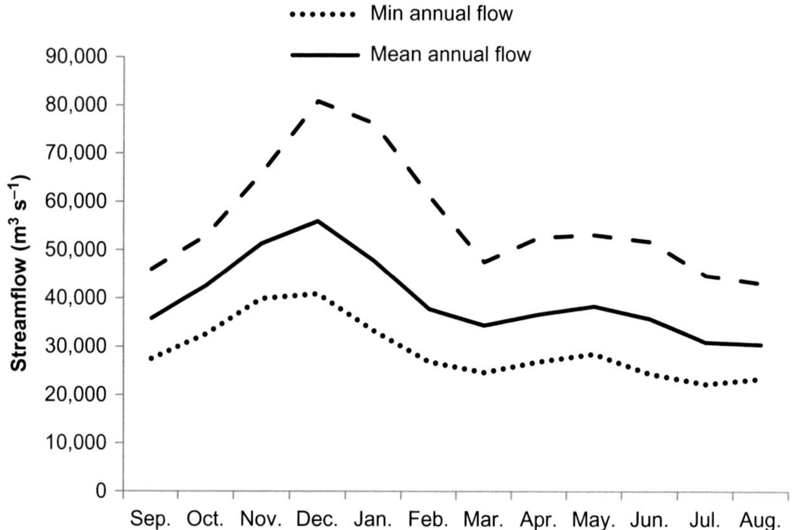

Fig. 8 Seasonal distribution of annual maximum, mean, and minimum flows (extremes high and low flow) for the main gaging site of the Congo Basin located at Kinshasa.

In general terms, there is a tendency towards increase in extreme high and low flows as observed for the Congo River Basin at the Kinshasa gaging site. The periods 1903–59 and 1982–94 appear homogenous, whereas the periods 1961–70 and 1995–2010 are characterized by major floods with regards to their annual maximum values.

3.2 A Wavelet Approach for Understanding Hydro Climate Processes in the Congo River Basin

Studies assessing the factors that modulate hydro-climate processes over the Congo River Basin have been variously challenged by the lack of data, and therefore application of consistent approaches that could allow deep analysis of the processes and causal relationships. A wavelet analysis is conducted here with the aim to assess the influence of the factors of climate variability on hydrological processes of the basin, to identify the main mode of variability in the basin, and eventually to provide an understanding of flood events in the basin. Streamflow data for the Kinshasa streamflow station for the period 1903–2012 and a rainfall grid from the CRU are used in this analysis. The North Atlantic Oscillation (NAO) is used for coherence analysis with streamflow and rainfall data. The wavelet technique is used to decompose a signal of streamflow time series on the basis of scaled and translated versions of a reference wavelet function. Each wavelet has a finite length (scale) and is always highly localized in time. The reference wavelet function comprises two parameters for time-frequency (or time scale) exploration, that is, a scale parameter a and a time-localization parameter b, so that

$$\psi_{ba}\left(t\right)=\frac{1}{\sqrt{a}}\psi\left(\frac{t-b}{a}\right) \tag{1}$$

A continuous wavelet transform of a signal $s(t)$ producing the wavelet spectrum is defined as

$$S\left(b,a\right)=\frac{1}{\sqrt{a}}\int_{-\infty}^{+\infty}\overline{\psi\left(\frac{t-b}{a}\right)}s\left(t\right)dt \tag{2}$$

This transform allows the decomposition in different frequency components that can be analyzed for a given time, thus allowing a better description of a non-stationary process (Schneider & Farge, 2006; Torrence & Compo, 1998). Literature provides several types of wavelet functions that can be used in the wavelet transform. In this study, the Morlet wavelet function is used, which offers a good resolution of frequency with a wave number so that wavelet scale and Fourier period should approximately be equal (De Moortel, Munday, & Hood, 2004). The global wavelet spectrum $G_f(a)$ is defined as follow and has been used to detect the dominant scales or modes of variability (Hudgins, Mayer, & Friehe, 1993; Meneveau, 1991).

$$G_f\left(a\right)=\int_{-\infty}^{+\infty}S_f\left(a,b\right)db \tag{3}$$

Fig. 9 shows the variability of streamflow for the Kinshasa station at different time scales. The *Y* axis is the frequency or periodicity, the *X* axis is the time of the streamflow series, and the *Z* axis is the energy that translates the variability in a given streamflow series. The Morlet wavelet analysis is performed for two signals: the mean monthly flow and the maximum annual flows for the period 1903–2012. The periodicity of the analysis ranges from 0.25 to 32 years, in which we clearly identify two main permanent oscillations throughout the period of analysis, and two temporary oscillations.

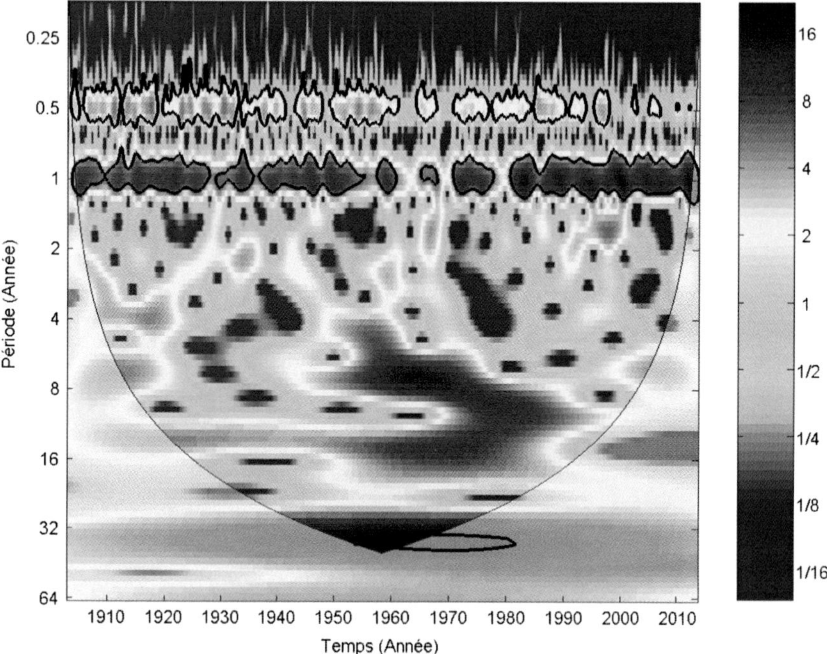

Fig. 9 Spectrums of the continuous wavelet analysis for the Congo River mean monthly flows at Kinshasa station. The dark contours show statistically significant fluctuations against a white noise (AR (1) = 0), at a confidence level of 90%.

At the signal of the mean monthly flows (Figs. 9 and 10), two permanent oscillations appear around the scale-periods of 0.5 and 1 year with the latter showing a very strong signal of variability. Two temporal oscillations are apparent, both only around 1950–80, with a peak markedly shown in 1960s, which historically corresponds to the year of maximum flood in the Congo River Basin. These two temporary oscillations are revealed for the scale-periods of 4–8 and 8–16 years.

At the signal of maximum annual flows (Fig. 11), we observe a more or less permanent and strong oscillation for the period scale of 13.1 years.

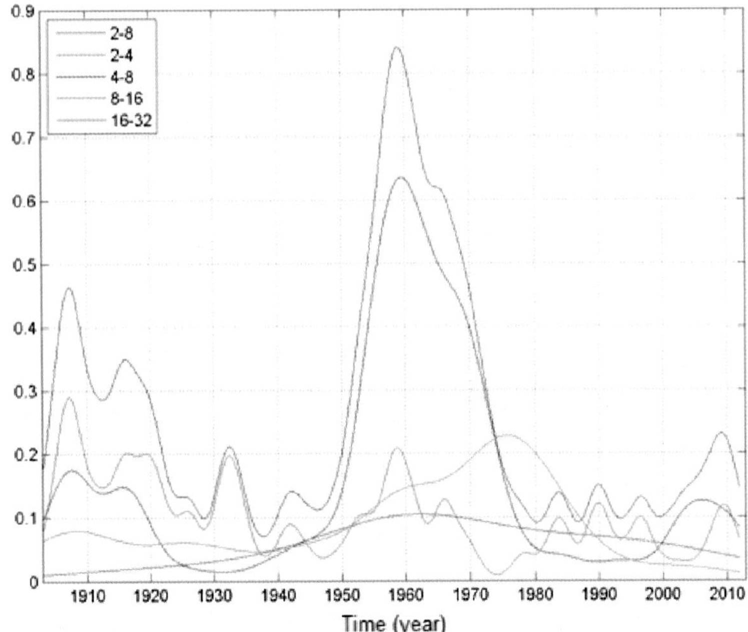

Fig. 10 Variance of streamflow (station: Kinshasa), average scale spectrum for the scale period of 2–8 and 8–16 years.

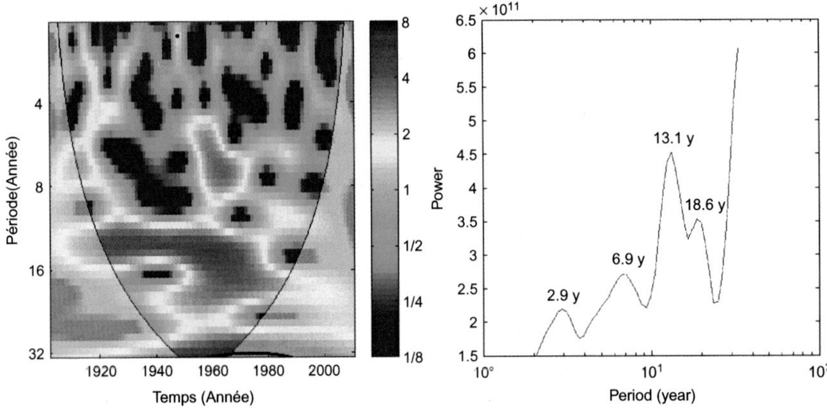

Fig. 11 Spectrums of the continuous wavelet analysis for the Congo River annual maximum flows at Kinshasa station. The dark contours show statistically significant fluctuations against a white noise (AR (1) = 0), at a confidence level of 90%.

Two other temporary oscillations appear around 1960 with the scale-period of 6.9 and 18.6 years.

Fig. 12 shows the average scale spectrum computed for the maximum annual flows for the scale periods of 2–4, 4–8, 2–8, and 8–16 years. It may be observed from this figure that there are different modes of variability.

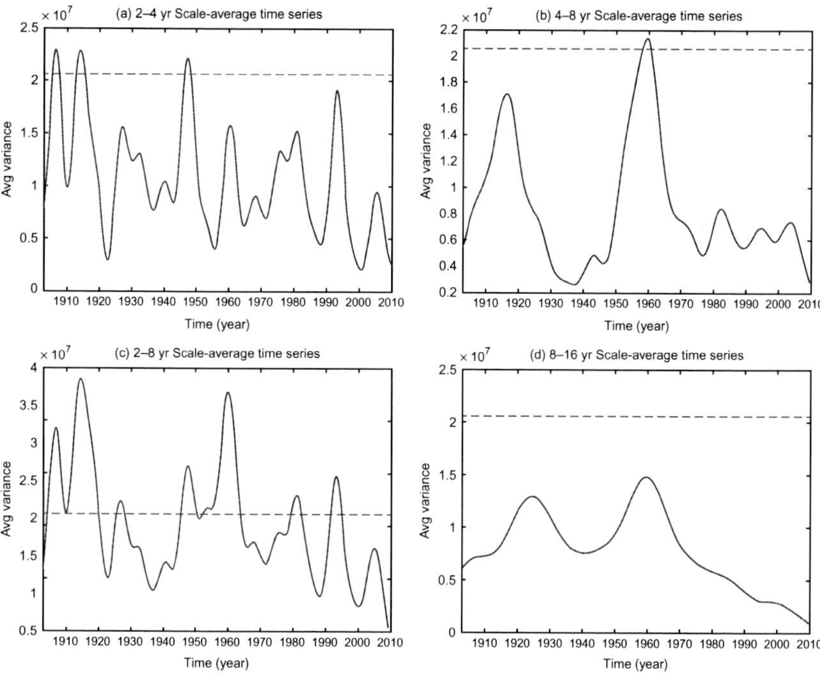

Fig. 12 Variance of annual maximum flows of the Congo River Basin at Kinshasa station, scale periods of 2–8 and 8–16 years.

Based on our wavelet analysis, the trend of streamflow discharge for the Kinshasa station, which represents about 98% of the basin drainage area and its main tributaries, is mainly structured into three modes of variability (Table 1), for the different scales used, namely annual, interannual, and decadal cycles. The high contribution to the variance is attributed to the annual cycle whereas the interannual (2–8 years) and decadal (8–16 years) cycles represent a very low contribution to the total variance with 7.68% and 2.7%, respectively. Most of the streamflow variability is associated with the events of very high frequency (≤1 year) and with seasonal cycle (1 year) for 67% of the total variance (cf. Fig. 10).

Although the interannual cycle shows a low contribution of about 7% to the variance of the streamflow, it appears that this cycle is linked to different floods observed in the Congo Basin. This strong contribution of the variance is observed for the periods 1903–20, 1960–70, and 2000–2010.

The main question here is: are the origins of these modes of variability of climate origin or are they internal to the Congo River Basin? In this regard, an attempt was made to carry out a comparative analysis using streamflow for the Kinshasa station, a regionalized precipitation grid, and some selected climate indices in order to detect possible teleconnections.

Table 1 Mapping of natural disasters in terms of events of occurrence for the nine countries of the Congo Basin (1964–2012).

Disaster type	**A**	**B**	**C**		**D**			**E**	**F**		**G**	**H**
Disaster subtype	*A1*	*B1*	*C1*	*C2*	*D1*	*D2*	*D3*	*E1*	*F1*	*F2*	*G1*	*H1*
DRC	2	3	23	41	18	2	7	5	1	3	3	2
Angola	7		2	15	24	1	6	1				
Burundi	6	1	1	12	11	5	9			3		
CAR	6		3		6	3	9			6	2	
Cameroon	4		1	25	10	2		1			3	
Republic of Congo		1	8	8	8		1	1				
Rwanda	6	2		11	9	1	2	3				
Tanzania	10	9	5	20	20	4	18	1	1	4		
Zambia	5		2	14	15		5	1				

The wavelet coherence was used, which describes a correlation measure between two signals (Labat, 2010) and can be used to characterize the correlation or linearity between two processes based on time frequency (Jiang, Gan, Xie, & Wang, 2013). The coherence analysis was therefore carried out between the mean monthly flow and mean monthly rainfall and the maximum annual flow (Fig. 13) with the NAO Index (Fig. 14). The coherence may therefore be defined as a module of spectrum crossed by wavelets and normalized by simple spectrums. Results of the coherence analysis range between 0 and 1, with 1 expressing a linear correlation between two signals at instant T of the scale a, and 0 expressing null correlation (Labat, 2010; Maraun & Kurths, 2004).

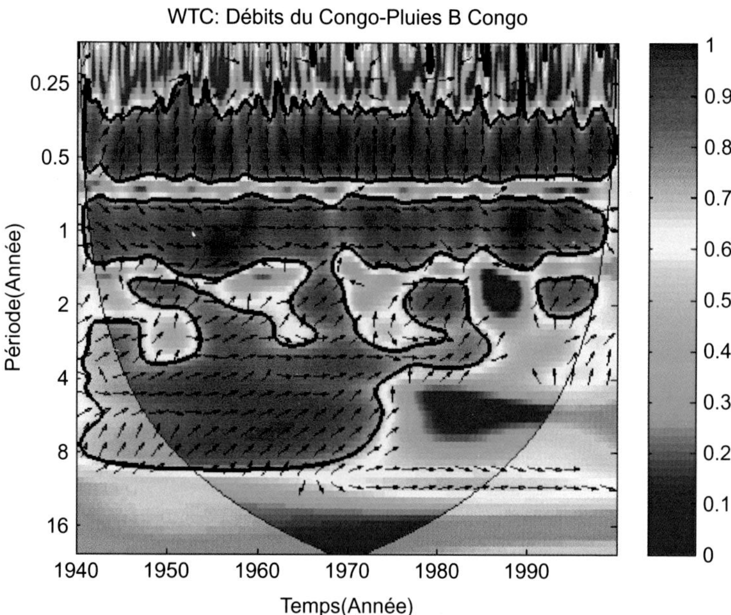

Fig. 13 Coherence analysis for the mean monthly flow of the Congo River Basin at Kinshasa and the regionalized rainfall for the basin.

It was previously observed that at the signal of mean monthly flows (Figs. 9 and 10), there were two permanent oscillations appearing around the scale-period of 0.5 and 1 year with the latter showing a very strong signal of variability. This is also portrayed in Fig. 15, where it is clear that the two permanent oscillations are a result of the rainfall regime within the basin.

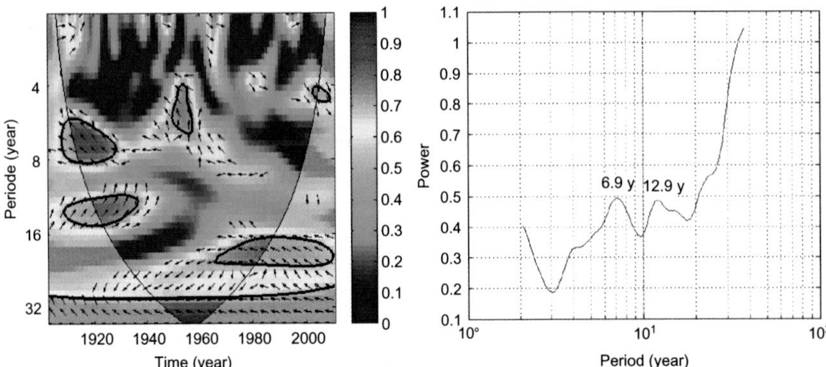

Fig. 14 Wavelet coherence and phase difference between the maximum annual flow and the annual NAO index. The arrows in the figures represent the phase difference of time periods and coherence bigger than 0.5 between two data series, with the phase pointing left and in phase pointing right.

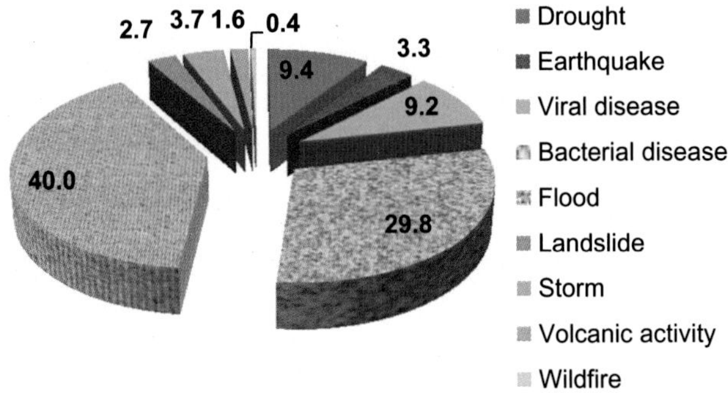

Fig. 15 Proportion of events of natural disasters in the Congo River Basin.

From Fig. 14, we observe the influence of the NAO index for the two modes of variability, inter annual and decadal cycles in the maximum annual flows.

The structure of variability in rainfall over the basin is similar to the structure of variability as observed in streamflow, with a break in the periodicity of 2–8 years from 1970. In all cases, the observed fluctuations in streamflow are strongly correlated with those of precipitation, thus suggesting that there is a strong influence of climate in the structure of streamflow variability.

It is observed that there is linearity between the annual maximum discharge of the Congo River Basin and the NAO index at the scale periods of 4–8, and 8–16 years until early 1940s and at the scale periods of 16–32 years from 1960 to 2010. A 13-year cycle is observed in both streamflow and NAO index.

3.3 Land Use Dynamics

The role of land cover in controlling catchment hydrological processes cannot be overemphasized (eg, Andrews & Bullock, 1994; Bosch & Hewlett, 1982; Edwards & Blackie, 1981; Mazvimavi, 2003). Advances in remote sensing technology have offered tremendous opportunities for improving knowledge of the dynamics of land cover and their characteristics at the catchment scale. The global land cover map (Bontemps et al., 2011) offers a relatively high level of reliability for environmental modeling of the Congo River Basin. Frequency distribution of Congo River Basin land cover (Tshimanga, 2012) shows that the major part of the central basin is covered by broadleaved evergreen or semi-deciduous forest, which constitutes the dominant land vegetation of the basin. This area is also characterized by high precipitation, low elevation, and low slopes. Anthropogenic activities, with sometimes remarkable consequences for land use change and natural variability of the climate systems, have induced major environmental changes which may have irreversible effects, at least at a certain time scale. Even though there has been an effort to understand the dynamics of land cover in the Congo Basin (de Wasseige et al., 2009), little has been done with regards to the impacts of the land use on its hydrological functioning. Various studies have demonstrated that the major impacts on water resources of the Congo Basin would stem from land cover and land use changes (Hoare, 2007; Ladel et al., 2008). Uncontrolled anthropogenic activities with potential impacts on water resources availability of the basin pose potential problems.

A United Nations' census for the period 2000–2005 (UN, 2007) shows a growth rate of 2.87% per year for the population living in the region of the Congo Basin, with the potential to double in 25–30 years. The majority of the population is characterized by low income, relying on subsistence agriculture for their livelihood. Rainfed agriculture is the main mode with slash burn, forest clearing, and shifting agriculture (de Wasseige et al., 2009). In addition, there have been increasing reports of uncontrolled large scale deforestation and mining which are known to impact on the patterns of hydrological behavior. Estimates for the deforestation with a focus on the evergreen forest zones of the basin for the period 1990–2000 show a net deforestation rate of 0.16% per year (de Wasseige et al., 2009). A loss of about 5% has been recorded in several catchments between these periods. These activities are sources of pressure on the basin water availability and their cumulative impacts could result in changes in the basin hydrological patterns, with increased flood frequency. Urban expansion which takes place in an uncontrolled way and population growth are the major factors that correlate well with flood occurrence and damages in the Congo Basin region. In many cases, the new urban areas lack adequate infrastructures to facilitate drainage of urban runoff.

A study conducted in 2009 on the urban expansion and population growth in the city of Kinshasa demonstrated that the population growth rate is 6.7% greater than the built areas of the urban expansion with 4.2%. Based on slope classes, more than 50% of urban expansion is observed in the areas of low slope (<10%) and about 30%in the areas of high slopes (>15%). About 15% of the expansion is observed in the slopes ranging between 10% and 15%.

3.4 Climate Change

From the Intergovernmental Panel on Climate Change, Assessment Report 4, it was concluded that the African continent has a high susceptibility, with low adaptation capability, towards stress caused by climate change. Scales and data issues have been among the major challenges faced by many studies carried out for climate change assessment over the Congo River Basin. Because of the difficulty to validate climate change scenarios at point or finer scales, many studies carried out in the basin have used a regionalization approach, consisting of averaging large-scale processes to obtain a lumped response. This approach to prediction, based on lumping large-scale processes, complicates the basin-wide development of water resources plans and there is a risk that the adaptation measures for future environmental changes will be based on a very large scale, which will undermine the possible impacts at smaller scales (Tshimanga & Hughes, 2012).

In a study of climate change scenarios for the Congo River Basin, Haensler, Saeed, and Jacob (2013) used a combination of 46 projections of SRES A2 and RCP 8.5 for high emission scenarios and 31 projections of SRES B1 and RCP 4.5 and 2.6 for low emission scenarios, with the aim to identify patterns and associated ranges of projected changes for the Congo Basin. The main finding from this assessment revealed warming tendency towards the end of the 21st century in the range of +1.5–3°C for the low emission scenarios, and +3.5–6°C for the high emission scenarios.

Projected changes in rainfall show both increasing and decreasing tendency for different geographical regions of the basin, with an increase in the intensity of heavy rainfall events as well as the frequency of dry spells during the rainy season. Change in the pattern of rainfall and temperature will result in changes in the hydrology of the Congo Basin, thus leading to an increase in runoff of up to 50%, especially in the wet season. The study concluded that flood risks will significantly increase in the future throughout the basin. Wet and dry seasons will show larger differences compared to the current climate, with more frequent wet and dry extremes.

4 TRENDS AND SOCIO-ECONOMIC IMPACTS OF FLOODS IN THE CONGO RIVER BASIN

The month of Dec. 2015 in Kinshasa, the capital city of DRC, was characterized by public emergencies due to wide spread damages that occurred following flooding in some of the major rivers that are used for urban water supply. About 2112 families were affected, resulting in 31 deaths and 8480 individuals being displaced. The disruption of the water supply plant led to water shortage for the main parts of the town. In the same time, the floods have also been reported in many parts of DRC, notably in the provinces of Central Congo, Bas Congo, Equateur, and Kisangani. Accounts report that the Congo River's annual high water season at Kisangani results in major inundations every 15–18 years.

In Nov. 1999, the Agence Française de Presse (AFP) reported that the "*Congo River 'flood of the century' hits capitals on both banks*," through which the AFP portrayed the picture of some important flood events that occurred in the Congo River Basin as well as the related impacts on socio-economic life. The flood occurred in Nov. and was reportedly expected to last until Jan. of the next year, thus approaching the two major flood events of the century, notably in 1903 and 1961–62. The 1999 flood occurred in both Kinshasa (DRC) and Brazzaville (Republic of Congo). The event affected tens of thousands of people in both Kinshasa and Brazzaville, and caused serious disruption of drinking water supply. The rising water level of the Congo River, which forms part of the border between the DRC and the Congo Republic, was the major cause of the observed impacts, and the rising water level was due to subsequent heavy rainfall.

Many of the natural disasters that occur across the world are related to the dynamics of hydro-climate processes, and flood disasters are the most prominent in terms of occurrence. Since the field measurement of flood events are very difficult to obtain, the Dartmouth Flood Observatory (Brakenridge & Anderson, 2004) and the emergency disaster database (EM-DAT, Guha-Sapir et al. www.emdat.be) identifies flood events through media reports with independent verification from satellite imagery. EM-DAT relies on disaster declarations and humanitarian assistance calls as well as media reports of fatalities or impacted populations to initiate a record of an event. Tables 1 and 2 show the types of natural disasters, in terms of number of events and total affected, that have been recorded for the region of the Congo River Basin, including Angola, Burundi, Cameroon, CAR, Congo Republic, DRC, Rwanda, Tanzania, and Zambia. The analysis is carried out for the period 1964–2012. The types of natural disasters recorded for the region are classified by the category: A = drought, B = earthquake, C = epidemic, D = flood, E = landslide, F = storm, G = volcanic activity, H = wildfire; and the sub-category: A1 = drought, B1 = earthquake, C1 = viral disease, C2 = bacterial disease, D1 = riverine flood, D2 = flash flood, D3 = other flood, E1 = landslide, F1 = tropical cyclone storm, F2 = convective storm, G1 = ash fall, H1 = wildfire. The analysis is here carried out in terms of the number of events and the total number of affected people.

Based on our analysis, of all the natural disasters documented for the region of the Congo River Basin, flood is the most reoccurring phenomena with 40% (196 events) of events recorded over the total number of all natural disaster events. Drought represents only 9.4% (46 EC) of events. However, in terms of number of affected people, drought impacts the most, with 69.6% of affected people compared to floods, which represent 23.8% of affected people (based on all of the documented natural disasters since 1964). Fig. 15 shows the proportion of events of the natural disasters for the region of the Congo River Basin as a whole.

From this analysis, it is demonstrated that floods occur in form of riverine floods, flash floods, and other combined types of flood, such that riverine floods are the most prominent type of flood with 121 events recorded since 1964 as observed in the region. Fig. 16 shows the proportion of events per type of flood and Fig. 17 shows the proportion of affected people per type of flood.

Taking into consideration only the hydro climate disasters including drought, flood, tropical cyclone storms, and convective storms (Fig. 18), it is clear that there is an increasing trend of events since 1964. This could probably also be an indication of improved means of reporting of the disaster events, in addition to the effects of environmental change.

Table 2 Mapping of natural disasters in terms of total number of people affected for the nine countries of the Congo Basin (1964–2012).

Disaster type	**A**	**B**	**C**		**D**			**E**	**F**		**G**	**H**
Disaster subtype	*A1*	*B1*	*C1*	*C2*	*D1*	*D2*	*D3*	*E1*	*F1*	*F2*	*G1*	*r*
DRC	800,000	21,266	545,857	165,374	147,704	81,750	45,328	2083	22,500	75,147	170,400	2895
Angola	4,443,900	0	665	137,928	1,088,608	595	109,484	0				
Burundi	3,062,500	120	8000	1,369,618	53,555	30,882	27,254			32,360		
CAR	3314	0	2490		77,990	15,873	94,469			27,440	835	
Cameroon	586,900	0	172	63,749	85,275	2305		100			13,447	
RC		1505	10,667	6080	131,114		42,000	668				
Rwanda	4,156,545	2286		7259	82,308	0	1,900,140	7937				
Tanzania	12,737,483	8991	1028	94,306	382,347	24,800	636,363	150	2500	6394		
Zambia	4,173,204	0	667	53,428	3,231,108		1,960,250	150				

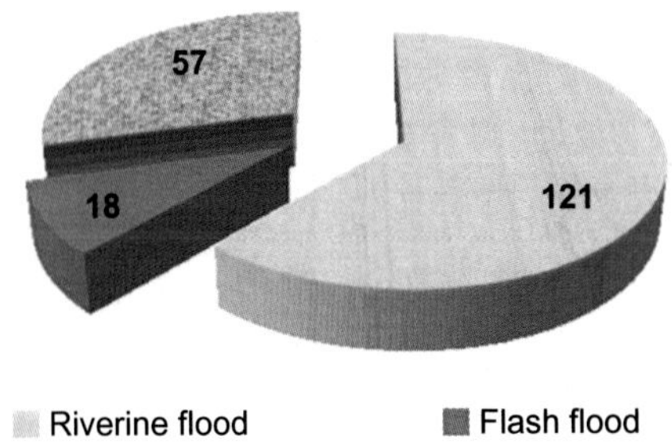

Fig. 16 Number of events per type of flood in the Congo River Basin.

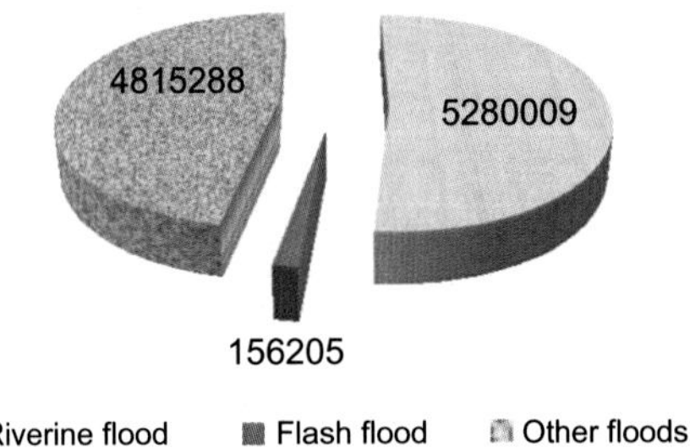

Fig. 17 Number of affected people per type of flood in the Congo River Basin.

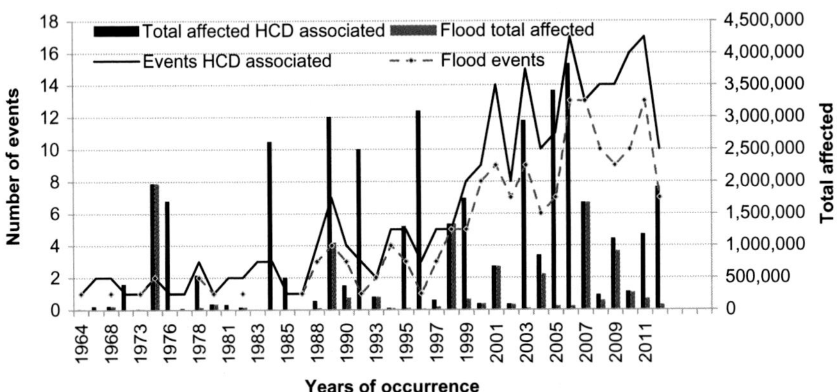

Fig. 18 Flood disaster events *(dotted red line)* and combined events of hydro climate disasters *(black line)*.

Floods have often happened in the Congo Basin with substantial socio-economic impacts. The geomorphologic nature of the basin hydrodynamic system, the rainfall pattern of the tropical region, and the pattern of land-use are among the drivers of the floods occurrence in the Congo River Basin. As a natural hazard, floods often occur when the river system carrying capacity is unable to convey the river flow especially after the heavy rain has occurred in the basin. This situation can worsen when the flood prone areas have been converted to socio-economic investments with no or little flood protection measures. In most cases, this leads to diversified socio-economic damages.

The Congo River main stem receives flow contribution from different tributaries of the Congo Basin. In many cases flood waves from the upper region of the Congo River system are attenuated through the Cuvette Centrale of the Congo including the wetland of the pool Malebo located downstream of the Cuvette Centrale. These natural depressions play an important role in the translation and attenuation processes of the flood waves in the Congo River Basin. When the Congo River main stem at Kinshasa reaches its carrying capacity, flood occurrences are observed mainly through the backwater effect in tributaries of adjacent catchments.

5 CURRENT STATUS OF FLOOD FORECASTING AND DISASTER MANAGEMENT SYSTEMS IN THE CONGO RIVER BASIN

The recognition of the need to ensure safety of the public and to protect the environment against the impacts of recurrent floods worldwide has propelled the rise of a relatively new discipline of hydrological science: flood forecasting. Flood forecasting can be defined as a process of estimating and predicting the magnitude, timing and duration of flooding based on known characteristics of a river basin, with the aim to prevent damages to human life, to properties, and to the environment. It allows evaluation of an event in real time leading to the issue of a general alert about hazardous conditions, and gives information on how flood disasters are likely to occur in the near future (minutes, hours, days, up to weeks), as well as the evolution in time and space. The system includes, therefore, an entire range of structural and nonstructural measures that encompass the determination of flood characteristics, the forecast as well as preparedness designed to safeguard lives and minimize losses dues to flood impacts.

An efficient flood forecasting system depends upon an adequate understating of hydro-climate processes and catchment characteristics driving the momentum of floods, the availability of data at scales (spatial and temporal) required for forecasting and management, the institutional and legal arrangements, and the technical and financial resources. Knowledge of rainfall pattern and streamflow events, catchment characteristics such as size, antecedent soil moisture, and land use are essential for reliability and lead time of flood forecasts.

5.1 Hydrometeorological Services and Monitoring

Central to an effective flood forecasting system is the ability to collect relevant data at the appropriate spatial and temporal scales. Three periods are essential to understanding hydrometeorological monitoring in the Congo Basin and these include the pre-colonial and post-colonial periods, which consist of ground-based monitoring network; and the period of satellite-based monitoring.

5.1.1 Ground-Based Monitoring

The year 1903 constituted a landmark period for hydrometeorological monitoring in the Congo River Basin, with the implementation of the Kinshasa gaging site (4.30S, 15.31E). Literature shows that over 400 streamflow gages as well as more than 150 synoptic stations and over 1000 rainfall posts were implemented between 1903 and the 1960s. After the 1960s, which was the period of political independence for many of the Congo Basin countries, we observe a decline in the implementation of hydrometeorological monitoring networks as well as the services related to data collection and dissemination. This is partly because these countries were not prepared or enabled in terms of capacities to carry out the duty, but also the objectives of hydrometeorological monitoring were not immediately the focus of political administrations that took power after the independence. It is here worth mentioning that even during the colonial era, the objectives of hydrometeorological monitoring were mainly related to the analysis of climate trends for agricultural production and control of water levels for navigation to facilitate export of goods and services. Fig. 19 draws a comparison of the network density of the hydrometeorological network before the 1960s and the current situation.

The services involved in hydrometeorological monitoring during the colonial period in the Congo Basin were organized through Service Météorologique de l'Afrique Equatoriale Française, Office de la Recherche Scientifique et Technique Outre Mer (ORSTOM), and Office National de Recherche pour le Développement (ONRD). Data collected during this period have been variously kept in different formats across different agencies in the world. After the 1960s, various hydrometeorological services

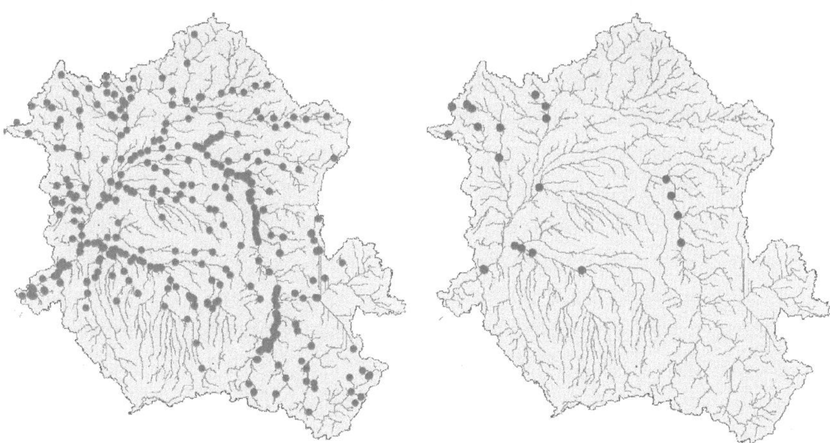

Fig. 19 Density of streamflow gages in the Congo River Basin before the 1960s *(left)* and the current situation *(right)*.

were created to continue the monitoring at national levels, notably: Agence Nationale de Météorologie et Télédétection par Satellite (METTELSAT), Régie des Voies Fluviales (RVF) and Congolaise des Voies Maritimes (CVM) in DRC; National Directorate of Meteorology (NDM), the Centre for Hydrologic Research (CRH) and HYDROMEKIN in Cameroon; Direction de la Météorologie Nationale, Service Météorologique (SMET), and Bureau de Coordination des Aérodromes Secondaires (BCAS) or Unité de Réseau Stations Météorologiques in CAR; La Direction de la météorologie and l'Institut de Recherche en Sciences Exactes et Naturelles (IRSEN) in Republic of Congo; and the Department of Water Affairs (DWA) and Zambia Meteorological Dept (ZMD) in Zambia. The services for other countries of the Congo River Basin are not reported here due to lack of information. Most of these hydrometeorological services rely on the use of the traditional types of gaging stations. During the last 2 decades, there has been major progress with regard to the technological development for hydrometeorological monitoring. The current trend is the use of automatic instruments to help overcome the challenge of access to remote areas. However, this remains a major challenge for the majority of the developing countries in the Congo Basin which are fraught with the issues of budget limitations, and in many cases, hydrometeorological monitoring is not part of the government priorities. There have been initiatives to establish hydrological forecasting using advanced technology of real-time data collection and transmission, but lack of maintenance did not enable continuous operation.

5.1.2 Satellite-Based Observation

Due to many challenges of inaccessibility to data, including the remoteness of the basin, satellite experiments have raised hope for improved environmental monitoring in the Congo River Basin. Studies conducted based on the use of satellite resources include, among others: temporal changes in river water levels derived from radar imagery (Rosenqvist & Birkett, 2002); seasonal and long-term trends of terrestrial water storage from gravity recovery and climate experiment (GRACE, Crowley, Mitrovica, Bailey, Tamisiea, & Davis, 2006); temporal changes in water surface elevations and connectivity between the main and floodplain channels evaluated from space-borne radar (Jung et al., 2010); evaluation of sources to sinks of the wetlands water using multiple remote sensing measurements such as GRACE, satellite radar altimetry, GPCP, JERS-1, SRTM, and MODIS (Lee et al., 2011); hydraulic characterization of the middle reach of the Congo River from Landsat imagery (O'Loughlin, Trigg, Schumann, & Bates, 2013); hydrological modeling of the Congo Basin using various sources of satellite-derived rainfall such as standard monthly satellite-based products from the Climate Prediction Centre Merged Analysis of Precipitation and the Global Precipitation Climatology Project (GPCP); monthly data from the CRU obtained using interpolation techniques; monthly TAMSAT satellite derived data; monthly CMORPH (CPC morphing technique) satellite-derived data; data from the Tropical Rainfall Monitoring Mission (TRMM); reanalysed data including NCEP/NCAR, CFSR and ECMWF (ERA-40 and ERA-Interim); and monthly data from Coupled Model Intercomparison Projects CMIP3 and CMIP5. The French Institute for Research and Development (IRD) has recently established a spatial altimetry data base for the Congo River Basin collected from various sources of satellite observation missions, including ENVISAT 1 and 2, JASON 2 and ALTICA (http://www.pecad.fas.usda.gov/cropexplorer/global_reservoir/, de l'USDA). Other initiatives include those provided by the Monitoring for Environment and Security in Africa project and the TIGER Capacity Building Facility, which aims at providing earth observation information to support decision-making in the various applications of environment and natural resources management.

While the benefits of these satellite-based products are of no question, it is still important to recognize the difficulty of validation and related uncertainties in some large areas of the Congo River Basin where there is a complete lack of ground-based historical observations (Washington, James, Pearce, & Pokam, 2013; Yin & Gruber, 2009).

5.2 Data Management

The main tasks carried out by the various hydrometeorological services within the region consist of data acquisition, transmission, processing, storage, and dissemination. Data are collected using regular reading from the existing hydrometeorological networks. The requirement for time interval (eg, synoptic hours) of data collection is generally the rule, but because of the difficulty associated with maintenance and operational budget as well as the lack of interest in short range events, data tend to be aggregated on a daily basis (over 24 h). Nevertheless, records with synoptic time interval are usually acquired from aeronautic services, which are mostly equipped with automatic stations. The various data collected are usually stored using papers (eg, Fig. 20).

Fig. 20 Historical data storage at one of the hydrological services in DRC.

5.3 Rainfall-Runoff Modeling and Forecasting

Flood forecasting systems involve predetermination of flood events necessary to help the implementation of structural measures for the mitigation of flood damages, regulation and operation of the multipurpose reservoirs with a focus on the control of incoming floods, and the evacuation of the affected people to the safer places. The use of hydrological models as well as GIS tools is of much benefit in this regard. Models are mathematical representations of the real dynamics of the basin hydrological processes and models are widely used for hydrological simulation and flood forecasting to support decision-making. Establishing a model for flood forecasting involves many issues of input data, parameters, structures, and the scales at which forecasting is needed.

Attempts to apply process-based hydrological models to simulate the hydrology of the Congo Basin on a large scale can be attributed to a few recent studies such as Asante (2000), Ducharne et al. (2003), Munzimi (2008), Chishugi and Alemaw (2009), Beighley et al. (2011), and Werth, Güntner, Petrovic, and Schmidt (2009). The above-mentioned studies have been variously challenged by problems related to a lack of appropriate data, as well as the lack of a thorough understanding of climate-hydrology processes, lack of integration of this understanding in models, and therefore lack of integrated and critical model assessment. This is true because of the complexity of runoff generation processes in the basin, which is compounded by the ungaged nature of the basin. Asante (Asante, 2000) used a "source to sink" modeling approach to compute flow routing parameters for continental-scale applications of watershed-based routing models. The approach was concomitantly applied with the hydrological modeling system in the Congo and Nile Basins, but only the gaging sites identified in the Nile Basin were used to compare the simulated flows from the models to the observed flows. Ducharne et al. (2003) applied the River-Transfer Hydrological Model over 11 large basins of the world, which included the Congo Basin. While the model results were successful in other basins, within the Congo Basin, the calibration results were characterized by a negative Nash coefficient of efficiency (Nash & Sutcliffe, 1970) due to a systematic overestimation of the simulated flow. Because of the lack of the observed data, the authors refrained from exploring the cause of the poor model performance for the Congo Basin. Munzimi's study (Munzimi, 2008) presents an attempt to predict the river flow of the Congo Basin using satellite-derived rainfall estimates. In this study, a Geospatial stream flow model was established for the whole Congo Basin and calibrated over 7 years, using the available observed streamflow data from the Global Runoff Discharge Center (GRDC). However, the preliminary simulation results did not adequately reproduce either the magnitude or timing of the observed flows. The causes of these discrepancies remain unknown as they were not discussed by the author who, however, recommended further exploration of the data input, model structure, and model parameters. Chishugi and Alemaw (2009) carried out a comprehensive study to simulate the hydrology of the Congo Basin, using a GIS-based hydrological water balance model. The main achievement of this study was the simulation of basin-wide mean annual soil moisture, evapotranspiration, and runoff. However, the authors failed to use the concurrent observed data to calibrate and validate the model, which limits the application of the model results for further studies and predictions in the basin. In fact, Chishugi and Alemaw (2009) attributed the failure to calibrate their model to the lack of the necessary observed data (Personal communication). In the same context, Werth et al. (2009) undertook a multi-objective calibration of the WaterGap Global Hydrology Model for the Amazon, Congo, and Mississippi Basins. In this study, the model was calibrated using both river discharge data and the total water storage change data from the Gravity Recovery and Climate Experiment (GRACE) satellite mission. The study results suggest that the model was able to produce improved simulations, with a good performance for the Amazon and Mississippi Basins. For the Congo Basin, the model calibration resulted in a much wider difference between the simulated and observed flows for the two objective functions used as performance criteria. The authors attributed the uncertainties in the calibration to the lack of consistent data and the particular characteristics of the rainfall distribution. At the same time, the authors recommended further studies to focus on the issues related to model structure, model input data, and the parameter space allowed for the model calibration. Similar disagreements are also reported by Papa, Guntner, Frappart, Prigent, and Rossow (2008). Beighley et al. (2011) noted challenges similar to other researchers. Beighley et al. (2011) used their Hillslope River Routing to route satellite-based rainfall across their model Congo Basin to produce river discharge, which they then compared to measured streamgage data. In this comparison, they found that TRMM, CMORPH, and PERSIANN yielded either too much rainfall across the entire basin or that rainfall from these products over smaller subbasins was poorly timed with observations.

Clearly, the discrepancies in the above-mentioned studies reveal the difficulty of modeling studies to represent properly the complexity of hydrological processes in the Congo Basin. The complexity is partly due to different response characteristics of the subbasins that compose the Congo River system (Laraque et al., 2001). Many of the observed data that are available for the Congo Basin are at the outlets of large subbasins, where it is very difficult to interpret the hydrological response characteristics because of the large scale of the basins and because of a multiplicity of interacting processes. These processes include surface and sub-surface response to rainfall at the small scale, but also include storage and attenuation effects of wetlands, floodplains, natural lakes, and the channel systems of large rivers.

Recent efforts in the application of process based hydrological models for the Congo Basin have led to the development of two conceptual rainfall-runoff models, SPATSIM (Hughes, Andersson, Wilk, & Savenije, 2006; Tshimanga & Hughes, 2014) and Water Evaluation and Planning (Cervigni, Rikard, James, & Kenneth, 2015; Yates, Sieber, Purkey, & Huber-Lee, 2005), with the aim to quantify the real hydrological processes at the basin scale and to use the models to assess the impacts of future environmental change in the basin. Fig. 21 shows simulation results obtained for both models at one of the main hydrometric station in the Congo Basin. Globally, the models show the ability to simulate historical and future hydrological response with high degree of confidence. Application of the models also shows the need for a finer scale in the modeling process to account for extreme events as well as modeling the downstream parts of the basin and specifically the routing of wet season flow volumes along the length of a very large river with adjacent floodplains. Currently, International Commission of Congo-Oubangui-Sangha (CICOS) is developing a Water Resources Planning tool under the MIKE HYDRO BASIN interface which could be of benefit for flood forecasting in the basin.

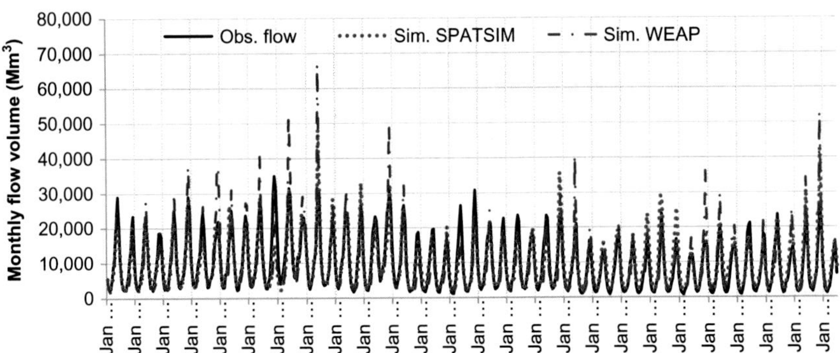

Fig. 21 Observed and simulated flows at the Bangui station, Oubangui River, using the PITMAN-GW rainfall-runoff module of SPATSIM and the soil moisture rainfall-runoff module of SPATSIM.

5.4 Institutional Arrangement and Flood Management Policy

A situational analysis of the current institutional arrangement of the hydrometeorological and disaster management services in most of the countries of the Congo River Basin shows several administrative and financial problems. In many departments, the human capacity to perform the tasks envisaged is lacking or very limited. The general policy framework in most countries is geared towards operation of hydrometeorological services and water departments for the general purpose of climate and water resources assessment, but with a very little focus on flood forecasting. This justifies the current stand of the observational network, which is mostly composed of a historical type of gaging stations as well as the absence of technical capacities and financial resources in the field of flood forecasting. There is a lack of coordination of institutions dealing with internal water resources (within the country) and external (transboundary water resources), and the disaster risk management. This is mainly due to insufficient technical capacity for the implementation of water sector development programs, and coordination difficulties in different sectors with different interests and priorities. For instance, many of the national hydrometeorological services in the region of the Congo River Basin belong to the ministries of transport, showing sometimes conflicting interests with other ministries in charge of water resources management. There is a lack of participation in policy formulation, legislation, and water resources management planning and development regarding the ownership and responsibilities of all stakeholders. Even in cases where policies and laws are in place, they are not well disseminated and applied, so less impact is recognized on the ground. There are overlapping responsibilities of institutions involved in the collection and use of hydrometeorological data, which reflects the lack of synergy and coordination of different stakeholders. The lack of a coherent legislative framework that can synergize the responsibilities of each institution should be noted.

The objectives of monitoring in terms of floods forecasting and management are very recent and are related to the current trend of awareness rising with regard to climate change, river basin development, and socio-economic impacts of hydro climate disasters. The advent of the integrated water resources management concept laid down a foundation for the implementation of many regional river basins organizations such as the Nile Basin Initiative (NBI), CICOS, and the Lake Tanganyika Water Authority (LTA). In addition, there are the water departments of the Southern African Development Commission (SADC) and the Economic Commission for Central African States (ECCAS). The implementation of these regional institutions has been valuable for increasing awareness about the various issues of water resources management, including specific programs on hydrometeorological monitoring and forecasting. Recently, a general framework for the coordination of the various measurements related to natural disasters including floods has been advocated in the basin, and efforts are being made in collaboration with various international agencies in order to increase the capacity of information and database management to support policy towards hydro climatic disasters management. The efforts envisaged aim at regulating the responsibilities of the departments or services which will be in charge of coordinating the measurements of the natural disasters.

6 CONCLUSION

The impacts of future environmental changes in the African continent have been highlighted in many studies, and it has become clear that water resource systems, including flowing rivers, wetlands, lakes, and groundwater will respond differently to the changing environmental conditions. Understanding the processes of changes and predicting the associated impacts, is therefore important for good management practices to enhance societal resilience. Flooding is a recurrent common phenomenon worldwide. The challenges of flood management are numerous at the global scale, but with a varying degree from place to place. This variation is associated with the inherent characteristics of the physical environment, the level of socio-economic development, and the social and political organization, which all have an influence on the level of vulnerability towards flood disasters.

In recognition of the requirements for effective flood forecasting and disaster management systems, analysis of the current status for the Congo River Basin shows that major efforts are needed, with regard to processes understanding, data collection, and enabling environment. Major challenges remain in terms of a lack of a clear flood management policy, a lack of adequate skills required for flood forecasting thus leading to uncertainties in quantification and prediction of flood events, a lack of operational hydrometric networks and state-of-the-art technology for provision of real-time or near-real-time data, inaccessibility to satellite-based products, and a lack of modeling tools for flood forecasting purpose. The complexity of hydro-climate processes in the basin and the lack of adequate data required at appropriate temporal and spatial scales are among the major factors that hinder an effective implementation of a flood forecasting system. This complexity is related to the heterogeneities of the landscape properties, which are compounded by the temporal and spatial variability.

Given the current status of flood forecasting systems in the basin, there is a need for the implementation of an operational framework of flood forecasting system, with flood forecasting centers which will have among its responsibilities the centralization of all information coming from meteorological and hydrological services in order to formulate the forecasts. In turn, the forecasts will be disseminated by the flood forecasting centers using different means of disseminations. Many of the networks exist at remote sites where communication capability is a real challenge. Improving the reliability of data transfer from the field is necessary for flood warning and flood disaster management. The transfer of data from field measurements to a forecasting center could be done through different means of communication. These can include telephone, satellite, landline, or automatic hydrometeorological stations connected directly to the forecasting center. The choice of any type of communication device will depend on local constraints.

Availability of hydrometeorological monitoring networks is necessary for flood forecasting and warning. In most cases, the operators of these networks work in different services with some times very little coordination between them. The lack of coordination is observed at national levels as well as at the basin-wide level. In many cases, the networks are not designed to acquire data during extreme events, or to provide data in real time and to common standards. The existing network has not been designed for a direct purpose of flood forecasting and they are lacking in flood-prone areas such as key urban centers and flood plain regions where forecasts are required. Therefore, strengthening and improving the existing network in terms of density and operational capability is a necessary step in the implementation of a flood forecast system.

Once the forecasts have been formulated by the flood forecasting center, the forecast is distributed to different government services in charge of disaster warnings. The government services in charge of disaster warnings are responsible for disseminating the flood forecast to all socio-economic sectors including press, radio, TV, police, defence, airports, civil authorities, railways authorities, etc. Hydrological models such as rainfall-runoff and flow routing models are useful tools in the support of forecasting systems. An inventory of existing models will also help to define the current capability within the individual services.

REFERENCES

Andrews, A. J., & Bullock, A. (1994). *Hydrological impact of afforestation in eastern Zimbabwe*. Wallingfornd: Institute of Hydrology. Overseas Development Report No. 94/5.

Asante, K. O. (2000). *Approaches to continental scale river flow routing*. PhD thesis, Austin: University of Texas at Austin. http://repositories.lib.utexas.edu/handle/2152/6800.

Balas, N., Nicholson, E. S., & Klotter, D. (2007). The relationship of rainfall variability in West Central Africa to sea-surface temperature fluctuations. *International Journal of Climatology, 27*, 1335–1349.

Beighley, E. R., Ray, L. R., He, Y., Lee, H., Schaller, L., Andreadis, M. K., et al. (2011). Comparing satellite derived precipitation datasets using the Hillslope River Routing (HRR) model in the Congo River Basin. *Hydrological Processes, 25*(20), 3216–3229.

Bosch, J. M., & Hewlett, J. D. (1982). A review of catchment experiments to determine the effect of vegetation changes on water yield and evaporation. *Journal of Hydrology, 55*, 3–23.

Brakenridge, G. R., & Anderson, E. (2004). *Satellite-based inundation vectors*. Hanover, NH: Dartmouth Flood Observatory, Dartmouth College.

Burke, K., & Gunnell, Y. (2008). The African erosion surface: A continental-scale synthesis of geomorphology, tectonics, and environmental change over the past 180 million years. *Geological Society of America Memoirs, 201*, 1–66.

Cervigni, R., Rikard, L., James, E., & Kenneth, M. (2015). *Enhancing the climate resilience of Africa's infrastructure: The power and water sectors. Overview booklet. License: Creative commons attribution CC BY 3.0 IGO*. Washington, DC: World Bank.

Chishugi, B. J., & Alemaw, F. B. (2009). The hydrology of the Congo River Basin: A GIS-based hydrological water balance mode. In *World environmental and water resources congress 2009: Great Rivers. ASCE conference proceedings*.

Crowley, J. W., Mitrovica, X. J., Bailey, C. R., Tamisiea, E. M., & Davis, L. J. (2006). Land water storage within the Congo Basin inferred from GRACE satellite gravity data. *Geophysical Research Letters, 33*, L19402. http://dx.doi.org/10.1029/2006GL027070.

De Moortel, I., Munday, S. A., & Hood, A. W. (2004). Wavelet analysis: The effect of varying basic wavelet parameters. *Solar Physics, 222*(2), 203–228.

de Wasseige, C., Devers, D., de Marcken, P., Ebaa Atyi, R., Nasi, R., & Mayaux, P. (2009). *Les forêts du Bassin du Congo- État des Forêts 2008*. Luxembourg: Office des publications officielles des Communautés européennes. p. 425.

Ducharne, A., Golaz, C., Leblois, E., Laval, K., Polcher, J., Ledoux, E., et al. (2003). Development of a high resolution runoff routing model, calibration and application to assess runoff from the LMD GCM. *Journal of Hydrology, 280*, 1–4.

Edwards, K. A., & Blackie, J. R. (1981). Results of the East African catchment experiments 1958–1974. In R. Lall & E. W. Russel (Eds.), *Tropical agriculture hydrology* (pp. 163–188). Chichester: Wiley.

Eltahir, E., Loux, B., Yamana, T., & Bomblies, A. (2004). A see-saw oscillation between the Amazon and Congo Basins. *Geophysical Research Letters, 31*(23), http://dx.doi.org/10.1029/2004GL021160.

Farnsworth, A., White, E., Williams, R. J. C., Black, E., & Kniveton, R. D. (2011). Understanding the large scale driving mechanisms of rainfall variability over Central Africa. *Advances in Global Change Research, 3*, 101–122.

Haensler, A., Saeed, F., & Jacob, D. (2013). Assessment of projected climate change signals over central Africa based on a multitude of global and regional climate projections. In A. Haensler, D. Jacob, P. Kabat, & F. Ludwig (Eds.), *Climate change scenarios for the Congo Basin*. Hamburg, Germany, ISSN: Climate Service Center. Climate Service Centre Report No. 11. ISSN: 2192–4058.

Hoare, A. L. (2007). *Clouds on the horizon: The Congo Basin's forests and climate change*. London: The Rainforest Foundation. http://www.rainforestfoundationuk.org.

Hudgins, L. H., Mayer, M. E., & Friehe, C. A. (1993). Fourier and wavelet analysis of atmospheric turbulence. In Y. Meye & S. Roques (Eds.), *Progress in wavelet analysis and applications* (pp. 491–498). Paris: Editions Frontieres.

Hughes, D. A., Andersson, L., Wilk, J., & Savenije, H. H. G. (2006). Regional calibration of the Pitman model for the Okavango River. *Journal of Hydrology, 331*, 30–42.

Hughes, R. H., & Hughes, J. S. (1987). *A directory of African wetlands: Zaire*. Tresaith, Wales: Samara House. http://www.iwmi.cgiar.org/wetlands/pdf/Africa/Region4.zaire.

Jiang, R., Gan, T. Y., Xie, J., & Wang, N. (2013). Spatiotemporal variability of Alberta's seasonal precipitation, their teleconnection with large-scale climate anomalies and sea surface temperature. *International Journal of Climatology, 34*, 2899–2917. http://dx.doi.org/10.1002/joc.3883.

Juarez, N., Robin, I., Li, W., Fu, R., & Fernandes, K. (2009). Comparison of precipitation datasets over the tropical South American and African continents. *Journal of Hydrometeorology, 10*, 289–299.

Jung, H. C., Hamski, J., Durand, M., Alsdorf, D., Hossain, F., Lee, H., et al. (2010). Characterization of complex fluvial systems using remote sensing of spatial and temporal water level variations in the Amazon, Congo, and Brahmaputra Rivers. *Earth Surface Processes and Landforms, 2010*(35), 294–304.

Kadima, E., Delvaux, D., Sebagenzi, S. N., Tack, L., & Kabeya, M. (2011). Structure and geological history of the Congo Basin: An integrated interpretation of gravity, magnetic and reflection seismic data. *Basin Research, 23*(5), 499–527.

Labat, D. (2010). Cross wavelet analyses of annual continental freshwater discharge and selected climate indices. *Journal of Hydrology, 385*, 269–278.

Ladel, J., Nguinda, P., Pandi, A., Tanania, K. C., Tondo, B. L., Sambo, G., et al. (2008). Integrated water resources management in the Congo Basin based on the development of earth observation monitoring systems in the framework of the AMESD programme in Central Africa. In *13th world water congress, Montpellier, France*.

Laraque, A., Mahé, G., Orange, D., & Marieu, B. (2001). Spatiotemporal variations in hydrological regimes within Central Africa during the XXth century. *Journal of Hydrology, 245*, 104–117.

Lee, H., Beighley, R. E., Alsdorf, D., Jung, H. C., Shum, C. K., Duan, J., et al. (2011). Characterization of terrestrial water dynamics in the Congo Basin using GRACE and satellite radar altimetry. *Remote Sensing of Environment, 2011*(115), 3530–3538.

Mahé, G. (1993). *Les écoulements fluviaux sur la façade atlantique de l'Afrique. Etude des éléments du bilan hydrique et variabilité interannuelle. Analyse de situations hydro climatiques moyennes et extrêmes*. Thèse de doctorat, collection études et thèses, Paris, France: ORSTOM.

Maraun, D., & Kurths, J. (2004). Cross wavelet analysis: Significance testing and pitfalls. *Nonlinear Processes in Geophysics, 11*(4), 505–514.

Mazvimavi, D. (2003). *Estimation of flow characteristics of ungauged catchments*. Unpublished PhD thesis, Enschede: Wageningen University and International Institute for Geo-Information and Earth Observation, ITC.

Meneveau, C. (1991). Dual spectra and mixed energy cascade of turbulence in the wavelet representation. *Physical Review Letters, 11*, 1450–1453.

Munzimi, Y. (2008). Satellite-derived rainfall estimates (TRMM products) used for hydrological predictions of the Congo River flow: Overview and preliminary results. START report. http://start.org/alumni-spotlight/yolande-munzimi.html.

Nash, J. E., & Sutcliffe, J. (1970). River flow forecasting through conceptual models part A: Discussion of principles. *Journal of Hydrology, 10*, 282–290.

Nicholson, E. S. (2009). A revised picture of the structure of the "monsoon" and land ITCZ over West Africa. *Climate Dynamics, 32*, 1155–1171.

O'Loughlin, F., Trigg, M. A., Schumann, G. P., & Bates, P. D. (2013). Hydraulic characterization of the middle reach of the Congo River. *Water Resources Research, 2013*(49), 5059–5070.

Paeth, H., & Friederichs, P. (2004). Seasonality and time scales in the relationship between global SST and African rainfall. *Climate Dynamics, 23*, 815–837.

Papa, F., Guntner, A., Frappart, F., Prigent, C., & Rossow, B. W. (2008). Variations of surface water extent and water storage in large river basins: A comparison of different global data sources. *Geophysical Research Letters*, *35*, L11401.

Poccard, I., Janicot, S., & Camberlin, P. (2000). Comparison of rainfall structures between NCEP/NCAR reanalyses and observed data over tropical Africa. *Climate Dynamics*, *16*, 897–915.

Rosenqvist, A. A., & Birkett, C. M. (2002). Evaluation of JERS-1 SAR mosaics for hydrological applications in the Congo River Basin. *International Journal of Remote Sensing*, *2002*(23), 1283–1302.

Runge, J. (2008). The Congo River, Central Africa. In A. Gupta (Ed.), *Large rivers: Geomorphology and management*. London: Wiley and Sons.

Schefuss, E., Schouten, S., & Schneider, R. R. (2005). Climatic controls on Central African hydrology during the past 20,000 years. *Nature*, *437*, 1003–1006.

Schneider, K., & Farge, M. (2006). On the long time behaviour of decaying two–dimensional turbulence in bounded domains. *Bulletin of the American Physical Society*, *51*(9), 98.

Torrence, C., & Compo, G. (1998). A practical guide to wavelet analysis. *Bulletin of the American Meteorological Society*, *79*, 61–78.

Tshimanga, R. M. (2012). *Hydrological uncertainty analysis and scenario-based stream flow modelling for the Congo River Basin*. PhD thesis, South Africa: Rhodes University repository. http://eprints.ru.ac.za/2937/.

Tshimanga, R. M., & Hughes, D. A. (2012). Climate change and impacts on the hydrology of the Congo Basin: The case of the northern sub-basins of the Oubangui and Sangha Rivers. *Physics and Chemistry of the Earth*, *50–52*(2012), 72–83.

Tshimanga, R. M., & Hughes, D. A. (2014). Basin-scale performance of a semi distributed rainfall-runoff model for hydrological predictions and water resources assessment of large rivers: The Congo River. *Water Resources Research*, *50*. http://dx.doi.org/10.1002/2013WR014310.

UN (2007). World Population and Housing Census Programme. http://unstats.un.org/unsd/demographic/sources/census/country_impl.htm.

Washington, R. R., James, H., Pearce, W. M., & Pokam, W. (2013). Moufouma-Okia. Congo Basin rainfall climatology: Can we believe the climate models? *Phil Trans. B*. *368*(1625), http://dx.doi.org/10.1098/rstb.2012.0296.

Werth, S., Güntner, A., Petrovic, S., & Schmidt, R. (2009). Integration of GRACE mass variations into a global hydrological model. *Earth and Planetary Science Letters*, *277*(1–2), 166–173.

Yates, D., Sieber, J., Purkey, D., & Huber-Lee, A. (2005). WEAP21 — A demand-, priority-, and preference-driven water planning model part 1: Model characteristics. *Water International*, *30*, 487–500.

Yin, X., & Gruber, A. (2010). Validation of the abrupt change in GPCP precipitation in the Congo River Basin. *International Journal of Climatology*, *30*, 110–119. http://dx.doi.org/10.1002/joc.1875.

CHAPTER 5

Flood Forecasting in Germany — Challenges of a Federal Structure and Transboundary Cooperation

N. Demuth*, S. Rademacher†
*State Environmental Agency Rhineland-Palatinate, Mainz, Germany
†German Federal Institute of Hydrology, Koblenz, Germany

1 INTRODUCTION

Water level forecasting has a long tradition in Germany, for instance, dating back to 1881 on the river Elbe. Since that time, various level relationship methods have been used to calculate the highest floodwater levels and the time at which the floodwaters will peak at certain water gauges. These methods were constantly improved, but the floods of the river Rhine in 1993 and 1995, of the river Oder in 1997, and along the rivers Danube and Elbe in 2002 contributed significantly to an increased awareness of the need for the most effective flood forecasting possible. This resulted in a process of continuous updating and improvement of forecasting techniques. Apart from the development of the actual calculation models, this also included data transmission methods as well as the entire organization of the flood alert service, including state-of-the-art communication technology.

Germany is a federal republic consisting of 16 constituent states — the Länder — with a total population of approximately 80 million. These federal states are legally responsible for flood forecasting in their own area. The German Weather Service (DWD) assists the states' water management services in performing this role. Above this, the DWD is also legally required to issue official warnings about extreme weather events that could pose a threat to public safety and order, in particular in relation to potential floods. Due to the administrative structures on the one hand, and the hydrological diversity of the natural landscapes on the other, a large number of regional and transregional flood warning centers have been founded (Fig. 1), which use different forecasting systems. Almost every state runs a State Flood Forecasting or Warning Centre supported by regional centers.

Flood Forecasting
http://dx.doi.org/10.1016/B978-0-12-801884-2.00005-0

Fig. 1 Flood forecasting in Germany — national and international networking.

Rivers do not stop at borders and many of the river basins in Germany are transnational due to Germany's position in the center of Europe. In order to obtain suitable flood forecasts in cross-border territories, the minimum requirement of an operational exchange of meteorological and hydrological data between the national centers is needed. This affects almost all of Germany's neighbors such as Poland, the Czech Republic, Austria, Switzerland, France, Luxembourg, and the Netherlands, as well as the individual federal states. Besides the data exchange, close cross-border cooperation, including joint development of forecasting systems, has developed in many areas.

In the following, the national and international networking between the flood forecasting and alert centers in transboundary river basins is demonstrated, taking the cross-border catchment area of the river Rhine as an example.

2 INTERNATIONAL COOPERATION — THE RHINE BASIN

The Rhine originates in the Swiss Alps and flows into the North Sea as a major river. Its catchment covers an area of 185,000 km^2 (Fig. 2) and is divided between nine countries; of these countries, only Switzerland, France,

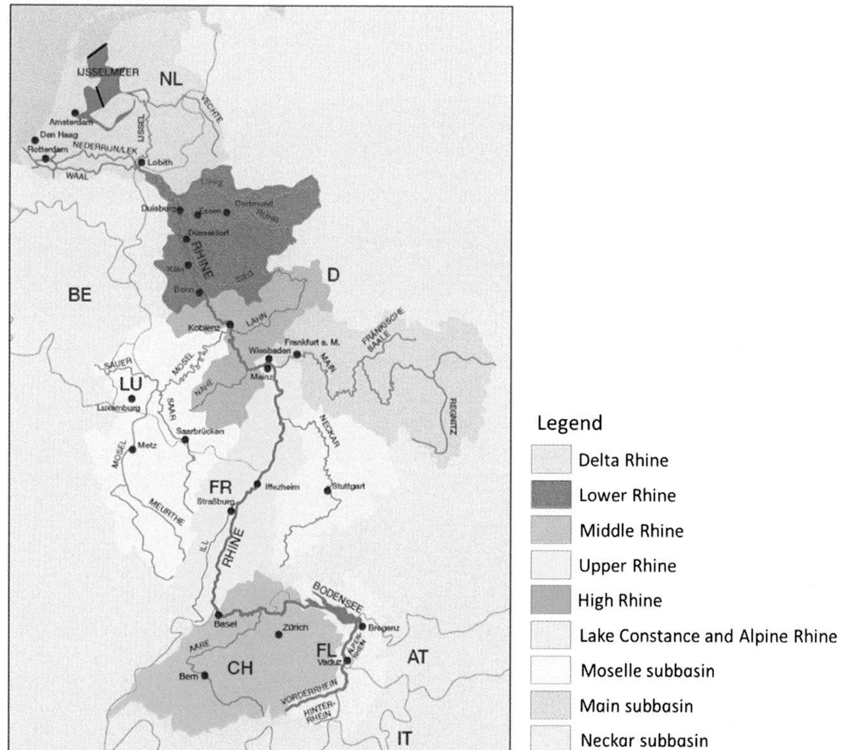

Fig. 2 The river Rhine basin.

Germany, and the Netherlands account for a great proportion of the area. The Rhine is one of the busiest waterways in the world, with about 60 million people living in its catchment area. The Upper Rhine was extensively regulated with weir systems between Basel and Iffezheim in the period between 1928 and 1977. The construction of 10 dams prevented further river bed erosion of the Rhine, while making it possible to generate hydroelectric power and notably improve its navigability. This development of the Rhine significantly increased the danger of flooding below the developed section of the river due to the loss of 130 km^2 natural inundation areas (Vieser, 1973). In 1982, France and Germany agreed in a treaty (Bundesgesetzblatt, 1984) to restore the flood control capacity that existed before the Rhine was developed, by building flood retention measures (Fig. 3).

Several national and regional forecasting centers are responsible for flood forecasting in the catchment area of the Rhine. The large number of centers primarily reflects the political structure of this area. For the Rhine itself there are five centers (Fig. 1), with others responsible for the Rhine's various tributaries:

- Flood forecasts department of the Federal Office for the Environment (FOEN) in Bern, Switzerland — from the source of the Rhine to Basel.

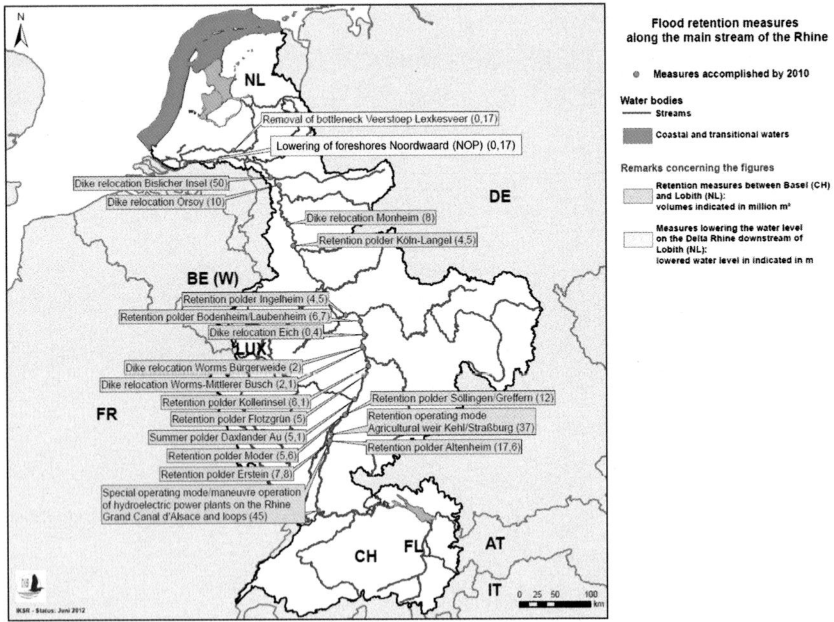

Fig. 3 Flood retention measures along the main stream of the river Rhine.

- Service de Prévision des Crues Rhin-Sarre in Strasbourg (DREAL Alsace-Champagne-Ardenne-Lorraine) — the Rhine, which forms the border between Germany and France between Basel and Lauterbourg and the French tributaries of the Rhine.
- Baden-Württemberg Flood Forecasting Centre (HVZ BW) in Karlsruhe — along the Rhine from the German-Swiss border to Mannheim and all tributaries of the Rhine in Baden-Württemberg.
- Flood Warning Centre Rhine (HMZ Rhein) — along the Rhine from Maxau to the German-Dutch border.
- Rijkswaterstaat Waterdienst in Lelystad (RWS) — the Rhine in the Netherlands.

Two centers are responsible for the German territory. The Baden-Württemberg Flood Forecasting Centre (HVZ BW) in Karlsruhe is responsible for forecasts along the Rhine downstream of Basel as far as Mannheim (Fig. 10), as well as for the entire part of the catchment area in Baden-Württemberg. The Flood Warning Centre Rhine (HMZ Rhein) in Mainz is responsible for flood information along the Rhine as far as the German-Dutch border. The HMZ Rhein is a joint institution of the state of Rhineland-Palatinate and the Federal Waterways and Shipping Administration (WSV). The WSV is responsible for maintaining Germany's large, navigable rivers. The cooperation between the centers has evolved over the years (Bürgi, Homagk, Prellberg, Sprokkereef, & Wilke, 2004), although the bilateral contacts initially originate from a specific need due to extreme floodings in 1993 and 1995. The cooperation was intensified due to the rapid developments in the field of information technology, and in most cases regulated by contractual agreements. In 1997, an international group of experts submitted an inventory of the reporting systems and proposals for improving flood forecasting in the Rhine basin (KHR, 1997). On Jan. 22, 1998, the International Commission for the Protection of the Rhine (ICPR) agreed a Flood Action Plan at the 12th Ministerial Conference on the Rhine, which included measures to improve flood forecasting, flood early warning systems, and international cooperation, and to extend forecast times (ICPR, 1998). In 2014, a working group of the federal and state water authorities developed recommendations for further improvements in flood forecasting in Germany (LAWA, 2014).

In addition to flood alerts, the communication between the centers also deals with hydrological and meteorological data, with the flow of information logically being primarily downstream. For stretches of the river where the

Rhine forms the border between the jurisdictions of the flood warning centers, the forecasts are generally only prepared by one of them. For the stretch of the Rhine downstream of Maxau as far as Worms (Fig. 10), both the HVZ BW and the HMZ Rhein prepare the forecasts. If there are significant differences between them, the centers agree on the forecasts before they are published. The forecasts by the HVZ BW and the HMZ Rhein are also part of the control criteria used for the flood retention measures along the Upper Rhine.

The flood forecasting centers along the Rhine organize annual meetings to discuss experiences, where the individual centers inform each other on current developments in their own areas. In addition to this, these meetings provide a venue to coordinate joint projects which aim to improve the flood forecasting systems — for example, on uniform determination of the forecast uncertainties, for publishing uncertainty bands, and for preparing joint reports following major flood events. The annual meetings of the flood forecasting centers take place at the ICPR and constitute part of the work of the flood working group.

3 MEASURED METEOROLOGICAL AND HYDROLOGICAL DATA

Hydrological and meteorological data is indispensable for flood forecasting. The hydrological monitoring networks are generally operated by the state water management agencies. A notable exception to this are the monitoring stations on the federal waterways (navigable waterways), which are the property of the Federal Government and are administered by the WSV. In the field of meteorology, Germany also has other monitoring networks, in addition to the monitoring network, operated by the national DWD, especially those operated by the states (some of which are also operated by the agricultural, environmental, and forestry administrations as well as the water management agencies, which have their own monitoring networks) and private meteorological services. The automation of data transmission has made it possible to implement data exchange between most of the monitoring network operators, meaning that there is a dense monitoring network available for flood forecasting, even though there are significant differences, both regionally as well as for the various measurement parameters. Fig. 4 shows the river level and precipitation measuring stations available for flood forecasting in the state of Rhineland-Palatinate (Germany).

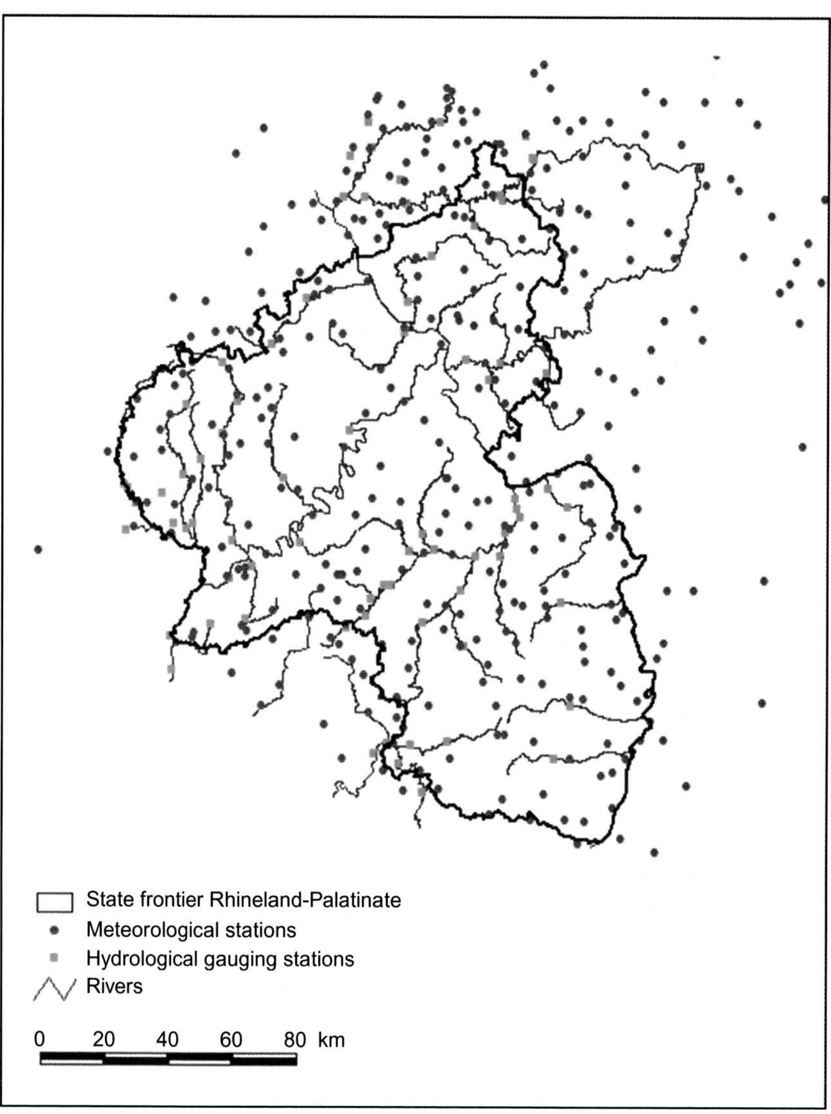

Fig. 4 Hydrological and hydrometeorological network in the State Rhineland-Palatinate (Germany).

As a general rule, the meteorological and hydrological data provided in real time consists of raw data. Since the quality of the input data is of central importance to the quality of the flood forecasts, a multi-stage check is performed before the data is fed into the model. When converting data, importing the data into the database, and exporting the input data for the forecasting model, data that is clearly wrong (negative precipitation values, for example) or data with "incorrect" time references or units is identified and in most cases replaced by error values or default values.

In the next stage, the meteorological input data for flood forecasting in Rhineland-Palatinate are checked for the forecasting model using the program NIKLAS (Einfalt, Gerlach, Podlasly, & Demuth, 2008; Einfalt & Podlasly, 2007). The NIKLAS program for checking the plausibility of meteorological data contains test routines for the following parameters: air temperature, relative air humidity, wind speed, global radiation, sunshine duration, air pressure, and precipitation. There are plans to extend the parameters to include snow depth and snow water equivalent as well as the hydrological variables of water level and runoff.

The program covers the following testing algorithms: completeness, limit value check (physically possible or plausible range), variability (temporal consistency), temporally constant values, internal consistency (the internal consistency refers to the behavior of a variety of parameters in one place to each other. Values for the same time and the same place, but of different parameters, have to be consistent with each other), and spatial consistency (during the spatial calibration, the values from different stations are compared to each other. Values for the same time and the same parameter from neighboring stations must not deviate too much from one another in this test).

The main emphasis is on testing the data in real time (measured on an hourly basis). There are, however, measurement errors, which only become apparent when tested over a prolonged period — which is why total amounts or mean values over a longer interval are taken into consideration for some parameters. For the test criteria implemented, a distinction is made between "hard" criteria (error status) and "soft" criteria (warning status). An error status excludes the data from further processing. A warning status marks the values, but leaves them unchanged. Many of the implausible values may not be readily or easily identifiable by checking the data automatically (warning status). Therefore data checking by a real-life hydrologist remains essential. This is normally done using the test reports generated by NIKLAS in combination with hydrograph visualization and maps of the interpolated values.

In addition to the measurements from the stations, quantitative radar precipitation data are also provided by the DWD for flood forecasting purposes. Since Jun. 2005, the DWD has provided gauge-adjusted radar data (RADOLAN data, DWD, 2004; Winterrath, Rosenow, & Weigl, 2012) from ground precipitation measurement stations with hourly precipitation levels at an intensity resolution of 0.1 mm and a spatial resolution of 1 km^2. The data is available almost in real time — within 30 min. In small catchment areas and regions with a low density of monitoring stations, the radar precipitation data is routinely used as precipitation input for flood forecasting. In areas where the density of the ground-level monitoring network is sufficient, however, the RADOLAN data is generally only used as an alternative to interpolated station data — in particular for convective heavy precipitation events. In spite of continuous product improvement, the RADOLAN product still contains incorrect data (eg, due to clutter, bright-band, and precipitation damping). These shortcomings could outweigh the advantages of the higher spatial resolution and the extensive availability, and result in errors in the flood forecasts. In this context, it is necessary to consider not only heavy rain that may cause flooding, but also incorrect precipitation input over a prolonged period, which may have a negative influence on the calculated hydrologic condition (soil moisture) when used in continuous water balance models.

For some transboundary catchment areas there is also data from foreign meteorological services available, for example, the radar product ANTILOPE (Champeaux et al., 2009) provided by the French meteorological service (Météo France) in the catchment areas of the Moselle and Saar.

4 NUMERICAL WEATHER PREDICTIONS

In many cases, the required forecasting periods exceed the hydrological response times of the catchment areas. The inclusion of numeric weather prediction (NWP) in operational flood forecasting has thus been commonly practice by many forecasting centers for many years now. This primarily makes use of the DWD's forecasting chain:

- The global model ICON with a mesh width of 13 km and a forecast range of 174 h (Zängl, Reinert, Ripodas, & Baldauf, 2015).
- The local model COSMO-EU with a mesh width of 7 km and a forecast range of 78 h (Schättler, Doms, & Schraff, 2012; Steppeler et al., 2003).
- The high-resolution regional model COSMO-DE with a mesh width of 2.8 km and a forecast range of 27 h (Baldauf et al., 2014).

In addition to the forecasts by the DWD, various other meteorological forecasts are also used for flood forecasting: AROME (Météo France, Seity et al., 2011), ARPEGE (Météo France, Pailleux et al., 2015), COSMO 7 (Météo Suisse, Schättler et al., 2012), ALARO-ALADIN (Zentralanstalt für

Meteorologie und Geodynamik, WMO, 1997), ECMWF (ECMWF, 2015), and GFS (Environmental Modeling Center, 2003). The output of the numerical weather forecast models is made available in the form of grid data. Depending on the type of flood forecasting model used, it is possible to use not only precipitation, but also other parameters (air temperature, relative air humidity, wind speed, global radiation, air pressure, and, where available, even snow water equivalent (see also Special forecasts)). With a few exceptions, the numerical weather forecasts are used as input for the hydrological models without any further processing (Model Output Statistics (MOS), for example). This can lead to problems if meteorological variables are subject to significant spatial variations or the NWP model systematically "misforecasts" certain variables or phenomena (luv/lee effects, air temperature over snow or inversion layers, for example). Fig. 5 shows an example of a runoff forecast with a temperature prediction that is too cold as the input. The forecast calculated on Dec. 8, 2010 underestimated the peak water level on Dec. 9, 2010 measured at Althornbach — which was above warning level 1 — significantly. Due to the predicted temperatures being too low, the snow on the ground in the catchment area did not melt in the model and the forecast precipitation fell as snow, rather than rain. A hindcast with the measured temperatures provides a realistic prediction.

Another problem with the integration of numerical weather forecasts is the transfer of the measured data to the forecast data (seamless prediction). If, for example, the progression of a front with heavy rain is earlier than predicted, the time lag between the time of the forecast and provision of the NWP may result in the precipitation already being recorded by the measurements from the stations, but being predicted as being in the future in the forecast and thus being doubled in the hydrological model. The same often applies to the temporal concatenation of forecasts, frequently used to extend the forecast range. These problems are a current topic of discussion not just between hydrologists and meteorologists, but also for several projects being conducted both by the DWD and flood forecasting centers.

Further to the deterministic numerical weather forecasts mentioned earlier, meteorological ensemble forecasts (EPS data) from various weather services have also been available to the flood forecasting centers for several years now:

- ECMWF Ensemble Prediction System ENS (15 forecast days, 52 members) (ECMWF, 2015; Molteni, Buizza, Palmer, & Petroliagis, 1996).
- COSMO-LEPS (132 h, 16 members) Regional ensemble developed by ARPA-SIM. Dynamical downscaling of the global ECMWF-EPS using the COSMO-Model (Marsigli, Boccanera, Montani, & Paccagnella, 2005).

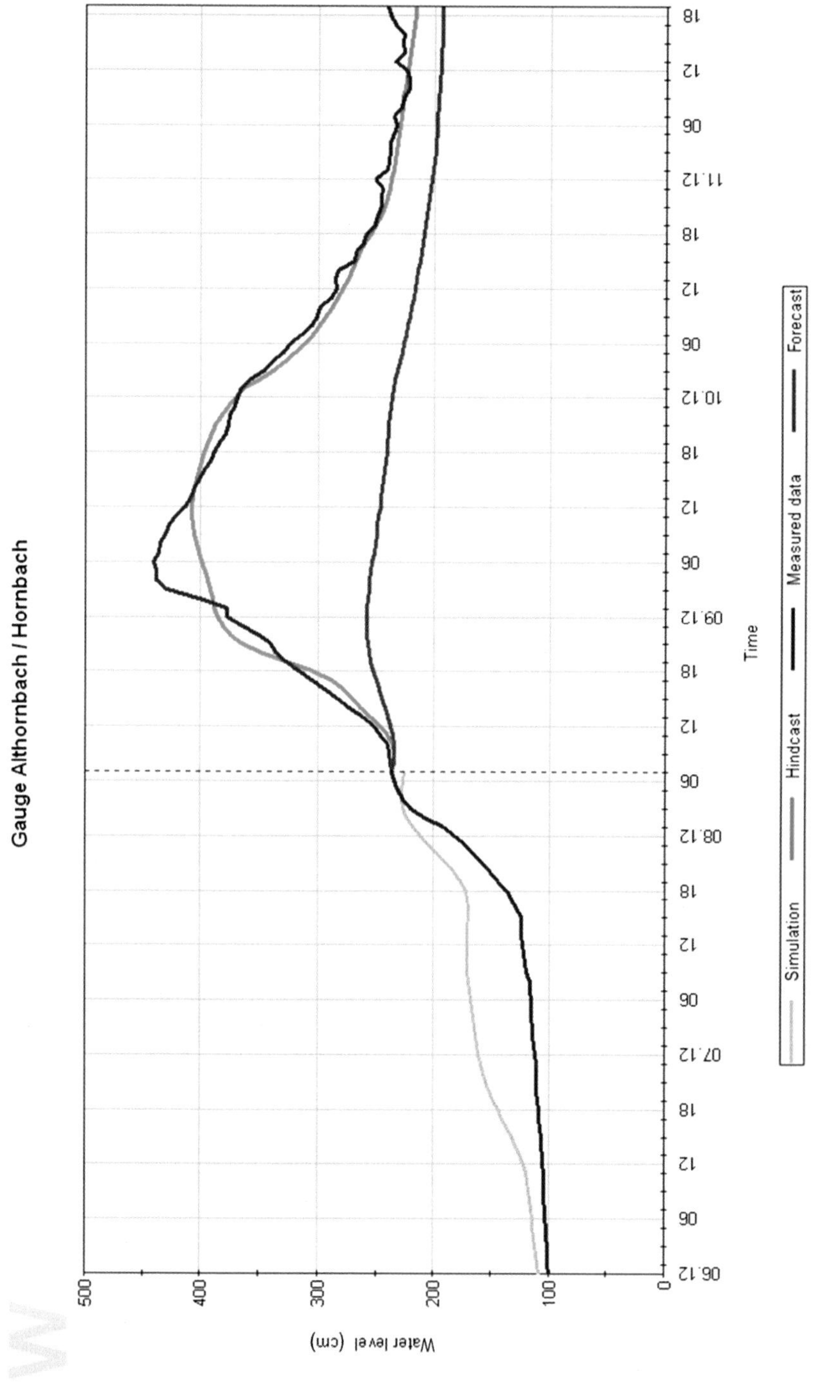

Fig. 5 LARSIM — flood forecast for gauge Althornbach/Hornbach.

- COSMO DE EPS (27 h, 20 members) Single model ensemble based on the convection permitting COSMO-DE model (Theis, Gebhardt, & Ben Bouallègue, 2014).
- SRNWP-PEPS — "Poor Man's Ensemble" of the operational Limited Area Models provided by the European national weather services in the framework of the Short Range Numerical Weather Prediction Program (SRNWP) (Heizenreder, Trepte, & Denhard, 2006).

Since just the calculation of the meteorological ensemble forecasts alone requires extensive adaptations of the operational systems, the integration of the EPS data into operational flood forecasting is still in its infancy in Germany (Johst & Demuth, 2015; Meißner, Klein, Lisniak, & Pinzinger, 2014). In addition to this, the use of EPS data heralds a paradigm shift from deterministic to probabilistic forecasting, which not only calls for new forecasting products and communication strategies (see forecast products), but which many hydrologists are also still skeptical about. The forecasting chain — that is, the transfer of the forecasts from one of the forecasting centers (upstream) to the next one (downstream) — is as yet not resolved in terms of the ensemble calculations, since the forecasting centers use different NWP products and EPS data, primarily due to the different forecasting ranges (short-term forecasts in the upper and middle course to long-term forecasts in the lower course).

Many of the forecasting centers have primarily used the meteorological ensemble forecasts qualitatively to evaluate the deterministic forecasts to date. For this purpose, the DWD provides probability maps, which show the likelihood of precipitation totals larger than 25, 30, 50, or 80 mm in a period of 12 or 24 h.

5 SNOWMELT FORECASTS

Whereas floods due to thawing snow alone are seldom in Germany, many floods in the Rhine basin are the result of thawing coinciding with heavy rain. A critical factor is a rapid succession of cold and warm fronts carrying a lot of snow and with high levels of precipitation. Especially in the low mountain ranges to the west, there can then be widespread melting of a blanket snow over a large area in a very short period of time.

Snow layering, snow aging, and snow melting are modeled directly by some hydrological forecasting models (eg, LARSIM, see later). Alternatively, the DWD also provides snow melt forecasts for Central Europe (Modell SNOW, Reich & Schneider, 2010). The SNOW model is based on a high spatial and temporal discretization of all of the energy and mass flows. For a grid with a mesh width of 1 km, SNOW provides data on the snow water equivalent and the precipitation yield (the sum of the rainfall and snowmelt) four times a day with hourly resolution, for the previous 30 h and for the following 72 h.

Because snow modeling involves great uncertainties, most of the models have processes for tracking the calculated snow water equivalents (eg, by optimizing the boundary limit temperature to distinguish between snow and rain, or for the assimilation of snow measurements and satellite data on snow cover). For many regions, the quality of the tracking is limited by an insufficient number of snow measurement stations.

6 REGIONAL ORGANIZATION AND TRANSBOUNDARY DATA EXCHANGE

The boundaries of the hydrological catchment areas correspond to the political borders in very few instances. Because floods do not stop at national borders, however, cooperation and exchange of data between upstream and downstream centers in transboundary catchment areas is indispensable for successful flood forecasting. This section describes the comprehensive exchange of data in the catchment area of the Moselle as an example. With a catchment area of 28,300 km^2, the Moselle is the Rhine's largest tributary. As the catchments are located in France, Luxembourg, Belgium, and the federal states of Rhineland-Palatinate, Saarland, and North Rhine-Westphalia, the Moselle is an international waterway. Due to the national and international borders, there are six forecasting centers responsible for flood forecasting in the Moselle basin. Following several major flood events in the mid-1980s, a Government Agreement between France, Germany, and Luxembourg concerning the flood reporting system in the Moselle basin was concluded in 1987 (Bundesgesetzblatt, 1988). After measurements from 6 gauges in France were initially sent to Germany and Luxembourg, the current exchange of data now encompasses almost 500 hydrological and meteorological stations (operated by 13 different monitoring network operators), radar precipitation data, numerical weather forecasts both from the German and the French meteorological service, as well as operating data from reservoirs and retention basins. A common exchange platform is used to ensure that the same measurements and forecast data — updated on an hourly basis — are available to all of the centers.

Due to the large number of data providers, the necessary agreements on data exchange and the usage rights as well as the various data formats, coordinate systems, measurement principles, monitoring network configurations, and reference and retrieval times pose a major, if not the greatest, challenge to running operational forecasting systems in transboundary catchment areas. Following the implementation of a new common flood forecasting system in the catchment areas of the Moselle (2004–08), it very soon became apparent that it is unrealistic to define uniform data standards and interfaces with all of the monitoring network operators. In order to be able to make use of as much of the available data as possible in operation, a large number of scripts and programs for data conversion — jointly funded by the forecasting centers — were written. In order to improve the cooperation between monitoring network operators, agreements were concluded to create win-win situations, where possible, which provide not only for the provision of data for use in flood forecasting, but also for a flow of data in the opposite direction. For instance, the maps of distribution of precipitation prepared in Rhineland-Palatinate for flood forecasting purposes are also used by the agrometeorological services.

7 RIVER FORECASTING MODELS

The German hydrological services use a range of forecasting models. This variety is not only a reflection of the federal structure, but is also caused by the hydrological diversity of the natural landscapes, ranging from alpine catchment areas in the south to catchment areas that react rapidly to changes in the low mountain ranges, from lowland rivers dominated by wave activity to the estuaries in the north with tidal influences. But the very varied flood damage potential (highly industrialized conurbations alongside sparsely populated agricultural regions) and the associated demands on flood forecasting have also contributed to this variety of models. Examples of two model systems that are used at several centers are presented in the following sections.

7.1 The LARSIM Water Balance Model

The LARSIM (Large Area Runoff Simulation Model) water balance model is a deterministic concept model, which allows process and spatially detailed simulation and forecasting of the terrestrial water cycle and the stream water temperatures (Fig. 6). LARSIM was developed as part of the BALTEX research project (Deutscher Wetterdienst BALTEX, 1995; Ludwig & Bremicker, 2006)

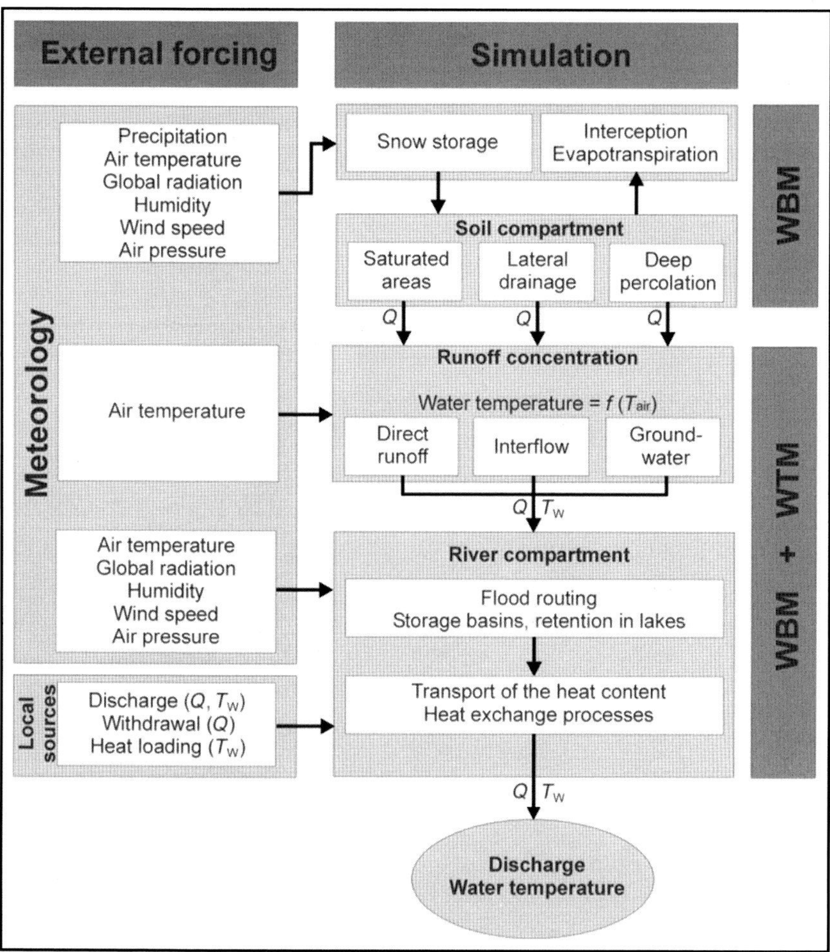

Fig. 6 LARSIM model-scheme.

on the basis of the FGMOD river basin model (Ludwig, 1982). However, LARSIM is also used in water management practice for researching a wide variety of aspects (eg, Bremicker, Casper, & Haag, 2011; Demuth, Haag, & Luce, 2010; Haag & Luce, 2008; Haag, Luce, & Gerlinger, 2006).

LARSIM can be used to simulate the following hydrological subprocesses (Ludwig & Bremicker, 2006):

- interception and interception evaporation;
- accumulation, metamorphosis, and melting of snow;
- infiltration, soil water balance, runoff formation, and deep drainage;
- evapotranspiration;

- runoff concentration (transfer from the surface into the gutter);
- translation and retention in the gutters;
- effect of lakes, reservoirs, and retention basins;
- inlets, outlets, and conduits; and
- thermal balance of watercourses and water temperatures (optional).

In addition to simulation of the hydrological processes, LARSIM also offers a lot of options specially developed for operational use, for example, for dealing with gaps in hydrometeorological input data or for the import and correction/manipulation of numeric weather forecasts. Of particular importance to flood forecasting are the automated processes for the assimilation of hydrological data and for tracking models (cf. Haag & Bremicker, 2013; Luce, Haag, & Bremicker, 2006). By comparing the measured and simulated outflows, it is possible to check how well the model represents the current water balance in day-to-day operation. In the event of major deviations between the measured and simulated outflows, it is possible to adjust and track the appropriate model variables so that the model represents the water balance at the current point in time as well as possible, and the forecasting quality is improved. The approaches to track the models differ in terms of the component in the model used to track the model (Komma, Drabek, & Blöschl, 2009; Refsgaard, 1997). It is possible for the user to track the input variables (model input), state variables, and model parameters or the calculation results (model output) himself (Fig. 7). In LARSIM, the following approaches can be activated (Haag & Bremicker, 2013; Luce et al., 2006):

Scheme of updating procedures

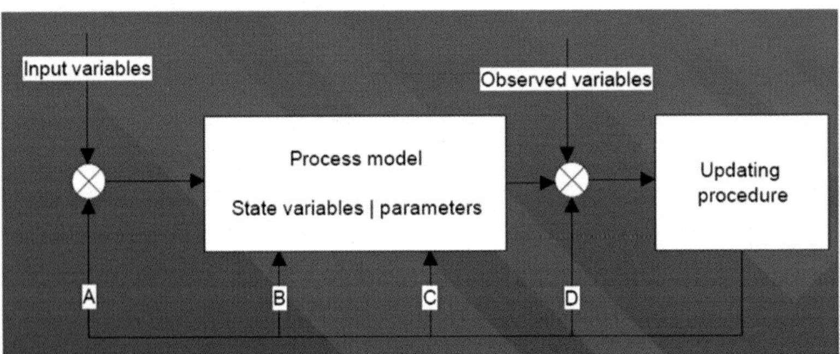

A: Input correction
B: Model state updating
C: Parameter correction
D: Output correction

Fig. 7 LARSIM — scheme of updating procedures.

- tracking the water resources (correction of the input variable);
- tracking the contents of the regional reservoir (correction of the system status);
- tracking the snow water equivalent and/or snow depths (correction of the system status);
- tracking the snow/rain boundary temperature (correction of the model parameters); and
- ARIMA correction (correction of the model output).

LARSIM models are used by the forecasting centers in the German states of Baden-Württemberg, Bavaria, Hesse, North Rhine-Westphalia, Rhineland-Palatinate, and Saarland, the Austrian states of Tyrol and Vorarlberg, in Luxembourg and the French regions of Alsace and Lorraine, and currently cover an area of more than 200,000 km^2. These high spatial resolution LARSIM models with zone areas of 0.25–10 km^2 are based on geodata (ground elevation, land use, ground reservoir volumes, gutter data, etc.) and are powered by meteorological data (precipitation, air temperature, etc.). In practical operation, various data available in real time, including meteorological measurements, data on discharge or water levels, and reservoir releases, are used for the model. Various different weather forecast models, reservoir control plans, and power plant utilizations are taken into account for the forecast period (Bremicker, Brahmer, Demuth, Holle, & Haag, 2013; Luce et al., 2006).

In addition to the actual flood forecasting, the high spatial resolution LARSIM models are also used operationally for early warning of floods, operational low water management, water temperature forecasts, and the optimization of reservoir releases. Furthermore, the same models are also used for nonoperational purposes such as for water resources planning, thermal load plans, assessment issues, and studies in the field of climate impact research (Bremicker et al., 2013).

Regular communication between the developers and users of the model and a transnational development community at the flood warning centers ensures that LARSIM is constantly improved and developed (Bremicker et al., 2013).

7.2 The WAVOS Water Level Forecasting System

For inland navigation on the free-flowing inland waterways there is centralized information on current and anticipated water levels, which is why the WSV also has its own forecasting service in addition to its water level reporting services. As well as boosting cost-effectiveness, water level forecasts contribute to an improvement in shipping safety and form a key building block of fairway information or navigation systems. Since the 1990s, the Federal Institute for Hydrology (BfG) has developed, updated, and optimized the WAVOS Water Level Forecasting System, including the integrated forecasting models, on behalf of the Federal Ministry of Transport and Digital Infrastructure (BMVI; Fröhlich, Heinz, Steinbach, & Wilke, 1998; Rademacher, Burek, & Eberle, 2004; Rademacher, Eberle, Krahe, & Wilke, 2004). The models basically cover the entire runoff spectrum, since the commercial shipping companies also need to be informed of the anticipated developments in water levels when levels are high. If defined maximum water levels are exceeded, speed restrictions and closures of waterways can cause considerable logistical constraints. However, since the federal states are also responsible for water level forecasting in the event of flooding for the federal waterways, no competing forecasts are published by the federal government in such instances (single-voice principle). Rather, WAVOS is also used by various flood forecasting centers (see Fig. 8).

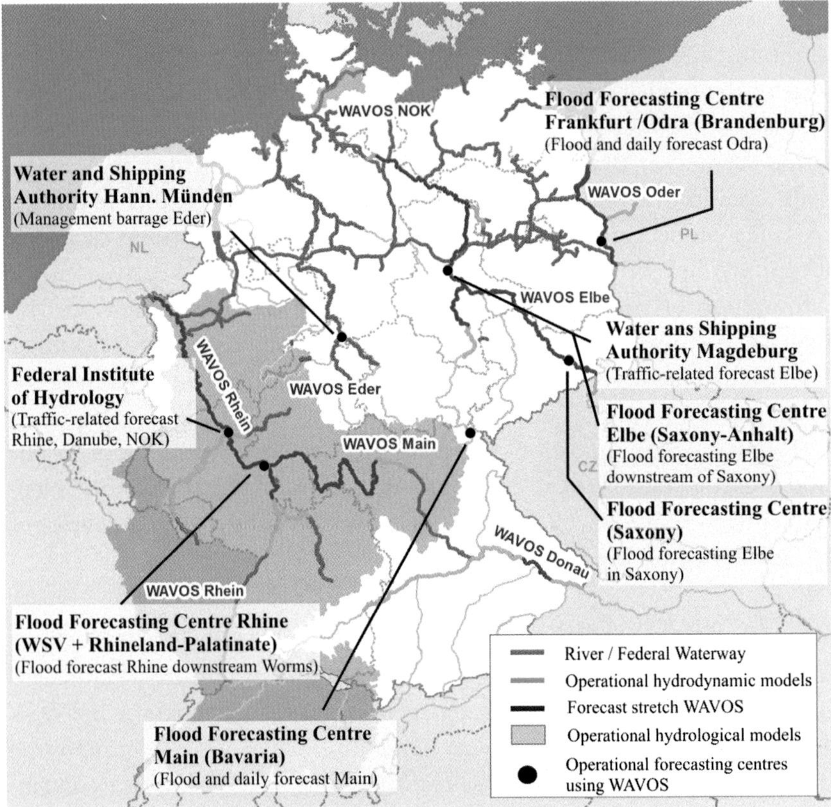

Fig. 8 Operational WAVOS-flood forecasting systems.

The calculation of the water level predictions for large waterways such as the federal waterways, in which the wave activity in main river channels and the interaction with the flood waves from the tributaries dominates, is primarily based on hydrodynamic models. Initially it is a one-dimensional hydrodynamic-numerical model WAVOS-1D, which was designed by the BfG especially for forecasting (Rademacher, Burek, et al., 2004; Rademacher, Eberle, Krahe & Wilke, 2004; Steinebach, 1999), and which is characterized by its stability and processing speed. In recent years, the commercial hydrodynamic-numerical modeling software SOBEK (Stelling & Duinmeijer, 2003) has seen increasing use within the WAVOS systems. These models are driven by the tributaries at the top, by the feeder rivers, and finally by the water level. For the preceding period, this data is based on measurements; in the forecast period, drainage flows calculated using rainfall runoff models are generally used (Meißner & Rademacher, 2010).

In addition to the forecasting models, WAVOS also provides preprocessors, which ensure smooth importing and processing of the wide range of input data, as well as postprocessors, which prepare the results quickly and meaningfully. The forecasting process is partly automated, for example, the data provision, and is partly supported by a user-friendly graphical user interface (eg, for checking of the data and results).

The WAVOS systems for the Rhine, Elbe, Oder, and Main have already been in operational use for flood forecasting by the competent flood forecasting centers in the respective federal states for many years. The BfG adapts the forecasting system to flood-specific features, such as the effect of flood retention measures (Burek, Rademacher & Wilke, 2006; Rademacher, Burek, et al., 2004; Rademacher, Eberle, Krahe & Wilke, 2004) in cooperation with the competent federal states.

7.3 Components of WAVOS Rhein

A key element of WAVOS Rhein, which is used for flood forecasting by HMZ Rhein, is a one-dimensional hydrodynamic model of the free-flowing Rhine (between Karlsruhe/Maxau and the German-Dutch border) as well as the Moselle from Trier onwards. In addition to the continuous updating of the model, the estuaries of the Neckar, Main, Nahe, Lahn, Sieg, and Lippe have been explicitly integrated as hydraulically linked model strands in the forecasting model. This makes it possible to take better account of both backwater effects as well as wave activity in the Rhine's major tributaries. In total the current model of the Rhine includes ~850 km of river. Beside all effective flood retention measures, the very last version includes 42 potential locations of dyke failures. This should provide a facility to calculate improved forecasts during extreme floods (see Fig. 9).

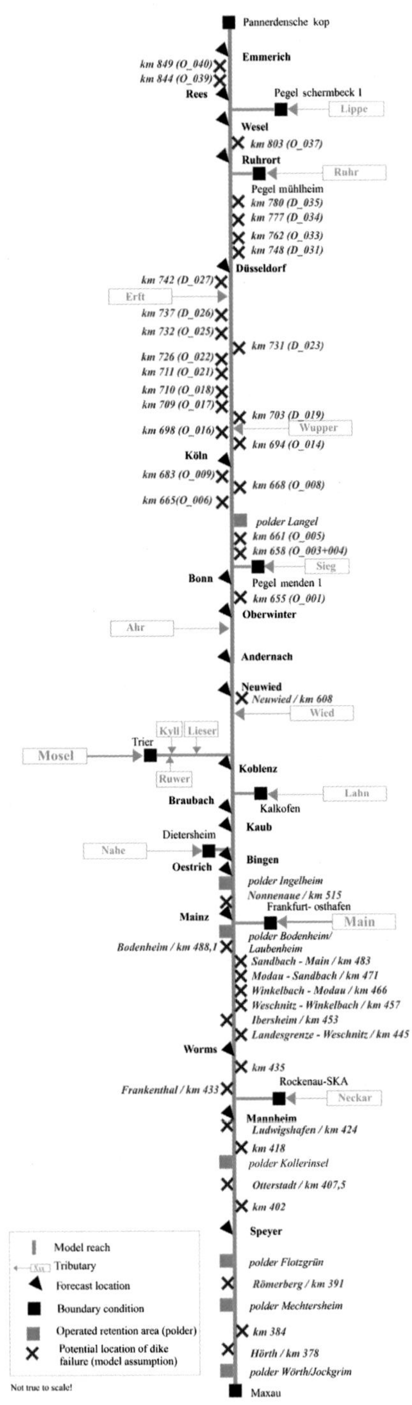

Fig. 9 Modell scheme of WAVOS Rhine.

For operation of the model, it is necessary to state the boundary conditions for the whole modeling duration — that is, the temporal development of the inflow at the top of the model (Karlsruhe/Maxau) and at the bottom (Pannerdensche Kop/NL) needs to be specified. Equally, the temporal development in the lateral inflows from the tributaries (Neckar, Main, Nahe, Moselle, etc.) into the model area is required. Whereas it is possible to take recourse to measurements to simulate the past, appropriate predictions are needed to calculate future water levels. There are different ways of determining the future boundary conditions in WAVOS. On the one hand it is possible to include (external) forecasts from other forecasting centers (HVZ BW, HMZ Main, HMZ Mosel); on the other hand it is possible to calculate internal inflow forecasts using hydrological or hydrodynamic models. WAVOS has thus already provided separate 1D hydrodynamic models for sections of the Main, Moselle, Sauer, and Saar for several years now.

8 FORECAST DISSEMINATION

Flood forecasting only makes sense if its results reach as many of the affected people as possible in a suitable form. This is the objective pursued by the state flood warning centers with some very different information on offer and communication strategies. The information on offer has developed over the years in accordance with the regional requirements and possibilities. With the demands of an increasingly mobile information society on the one hand and the increasing performance of the IT systems on the other, the information offered by the flood warning centers is becoming increasingly well-coordinated, and particularly the information available online including mobile applications is being continuously updated. A transnational flood portal, operated jointly by the German federal states, gives an overview of the present flood situation in the whole of Germany and in neighboring countries. In addition to the discharge situation at the flood gauges, it provides a summary of the current flood situation and enables easy access to the regional, detailed information offered by the flood warning centers. The next section describes the information offered by the flood alert service in Rhineland-Palatinate, as an example.

The state of Rhineland-Palatinate operates a flood alert with significant involvement by the Federal Water and Shipping Administration for the federal waterways Rhine, Lahn, Moselle, and Saar. Three regional flood monitoring centers inform the general population and the affected districts and communities about the development and the course of flood events on the major waterways by way of regularly updated flood forecasts, flood water level reports data, and status reports. The flood forecasts for the major waterways based on water levels gauge data are augmented by a regional flood early warning system for small catchment areas (<300 km^2) covering the total area of Rhineland-Palatinate. Details on the flood alert service in the state flood monitoring centers are given in the regional flood alert plan, which is updated regularly.

The responsible flood monitoring center triggers the flood alert service if the critical water level (reporting level) at one of the gauges is exceeded. With an initial warning (status report), the affected districts and communities and the general population are given early warning of a flood wave, so that local defensive measures can be initiated in good time. After the flood alert service has been triggered, all further flood reports are dependent on the current flood situation. If the water levels given in the tables are exceeded at the gauges, current flood reports are issued:

- at MARK 1: between 7:00 am and 3:00 pm;
- at MARK 2: between 7:00 am and 9:00 pm; and
- at MARK 3: between 12:00 am and midnight.

The level measurement values are updated once an hour and the forecasts every 3 h. The forecasts normally cover a more reliable period (see Fig. 5, red line) as well as an assessment of the subsequent outlook. Depending on the river basin and level, the more reliable part may cover periods between 6 h and at most 24 h. In the event of flooding along the Rhine, the subsequent estimated period is maximum 48 h at most and 24 h for the tributaries.

Due to the high degree of uncertainty in small catchment areas, specific local forecasts accurate to the centimeter are impossible. Instead, warnings are issued for regions (Demuth, 2008) if the predicted runoff from several stretches of the river exceeds a certain threshold. The regions for which warnings are issued are the towns and rural districts. The warning is issued in five warning classes and refers to the annuality of the maximum predicted runoff in the next 24 h. The regional flood early warnings for small catchment areas are updated between two and eight times a day, depending on the runoff situation.

The flood monitoring centers summarize the flood situation on a daily basis, in at least one flood report per day, from the time the flood alert service is triggered until it ceases. The flood report is sent to regional and local authorities as well as other recipients connected with the flood alert service. It should be borne in mind that the flood monitoring centers are only obliged to issue the initial warning and the flood reports (PUSH). For all other flood alerts, the obligation is on those affected (PULL).

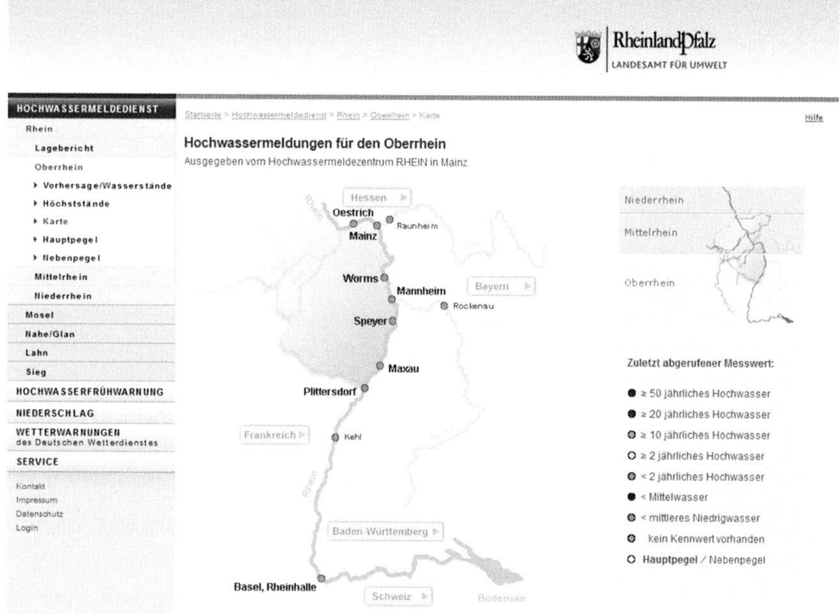

Fig. 10 Rhineland-Palatinate flood warning service — Internet presence for the Upper Rhine (www.hochwasser-rlp.de).

The flood alerts are published via various channels:

- on the Internet (Fig. 10);
- on teletext;
- on the radio; and
- on telephone answering machine at gauges (only water levels).

9 FLOOD PARTNERSHIPS

As part of the implementation of the European Flood Risk Management Directive, the Rhineland-Palatinate water management agency initiated flood partnerships (HPI, 2013). Flood partnerships are voluntary collaborations between towns and cities, communities, associations of cities, and rural districts that cooperate, sometimes even across regional and national borders, to provide flood protection and prevention together. The aim of the flood partnerships is to maintain the awareness of the danger of all those concerned through local networking and cooperation, and to ensure the best possible protection against the next flood. In workshops, the stakeholders develop realistic objectives and measures to this end in cooperation with the specialist authorities. On the topic of flood forecasting and early warning, the information offered by the flood monitoring center as well as the limits of flood forecasting are discussed. Specific issues, problems, and proposals are collected and measures are taken to improve the forecasting and early warning system, both in terms of providing better information to the general public as well as for the responses of the emergency and rescue services in the communities. The flood partnerships are thus a key component in the user-driven development of the flood forecasting and early warning systems.

10 SUMMARY

In Germany, several regional and transregional services are responsible for flood forecasting. The various flood warning centers tasks and organizational structures can vary quite considerably. They use different forecasting systems and offer different information services. This variety is not only a reflection of the federal structure, but is also caused by the hydrological diversity of the natural landscapes, ranging from alpine catchment areas in the south to catchment areas that react rapidly to changes in the low mountain ranges, from lowland rivers dominated by wave activity to the estuaries in the north with tidal influences. But the substantial regional variations in flood damage potential and the related demands on flood forecasting have also contributed to this diversity. With the demands of an increasingly mobile information society on the one hand and the increasing performance of the IT systems on the other, the information offered by the flood warning centers is becoming increasingly well-coordinated. Flood forecasting in the catchment area of the Rhine and the flood alert service in the federal state of Rhineland-Palatinate are explained as examples.

The inclusion of meteorological ensemble forecasts in flood forecasting is still in its infancy in Germany. It requires extensive adaptations of the operational systems, agreements on the forwarding of ensemble forecasts from one forecasting center (upstream) to the next one (downstream), as well as the development of new forecasting products and communication strategies.

REFERENCES

Baldauf, M., Förstner, J., Klink, S., Reinhardt, T., Schraff, C., Seifert, A., et al. (2014). Kurze Beschreibung des Lokal-Modells Kürzestfrist COSMO-DE (LMK) und seiner Datenbanken auf dem Datenserver des DWD, version 2.3. In Offenbach: Deutscher Wetterdienst.(pp. 86). http://www.dwd.de/SharedDocs/downloads/DE/modelldokumentationen/nwv/cosmo_de/cosmo_de_dbeschr_version_2_3_201406.pdf?__blob=publicationFile&v=4 Accessed 04.02.16.

Bremicker, M., Brahmer, G., Demuth, N., Holle, F.-K., & Haag, I. (2013). Räumlich hoch aufgelöste LARSIM Wasserhaushaltsmodelle für die Hochwasservorhersage und weitere

Anwendungen. *KW Korrespondenz Wasserwirtschaft, 6*(9), 509–514. http://dx.doi.org/10.3243/kwe2013.09.004.

Bremicker, M., Casper, M. C., & Haag, I. (2011). Extrapolationsfähigkeit des Wasserhaushaltsmodells LARSIM auf extreme Abflüsse am Beispiel der Schwarzen Pockau. *KW Korrespondenz Wasserwirtschaft, 4*(8), 445–451. http://dx.doi.org/10.3243/kwe2011.08.002.

Bundesgesetzblatt. (1984). Vereinbarung zur Änderung und Ergänzung der Zusatzvereinbarung vom 16. Juli 1975 zum Vertrag vom 4.07.1969 zwischen der Bundesrepublik Deutschland und der Französischen Republik über den Rheinausbau zwischen Kehl/Straßburg und Neuburgweier/Lauterburg. Bundesgesetzblatt 1984. part II (pp. 268–275).

Bundesgesetzblatt. (1988). Übereinkommen der Regierung der Bundesrepublik Deutschland, der Regierung der Französischen Republik und der Regierung des Großherzogtums Luxemburg über das Hochwassermeldewesen im Moselgebiet. Bundesgesetzblatt 1988. part II (pp. 93 ff).

Burek, P., Rademacher, S., & Wilke, K. (2006). The operational flood forecasting system WAVOS for the river Elbe. In J. Alphen, E. van Beek, & M. Taal (Eds.), Floods, from defence to management: Proceedings of the third international symposium on flood defence, Nijmegen, Netherlands, 25–27 May 2005.

Bürgi, T., Homagk, P., Prellberg, D., Sprokkereef, E., & Wilke, K. (2004). Das internationale Hochwasservorhersagesystem am Rhein. *Wasserwirtschaft, 12,* 12–19. http://dx.doi.org/10.1007/BF03243615.

Champeaux, J.-L., Dupuy, P., Laurantin, O., Soulan, I., Tabary, P., & Soubeyroux, J.-M. (2009). Rainfall measurements and quantitative precipitation estimations at Météo-France: inventory and prospects [Les mesures de précipitations et l'estimation des lames d'eau à Météo-France: état de l'art et perspectives]. *La Houille Blanche, 5,* 28–34. http://dx.doi.org/10.1051/lhb/2009052.

Demuth, N. (2008). Hochwasserfrühwarnung in Rheinland-Pfalz. In R. Jüpner, V. Lüderitz, & A. Dittrich (Eds.), *Reihe der Berichte des Fachgebietes Wasserbau und Wasserwirtschaft der TU Kaiserslautern: Vol. 19.* Beiträge zum Fachkolloquium "Extremwerte in der Wasserwirtschaft". Kaiserslautern. (pp. 23–28).

Demuth, N., Haag, I., & Luce, A. (2010). Integrating spatially distributed information on dominant runoff processes into the hydrological model LARSIM. In L. Pfister, et al. (Eds.), Looking at catchments in colors — Debating new ways of generating and filtering information in hydrology, EGU Leonardo topical conference.(pp. 94).

Deutscher Wetterdienst BALTEX. (1995). Baltic Sea experiment BALTEX: Initial implementation plan. Geesthacht: International BALTEX Secretariat at the GKSS Research Center Publication no. 2.

DWD. (2004). *DWD-LAWA Project RADOLAN: Routine method for the online adjustment of radar precipitation data with the help of automatic surface precipitation stations (Ombrometer).* Final report, Offenbach, 111 pp. (in German).

ECMWF. (2015). European Centre for medium-range weather forecasts. http://www.ecmwf.int/ Accessed 04.02.16.

Einfalt, T., Gerlach, N., Podlasly, C., & Demuth, N. (2008). Rainfall and climate data quality control. In: Proceedings of 11th ICUD, Edinburgh.

Einfalt, T., & Podlasly, C. (2009). NIKLAS — Program for the plausibility check of meteorological station measurements, version 1.2.4, user manual, 40 pp.

Environmental Modeling Center. (2003). The GFS atmospheric model. *NCEP office note 442.* Camp Springs, MD: Global Climate and Weather Modeling Branch, EMC.

Fröhlich, W., Heinz, M., Steinbach, G., & Wilke, K. (1998). Water level forecasts for navigation on the river Elbe and river Rhine — A contribution to an intelligent waterway. In: Proceedings of 29th PIANC International Navigation Congress The Hague 1998, section I, subject 3.International Navigation Association.(pp. 55–60).

Haag, I., & Bremicker, M. (2013). Möglichkeiten und Grenzen der Schneesimulation mit dem Hochwasservorhersagemodell LARSIM. *Forum für Hydrologie und Wasserbewirtschaftung, 33,* 47–58.

Haag, I., & Luce, A. (2008). LARSIM-WT: An integrated water-balance and heat-balance model to simulate and predict stream water temperatures. *Hydrological Processes, 22*(7), 1046–1056. http://dx.doi.org/10.1002/hyp.6983 Accessed 04.02.16.

Haag, I., Luce, A., & Gerlinger, K. (2006). Effects of conservation tillage on storm flow: A model-based assessment for a mesoscale watershed in Germany. In M. Sivaplan, et al. (Eds.), In *Predictions in ungagged basins: Promises and progress: Vol. 303*. (pp. 342–350). IAHS-Publications.

Heizenreder, D., Trepte, S., & Denhard, M. (2006). SRNWP-PEPS: A regional multi-model ensemble in Europe. *The European Forecaster, Newsletter of the WGCEF, 11.*

Hinterding, A. (2003). Entwicklung hybrider Interpolationsverfahren für den automatisierten Betrieb am Beispiel meteorologischer Größen, ifgi-Prints — Schriftenreihe des Instituts für Geoinformatik, Westfälische Wilhelms. Dissertation Universität Münster 19.

HPI. (2013). International Support Centre for Flood Partnerships, HPI. Flood partnerships — Transboundary cooperation on flood risk management via community-based networksbrochure, 40 pp. http://213.139.130.160/servlet/is/84593/200429_Flood_Partnerships_EN_web.pdf?command=downloadContent&filename=200429_Flood_Partnerships_EN_web.pdf Accessed 04.02.16.

ICPR. (1998). Action plan on floods. Koblenz: International Commission for the Protection of the Rhine.

Johst, M., & Demuth, N. (2015). Verwendung von COSMO-DE-EPS Wettervorhersagen in der operationellen Hochwasservorhersage. *Newsletter Hydrometeorologie DWD 13,* 8–11.

KHR. (1997). Internationale Kommission für die Hydrologie des Rheingebietes und Internationale Kommission zum Schutze des Rheins: Bestandsaufnahme der Meldesysteme und Vorschläge zur Verbesserung der Hochwasservorhersage im Rheingebiet, KHR-Bericht Nr. II-12, Lelystadt. 139 pp.

Komma, J., Drabek, U., & Blöschl, G. (2009). Aktuelle Methoden der Hochwasservorhersagen. *Wiener Mitteilungen, Hochwässer — Bemessung, Risikoanalyse und Vorhersage, 216,* 181–212.

LAWA. (2014). Bund/Länder-Arbeitsgemeinschaft Wasser — Handlungsempfehlungen zur weiteren Verbesserung von Grundlagen und Qualität der Hochwasservorhersage an den deutschen Binnengewässern http://www.lawa.de/documents/W-Vorhersage_Handlungsempfehlungen_de1.pdf Accessed 04.02.16.

Luce, A., Haag, I., & Bremicker, M. (2006). Einsatz von Wasserhaushaltsmodellen zur kontinuierlichen Abflussvorhersage in Baden-Württemberg. *Hydrologie und Wasserbewirtschaftung, 50*(2), 58–66.

Ludwig, K. (1982). The program system FGMOD for calculation of runoff processes in river basins. *Zeitschrift für Kulturtechnik und Flurbereinigung, 23,* 25–37.

Ludwig, K., & Bremicker, M. (2006). *The water balance model LARSIM — Design, content and applications*. Freiburger Schriften zur Hydrologie. Institut für Hydrologie der Universität Freiburg i. Br., 22. 130 pp.

Marsigli, C., Boccanera, F., Montani, A., & Paccagnella, T. (2005). The COSMO-LEPS ensemble system: Validation of the methodology and verification. *Nonlinear Processes in Geophysics, 12,* 527–536.

Meißner, D., Klein, B., Lisniak, D., & Pinzinger, D. (2014). Probabilistic flow and water-level forecasts — Communication strategies and potential uses for inland navigation. *Hydrologie und Wasserbewirtschaftung, 58*(2), 119–127. http://dx.doi.org/10.5675/HyWa_2014,2_7.

Meißner, D., & Rademacher, S. (2010). Die verkehrsbezogene Wasserstandsvorhersage für die Bundeswasserstraße Rhein —Verlängerung des Vorhersagezeitraums und Steigerung der Vorhersagequalität. KW Korrespondenz Wasserwirtschaft. *9,* 531–537. http://dx.doi.org/10.3243/kwe2010.09.004.

Molteni, F., Buizza, R., Palmer, T. N., & Petroliagis, T. (1996). The ECMWF ensemble prediction system: Methodology and validation. *Quarterly Journal of the Royal Meteorological Society, 122*(529), 73–119. http://dx.doi.org/10.1002/qj.49712252905.

Pailleux, J., Geleyn, J.-F., El Khatib, R., Fischer, C., Hamrud, M., Thépaut, J.-N., et al. (2015). Les 25 ans du système de prévision numérique du temps IFS/Arpège. *La Météorologie, 89,* 18–27. http://dx.doi.org/10.4267/2042/56594.

Rademacher, S., Burek, P., & Eberle, M. (2004). Niedrigwasser — Vorhersage an Bundeswasserstraßen — Entwicklung und operationeller Betrieb. In H.-B. Kleeberg & G. Koehler (Eds.), Niedrigwassermanagement — Beiträge zum Seminar am 11/12 November 2004 in Koblenz. Hydrologische Wissenschaften — Fachgemeinschaft der ATV-DVWK.

Rademacher, S., Eberle, M., Krahe, P., & Wilke, K. (2004). Enhancing the operational forecasting system of the River Rhine. In D. Malzahn & T. Plapp (Eds.), Disasters and society — From hazard assessment to risk reduction (pp. 191–198). Karlsruhe.

Refsgaard, J.Ch. (1997). Validation and intercomparison of different updating procedures for real-time forecasting. *Nordic Hydrology, 28*(2), 65–84. http://dx.doi.org/10.2166/nh.1997.005.

Reich, T., & Schneider, G. (2010). SNOW — Ein Modell zur Analyse und Vorhersage der Schneedeckenentwicklung. *Newsletter Hydrometeorologie, 3,* 4–14. DWD http://www.dwd.de/DE/Leistungen/snow/snow.html?nn=353182 Accessed 04.02.16.

Schättler, U., Doms, G., & Schraff, C. (2012). A description of the nonhydrostatic regional COSMO-Model. Part VII: User's guide, consortium for small-scale modelling (COSMO)192 pp. http://www.cosmo-model.org/content/model/documentation/core/cosmoUserGuide.pdf Accessed 04.02.16.

Seity, Y., Brousseau, P., Malardel, S., Hello, G., Bénard, P., Bouttier, F., et al. (2011). The AROME-France convective-scale operational model. *Monthly Weather Review, 139,* 976–991. http://dx.doi.org/10.1175/2010MWR3425.1.

Steinebach, G. (1999). Using hydrodynamic models in forecast systems for large rivers. In K. P. Holz, W. Bechteler, S. S.Y. Wang, & M. Kawahara (Eds.), In *Proceedings of advances in hydro-science and engineering, Cottbus: Vol. 3.* incl. CD-ROM.

Stelling, G. S., & Duinmeijer, S. P. A. (2003). A staggered conservative scheme for every Froude number in rapidly varied shallow water flows. *International Journal for Numerical Methods in Fluids, 43*(12), 1329–1354. http://dx.doi.org/10.1002/fld.537.

Steppeler, J., Doms, G., Schättler, U., Bitzer, H. W., Gassmann, A., Damrath, U., et al. (2003). Meso-gamma scale forecasts using the nonhydrostatic model LM. *Meteorology and Atmospheric Physics, 82*(1), 75–96. http://dx.doi.org/10.1007/s00703-001-0592-9.

Theis, S., Gebhardt, C., & Ben Bouallègue, Z. (2014). Beschreibung des COSMO-DE-EPS und seiner Ausgabe in die Datenbanken des DWD, Version 2.0. Offenbach: Deutscher Wetterdienst. 77 pp. http://www.dwd.de/EN/research/weatherforecasting/num_modelling/04_ensemble_methods/ensemble_prediction/cosmo-de-eps-beschreibung.pdf?__blob=publicationFile&v=1 Accessed 04.02.16.

Vieser, H. (1973). Folgen der Ausbaumaßnahmen am Oberrhein auf den Hochwasserabfluß. In Dt. gewässerkundl. Mitteilungen Sonderheft (pp. 42–50).

Winterrath, T., Rosenow, W., & Weigl, E. (2012). On the DWD quantitative precipitation analysis and nowcasting system for real-time application in German flood risk management. *Weather Radar and Hydrology, 351,* 323–329 (Proceedings of a symposium held in Exeter, UK, April 2011), IAHS.

WMO. (1997). The ALADIN project: Mesoscale modelling seen as a basic tool for weather forecasting and atmospheric research. *WMO Bulletin, 46*(4), 317–324.

Zängl, G., Reinert, D., Ripodas, P., & Baldauf, M. (2015). The ICON (ICOsahedral Non-hydrostatic) modelling framework of DWD and MPI-M: Description of the non-hydrostatic dynamical core. *Quarterly Journal of the Royal Meteorological Society, 141*(687), 563–579. http://dx.doi.org/10.1002/qj.2378.

CHAPTER 6

Operational Flood Forecasting in Israel

A. Givati*, E. Fredj†, M. Silver‡
*Israeli Hydrological Service, Jerusalem, Israel
†The Jerusalem College of Technology, Jerusalem, Israel
‡Ben Gurion University, Beersheba, Israel

1 INTRODUCTION

Floods are among the most dangerous natural hazards in the Mediterranean region due to the number of people affected and the relatively high frequency by which human activities and goods suffer damages and losses (Llasat-Botija, Llasat, & López, 2007). Flood damages are especially severe in semiarid regions such as in the Middle East, where estimation and prediction of highly variable precipitation during the rainy season is problematic. Additionally, it is widely expected that climate change will increase the occurrence of extreme rainfall events in many regions around the world (Andersen & Marshall Shepherd, 2013; Kundzewicz, Hirabayashi, & Kanae, 2010; Milly, Wetherald, Dunne, & Delworth, 2001; Trenberth, 2011; Wagener et al., 2010; Zwiers et al., 2013). In the Mediterranean basin, the effects could be increasing droughts on the one hand (Dai, 2011; Hoerling et al., 2012; Smiatek, Kaspar, & Kunstmann, 2013; Smiatek, Kunstmann, & Heckl, 2011, 2014; Törnros & Menzel, 2014) and intensified flood events on the other hand (Samuels, Smiatek, Krichak, Kunstmann, & Alpert, 2011). Land use changes and increasing urbanization also enhance flood intensity and frequency (eg, Bronstert, Niehoff, & Bürger, 2002; Chang & Franczyk, 2008; Delgado, Llorens, Nord, Calder, & Gallart, 2010; Githui, Mutua, & Bauwens, 2010; Kalantari et al., 2014; Li, Feng, & Wei, 2013), and flood prediction tools must be able to incorporate dynamically evolving atmospheric and land-surface conditions.

Advanced warning systems for floods can reduce flood risk and allow emergency response personnel to be better prepared for and mitigate damages. The accuracy of flood forecasts is highly determined by the skill of quantitative precipitation forecasts and its spatial distribution (Cloke & Pappenberger, 2009; Sin Shiha, Cheng-Hsin, & Tsyh Yeh, 2014;

Flood Forecasting
http://dx.doi.org/10.1016/B978-0-12-801884-2.00006-2

Younis, Anquetin, & Thielen, 2008). Most modern hydrological models can use precipitation input from various sources like rain gauges, radar, remote sensing, or simulated precipitation from numerical weather models. Operational weather forecast centers routinely provide relatively coarse (16–27 km grid) precipitation forecasts, which are typically incapable of resolving the necessary details of complex, intense precipitation structures that are forced by mesoscale orography, land-surface heterogeneities, and land-water contrasts (Fiori et al., 2014). In the eastern Mediterranean region, strong sea-air interaction and orographic forcing produce precipitation with very strong gradients that are generally missed by the coarse-grid operational models. In Israel, the precipitation patterns are particularly complex, and large precipitation gradients occur over a relatively small geographical distance (2–10 km). These large climatological precipitation gradients in Israel are caused by the preferred tracks of extra-tropical cyclones, the complex orography, and the shape of the coastline (Saaroni, Halfon, Ziv, Alpert, & Kutiel, 2009). To address those problems, Givati, Lynn, Liu, and Rimmer (2012) used the Weather Research and Forecasting (WRF) model to provide high-resolution precipitation forecasts during the 2008–09 and 2009–10 winters (wet seasons) for Israel and the surrounding region where complex terrain dominates. They showed that by using high-resolution (1.3–4 km) grids, the WRF model was capable of forecasting daily precipitation amounts and structures in northern Israel reasonably well, and in doing so were able to improve daily hydrological simulations for the upper Jordan River.

2 CLIMATE AND HYDROLOGICAL CHARACTERISTICS OF ISRAEL

Israel is characterized by various climate and hydrological regimes. It is located in the eastern part of the Mediterranean, between latitude 29 degree and 33 degree north. The northern part of the country is relatively wet with annual precipitation around 800–1000 mm at the higher elevations, which provide the majority of the water resources in the country. The central parts of the country are drier with annual precipitation around 550–700 mm. The southern and eastern regions are arid areas with annual precipitation less than 200 mm, as can be seen in Fig. 1. Most of the precipitation occurs during the winter, between Dec. and Mar. (over 80% of the annual precipitation). More than 90% of the rainfall in Israel is caused by cold fronts and moist air masses, which are associated

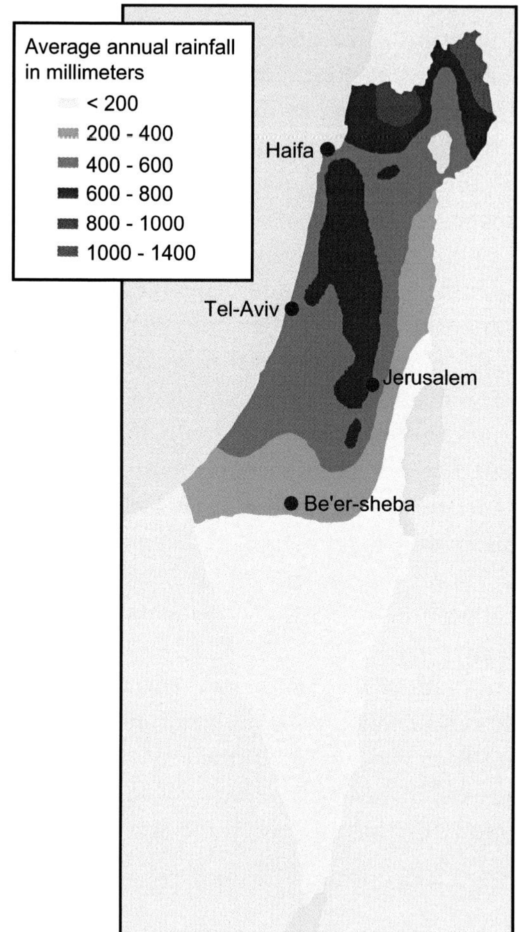

Fig. 1 Average annual rainfall in Israel.

with extra-tropical cyclones that pass through the northeastern corner of the Mediterranean Sea (Goldreich, 2003).

The strong spatial variability in precipitation between the different parts of the country: north-south, west-east, which can be observed in Fig. 1 are due to the following factors: (a) latitude-precipitation tends to decrease towards the south; (b) distance from the sea-precipitation tends to decrease away from the sea; (c) terrain elevation (orography) — precipitation tends to increase at higher elevations; (d) mountain downwind — precipitation tends to decrease when a mountain range is parallel to the coast. As noted by Goldreich (2003), these four factors similarly affect almost any geographic area, but the first one is particular to Israel's geographic location at the southeastern Mediterranean: the preferred tracks of the cyclones associated with the cold season precipitation are oriented towards the northeastern Mediterranean (Alpert, Neeman, & Shay-El, 1990; Schädler & Sasse, 2006; Trigo, Davies, & Bigg, 1999).

Fig. 2 displays the main basin network in Israel. While the streams in the western drainage area flow into the Mediterranean Sea, the streams in the eastern slopes of the Judea and Samaria hills (the lee side, rain-shadow desert) drain into the Lake of Galilee and the Dead Sea. The Eastern streams are actually wadis, rocky slopes (Fig. 2) with an elevation change from +800 m in the northwest to −425 m in the southeast. Flash floods develop in the area almost every year, mostly in the autumn and spring, as a result of strong convective conditions which bring moisture from the south and the east. Due to the high rain intensity (commonly thunderstorms), and the relatively impervious marl and clay formations which outcrop throughout the area and lack soil horizons, floods are sudden and intense. The runoff-precipitation ratio in this area is very high (up to 50%). A typical wadi in the Judea and Samaria eastern slopes, the Wadi Zin, can be viewed in Fig. 3.

At the western drainage area, significant changes in land use occurred during recent decades, mainly across the coastal areas. Ohana-Levi, Karnieli, Egozi, Givati, and Peeters (2015) quantified the changes in the urban areas at the Yarqon-Ayalon basin between 1989 and 2009, and showed the significant increase in the urban areas in the late period in respect to the early one.

3 THE FLOOD FORECASTING MODELING SYSTEM IN ISRAEL

3.1 Flood Forecasting at the Israeli Domain Using the WRF-Hydro Model

Operational flood forecasting is carried out in Israel by the Israel Hydrological Service (IHS). The IHS, part of the Water Authority, is responsible for monitoring all the water resources in the county. The IHS operates approximately 120 hydrometric stations in the country, half of which are online telemetric stations. The hydrometric network in Israel is relatively dense, especially in the south, extremely arid areas in the country, in respect to other places in the world.

Since 2013, the IHS has adopted and applied the WRF-Hydro model as its main model for flood forecasting. The WRF-Hydro model was developed

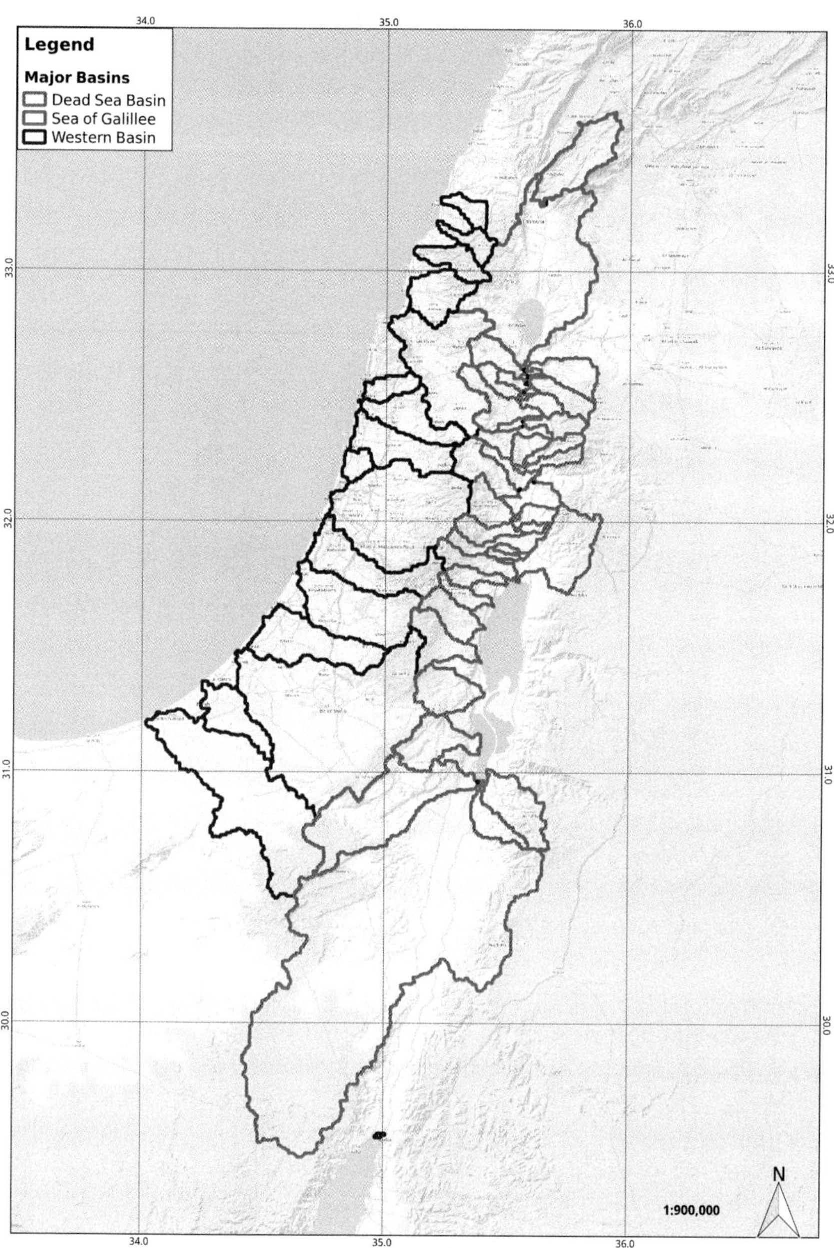

Fig. 2 The western and eastern drainage areas.

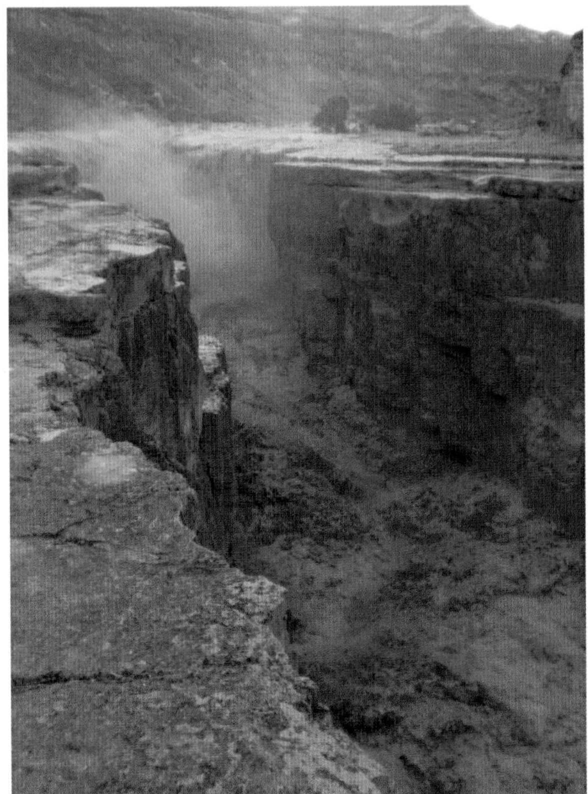

Fig. 3 The Zin basin at the eastern drainage areas.

by the National Center for Atmospheric Research (NCAR) to provide hydrological model components to the WRF atmospheric model. The model improves the representation of terrestrial hydrologic processes related to the spatial redistribution of surface, subsurface, and channel waters across the land surface. A suite of terrestrial hydrologic routing physics is contained within the WRF-Hydro. Detailed description regarding the model parameters and calibration can be found in Yucela, Onena, Yilmazb, and Gochisc (2015) and (Fredj, Silver, & Givati, 2015).

The WRF-Hydro domain covers areas in southern Lebanon, Israel, West Jordan, and South Egypt. Fig. 4 shows the WRF-Hydro modeling domains, at 9 and 3 km resolution, respectively. Fig. 5 displays the channel network at the Israeli WRF-Hydro domain and the locations of the model forecast points. The model was set up to extract predicted streamflow values for 170 different locations within the domain, where approximately

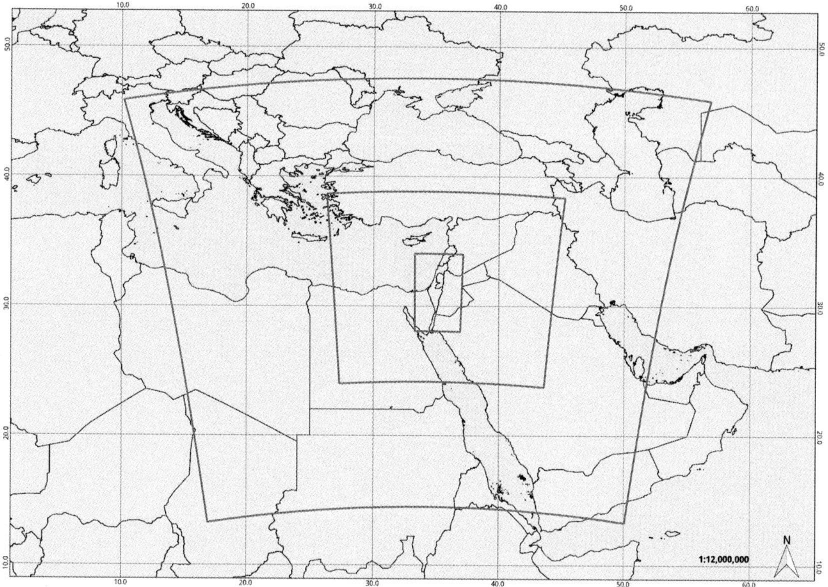

Fig. 4 Map of the three nested domain grids of WRF simulations at the east Mediterranean (Domain 1 = 27 km, Domain 2 = 9 km, Domain 3 = 3 km).

100 locations are equipped with active hydrometric stations operated by the IHS and used for the model calibration.

The model runs operationally twice a day, based on 00:00 and 12:00 GMT GFS data, with a forecast period of 72–84 h in advance (this model time lead can be determined according to the changing needs). The model results are uploaded to the IHS website where flood alerts appear using a color-coded legend where each color represents different flood magnitudes and return periods, as can be seen in Fig. 6. In addition, the website presents animations of expected precipitation (total storm accumulated precipitation and 3 h precipitation) and hourly animated runoff forecasts across the domain, as can be seen in Fig. 7.

3.2 Flood Forecasting for the Tel Aviv Metropolis Using the HEC-HMS Model

The IHS and regional Yarqon Drainage Authority chose to adopt a special monitoring and forecast system for the Ayalon basin in central Israel. This watershed is a subbasin of the Yarqon basin, draining 800 km^2 (Fig. 8A and B). The Ayalon river headwaters are in the Judea and Samaria Mountains at elevation of 800–900 m, and drain west towards the Mediterranean Sea, crossing the city of Tel Aviv. This watershed,

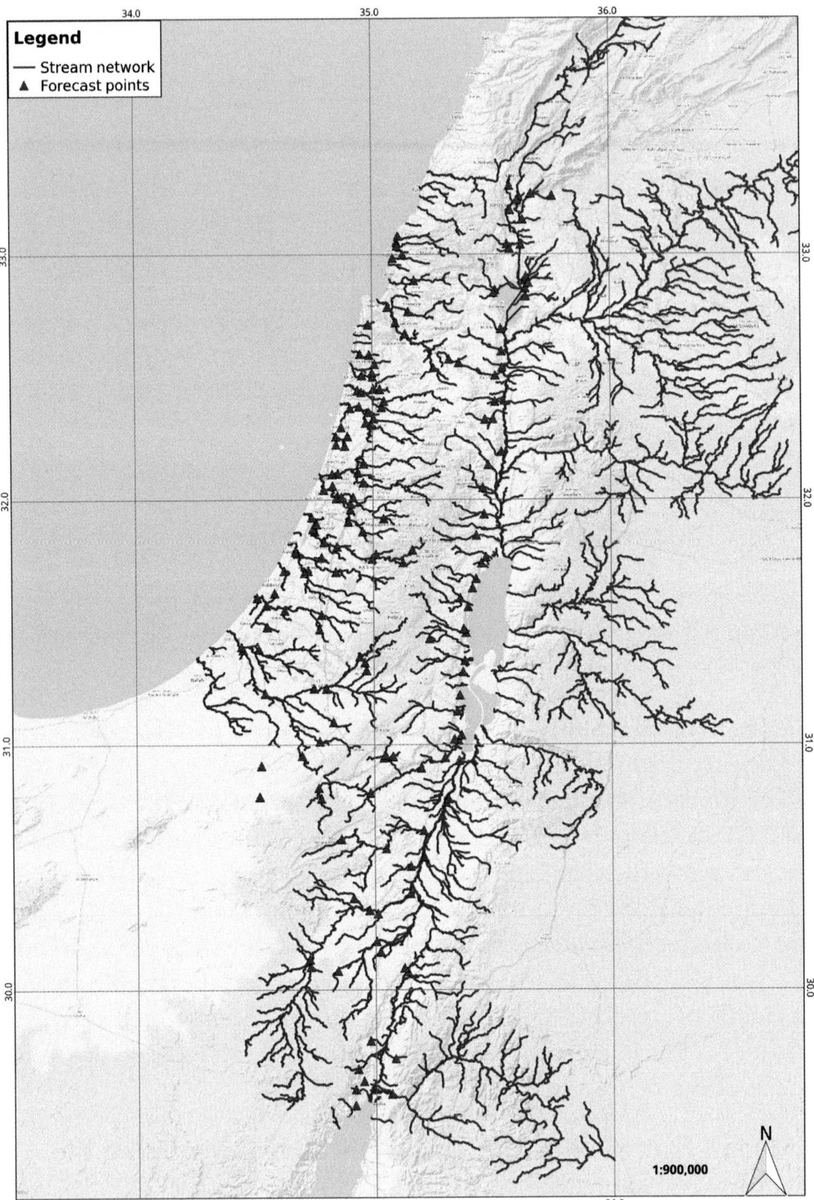

Fig. 5 The channel network at the Israeli WRF-Hydro domain and the locations of the model forecast points.

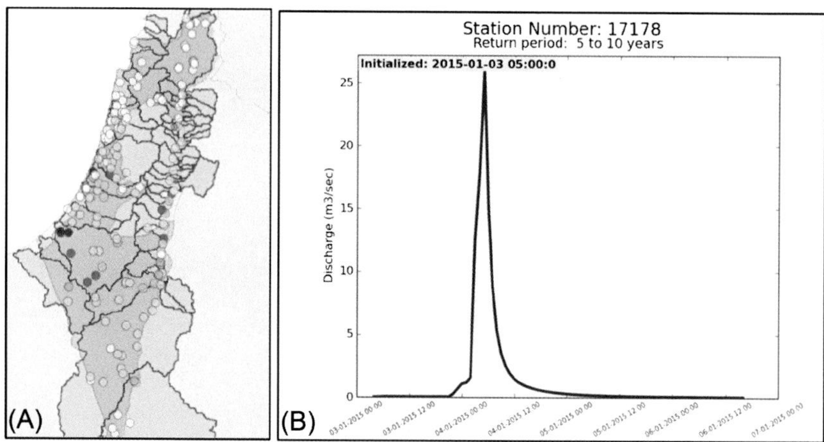

Fig. 6 Flood alerts map based on the WRF-Hydro forecast. (A) Each colored dot represents a different flood magnitude according to the return period. (B) Predicted hydrograph for a single location in the domain.

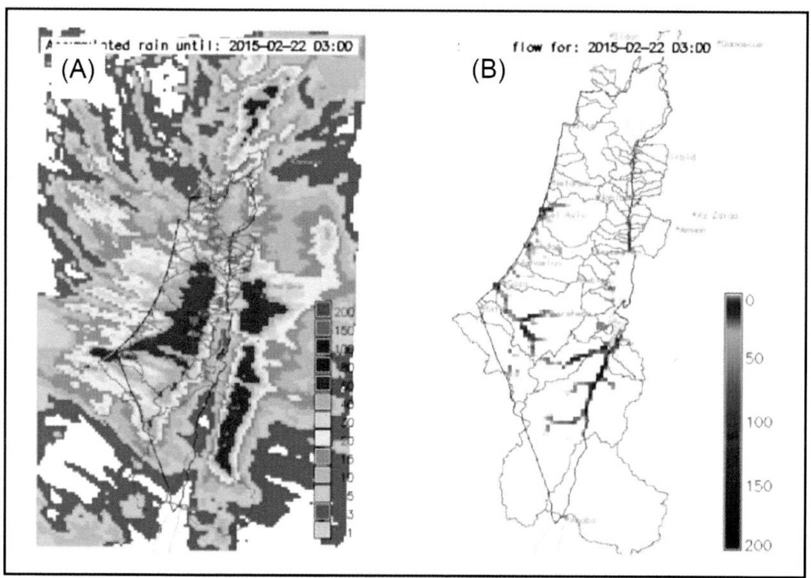

Fig. 7 Forecast maps for (A) accumulated 72h WRF-Hydro precipitation, 3 days in advance; (B) stream flow forecast (cubic meter per second) for the domain.

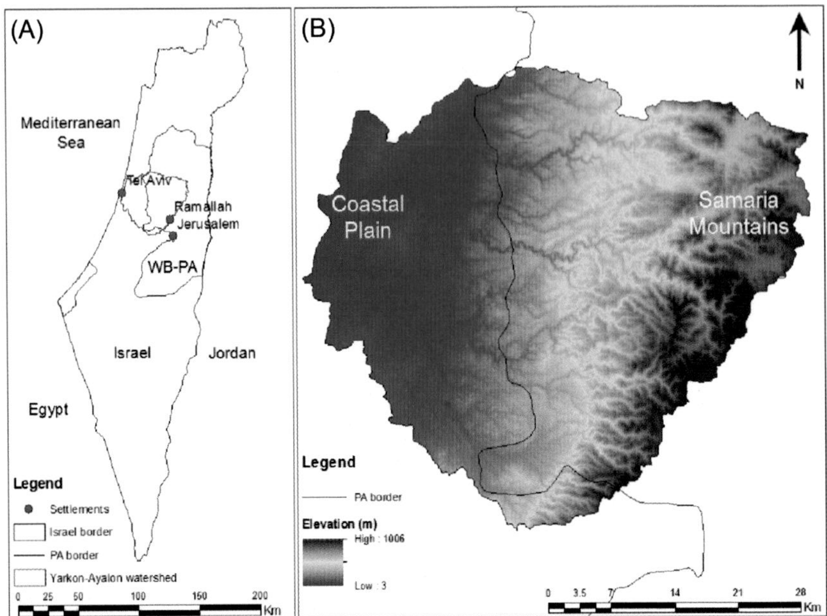

Fig. 8 (A) The location of the Yarqon basin in Israel. (B) The topography at the Yarqon-Ayalon basin.

located in a semiarid climate zone with a mean annual precipitation of about 580 mm/year, contains the most densely populated region of the country, the Tel Aviv metropolis.

During extreme flood events, the water level at the Ayalon River rises more than 5.5 m (17.5 m above sea level) and floods the Ayalon Highway (the busiest road in the country), as well as the railway and neighborhoods in south Tel Aviv. When severe flooding is expected, the IHS creates flood forecasts based on the Hydrologic Modeling System (HEC-HMS) only for the small domain that includes the Ayalon basin. The HEC-HMS is capable of representing the hydraulic conditions of the channels and takes into account the dynamic operation in the basins: dams, reservoirs, and other factors that affect the flow in addition to rainfall-runoff relationships. The model is comprised of a network of more than 200 locations representing hydrological intersections, as seen in Fig. 9. The input for the HEC-HMS model is a simulated grid of precipitation at 3 km resolution, which is extracted either from the operational WRF-Hydro runs or from observed precipitation at rain gauges for short range forecasts of a few hours.

Fig. 10 displays the HEC-HMS simulated versus observed hydrograph from the hydrometric station at the outlet of the Ayalon basin during the

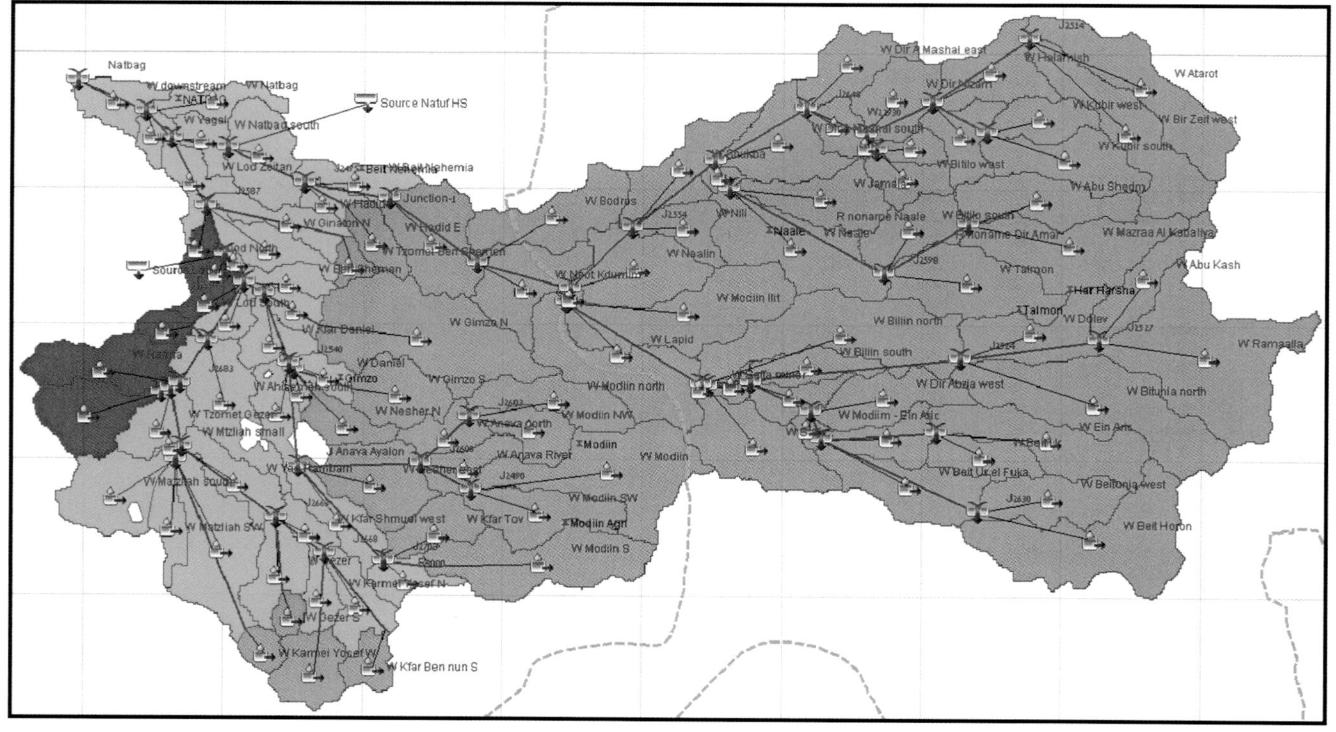

Fig. 9 The location of the forecast points in the Ayalon basin that represent hydrological intersections.

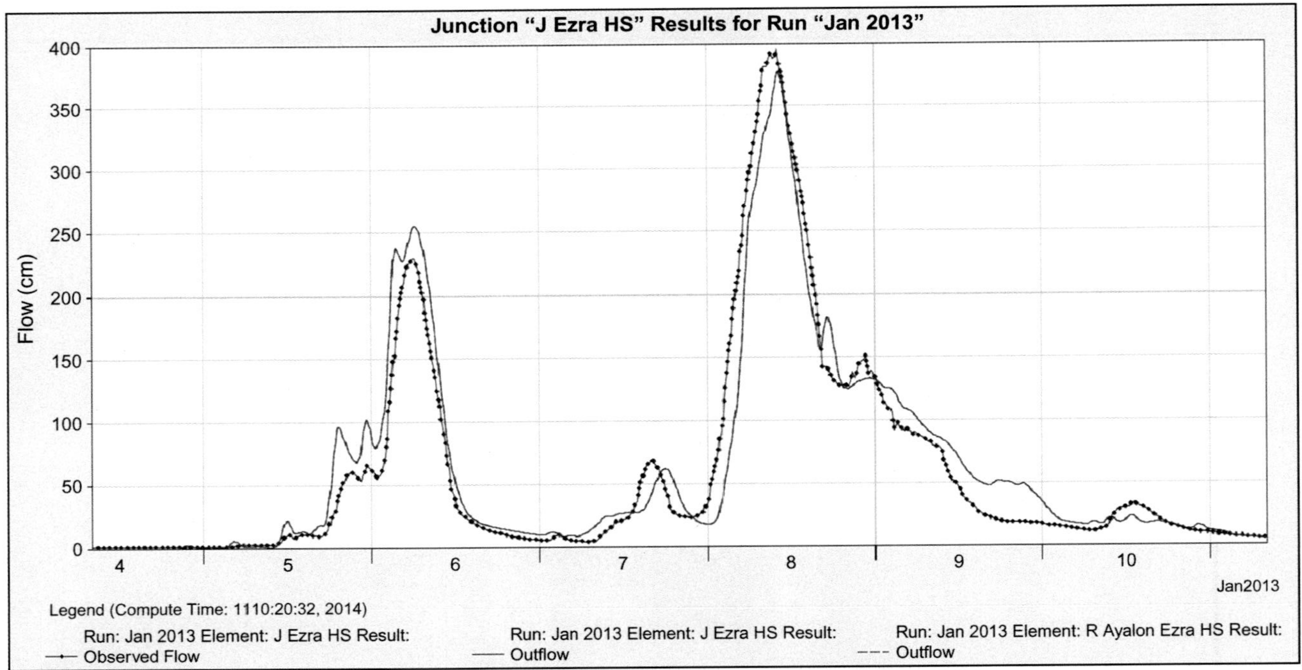

Fig. 10 HEC-HMS simulated (in *blue*) versus observed (in *black*) hydrograph at the hydrometric station located at the bottom of the Ayalon basin.

08/01/2013 flood event. This 2% probability severe flood caused significant damage and the model simulation, which was based on observed precipitation, nicely matched observed flow regarding both peak discharge and total water volume.

4 SUMMARY

Operational flood forecasting in Israel continues to be developed and improved. While we report encouraging results from numerical models like WRF-Hydro, there is still a need for further calibration and reducing bias by using the best precipitation input possible. We recognize a need to isolate the meteorological from the hydrological factors that cause bias in the forecast. We need to find the best forecast lead time that will allow satisfactory advanced warning, yet with sufficient skill and accuracy in the forecast. Close collaboration between meteorologists and hydrologists may improve the input precipitation data and allow the running of the hydrological model with an ensemble of different Numerical Weather Prediction datasets. Previous work showed the advantage of using data assimilation for short-term flood forecasting. The IHS plans to incorporate observations like sea surface temperature, convection from satellites, wind from radar, and meteorological stations in order to improve the simulation of precipitation.

The challenge of hydrometeorological flood forecasting presents a complicated task but promises significant gain when successful. The hydrometeorological community should redouble research and operational efforts to improve the flood alert and warning systems, especially in an era of rapid land use and climate changes.

REFERENCES

Alpert, P., Neeman, B. U., & Shay-El, Y. (1990). Climatological analysis of Mediterranean cyclones using ECMWF data. *Tellus, 42A*, 65–77.

Andersen, T. K., & Marshall Shepherd, J. (2013). Floods in a changing climate. *Geography Compass,* 7, 95–115. http://dx.doi.org/10.1111/gec3.12025.

Bronstert, A., Niehoff, D., & Bürger, G. (2002). Effects of climate and land-use change on storm runoff generation: Present knowledge and modeling capabilities. *Hydrological Processes, 16*, 509–529. http://dx.doi.org/10.1002/hyp.326.

Chang, H., & Franczyk, J. (2008). Climate change, land-use change, and floods: Toward an integrated assessment. *Geography Compass, 2*, 1549–1579. http://dx.doi.org/10.1111/j.1749-8198.2008.00136.x.

Cloke, H. L., & Pappenberger, F. (2009). Ensemble flood forecasting: A review. *Journal of Hydrology, 375*, 613–626.

Dai, A. (2011). Drought under global warming: a review. *WIREs Clim Change, 2*, 45–65. http://dx.doi.org/10.1002/wcc.81.

Delgado, J., Llorens, P., Nord, G., Calder, I. R., & Gallart, F. (2010). Modelling the hydrological response of a Mediterranean medium-sized headwater basin subject to land cover change: The Cardener River basin (NE Spain). *Journal of Hydrology, 383*(1-2), 125–134.

Fiori, E., Comellas, A., Molini, L., Rebora, N., Siccardi, F., Gochis, D. J., et al. (2014). Analysis and hindcast simulations of an extreme rainfall event in the Mediterranean area: The Genoa 2011 case. *Atmospheric Research, 138*, 13–29.

Fredj, E., Silver, M., & Givati, A. (2015). An integrated simulation and distribution system for early flood warning. *International Journal of Computer and Information Technology, 4*(3), 517–526.

Githui, F., Mutua, F., & Bauwens, W. (2010). Estimating the impacts of land-cover change on runoff using the soil and water assessment tool (SWAT): Case study of Nzoia catchment, Kenya. *Hydrological Sciences Journal, 54*(5), 899–908.

Givati, A., Fredj, E., Rummler, T., Gochis, D., Silver, M., Kunstmann, H., et al. (2015). Using the fully coupled WRF-Hydro model for extreme events flood at Mediterranean basins. *Journal of Hydrology,* (in review).

Givati, A., Lynn, B., Liu, Y., & Rimmer, A. (2012). Using the WRF model in an operational streamflow forecast system for the Jordan River. *Journal of Applied Meteorology and Climate, 51*(2), 285–299.

Goldreich, Y. (2003). *The climate of Israel — Observations, research and applications.* New York: Kluwer Academic/Plenum Publishers. 270 pp.

Hoerling, M., Eischeid, J., Perlwitz, J., Xiaowei, Q., Zhang, T., & Pegion, P. (2012). On the increased frequency of Mediterranean drought. *Journal of Climate, 25*, 2146–2161.

Kalantari, Z., Lyon, S. W., Folkeson, L., French, H. K., Stolte, J., Jansson, P. E., et al. (2014). Quantifying the hydrological impact of simulated changes in land use on peak discharge in a small catchment. *Science of the Total Environment, 466*, 741–754.

Kundzewicz, Z. W., Hirabayashi, Y., & Kanae, S. (2010). River floods in the changing climate — Observations and projections. *Water Resources Management, 24*(11), 2633–2646.

Li, J., Feng, P., & Wei, Z. (2013). Incorporating the data of different watersheds to estimate the effects of land use change on flood peak and volume using multi-linear regression. *Mitigation and Adaptation Strategies for Global Change, 18*(8), 1183.

Llasat-Botija, M., Llasat, M. C., & López, L. (2007). Natural hazards and the press in the western Mediterranean region. *Advances in Geosciences, 12*, 81–85.

Milly, P. C. D., Wetherald, R. T., Dunne, K. A., & Delworth, T. L. (2001). Increasing risk of great floods in a changing climate. *Nature, 415*, 514–517.

Ohana-Levi, N., Karnieli, A. M., Egozi, R., Givati, A., & Peeters, A. (2015). Modeling the effects of land-cover change on rainfall-runoff relationships in a semi-arid, Eastern Mediterranean watershed. *Advances in Meteorology,* (in press).

Saaroni, H., Halfon, N., Ziv, B., Alpert, P., & Kutiel, H. (2009). Links between the rainfall regime in Israel and location and intensity of Cyprus lows. *International Journal of Climatology, 30*(7), 1014–1025. http://dx.doi.org/10.1002/joc.1912.

Samuels, R., Smiatek, G., Krichak, S., Kunstmann, H., & Alpert, P. (2011). Extreme value indicators in highly resolved climate change simulations for the Jordan River area. *Journal of Geophysical Research, 116*. http://dx.doi.org/10.1029/2011JD016322.

Schädler, G., & Sasse, R. (2006). Analysis of the connection between precipitation and synoptic-scale processes in the eastern Mediterranean using self-organizing maps. *Meteorologisch Zeitschrift, 15*, 273–278.

Sin Shiha, D., Cheng-Hsin, C., & Tsyh Yeh, G. (2014). Improving our understanding of flood forecasting using earlier hydro-meteorological intelligence. *Journal of Hydrology, 512*, 470–481.

Smiatek, G., Kaspar, S., & Kunstmann, H. (2013). Hydrological climate change impact analysis for the Figeh Spring in Damascus area, Syria. *Journal of Hydrometeorology, 14*(2), 577–593. http://dx.doi.org/10.1175/JHM-D-12-065.1.

Smiatek, G., Kunstmann, H., & Heckl, A. (2011). High resolution climate change simulations for the Jordan River area. *Journal of Geophysical Research — Atmosphere, 116*, D16111. http://dx.doi.org/10.1029/2010JD015313.

Smiatek, G., Kunstmann, H., & Heckl, A. (2014). High-resolution climate change impact analysis on expected future water availability in the upper jordan catchment and the Middle East. *Journal of Hydrometeorology, 15*(4), 1517–1537. http://dx.doi.org/10.1175/JHM-D-13-0153.1.

Törnros, T., & Menzel, L. (2014). Addressing drought conditions under current and future climates in the Jordan River region. *Hydrology and Earth System Sciences, 18*, 305–318.

Trenberth, K. E. (2011). Changes in precipitation with climate change. *Climate Research, 47*(1), 123.

Trigo, I. F., Davies, T. D., & Bigg, G. R. (1999). Objective climatology of cyclones in the Mediterranean region. *Journal of Climate, 12*, 1685–1696.

Wagener, T., Sivapalan, M., Troch, P. A., McGlynn, B. L., Harman, C. J., Gupta, H. V., et al. (2010). The future of hydrology: An evolving science for a changing world. *Water Resources Research, 46*, W05301. http://dx.doi.org/10.1029/2009WR008906.

Younis, J., Anquetin, S., & Thielen, J. (2008). The benefit of high-resolution operational weather forecasts for flash flood warning. *Hydrology and Earth System Sciences, 12*, 1039–1051. http://dx.doi.org/10.5194/hess-12-1039-2008.

Yucela, I., Onena, A., Yilmazb, K. K., & Gochisc, D. J. (2015). Calibration and evaluation of a flood forecasting system: Utility of numerical weather prediction model, data assimilation and satellite-based rainfall. *Journal of Hydrology, 523*, 49–66.

Zwiers, F. W., Alexander, L. V., Hegerl, G. C., Knutson, T. R., Kossin, J. P., Naveau, P., et al. (2013). *Climate extremes: Challenges in estimating and understanding recent changes in the frequency and intensity of extreme climate and weather events.* In *Climate Science for Serving Society* (pp. 339–389). Netherlands: Springer.

CHAPTER 7

Operational Hydrologic Forecast System in Russia

S. Borsch, Y. Simonov
Hydrometeorological Research Centre of Russian Federation, Moscow, Russia

1 INTRODUCTION

Hydrological forecast information about the hydrological regime of water objects is a necessary condition for solving different tasks of water issues in the Russian Federation. A sharp need in hydrological forecasts arises as a basis for stable and effective use of water recourses in different fields of economy (hydro energy, river transport, agriculture, industry, etc.). Another issue is safety from water-related hazards as these are one of the most challenging events for the Russian economy: water-related hazards significantly threaten the social and economic state of the country. More than 50% of total damage from natural disasters in Russia comes from water-related hazards (Borshch, Asarin, Bolgov, & Polunin, 2012). All types of water-related hazards occur in Russia: floods, low water, ice-jams, mudslides, etc. The greatest economic losses (around 30% of all natural disasters) in Russia come from floods: the total area of periodical flooding is around 400,000 km^2, and floods threaten more than 700 cities and over several thousand locations (Atlas, 2010).

The Russian Federal Service for Hydrometeorology and Environmental Monitoring (Roshydromet) is an executive authority in Russia in the field of hydrometeorology and its adjacent fields, environmental monitoring, governmental supervision on environment pollution, and active impact on its geophysical state. Organizations within Roshydromet are producing a wide range of information and analyses, hydrological forecasts of runoff characteristics, including inflow into water river reservoirs, maximal levels of spring floods, minimal critical levels that limit river transport, and ice phenomena of river, lakes, and water river reservoirs in support to operations of water users. This chapter briefly touches on the hydrologic forecast system of the Roshydromet: hydrological phenomena to forecast; forecasting techniques and models used operationally; the hydrometeorological data network; and automated forecast systems.

Flood Forecasting
http://dx.doi.org/10.1016/B978-0-12-801884-2.00007-4

2 HYDROLOGICAL PRACTICES IN RUSSIA

To monitor and forecast the state of environmental objects including rivers, lakes, and river reservoir is the task of the Roshydromet. Its operational divisions issue all known forecasts, with 2000 forecasts being issued annually by the Hydrometcentre of Russia (the leading scientific and operational center of Roshydromet). The main phenomena to forecast are listed in the following text and are connected with the forecast time frame. The classification of forecasts according to the time range differs slightly from World Meteorological Organization's classification (World Meteorological Organization, 2009). The main forecasted phenomena and forecast range are as follows:

- short-range forecasts (up to 5 days' lead time):
  - daily levels (discharges) during spring flood or rain flood events;
  - daily inflow into river reservoirs;
  - dates of ice phenomena, ice thickness, ice destruction, ice-jam formation;
- medium-range forecasts (6–15 days' lead time):
  - maximal level (discharge) of the spring flood;
  - water inflow into reservoir during decade (10 days);
  - spring flood hydrograph;
  - dates of ice phenomena formation on rivers and reservoirs, including ice-jams;
- long-range forecasts (more than 15 days' lead time):
  - maximal level (discharge) of the spring flood;
  - inflow into river reservoir during a month, quarter of a year, spring flood duration;
  - vegetation period's river runoff in agricultural regions;
  - dates of ice phenomena formation on rivers and ice cover formation on reservoirs, ice break-up dates;
  - probability of ice-jam formation and ice-jam water level;

The methodological database of operational hydrological forecasts in Russia is greatly determined by the spatial scale (and degree of local processes influence) of hydrological processes and their significant spatial variability. This leads to mostly empirical forecasting methods being used in hydrological practice. Even if a theoretical model is used as the basis, an empirical determination of the model coefficients and parameters is applied. This results in a diversity of forecasting technique determination methods being used in practice (Bel'chikov, Koren', & Nechaeva, 1992; Borsch, Samsonov, Simonov, & Lvovskaya, 2013; Borsch, Leonteva, & Simonov,

2014; Ginzburg, 1984; Guide to Hydrological Forecasting, 1989). The most widespread operational forecast techniques are based on a physics-statistical method — that is, based on an approximation of water and heat balance equations, equations of motion and continuity, and estimates of different types of water losses, accounting for water supply in river networks. The most important types of forecast based on such technique are forecasts of spring flood phenomena (as discussed later).

Techniques based on solving equations of water movement in river channels found wide applications in hydrological forecast practice mainly for average to large rivers in Russia. The physical bases of such techniques are the theory of flood wave movement, water balance of a river section, and flood water inundation.

Another group of techniques is based on using runoff formation models on the watershed. Such techniques are applied in operational practice mainly for small plain watersheds and for mountain rivers to account for runoff from different elevation zones. Forecasting techniques in large river networks often utilize both of these groups of methods — discharge forecasting in upper catchments outlets using the runoff formation model (snow- and rainfall-runoff) and then routing along the main river reach using the hydrodynamic model.

3 THE HYDROLOGIC FORECASTING SYSTEM OF THE ROSHYDROMET

Among the numerous tasks of the Roshydromet, monitoring and forecasting of inland waters is of vital importance, as it helps to ensure the hydrological safety of the country. A hydrological forecasting system (HFS) operates within the Roshydromet framework to meet the demands of the economy and population in complete, effective, and operational forecasting products of inland waters' hydrological state. The Roshydromet's HFS is a part of the national hydrometeorological monitoring and forecasting system, and thus all data flow in hydrology and meteorology circulate within one organization, making it less complex to manage and operate at a high rate. A peculiarity of Roshydromet operations is that operational hydrological forecasts are issued in different seasons of a year for dramatically different water objects (in terms of natural conditions, basin area, hydrological regime, etc.). Elements of the water regime are forecasted with different lead times that are regulated by water users' demands and hydrological forecasting techniques' limitations. In the case of long-range forecasts, corrections are also issued.

The structure of the Roshydromet HFS is based on an administrative principle (Fig. 1). Local-level centers for hydrometeorology operate within administrative subjects' border, with their main tasks being to support observational network, process data, issue local short-range forecasts and warnings, and communicate with local administrations and other users on a local level.

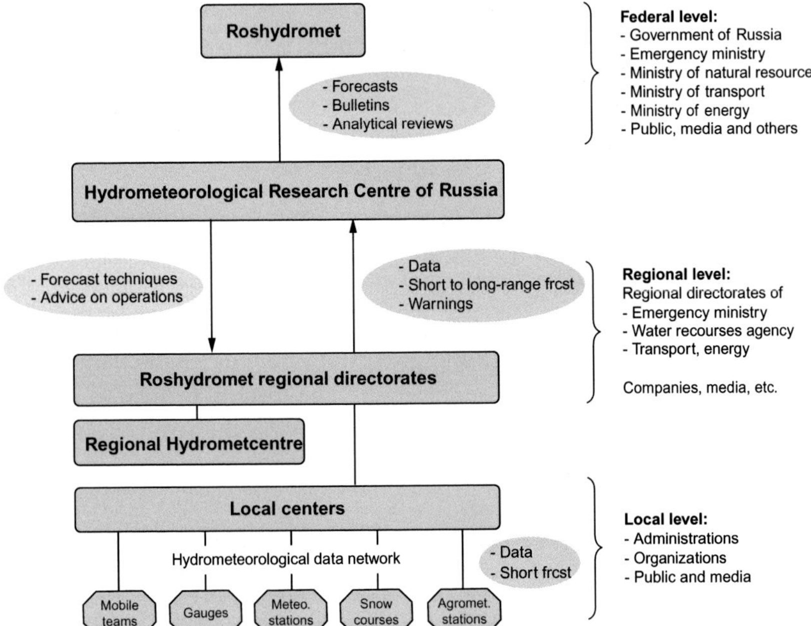

Fig. 1 Functional structure of the HFS of the Roshydromet (chain of operational bodies of the Roshydromet, *center*; and forecast users, *right*).

Local centers for hydrometeorology are affiliated with the regional directorate of the Roshydromet within a large region, which may also include the scientific branch (hydrometeorological center) for forecast techniques development and running in operational regime. Regional directorates of the Roshydromet are responsible for hydrological forecasts and warnings, and other related analytical documents on a regional level, with their further transmission to regional directorates of ministries of emergency situations, transport, agriculture, water resources agency, and others (Fig. 1).

Listed above the centers are the operational bodies of the Roshydromet; there are also a number of scientific institutes of the Roshydromet offering scientific support to HFS. The most important one in terms of HFS operations is the Hydrometeorological Research Center of Russia, which has both scientific (forecasting methods development) and operational (forecast and warnings issued collaboratively with regional centers, analytical products production) functions. The Hydrometeorological Research Center has a direct link to Roshydromet central quarters with top governmental hydrologic advice functions.

Thus the HFS is a part of the national hydrometeorological monitoring and forecasting system, which leads to advantage in terms of operational data, forecasts, and informational exchange as meteorological, hydrological, agriculture soil data, etc. are flowing within a single organization — the Roshydromet.

4 HYDROMETEOROLOGICAL DATA

Informational support is one of the basic components of the HFS, which determines the success of its operation. The hydrological network in Russia is one of the oldest in the world — it accounts for more than 100 years of development and operation. Hydrological gauges measure and transmit comprehensive sets of data about the factual status of rivers, lakes, and reservoirs. The modern period of hydrologic network development is devoted to network increasing (after its recession in 1990s) and the technical modernization of its state, which is at the moment quite uneven in different regions — from full automatization in a selected basin to manual measurements being taken several times a day. The size of the hydrological station network was reduced in the 1990s to 30% compared with its optimal size (in 1985 in the USSR). Nowadays, around 3100 hydrological gauges are operating and their number will be increased.

Measurements of snow and soil parameters (moisture content and freeze depth) are of great importance for long-range spring flood forecasts, as they determine the initial conditions of spring flood phenomena. A network of snow courses (Fig. 2) provides direct measurements of snow height, weight, and density (to obtain snow water equivalent (SWE) values) every 5 days during winter in field and forest areas separately. Information on agrometerological stations soil moisture content (Fig. 2) on different layers and soil freeze depth data are assimilated together as initial conditions for spring flood long-range forecasts.

Satellite data of snow cover extent are often used in a number of river basins in order to adjust long-range spring flood forecasts. Forecast techniques of spring flood forecasts of water inflow into reservoir are based on conceptual models operation and input snow states (areal snow cover) correction in different altitude zones of a mountain basin based on snow

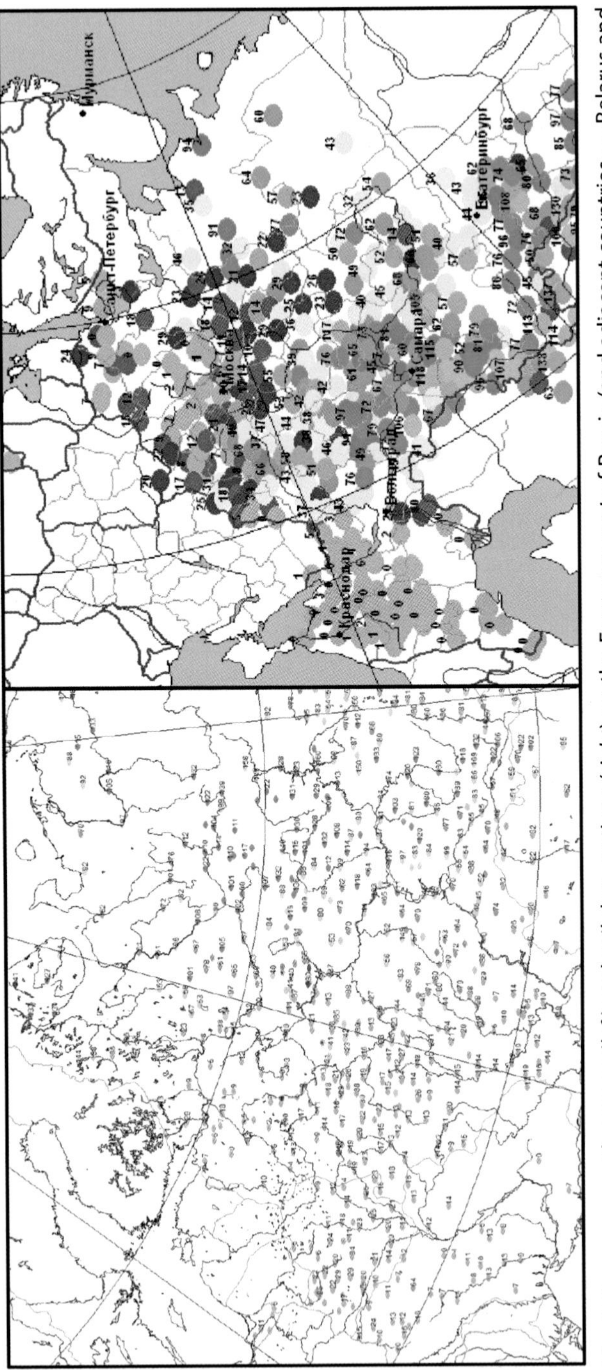

Fig. 2 Snow courses' locations (*left*) and soil data stations (*right*) on the European part of Russia (and adjacent countries — Belarus and Ukraine).

cover extent data from satellites (Burakov, 2009). In some cases, such forecast correction provides a significant increase in forecast quality.

The main numerical weather prediction (NWP) model used in HFS short-range forecasting is the consortium for small-scale modeling-Russia (COSMO-RU) model (Vilfand, Rivin, & Rozinkina, 2010), which is running in the Hydrometeorological Research Center of Russia in different modes for different regions (Fig. 3). For example, in the Sochi region, resolution 1-km meteorological forecast grids of temperature and precipitation from COSMO-RU operational NWP are available four times a day with 36 h lead time. In the far east of Russia, outputs from several models are used — COSMO-RU (13 km), weather research and forecasting (WRF) and advanced research WRF (ARW) (WRF-ARW), and others.

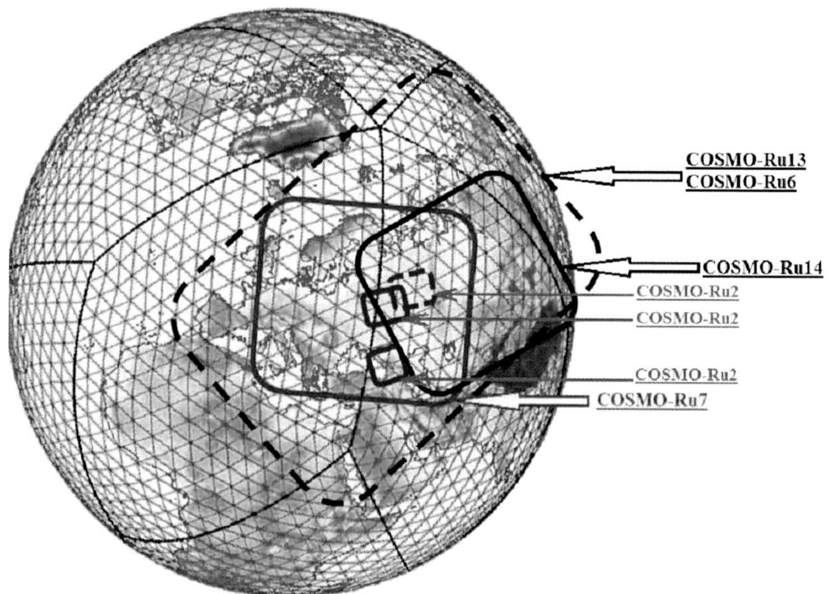

Fig. 3 Modes of COSMO-RU NWP operational model for different regions. Adapted from Vilfand, R. M., Rivin, G. S., & Rozinkina, I. A. (2010). COSMO-RU system of non-hydrostatic mesoscale short-range weather forecast in Hydrometcentre of Russia: First step of modernization and development. *Russian Meteorology and Hydrology, 35*(8), 503–514.

Long-range hydrological forecasts assimilate mainly long-range meteorological ensemble predictions, issued by several scientific bodies of the Roshydromet (Bundel et al., 2011). Nowadays research is being carried out on using weather generator outputs together with hydrologic model in long-range predictions. This method is not applied operationally, but has significant potential in the future.

Methods used in the Roshydromet HFS operational regime differ significantly (as described earlier) — from statistical to watershed process modeling, depending on a basin's peculiarities, hydrological phenomena to forecast, and available data. Brief descriptions of the most important forecast techniques are discussed later.

5 LONG-RANGE SPRING FLOOD PREDICTIONS

During the spring flood period, 50–90% of annual river runoff passes river reaches (spring flood is the main phase of annual flow distribution). Such unevenly distributed flow indicates that forecasts of spring hydrological phenomena are the most important of all forecasts in Russia — spring flood inflow into reservoirs defines further hydropower plants' regime up to next spring, and agriculture and fisheries and many others are also highly dependent on spring hydrological conditions. That is why long-range hydrological forecasts of spring flood are of great interest of water-dependent sectors of the economy. Long-range forecasts of the following hydrological phenomena are of great importance: volume of spring flood, maximal levels, ice phenomena (break-up dates and ice-jams formation). Different models and approaches are used to predict such values depending on a river basin's characteristics, but the most widespread are physically based statistical approaches, which account for spatial distribution of main spring flood factors like maximal SWE, soil freeze depth, and soil moisture. Such methods allow the issuing of long-range forecasts of spring flood phenomena with sufficient quality and lead time from 3 to 6 months.

Influence of the unknown at the forecast issue date factors (future meteorological conditions) on the spring flood formation is dependent along the country with variations of climatic conditions — all of which determine the forecast's uncertainty, its possible time range and the technique used to issue the forecast. There are suitable regions where the main source of spring floods is determined by snow (more than 90%) accumulated during winter (mainly mountainous regions, watersheds with a long melting season), where lead time may exceed 6 months. Plainly watersheds with mainly snow source (but also spring rainfalls) allow forecast issue lead times of up to 2–4 months, but possibilities of long-range forecasts for such rivers are limited by more significant meteorological factors during spring (rainfalls, etc.).

Initial conditions on the watershed before melt starts are one of the greatest factors that determine the forecast quality — assimilating SWE, soil freeze depth, and soil moisture data into models is of great importance. An intensive network of measurement stations for these elements is used (Fig. 2).

There are three main types of methods applied as the basis for long-range hydrological predictions: the direct water balance method (for a few river basins), the physical-statistical method (the most widespread operational method), and the physically based distributed hydrological model together with long-range weather scenarios (under testing for capabilities to run operationally).

Direct water balance methods are based on direct calculation of water balance elements of the basin — these methods have limited applicability due to difficulties in direct accounting for runoff losses, and are applied mainly for watersheds with peculiarities of water balance regime (not significant or not variable factors like infiltration in regions with permafrost). Physical-statistical methods are the basis for long-range operational forecasting techniques. Such methods link spring runoff phenomena with its main factors with assumption of their certain spatial distribution law along the river basin. Such dependencies are constructed for every watershed using long-term measurement data together with statistical and optimization techniques. Some factors are estimated using their indexes — for example, soil detention capacity is assessed using soil freeze depth and moisture level data. Several physically based hydrologic models are used together with differently constructed meteorological ensembles (climatologic, stochastic weather generator) to receive probabilistic forecasts of water inflow into the Volga river reservoir in test mode. Experiments showed close efficiency between "traditional" methods (physical-statistical) and watershed modeling; however, the probabilistic form of forecast presentation has a vital role in the practical utilization in the water sectors of the economy, in particular for long-range planning. The forecast form is presented as a confidence interval, which includes a forecasted value with certain probability.

Long-range forecasts of spring floods are issued in very beginning of spring (Mar.–Apr.) depending on watershed climate conditions. Correction of the forecasts may be issued (if there is a significant difference in the estimation of some factors) in the one-third of the forecast lead time using snow cover estimations with satellites, current rainfall amount, and its updated forecast.

6 EARLY WARNING FLOOD FORECASTING SYSTEMS

According to the Atlas of natural and manmade hazards in Russia (Atlas, 2010), almost 30% of economic losses come from floods. There are typically several major reasons for flooding in Russia, which hold potential damage to population: flooding from spring flood wave originated from snowpack melting and floods from heavy rainfalls in the summer-autumn period of a year. The first phenomenon from this list happens on the majority of Russian river basins; it has a relatively slow-developing nature and is covered by long-range forecasts of maximal levels. Economy and population are often prepared for such floods and they pass without human victims. Another situation happens with heavy rain floods especially in mountainous rivers — such floods happen almost every summer and cause great damage and take human lives. The rapid response of rainfall induced floods allows us to characterizes these as flash floods. The river basins affected by this type of flooding are located mostly in mountainous areas where convective heavy storms generate during the warm period of a year—for example, in the Caucasian Mountains region (southwest of Russia) — the Kuban river and its tributaries and other rivers flowing down from the mountain range to the Black Sea. This area suffers from heavy floods almost every year (the 2013 flood took more than 150 lives, causing great economic damage to the city of Krimsk). Such basins were chosen as number-one priority basins for developing and implementing early warning flood forecasting systems (EWFFS) that incorporate all hydrometeorological data flow—from measurement to forecast issue and forecasting products generating and their further transmission to forecast users.

The structure of the EWFFS incorporates three main modules: hydrometeorological data module (observations and meteorological forecast processing), forecast module (flood forecast calculation, its correction and issue), and final products module (visualization of flood forecast in geographic information systems (GIS)) (Fig. 4). The data processing module includes operational hydrometeorological database and numerous routines for receiving, decoding, and checking incoming observations data and meteorological forecast data from four NWP models (including the COSMO as a basic one). The database also contains landscape characteristics derived with the help of GIS for further parameterizations of hydrologic models. Such a database is operating 24/7 in the Hydrometcentre of Russia. The database is large basin scale base; it is generated for primary rivers, and thus it incorporates data for all tributaries of the river. The flood forecasting block is the calculation heart of the system—it contains forecasting models for different parts of the basin together with forecast updating techniques. The final products generation module is based on GIS and web technologies, and provides users with forecast data in an appropriate format and interface.

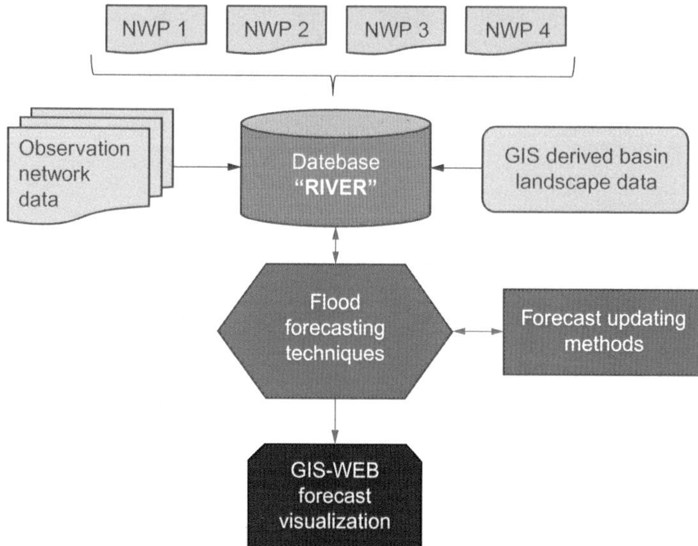

Fig. 4 Structure of the Hydrometcentre EWFFS.

Hydrological short-range forecast techniques are based generally on implementing conceptual models and physics-statistical models driven by observed and forecasted meteorological data. Conceptual models are implemented for mountainous watersheds to account for water balance elements in elevation and vegetation (field and forest) zones. The main blocks of a model are snow dynamic and soil moisture accounting block. Physics-statistical models use quite simple equations that account for water volume in upstream river network and meteorological conditions. They are applied in the lower part of a basin with relatively plain relief and hydrologic conditions. Forecasts are issued with 72 h lead time. Final products are generated using GIS technologies; they are presented in several ways: as a map with forecasted hydrologic situation (Fig. 5A), as an observed and forecasted hydrograph (Fig. 5B), and as data in tabular form. The latest updates of the forecast presentation technologies include web-based products, available online for Roshydromet forecast users.

There are several large river basins with EWFFS implemented — two of them are the Kuban River basin and the Amur River basin. The presented EWFFS is operating in the Hydrometeorological Centre of Russia in an operational mode for several river basins. There are several vast watersheds on the way (the Volga River and the Yenisei River) for this system's implementation.

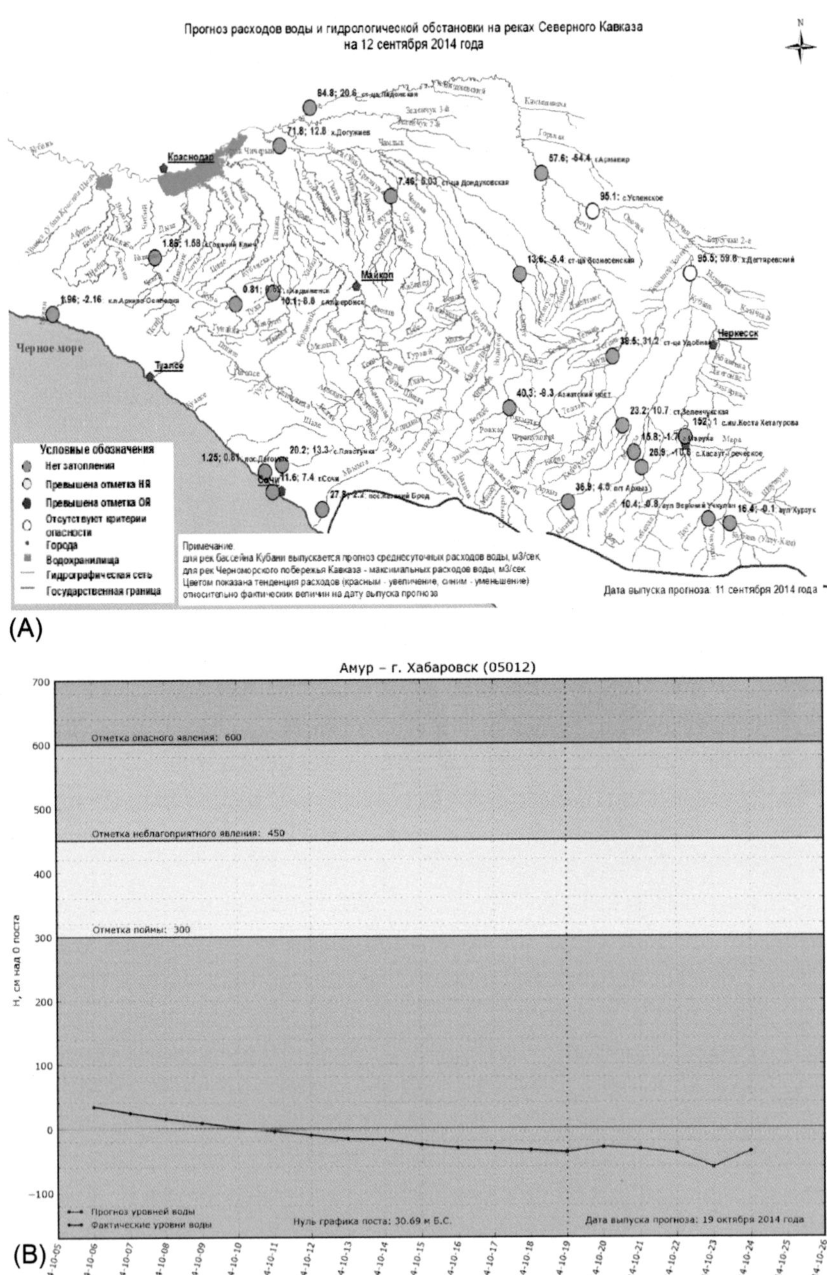

Fig. 5 (A) Short-range forecast (24 h ahead) of flood hazard level (*green*, no flood) on the mountainous river in the southwest part of Russia (the Caucasian Mountains region). (B) Combined hydrograph (*blue*, observed levels; *red*, forecasted levels) for the Amur River gauge (the city of Khabarovsk).

7 CONCLUSIONS

The hydrologic forecast system in Russia operates within the Roshydromet framework and includes operational and scientific branches. Issued hydrological forecasts cover all known hydrologic phenomena and all time-ranges to meet demand from forecast users in different regions of Russia, with its peculiarities in hydrologic regime, climate conditions, and data availability. The modern tendency of the hydrologic forecast system incorporates the following processes: automatization of developed forecasting techniques (from short- to long-range), incorporating them into a developed automatic forecast system structure, and developing web-based interfaces for the main users of forecasts. Further development of Roshydromet hydrologic forecast techniques is highly correlated with network modernization process (including automatization of gauging stations and weather radar network installation), which is now taking place in Russia.

REFERENCES

Atlas of natural and man-made hazards and risks from emergency situations in Russian Federation. (2010). Moscow. 696 pp.

Bel'chikov, V. A., Koren', V. I., & Nechaeva, N. S. (1992). *Avtomatizirovannye kratkosrochnye prognozy raskhodov i urovney vody dlia rechnoy sistemy Severnoy Dviny [Automated short term forecasts of discharges and water levels for the Northern Dvina river system].*

Borsch, S.V., Leonteva, E. A., & Simonov, Yu. A. (2014). GIS-based representation of operational hydrological forecasts. *Proceedings of Hydrometcentre of Russia, 351*, 141–153.

Borsch, S. V., Samsonov, T. E., Simonov, Y. A., & Lvovskaya, E. A. (2013). Visualization of hydrological phenomena in large river basins using GIS technologies. *Proceedings of Hydrometcentre of Russia, 349*, 47–62.

Borshch, S. V., Asarin, A. E., Bolgov, M. V., & Polunin, A. Ya. (2012). *Floods.* In *Assessment methods of climate change consequences for physical and biological systems.* (pp. 87–125). Moscow. (in Russian).

Bundel, A.Yu., Kryzhov, V. N., Min, Young-Mi, Khan, V. M., Vilfand, R. M., & Tishchenko, V. A. (2011). Assessment of probability multimodel seasonal forecast based on the APCC model data. *Russian Meteorology and Hydrology, 36*(3), 145–154.

Burakov, D. A. (2009). Long-term forecasting of snowmelt runoff. In Lev S. Kuchment & Vijay P. Singh (Eds.), *Encyclopedia of Life Support Systems (EOLSS), Jun 18, 2009: Vol. 1* (364 pages).

Ginzburg, B. M. (1984). *Gidrologicheskie prognozy [Hydrological forecasts]* (439 p).

Guide to hydrological forecasting (Vols. 1–3). (1989). Leningrad: Hydrometeoizdat.

Vilfand, R. M., Rivin, G. S., & Rozinkina, I. A. (2010). COSMO-RU system of non-hydrostatic mesoscale short-range weather forecast in Hydrometcentre of Russia: First step of modernization and development. *Russian Meteorology and Hydrology, 35*(8), 503–514.

World Meteorological Organization. (2009). Management of water resources and application of hydrological practices. WMO-No.168, *Guide to hydrological practices:* Vol. 2. Geneva: World Meteorological Organization.

CHAPTER 8

Increasing Early Warning Lead Time Through Improved Transboundary Flood Forecasting in the Gash River Basin, Horn of Africa

G. Amarnath*, N. Alahacoon*, Y. Gismalla†, Y. Mohammed†, B.R. Sharma‡, V. Smakhtin*

*International Water Management Institute (IWMI), Colombo, Sri Lanka
†Hydraulic Research Station (HRS), Wad Madani, Sudan
‡International Water Management Institute (IWMI), New Delhi, India

1 INTRODUCTION

Floods are major natural disasters that affect many regions around the world, causing loss of life, and damaging economies and harming human health. More than one-third of the world's land area is prone to flooding which affects about 82% of the global population (Dilley et al., 2005). According to EM-DAT (2015), out of all the natural disasters, nearly half of the deaths and one-third of all economic losses are due to flooding. Such problems vary, according to the sources of flooding and the ambient developed landscapes. In the last three decades, several devastating floods were recorded in 1975, 1983, 1988, 1993, 1998, 2003, and 2007. The most damaging of these was the floods of 2003, where almost half of Kassala City was washed out and affected approximately 60–70% of the city's population (OCHA, 2003). The total loss is estimated to be USD 150 million (Artan et al., 2007). Another big flood event in 2007 (Fig. 1) took the lives of 20 people, damaged 16,300 houses, and affected more than 20,000 people (IFRC, 2007).

In recent years, satellite technology has become extremely important in providing cost-effective, reliable, and crucial mechanisms for preparedness, damage control, and relief management of flood disasters. A variety of mitigation measures can be identified and implemented to reduce or minimize the impact of flooding. Such mitigation measures include flood forecasting and warning, adopting proper land-use planning, and flood-prone area zoning and management. Improving regional flood-risk forecasting requires extensive flood-related information. Traditionally, gathering and analyzing

Flood Forecasting
http://dx.doi.org/10.1016/B978-0-12-801884-2.00008-6

Fig. 1 Aerial view of the 2007 Gash River flooding in Kassala, Eastern Sudan (photo: Mohamed Nureldin Abdallah/Reuters).

hydrologic data related to floodplains and flood-prone areas have been a time-consuming effort which requires extensive field observations and calculations. With the development of remote sensing (RS) and computer analysis techniques, traditional techniques can now be supplemented with these new methods of acquiring quantitative and qualitative flood hazard information.

2 BUILDING A FLOOD FORECAST AND EARLY WARNING SYSTEM

The simple and commonly used practice is to record real-time rainfall data using rain gauges, and then apply the measured rainfall to forecast discharge at the point of interest using a rainfall-runoff model (Amarnath, Shrestha, & Islam, 2016; Rao, Bhanumurthy, & Roy, 2009, Rao, Rao, Dadhwal, Behera, & Sharma, 2011). However, for small catchments with little hydrological response time, flood forecasting with real-time rainfall data is not sufficient for disseminating flood warnings at the required lead time, so that necessary precautions can be taken. In such instances, the use of the forecasted rainfall, instead of the measured rainfall, in the rainfall-runoff model provides additional lead time for early warning (Maharjan, 2013).

Runoff prediction and river flow forecasting are important hydrological studies for water resources development, planning, and management, among other uses. Rainfall-runoff models, which aim at simulating the catchment response and flow hydrograph, are extensively used to support flood forecasting (Nayak, Venkatesh, Krishna, & Jain, 2013) and water resources planning (Laurent, Jobard, & Toma, 1998). The need for predicting runoff from specific catchments remains large, especially in densely populated and flood-prone areas (Rao et al., 2011) such as the Gash River Basin in Eastern Africa (Bashar, Abdo, & Gadain, 2005). Its hilly and mountainous catchment with high rainfall intensity and relatively sparse vegetation cover makes Gash more prone to flooding (Wheater, Sorooshian, & Sharma, 2007). Forecasting of transboundary flooding in the lower region of Gash River Basin, in particular Eastern Sudan, remains notoriously difficult when using conventional modeling approaches that rely on extensive and real-time in situ data. Also, in transboundary river basins, information exchange and sharing of data between riparian countries is regarded as a first and essential step towards fostering cooperation and trust (Gerlak, Lautze, & Giordano, 2013; Khan, Hong, et al., 2011). In the Gash Basin, there are limited known events of regional data and information exchanges, although bilateral processes already exist to a certain extent. Cooperation on sharing water-level information in the Gash River is limited between Sudan and Eretria. More importantly, since 1986, the Gash River Training Unit (GRTU) of the Hydraulic Research Centre (HRC) under the Ministry of Water Resources and Electricity (MoWRE) is actively engaged in flood forecasting and a warning system on the Gash River, by maintaining five hydrometric sites in Sudanese territory. However, the key challenge in the Gash Basin is the transboundary nature of the river and the absence of a data sharing mechanism. Most of the river catchment lies in Eretria, but there is no information exchange and sharing of data, and there is often hesitance and caution to share any kind of hydrological information.

RS can potentially close some of the gaps in data availability (Stisen, Jensen, Sandholt, & Grimes, 2008). Satellite-based rainfall monitoring is widely used because of its increasing global coverage. It has great importance for operational purposes in data-scare regions such as Africa. It also has potential benefits, such as providing input to hydrological models, due to its real-time availability, low cost, and good spatial coverage (Ashouri et al., 2015). Radar has also played an increasingly important role in technologically advanced countries, particularly with regard to provision of data in real time (Manjusree, Prasanna Kumar, Bhatt, Rao, & Bhanumurthy, 2012). However, selecting an appropriate precipitation dataset for hydrological modeling is quite challenging. Users often face a dilemma in selecting an appropriate model from a large pool of models or even a suitable method to represent a particular process within one hydrologic model.

These types of studies are relatively limited in developing nations (Verma, Jha, & Mahana, 2010).

The availability of technologically advanced geographic information systems (GIS), remotely sensed data, decision support systems, innovative analytical tools, and web-based communication materials in the public domain can be the catalyst for bringing about change in the current level of transboundary information exchange in the Gash River Basin (Fig. 2) (Gerlak et al., 2013; Khan, Dawe, Ali, & Puestow, 2011; Nishat & Rahman, 2009; Stimson Center, 2010). The figure illustrates how geospatial datasets are combined with hydrological models to provide 7–10 days of lead time in regional flood early warning systems. For example, a wide range of satellite-based rainfall estimates from the Climate Prediction Center (CPC) morphing technique (CMORPH), Precipitation Estimation from Remotely Sensed Imagery Using Artificial Neural Networks (PERSIANN) (Ashouri et al., 2015), and Tropical Rainfall Measuring Mission (TRMM)-based 3B42RT and the recent product from the National Aeronautics and Space Administration (NASA) and Japan Aerospace Exploration Agency (JAXA) on Global Precipitation Measurement (GPM) have the best potential in advanced modeling, forecasting, communications, and warning systems. A

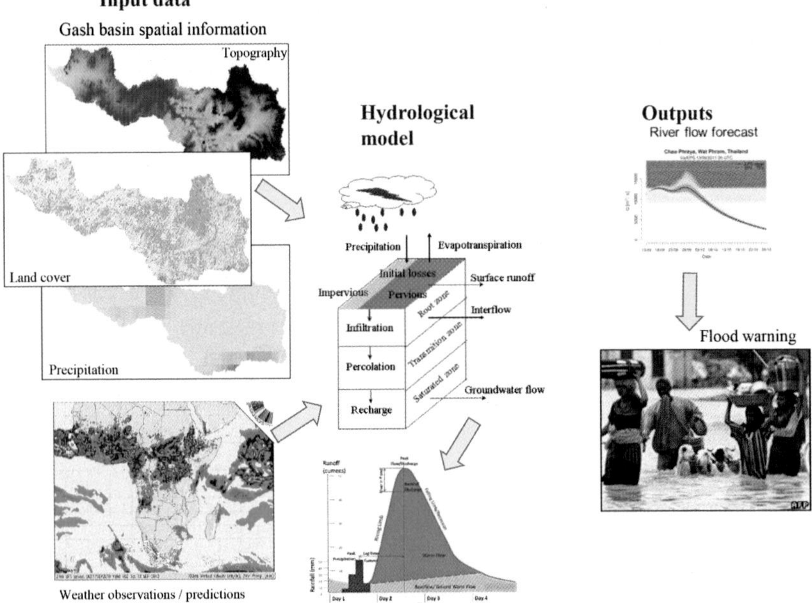

Fig. 2 Example of an illustrative flood forecasting system by combining weather forecast inputs, geospatial data, and hydrological models for streamflow predictions.

shared information base for basin planning will help countries seize opportunities in the basin and manage its risks.

Jaun and Ahrens (2009) studied the limitation of a probabilistic forecasting system, which is based on a hydrometeorological ensemble prediction approach. Thielen, Bartholmes, Ramos, and de Roo (2009) presented the development of the European Flood Awareness System (EFAS), which aims at increasing preparedness for floods in transnational European river basins by providing local water authorities with medium-range, probabilistic flood forecasting information 3–10 days in advance. Rao et al. (2009) have worked on developing a medium-range flood forecasting model for the Brahmaputra River Basin. They demonstrated the scope of using satellite-based rainfall products in flood forecasting and its limitations. Bogner and Kalas (2008) tested for adjusting the ensemble traces using a transformation derived from simulated and observed flows in the Upper Danube River. Their work involved the combination of state-space models and wavelet transformations, in order to update errors between the simulated (forecasted) and observed discharges.

De Roo et al. (2003) developed a prototype flood forecasting system for European test basins. They used real-time rainfall data and rainfall forecast grids of coarse resolution for flood forecasting at basin scale, by integrating hydrological models with weather forecasts. A flood forecasting system has been developed by the Danish Hydraulic Institute in collaboration with the Bangladesh Water Development Board for the part of the Brahmaputra River that falls in Bangladesh territory. Tele-meteorological rainfall data were used in this study. Runoff was computed for the catchment area within Bangladesh and added to the inflow discharges measured at upstream sections of the river within the Bangladesh area (Jorgensen & Host-Madsen, 1997). Considering the current state of research in the field and the need for flood forecasting, this chapter focuses on the development of a flood forecasting model for the Gash River Basin in HEC-HMS and HEC-GeoHMS modeling environment (U.S. Army Corps of Engineers, 2000, 2001, 2003). The model is validated with 3-h, real-time hydrometeorological data from 2013.

3 GEOGRAPHICAL SETTING OF THE GASH RIVER BASIN

The Gash River Basin is a transboundary basin shared between Ethiopia, Eritrea, and Sudan (Fig. 3). The river originates from the Eritrean Highlands and Ethiopian Plateau in an area characterized by steep slopes. The upper course of the river in Eritrea is known as the Mareb River (Artan et al., 2007). Historically, the basin used to be part of the Nile River system.

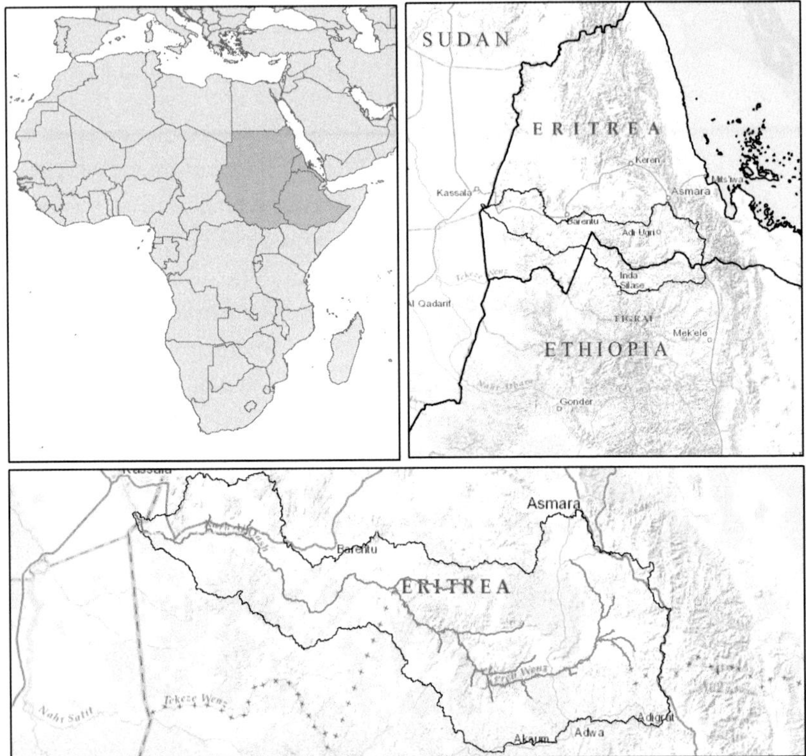

Fig. 3 Location map of the Gash River Basin in Eastern Africa.

However, tectonic activities, sedimentation, and other morphological developments have dramatically changed the course of the river (Elsheikh, Khalid, & Shaza, 2011). The Gash Spate Irrigation Scheme (GSIS) area is characterized by a semiarid climate with two notable seasons (winter and summer). The maximum temperature may exceed 45°C in the summer, and drops to an average of 25°C in the winter (Elsheikh et al., 2011). The Gash River, water source of GSIS, travels about 121 km from the border with Eritrea down to the Gash Die (the end of the delta). The total catchment area of the river is 21,000 km^2 (Anderson, 2011). It is a seasonal river, which flows between late Jun. and Oct., and high flows occur between Jul. and Sep. The maximum annual discharge recorded in 1983 was 1430 million cubic meters (Mm^3), and an annual minimum flow of 140 Mm^3 was recorded in 1921 (Anderson, 2011). The average annual discharge is 1056 Mm^3 at El-Gera upstream gauge station and 587 Mm^3 at Salam-Alikum downstream gauge station (Elsheikh et al., 2011). The topography of the basin varies from 531 m above mean sea level (amsl) to 3259 m amsl.

4 THE MODELING APPROACH

In a distributed modeling approach, the spatial variations of topographic and hydrometeorological parameters are considered, and the runoff is computed in the spatial domain. In this approach, there are sources, sinks, and boundary conditions besides other geospatial data; a GIS platform with the integration of satellite data offers an excellent solution. HEC-HMS and HEC-Geo HMS are used as a modeling environment for developing the flood forecasting model for the Gash River Basin. The base model setup consists of two main modules — the topographic model and the hydrometeorological model.

4.1 Topographic Model Setup

The physical representation of watersheds or basins and rivers is configured in the topographic model. Hydrological elements are connected in a dendritic network to simulate runoff processes (Fig. 4). Available elements in the topographic model are subbasin, reach, junction, reservoir, diversion, source, and sink. Computation in the model proceeds from upstream elements in a downstream direction. Various thematic layers — such as land use, soil texture, subbasins, etc. — which are required for the topographic model are prepared in an ArcGIS environment.

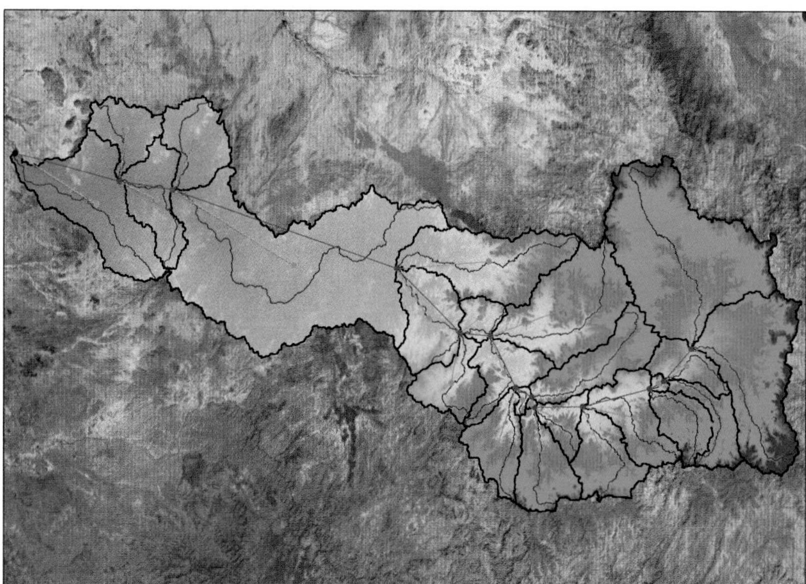

Fig. 4 Delineation of catchments and river nodes for the Gash River Basin using the HEC-GeoHMS model.

4.2 Spatial and Nonspatial Databases

The study area is covered by two satellite scenes of the Landsat images, which are radiometric and geometrically corrected, provided by the United States Geological Survey (USGS) Earth Resources Observation and Science (EROS) Center. The images were used to map bare areas that were applied in the HEC-HMS model as a percentage of impervious areas. Land use/land cover (LULC) is a very important parameter in hydrological modeling. Evapotranspiration, interception, and catchment characteristics are mainly dependent on this input. Considering its hydrological characteristics, land use is further reclassified into a hydrological LULC map. Land-use type with hydrologically poor conditions indicates more runoff potential and vice versa. Parameters derived from the land-use map are used for runoff estimation and hydrodynamic flow routing.

In the hydrological cycle, infiltration is a major component. Infiltration depends on soil texture, which, in conjunction with land use, provides various basin parameters for modeling. A soil texture map of the study area at 1:250,000 scale was obtained from the Food and Agriculture Organization of the United Nations (FAO). The Digital Elevation Model (DEM) is the main input for topographic parameter extraction. Runoff within a watershed and in a channel depends on the slope of the watershed and channel, respectively. The Shuttle Radar Topographic Mission (SRTM) DEM of 90 m resolution is used to extract various topographic and hydraulic parameters of the basin, such as subbasin and channel slopes, Manning's coefficients, lag time, time of concentration, etc. Subbasins and the drainage network are also delineated using the DEM through an automated process.

4.3 Terrain Processing

Terrain pre-processing is a series of steps carried out to derive various topographic and hydraulic parameters. These steps consist of computing the flow direction, flow accumulation, stream definition, watershed delineation, watershed polygon processing, stream processing, and watershed aggregation. Computations were carried out using a step-by-step procedure or in a batch manner. The basin model file contains the hydrologic data structure, which includes the hydrologic elements, their connectivity, and related parameters. All topographic and hydraulic parameters are computed in the terrain processing stage using land use and soil texture information, and DEM, and are exported to the topographic model.

Main streams were digitized using the satellite data and were fused on the SRTM DEM using the "stream burning" technique. This technique facilitates delineating subbasins and streams in a flat topography more accurately. The burned DEM was used to calculate flow direction and flow accumulation. Upstream drainage area at a given cell can be calculated by multiplying the flow accumulation value by the cell area. Stream definition classifies all cells with flow accumulation greater than the user-defined threshold as cells belonging to the stream network. Streams in the basin have been delineated using this automated technique. Stream segmentation divides the stream into segments. Stream segments are links that connect two successive junctions, a junction and an outlet, or a junction and the drainage divide. The subbasin delineation process delineates a basin or watershed for every stream segment. Keeping in mind the spatial extent of the basin, computational time, and the desired accuracy, the basin is divided into 25 subbasins.

4.4 Hydrological Parameter Extraction

Topographic characteristics of streams and watersheds have been computed using a model pre-processor. These characteristics are useful in estimating hydrological parameters of basins and for comparing the basins. Physical characteristics of all streams and basins are stored in the attribute tables that can be exported to the model for further modeling processes. The physical characteristics that are extracted for the streams and subbasins are river length, river slope, basin centroid, longest flow path, centroidal flow path, etc. When the physical characteristics of streams and subbasins are extracted, hydrological parameters can be derived easily. Infiltration rate is estimated as grid-based quantities that are based on land use and soil types. Other hydrological parameters, such as time of concentration, lag time, and Muskingum routing parameters, are computed from the terrain characteristics. All the abovementioned hydrological parameters are extracted for all the subbasins of the study area in an ArcGIS environment and fed into the model.

4.5 Hydrometeorological Model Setup

The response of a watershed is driven by the precipitation it receives and evapotranspiration. The precipitation may be the observed rainfall from a historical event or a frequency-based hypothetical rainfall event. Historical data may be useful for calibration and validation of model parameters. Rainfall data from nearly five stations covering the upstream catchment were obtained for the period 2007–12 from the National Meteorology Agency, Ethiopia. These data were used in the calibration and validation of the complete model. The main data sources that were incorporated in the rainfall-runoff model are the daily rainfall estimates from NASA TRMM 3B42 data. Bias correction of the Satellite Rainfall Estimates (SRE) was carried out using existing rainfall data. However, it was not possible to obtain the measured data for the catchment covering Eretria and part of Sudan from the meteorological station. The uncertainties were not estimated for this sub-catchment. Fig. 5 clearly shows the enhancement of the SRE TRMM product for the 2009 Humera station. For real-time validation, 3-hourly NASA TRMM meteorological data from 2013 were used for computation of the operational flood hydrograph.

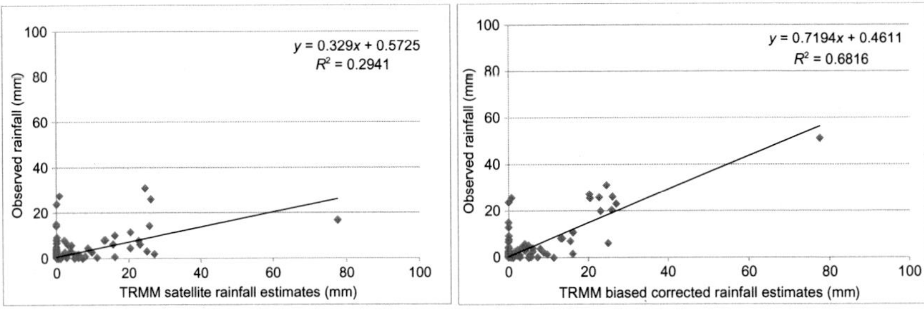

Fig. 5 Comparison of observed and TRMM SRE (left) and corrected TRMM SRE (right) using the distribution method for the 2009 Humera station, Ethiopia.

4.6 Model Setup and Simulation Run

In this study, the distributed modeling approach is adopted for computation of the flood hydrograph. The methodology involved in computing the flood hydrograph of the basin at the outlet can be broadly divided into five stages — computing runoff volume (excess rainfall), modeling direct runoff, flood routing, calibration of the model, and model validation. The surface runoff (excess rainfall) in each pixel will drain towards the outlet of the subbasin/basin. The process is referred to as the "transformation" of excess precipitation into point runoff, ie, computation of the flood hydrograph at each sub-watershed outlet, and it is measured with reference to time. The Soil Conservation Service (SCS) Curve for loss estimation and SCS-Unit Hydrograph techniques are used for estimation of direct runoff. These models are selected based on the size, shape, and slope of the subbasin. In this study, the straight-line method, a popular method used for the computation of baseflow in each sub-watershed, is adopted.

Once direct runoff in each watershed is calculated, it has to be routed to the main outlet. A flood wave is attenuated by friction and channel storage as it passes through a reach. The process of computing the travel time and attenuation of water flowing in the reach is often called routing, which is implemented using the TR55 methodology available in HEC-GeoHMS. Travel time and attenuation characteristics vary widely between different streams. The travel time is dependent on characteristics such as length, slope, friction, and flow depth. Attenuation is also dependent on friction, in addition to other characteristics such as channel storage. Direct runoff in various subbasins has been routed to the main outlet at Kassala Bridge using Muskingum and modified SCS lag methods.

After completing the model setup, trial runs were conducted to obtain results. Each run combines a topographic model, hydrometeorological model, and control specification components with run options. All errors, such as missing sink and source nodes, channel connectivity, etc., in the model setup were rectified during trial runs. Runs can be re-executed at any time to update results, when data in the components are changed.

4.7 Model Calibration and Validation

Model calibration is the process of adjusting model parameter values until model results match historical data. The process can be completed using engineering judgement by repeatedly adjusting parameters, and computing and inspecting the goodness-of-fit between the computed and observed hydrographs. Significant efficiency can be realized with an automated procedure (U.S. Army Corps of Engineers, 2001). The quantitative measure of the goodness-of-fit is the objective function. An objective function measures the degree of variation between computed and observed hydrographs. The key to automated calibration is a search method for adjusting parameters to minimize the objective function value and to find optimal parameter values.

A hydrograph is computed at the target element (outlet) by computing all the upstream elements and by minimizing the error (minimum deviation with the observed hydrograph) using the optimization module. Parameter values are adjusted by the search method; the hydrograph and objective function for the target element are recomputed. The process is repeated until the value of the objective function reaches the minimum to the best possible extent. During the simulation run, the model computes direct runoff of each watershed, and the inflow and outflow hydrograph of each channel segment. The model computes the flood hydrograph at the outlet after routing flows from all subbasins to the basin outlet. The computed hydrograph at the outlet is compared with the observed hydrograph at Kassala Bridge stations.

After computing the exact value of the unknown variable during the calibration process, the calibrated model parameters are tested for another set of field observations to estimate accuracy of the model. In this process, if the calibrated parameters do not fit the data of validation, the required parameters have to be calibrated again. Thorough investigation is needed to identify the parameters to be calibrated again. In this study, hydrometeorological data of 2007 and 2012 were used for model validation, because floods occurred in those years.

In order to measure model performance of simulated and observed flow, the objective function "*Y*" can be effectively used (Akhtar, Ahmad, & Booij, 2009). The *Y* value is defined by combining the Nash-Sutcliffe coefficient (*NS*) (Eq. (1)) and Relative Volume Error (*RVE*) through an equation (Eqs. (2), (3)). The discharge data of Kassala gauge station in 2007, 2011, and 2012 were used to calculate *NS*, *RVE*, and *Y* values.

$$NS = 1 - \frac{\sum_{i=1}^{i=N}\left[Vs(i) - Vo(i)\right]^2}{\sum_{i=1}^{i=N}\left[Vo(i) - \overline{Vo(i)}\right]^2} \quad (1)$$

where i = time steps, N = total number of time steps, $Vs(i)$ = simulated flow, $Vo(i)$ = observed flow, and $\overline{Vo(i)}$ = average of the observed flow.

Interpretation of *NS* (Eq. (1)) together with *RVE* (Eq. (2)) provides frank results. When *NS* is equal to 1 and *RVE* is equal to 0 (zero), model performance is very good. With respect to *RVE*, ±5% value represents very good performance, and ±10% represents satisfactory performance. Once *NS* becomes 0.8 and *RVE* is less than 15%, simulated flow matched observed flow at an acceptable level.

$$RVE = \frac{\sum_{i=1}^{i=N}\left[Vo(i) - Vs(i)\right]}{\sum_{i=1}^{i=N} Vo(i)} \quad (2)$$

where i = time steps, N = total number of time steps, $Vs(i)$ = simulated flow, and $Vo(i)$ = observed flow.

$$Y = \frac{NS}{1 + |RVE|} \quad (3)$$

5 RESULTS AND DISCUSSION

Prior to using NASA TRMM rainfall estimates in computing the hydrograph, these data were validated with rainfall stations obtained from the National Meteorology Agency, Ethiopia. Rain gauge observations from five stations were obtained for bias correction, and were adjusted to SRE and subsequently integrated into the hydrological model for development of the Gash River Basin flood forecasting operations. The computed hydrograph during the validation process and observed hydrograph at Kassala Bridge stations is shown in Fig. 6. This figure indicates that the computed hydrographs match well with the observed hydrographs. Real-time flood forecasting was provided by continuous simulation of flood hydrographs using the 3-hourly NASA TRMM SRE data of the 2013 flood season (Fig. 7).

Table 1 presents *NSE*, *RVE* and *Y* values for three different years during the calibration and validation periods of the Gash catchment. The model shows that, during the calibration period, *RVE* is very small and indicates that the average simulated and observed discharge are close to each other. General testing of conceptual models (Rango, 1992) has shown that an *NS* value higher than 0.8 is above average for runoff modeling. Therefore, *NSE* values obtained during calibration are satisfactory for the Gash catchment, and the highest value is achieved by the 2011 flood season (0.79). During the calibration period, the peak values are generally underestimated, and discharge during periods of low flow is well simulated by the HEC-HMS model. During the calibration period, efficiency (*Y*) values and visual inspection of the hydrograph show that performance of the 2011 model is satisfactory.

During validation, *RVE* and *Y* values show a reasonable agreement with the calibration phase. However, the validation period is somewhat less compared to the calibration, mainly due to the volume errors. The comparison of *Y* values between different years has to be considered carefully, because this statistical measure is strongly influenced by runoff variability, upstream abstraction, losses from evapotranspiration, and groundwater recharge.

The model could forecast most of the flood peaks exactly. Accuracy in computing peak discharge during the flood events was 79% when compared to the observed flows. Model computations are 13% higher than the observed discharges. This error could be due to several factors, including upstream abstraction, loss due to evapotranspiration, and groundwater recharge, which the model could not take into account. Flood forecasting lead time is increased

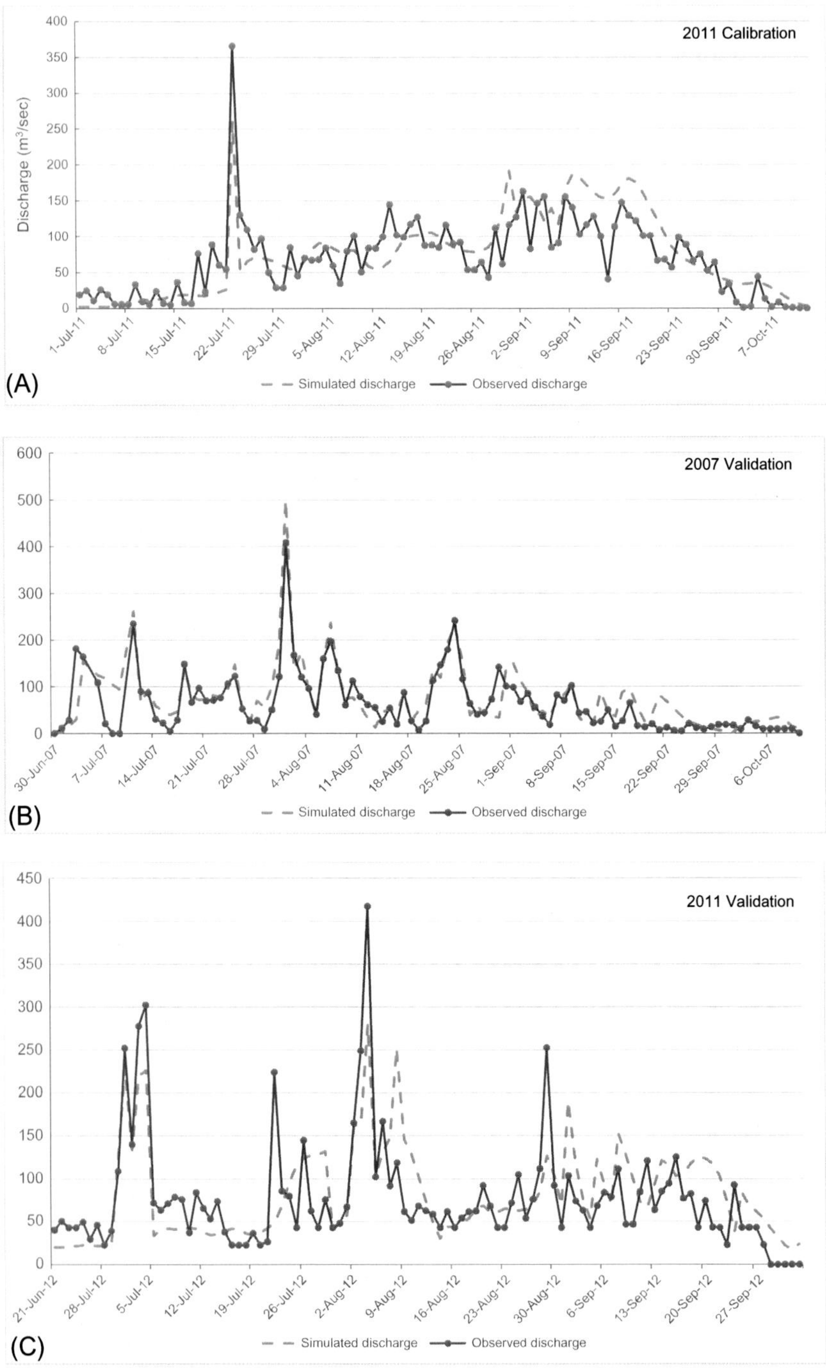

Fig. 6 Observed and simulated discharge (m³/s) for the Kassala Bridge station during the (A) 2011 calibration, and (B) and (C) 2007 and 2012 validation periods.

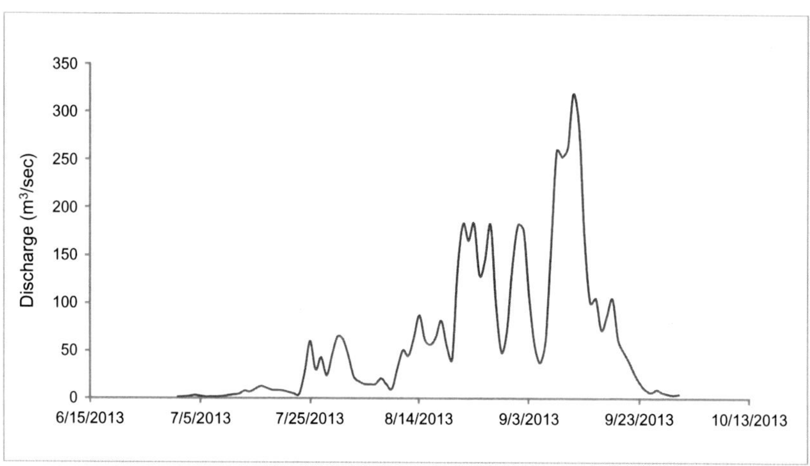

Fig. 7 Near real-time simulated flood hydrographs at Kassala Bridge for the 2013 flood season.

Table 1 Model calibration and validation for the Kassala Bridge station, Gash River Basin

		NSE	*RVE%*	*Y*
Calibration	2011	0.79	−0.05	0.94
Validation	2007	0.72	−0.17	0.87
	2012	0.71	−0.06	0.65

by 12h compared to the present conventional method. Due to the hydrological modeling technique, accuracy of discharge computations has improved. Once the flood forecasting model is developed, it can forecast floods at any river confluence. This will help in forecasting flood discharges at intermediate river junctions. Discharge in any subbasin of the study area can be predicted separately with the adoption of this hydrological modeling approach.

6 CONCLUSION

The simulations show that the computed hydrographs match well with the observed hydrographs. These hydrographs match very well when the discharge in the river is less than 200 m^3/s. There will be a slight increase in the error when discharge is beyond this range, due to the impact of water resources and evapotranspiration. With this hydrological modeling approach, the accuracy in discharge computations has improved when compared to conventional methods, flood forecasting can be carried out at any river confluence, and the influence of any tributary can be examined separately. In the near real-time application, the flood forecast model improved the lead time by 12h compared to conventional methods of forecasting. We need to improve further on the real-time operations to link with automatic gauge stations to assess model accuracy. Further impacts of land-use changes on the water balance, abstraction in the upstream areas, and evapotranspiration need to be augmented in the model, in order to generate real-time flood forecasting to support the GSIS in better planning and improving water productivity.

ACKNOWLEDGMENTS

The authors would like to thank the International Fund for Agricultural Development (IFAD) and the CGIAR Research Program on Water, Land and Ecosystems (WLE) for providing support with funding. Meteorological data provided by the National Meteorological Agency, and water level and discharge data provided by the Gash River Training Unit (GRTU), Sudan, are greatly appreciated. The technical assistance and cooperation provided by the Hydraulic Research Centre (HRC), and the Ministry of Water Resources and Electricity (MoWRE), Sudan, are gratefully acknowledged.

REFERENCES

Akhtar, M., Ahmad, N., & Booij, M. J. (2009). Use of regional climate model simulations as input for hydrological models for the Hindukush-Karakorum-Himalaya region. *Hydrology and Earth System Sciences, 13*(7), 865–902.

Amarnath, G., Shrestha, M. S., & Islam, A.K.M.S. (2016). Monitoring and assessment of floods and drought in Ganges basin. In L. Bharati, B. R. Sharma, A. S. Mandal, & V. Smakhtin (Eds.), *Earthscan/IWMI series on major river basins of the world: The Ganges river basin: Status and challenges in water, environment and livelihoods.* London: Routledge Publishers (352 pp.).

Anderson, I. M. (2011). *Technical paper on main findings and recommendations: The Eastern Sudan Rehabilitation and Development Fund.*

Artan, G., Gadain, H., Smith, J., Asante, K., Bandaragoda, C., & Verdin, J. (2007). Adequacy of satellite derived rainfall data for stream flow modeling. *Natural Hazards, 43*(2), 167–185. http://dx.doi.org/10.1007/s11069-007-9121-6.

Ashouri, H., Hsu, K. L., Sorooshian, S., Braithwaite, D. K., Knapp, K. R., Cecil, L. D., et al. (2015). PERSIANN-CDR: Daily precipitation climate data record from multisatellite observations for hydrological and climate studies. *Bulletin of the American Meteorological Society, 96*(1), 69–83.

Bashar, K. E., Abdo, G. M., & Gadain, H. (2005). Use of space technology in the management of Wadi water resources. In *Paper presented at the proceeding of the third international conference on Wadi hydrology, Sanaa, Yemen.*

Bogner, K., & Kalas, M. (2008). Error correction methods and evaluation of an ensemble based hydrological forecasting system for the upper Danube catchment. *Atmospheric Science Letters, 9*(2), 95–102.

De Roo, A., Gouweleeuw, B., Thielen, J., Bates, P., Horritt, M., et al. (2003). Development of European flood forecasting system. *International Journal of River Basin Management, 1*(1), 49–59.

Dilley, M., Chen, R. S., Deichmann, U., Lerner-Lam, A. L., Arnold, M., Agwe, J., et al. (2005). *Natural disaster hotspots: A global risk analysis*. Washington, DC: International Bank for Reconstruction and Development/The World Bank and Columbia University.

Elsheikh, A. E. M., Khalid, A. E. Z., & Shaza, A. E. (2011). Groundwater budget for the upper and middle parts of the River Gash Basin, eastern Sudan. *Arabian Journal of Geosciences, 4*, 567–574.

EM-DAT. (2015). *The OFDA/CRED international disaster database*. Brussels, Belgium: Centre for Research on the Epidemiology of Disasters (CRED), Université catholique de Louvain. Available at http://www.emdat.be. Accessed 31.07.15.

Gerlak, A. K., Lautze, J., & Giordano, M. (2013). *Greater exchange, greater ambiguity: Water resources data and information exchange in transboundary water treaties' GWF Discussion Paper 1307*. Canberra, Australia: Global Water Forum.

IFRC. (2007). *Floods DREF bulletin no. MDRSD004*. http://www.ifrc.org/docs/appeals/07/MDRSD004.pdf.

Jaun, S., & Ahrens, B. (2009). Evaluation of a probabilistic hydrometeorological forecast system. *Hydrology and Earth System Sciences, 13*, 1031–1043.

Jorgensen, G. H., & Host-Madsen, J. (1997). Development of a flood forecasting system in Bangladesh. In *Proceedings of conference on operational water management, 3–6 September 1997* (pp. 137–148). Copenhagen: AA Balkema.

Khan, H., Dawe, P., Ali, K. A., & Puestow, T. (2011). Innovative approaches to monitoring for transboundary water governance. In *International conference on environment science and engineering IPCBEE (Vol. 8)*. Singapore: IACSIT Press.

Khan, S. I., Hong, Y., Wang, J., Yilmaz, K. K., Gourley, J. J., Adler, R. F., et al. (2011). Satellite remote sensing and hydrological modeling for flood inundation mapping in Lake Victoria Basin: Implications for hydrologic prediction in ungauged basins. *IEEE Transactions on Geoscience and Remote Sensing, 49*, 85–95. http://dx.doi.org/10.1109/TGRS.2010.2057513.

Laurent, H., Jobard, I., & Toma, A. (1998). Validation of satellite and ground-based estimates of precipitation over the Sahel. *Atmospheric Research, 47–48*, 651–670. http://dx.doi.org/10.1016/S0169-8095(98)00051-9.

Maharjan, R. (2013). *Coupling hydrological model with Delft-FEWS for flood forecasting in Bagmati Basin, Nepal*. (M.Sc. thesis). Delft, Netherlands: UNESCO-IHE Institute for Water Education.

Manjusree, P., Prasanna Kumar, L., Bhatt, C. M., Rao, G. S., & Bhanumurthy, V. (2012). Optimization of threshold ranges for rapid flood inundation mapping by evaluating backscatter profiles of high incidence angle SAR images. *International Journal of Disaster Risk Science, 3*(2), 113–122. http://dx.doi.org/10.1007/s13753-012-0011-5.

Nayak, P. C., Venkatesh, B., Krishna, B., & Jain, S. K. (2013). Rainfall-runoff modeling using conceptual, data driven, and wavelet based computing approach. *Journal of Hydrology, 493*, 57–67. http://dx.doi.org/10.1016/j.jhydrol.2013.04.016.

Nishat, B., & Rahman, S. M. (2009). Water resources modeling of the Ganges-Brahmaputra-Meghna river basins using satellite remote sensing data. *Journal of American Water Resources Association, 45*(6), 1313–1327. http://dx.doi.org/10.1111/j.1752-1688.2009.00374.x.

OCHA. (2003). *Flooding in Kassala state information bulletin no. 1. ReliefWeb*. Retrieved from, http://reliefweb.int/report/sudan/sudan-flooding-kassala-state-information-bulletin-no-12003.

Rango, A. (1992). Worldwide testing of the snowmelt runoff model with applications for predicting the effects of climate change. *Nordic Hydrology, 23*, 155–172.

Rao, K. H. V. D., Bhanumurthy, V., & Roy, P. S. (2009). Application of satellite based rainfall products and SRTM DEM in hydrological modelling of the Brahmaputra basin. *Journal of Indian Society of Remote Sensing, 37*(4), 539–552.

Rao, K. H. V. D., Rao, V. V., Dadhwal, V. K., Behera, G., & Sharma, J. R. (2011). A distributed model for real-time flood forecasting in the Godavari Basin using space inputs. *International Journal of Disaster Risk Science, 2*(3), 31–40.

Stimson Center. (2010). *Fresh water futures: Imagining responses to demand growth, climate change, and the politics of water resource management by 2040.*

Stisen, S., Jensen, K. H., Sandholt, I., & Grimes, D. I. F. (2008). A remote sensing driven distributed hydrological model of the Senegal River basin. *Journal of Hydrology, 354*(1–4), 131–148. http://dx.doi.org/10.1016/j.jhydrol.2008.03.006.

Thielen, J., Bartholmes, J., Ramos, M.-H., & de Roo, A. (2009). The European flood alert system — Part 1: Concept and development. *Hydrology and Earth System Sciences, 13*(2), 125–140.

U.S. Army Corps of Engineers. (2000). *Hydrological modelling system HEC-HMS technical reference manual.* Davis, CA: U.S. Army Corps of Engineers, Hydrologic Engineering Centre.

U.S. Army Corps of Engineers. (2001). *Hydrological modelling system HEC-HMS user's manual.* Davis, CA: U.S. Army Corps of Engineers, Hydrologic Engineering Centre.

U.S. Army Corps of Engineers. (2003). *Geospatial hydrological modelling extension HEC GeoHMS, User's manual.* Davis, CA: U.S. Army Corps of Engineers, Hydrologic Engineering Centre.

Verma, A., Jha, M., & Mahana, R. (2010). Evaluation of HEC-HMS and WEPP for simulating watershed runoff using remote sensing and geographical information system. *Paddy and Water Environment, 8*(2), 131–144. http://dx.doi.org/10.1007/s10333-009-0192-8.

Wheater, H., Sorooshian, S., & Sharma, K. D. (2007). *Hydrological modelling in arid and semi-arid areas.* Cambridge: Cambridge University Press.

CHAPTER 9

Flood Forecasting — A National Overview for Great Britain

C. Pilling*, V. Dodds†, M. Cranston‡, D. Price*, T. Harrison§, A. How¶
*Flood Forecasting Centre, Met Office, Exeter, United Kingdom
†Bureau of Meteorology, Brisbane, QLD, Australia
‡Scottish Flood Forecasting Service, SEPA, Perth, United Kingdom
§Flood Incident Management, Environment Agency, Solihull, United Kingdom
¶Flood and Operational Risk Management, Natural Resources Wales, Cardiff, United Kingdom

1 BACKGROUND AND CATALYSTS FOR CHANGE

1.1 Introduction

Great Britain comprises a diverse range of river catchments. These include highly regulated lowland basins such as the River Thames in the southeast of England; rural, upland catchments such as the River Tay in Scotland (the longest river in Britain); steep, fast-responding coastal catchments in Wales and the southwest of England; and canalized channels in urban areas.

There are different sources of natural flooding which are categorized as: coastal, river, surface water, and groundwater. Often the primary cause of flooding is exacerbated by another source of flooding, and the meteorological and hydrological drivers behind significant flood events can be complex.

Orography, which is highest in the west, combined with weather systems that primarily track west to east, result in an uneven distribution of rainfall across Britain. As Fig. 1 shows, annual average rainfall amounts range from 400 mm in the southeast of England, up to 4700 mm in the western Highlands of Scotland.

Large fluvial floods are often caused by a succession of synoptic-scale low pressure systems tracking in from the Atlantic during the autumn and winter, moving west to east. Rainfall totals are often enhanced across the windward facing slopes in the west, or where elements of convection develop within the frontal systems associated with the low pressure systems. Rainfall from these weather systems can quickly saturate river catchments and generate significant runoff, with flood peaks taking several days to propagate downstream.

There have been several notable fluvial floods across Great Britain in recent years, and those that affected the River Thames are summarized in Case Study 1.

Flood Forecasting
http://dx.doi.org/10.1016/B978-0-12-801884-2.00009-8

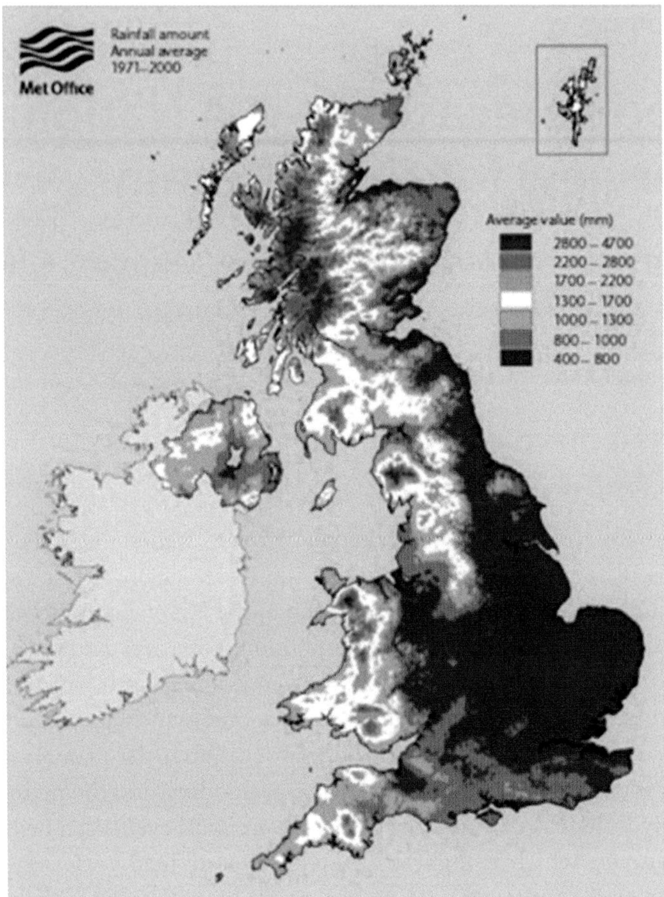

Fig. 1 Annual average rainfall across Great Britain and Northern Ireland, 1971–2000 (Met Office, 2013)

Case Study 1: Old Father Thames

The River Thames, a lowland river basin, is one of the most well-known rivers in the world. It is the longest river in England at over 382 km and drains an area of 9950 km^2 above Teddington weir in London (CEH, 1983).

The largest flood on record on the lower Thames was in 1894, with an estimated peak flow of 1059 cumecs (CEH, 1983). In recent years there have been a number of major floods, notably in 2000, 2004, 2007, 2012, and 2013/14, although in these events the peak flow has not exceeded 500 cumecs.

The source of the Thames is at Thames Head (Gloucestershire) in the west of the catchment; a rural landscape characterized by undulating hills and farmland, and picturesque villages and towns. As the river meanders further east the catchment becomes more urbanized with major population centers including Oxford, Reading, and London before it drains into the Thames estuary and then the North Sea. The Thames catchment receives on average 690 mm of rainfall per year, making it one of the drier areas in Britain (Environment Agency, 2009a).

There are a number of major water supply abstractions, particularly in the lower Thames, as well as 44 locks and weirs between Cricklade and Teddington (Environment Agency, 2009a).

The underlying geology varies from chalk, limestone, gravel, sand, and clay (Environment Agency, 2009a). There are significant groundwater resources within the catchment associated with the limestone and chalk aquifers in the Cotswolds, the Berkshire Downs, and the Chilterns. These groundwater resources also result in "underground flow out of the Thames catchment" into adjacent river basins (CEH, 1983).

There are 38 main tributaries of the Thames including the Windrush, Thame, Kennet, Loddon, Mole, and Wey, which all drain upstream of Teddington, and the Crane, Brent, Ravensbourne, and Wandle, which drain into the Thames estuary. The interaction of flood peaks from all the different tributaries adds a level of complexity to flood forecasting for the River Thames. As a result, sophisticated hydrodynamic models, hosted in the National Flood Forecasting System (NFFS: see Section 2), have been calibrated and greatly assist hydrologists with flood forecasting.

Approximately 135,000 properties have more than a 1% chance of river flooding in any 1 year in the Thames catchment, with a further 300,000 properties at risk from tidal flooding (Environment Agency, 2009a). The majority of properties at risk can be found in London or the lower Thames, although they are protected by the tidal defences and in particular the iconic Thames Tidal Barrier, which protects the capital from a tidal surge.

Convective scale rainfall events associated with sea-breeze convergence, squall lines, or meso-scale convective systems typically occur in the summer months. They can result in intense downpours in a short period of time and as a consequence surface water flood events or flash floods in steep upland river catchments can happen.

High astronomical tides, or "spring tides" coinciding with a storm surge, strong onshore winds, and large, high-energy waves, can give rise to significant coastal flood events.

The presence of aquifers and porous rock such as chalk, for example, in the south and east of England, can result in significant groundwater flood events after prolonged periods of above average rainfall. In addition, water can permeate through gravels, as happened along parts of the River Thames in 2014, resulting in the combination of groundwater and fluvial flooding.

1.2 Flood Risk in England, Wales, and Scotland

All of the leading flood agencies in England, Wales, and Scotland have adopted a risk-based approach to managing natural floods (coastal, river, surface water, and groundwater flooding) and offer a flood warning service for all main rivers.

The Environment Agency has the strategic overview role for flood risk management for all sources of flooding in England, and its responsibilities include forecasting flood risk and providing warnings to the public and emergency response community (Environment Agency, 2009b) for river and coastal flooding. According to the Environment Agency's national assessment of flood risk, there are around 5.2 million properties in England (1 in 6) at risk of flooding from rivers, the sea, and surface water (Environment Agency, 2009b). In addition to this, the British Geological Survey and the Environment Agency estimate the number of properties at risk of groundwater flooding in England is between 122,000 and 290,000 (McKenzie & Ward, 2015, p. iv).

Within Wales, the responsibility for flood risk forecasting as well as the operation of the flood warning service falls within the remit of Natural Resources Wales. Natural Resources Wales was formed in 2013 by the merger of the Countryside Council for Wales, Forestry Commission Wales, and Environment Agency Wales, creating a single body responsible for maintaining and enhancing the environment and natural resources of Wales. The latest estimates in 2014 suggests about 208,500 properties are at risk from rivers and/or the sea, accounting for 11% of all properties in Wales. There are 163,000 properties at risk from surface water flooding (NRW, 2014, p. 3).

In 2009, the Flood Risk Management (Scotland) Act was given Royal Assent. The Act provides a modern and sustainable framework for the management of flood risk in Scotland and provided the Scottish Environment Protection Agency (SEPA) with a host of new responsibilities. Approximately 1 in 22 of all residential and 1 in 13 of all non-residential properties are at risk of flooding in Scotland, with average annual damages estimated to be between £720 and £850 million (SEPA, 2011).

As well as properties, there are also a number of critical national infrastructures at risk of flooding, such as transport networks, energy suppliers, and utilities across all three countries.

1.3 Flood Forecasting and Warning Landscape 1998–2009

The flood forecasting and warning landscape in Britain has changed significantly in the last 25 years and has undergone a paradigm shift from flood defence to the current philosophy based on flood risk management. Before the formation of the Environment Agency for England and Wales in 1996, numerous reorganizations saw the responsibility for flood forecasting and warning change from "river authorities to water authorities to the National Rivers Authority (NRA) in 1989" (Bye & Horner, 1998). The NRA and its predecessors generally took the lead for preparing flood warnings, and the police took a prominent role in the dissemination of these warnings to the public (Bye & Horner, 1998).

Following the Environment Act 1995, the Environment Agency, a non-departmental public body, was formed in 1996 and became the lead organization for flood defence (Environment Agency, 2001). Its responsibilities included the "power to provide and operate flood warning systems" (Bye & Horner, 1998), although its powers were deemed to be permissive and not a statutory duty (Bye & Horner, 1998). Forecasting was provided at the regional level (seven Environment Agency regions plus Environment Agency Wales), while a flood warning service was provided at an Area level (20 Areas) for all main watercourses and many coastal communities.

The Easter 1998 floods were a real test for the newly formed Environment Agency, as they affected large swathes of England and Wales. Five people lost their lives, 4500 families lost their homes, and damages were estimated to exceed £300 million (Bye & Horner, 1998). After the floods, an independent review was commissioned by the government, and this was conducted by Bye and Horner (1998). It concluded that the "Agency's performance on issuing warnings was, on average, 'unsatisfactory' and 'nationally' inconsistent and inadequate procedures and systems resulted in poor overall performance" (Bye & Horner, 1998). At the time, the Government Minister for Fisheries and the Countryside, Elliot Morley, called for a "seamless and integrated service of flood forecasting, warning and response" (Environment Agency, 2001).

In autumn 2000, a succession of weather events resulted in further flooding of around 10,000 homes and businesses across England and Wales. The Environment Agency's Lessons Learned report (Environment Agency, 2001) reiterated many of the recommendations to come out of the Bye Report, although it also acknowledged that there had been some significant improvements. Recognized improvements included the introduction of new flood warning codes (Flood Watch, Flood Warning, Severe Flood Warning, All Clear) and a successful flood awareness campaign (Environment Agency, 2001).

The Bye Report and the Environment Agency's Autumn 2000 Lessons Learned report were key drivers for change and led to a more nationally consistent approach to flood forecasting and warning in England and Wales. Over the next 10 years, numerous legacy flood forecasting and warning systems were decommissioned across the regions as the Environment Agency embraced the latest technological and scientific advances. For example, by 2006 the NFFS, which adopted the Delft-FEWS (Flood Early Warning System), became the primary operational forecasting system across all regions of the Environment Agency and was subsequently adopted by SEPA as the FEWS Scotland system (Werner, Cranston, Harrison, Whitfield, & Schellekens, 2009). In conjunction with this, improved dissemination tools and systems such as Floodline Warnings Direct (FWD) were rolled out, and this in turn was supported by nationally consistent procedures and documentation (Andryszewski et al., 2005).

One of the earliest examples of coordinated flood forecasting in the UK involved coastal flooding. The UK Coastal Monitoring and Forecasting Service, formerly known as the Storm Tide Forecasting Service, was set up following the devastating east coast surge and flood of 1953. This provides a collaborative service that monitors and provides forecasts of tide, surge, and wave conditions (Environment Agency, 2009c). Principally established for England and Wales, the service was extended to include partners across Britain including SEPA following the Western Isles Storm of 2005 (Cranston & Tavendale, 2012). Further details are provided in Case Study 3.

While the flood forecasting and warning service had undergone significant improvements at the local and regional level, there remained "gaps" at a strategic level, and these became evident in the summer 2007 floods.

1.4 The Catalyst for Change: Summer 2007 Floods

In Jun. and Jul. of 2007, exceptionally heavy summer rainfall resulted in severe floods affecting Hull and Sheffield (Jun. 24–25) and large areas of central southern England and Wales (Jul. 19–20). The record-breaking rainfall totals were in part attributed to the strength and position of the polar

Fig. 2 Demountable flood defences providing protection in Bridgwater, England, during Feb. 2014.

front jet stream, which was stronger and further south than usual and which consequently steered a succession of Atlantic storms across southern and central parts of the United Kingdom. This, combined with very warm and moist air, resulted in unprecedented rainfall totals across many areas (Pitt, 2008) (Fig. 2).

The highest recorded rainfall total was 157.4 mm in 48 h at Pershore College (Worcestershire) in Jul. (Met Office, 2013). Fig. 3 shows the rainfall distribution that led to the severe flooding in 2007.

River and surface water flooding or a combination of both characterized the floods, with some estimates suggesting that nearly two-thirds of properties flooded as a result of surface water. However, in 2007, no agency was responsible for the provision of a surface water flood forecasting service (Werner et al., 2009).

The floods brought parts of the country to a standstill, with 13 fatalities and 55,000 properties flooded. Transport and utilities were significantly disrupted as the floods resulted in the "largest loss of essential services since World War II, with almost half a million people without mains water or electricity" (Pitt, 2008). Notably, despite over 200 major floods worldwide during 2007, these UK floods ranked as the most costly in the world that year, with over £3 billion in damages (Pitt, 2008). The 2007 floods and the reviews that followed resulted in another step change in the quality of flood forecasting and warning services in England and Wales.

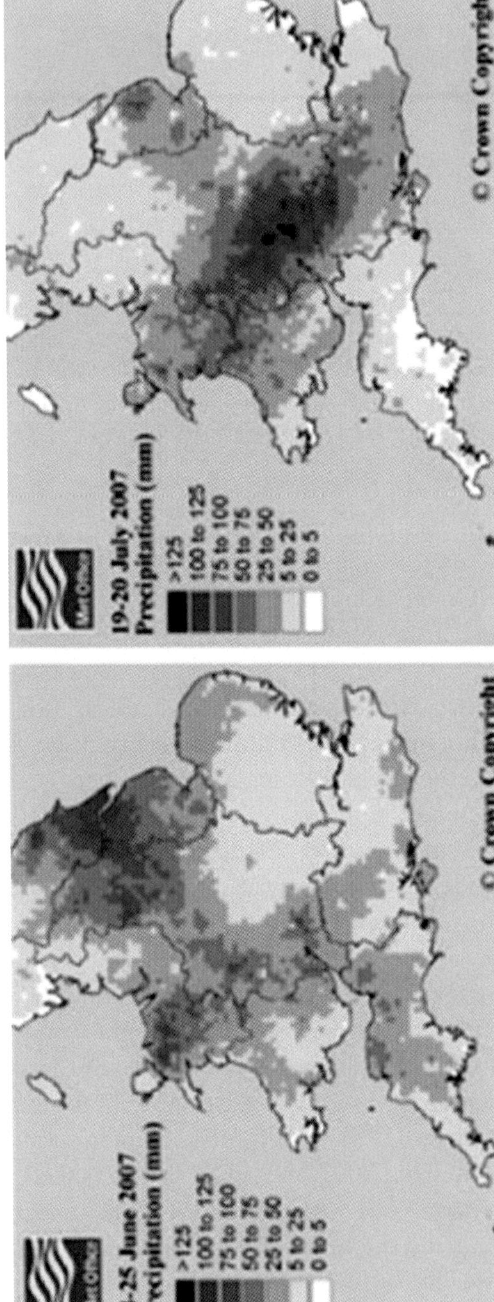

Fig. 3 Precipitation amounts across England and Wales during Jun. 24–25 *(left)* and Jul. 19–20, 2007 *(right)*. *Source:* Pitt, M. (2008). *Learning Lessons from the 2007 Floods*. Available from: http://webarchive.nationalarchives.gov.uk/20080906001345/http://cabinetoffice.gov.uk/thepittreview/final_report.aspx.

1.5 The Pitt Review and the Establishment of the Flood Forecasting Centre

As a result, the government instigated the Pitt Review, a comprehensive review of the event and performance of all agencies and organizations involved. This Review was published in Jun. 2008 and was one of the "widest ranging policy reviews ever carried out in the UK" (Pitt, 2008). Key areas addressed included forecasting and warning, the role of the emergency responder community, and the protection of critical infrastructures and utilities. It also considered which organizations should have overall strategic responsibility for flooding. The government's response to the Pitt Review stated that it supported all of the recommendations in the review. Subsequently, the Flood and Water Management Act 2010 has introduced legislation to address many of the issues identified in the Pitt Review, including giving the Environment Agency the strategic overview for all sources of flooding. From a flood forecasting and warning perspective, two of the key recommendations from the Pitt Review were:

> *The Environment Agency (EA) and the Met Office (MO) should work together, through a joint centre to improve their technical capability to forecast, model and warn against all sources of flooding (Recommendation 6).*

> *The Met Office and the Environment Agency should issues warnings against a lower threshold of probability to increase preparation lead times for emergency responders (Recommendation 34).*

These recommendations resulted in the establishment of the Flood Forecasting Centre (FFC), a joint center combining Environment Agency and Met Office expertise. The Centre brought together staff from its parent organizations (Met Office and Environment Agency) to a single physical location, working towards a common purpose of improving the flood forecasting capabilities of England and Wales. It was responsible for issuing longer lead-time guidance to the emergency response community to allow a proportionate and risk-based response, and also filled a gap at the strategic level for flood forecasting and warning.

The FFC was officially opened by the Secretary of State for the Environment (Hilary Benn) in London in Apr. 2009. Initial funding was provided for the first 2 years by the Department for Environment, Food and Rural Affairs, as well as the Met Office and the Environment Agency.

Between 2009 and 2011, flood operations within the FFC were performed by Met Office meteorologists sitting alongside Environment

Agency hydrologists on a 24/7 basis. The benefits of flood hydrologists and operational meteorologists working closely together was immediately clear.

In its first year of operation, the FFC performed well and was praised by Hilary Benn for its provision of timely advice to Category 1 and 2 responders during the Cumbrian floods in Nov. 2009 (Sibley, 2010). In May 2010, a long-term mandate for the continuation of the FFC was agreed and has subsequently been accepted as an essential and permanent part of flood forecasting in England and Wales.

After 2 years of flood operations, the benefit was recognized in going one step further and combining the disciplines of hydrology and meteorology to create the role of operational hydrometeorologist. Prior to 2011, this role did not exist in Great Britain. A structured and assessed technical development framework was established to provide skills assurance for this new role and the vocational qualification has recently been updated to a "Diploma in Operational Hydrometeorology and Flood Forecasting" (PAA\VQSET, 2014).

1.6 Establishment of the Scottish Flood Forecasting Service

This approach for much closer collaboration between meteorologists and hydrologists was also considered in Scotland, and in 2011 the Scottish Flood Forecasting Service (SFFS) was established. The aim of the SFFS is to improve resilience to flooding through developing a technical capability to forecast, model, and warn against all sources of flooding. This approach to closer collaboration was based on international best practice such as the FFC and the Service Central d'Hydrometeorologie et d'Appui a la Previsions des Inondations (SCHAPI) in France, and was built on respective meteorological and hydrological centers in Aberdeen and Perth (Cranston et al., 2012).

The primary benefits of such a new approach identified by Cranston and Tavendale (2012) included:

- a combined flood forecasting service for Scotland, fully integrating meteorological and hydrological aspects for the first time, providing knowledge transfer through joint working between SEPA and the Met Office, and leading to the development of an integrated hydrometeorological skills set;
- the provision of regular, consistent information on flood threat to emergency responders in an accessible format;
- working towards the provision of a single integrated flood advisory service for emergency responders;

- improved accuracy and timeliness of flood forecasting; and
- a vehicle for the development of flood forecasting for all forms of flooding in Scotland.

The introduction of the new service has certainly contributed to improved resilience to flooding, with praise from the Minister for Environment and Climate Change: "I was extremely impressed by how well prepared all the authorities were — aided by the timely and accurate flood forecasts from the Scottish Flood Forecasting Service" (BBC News, 2014). This led to a pull-through of science capabilities to improve countrywide flood forecasting and new capabilities in the prediction of surface water flooding. However, as approaches to flood risk management are moving away from an era of flood prevention to more sustainable methods (Tavendale, 2009), this in turn is placing greater emphasis on flood forecasting and warning linked to measures such as property-level protection.

2 COUNTRYWIDE FLOOD FORECASTING MODELING APPROACH

2.1 Drivers, History, and Context for a Countrywide Flood Forecasting Approach

The requirement for a countrywide flood forecasting capability was a direct response to the recommendations set out in the Pitt Review (Pitt, 2008), which identified the need for a coherent service delivered at the national level and commensurate with the needs of national stakeholders. Flood forecasting is now delivered at both a national and a local level, with these services complementing each other to support the delivery of consistent flood guidance and flood warning.

Historically within the flood forecasting agencies (Environment Agency, Natural Resources Wales, SEPA), a number of flood forecasting techniques have been used (Moore, Bell, Cole, & Jones, 2007; Werner et al., 2009). For river forecasting, these have ranged from simple upstream level exceedance triggers, and level-to-level correlations; to more complex coupled hydrological and hydrodynamic models. There are numerous examples of catchments for which hydrological and hydrodynamic flood forecasting models have been developed and which are used to provide an operational flood forecasting service. In these examples, rainfall runoff models (typically PDM, Moore, 2007) are coupled with routing and/or hydrodynamic models (typically Flood Modeller, formally known as ISIS, Lin, Wicks, Falconer, & Adams, 2006) to provide forecasts of level and flow that can be assessed against known thresholds, to cause the issue of warnings at the community scale. Although PDM and Flood Modeller are the most commonly used modeling techniques, many other forecasting methodologies have also been implemented operationally (Moore et al., 2007; Werner et al., 2009).

Coastal flood forecasting has also seen various modeling techniques used. From trigger levels on observed tidal gauges and simple chainage calculations translating sea level along the shore, to complex wave transformation modeling and neural network computations of wave overtopping of defences.

All flood forecasting models have been integrated and configured onto a common forecasting platform, Delft-FEWS, which has enabled the best use of real-time observed and forecast (rainfall) data sets. This is referred to as the NFFS in the Environment Agency and Natural Resources Wales, and FEWS Scotland in SEPA.

The next section provides an overview of the approaches used for fluvial, surface water, coastal, and groundwater flood forecasting. In line with the emphasis of this chapter, this section focuses on the developments in countrywide flood forecasting modeling approaches delivered at a national level, rather than those delivered at the local level.

2.2 Countywide Fluvial Flood Forecasting

Even before the Pitt Review, the Environment Agency had embarked on a Research and Development project to find improved and more appropriate modeling techniques to transform high-resolution numerical weather prediction (NWP) rainfall forecasts into accurate flood forecasts (Environment Agency, 2010). This project tested and led to further development of the Centre for Ecology and Hydrology's (CEH) Grid-to-Grid (G2G) model, which has since become the FFC's primary hydrological forecasting tool (Price, Hudson, et al., 2012).

The G2G model offers a number of significant advantages over the more commonly used lumped catchment models, not least its ability to capture the spatial and temporal structure of rainfall events (Cole & Moore, 2009; Price, Pilling, et al., 2012). In addition, pre-existing spatial data sets covering terrain, soil/geology, and land cover properties are used in the configuration and parameterization of the model.

The G2G model was commissioned by the FFC in 2009, and shortly after by the SFFS, to support the countrywide delivery of a flood guidance service. The G2G is a physical-conceptual distributed hydrological model (Bell, Kay, Jones, Moore, & Reynard, 2009; Moore et al., 2007) with runoff and routing components. It is configured across England, Wales, and Scotland on a 1 km^2 resolution grid and runs on a 15-min timestep. This ensures the model can utilize high-resolution rainfall data available from the Met Office, and that this can be better reflected in catchment (hydrological) response.

The G2G uses a simple runoff production scheme to generate surface and subsurface runoff from inputs of gridded rainfall and potential evapotranspiration (see Fig. 4). The runoff production is controlled by the soil characteristics of each grid cell, including water holding capacity, and these are specified directly using imported soil property data. Variation in water holding capacity within grid squares is represented in a probability distributed way (Moore, 1985). Lateral and vertical drainage within and between model cells is represented, and specified, through imported soil property data. Land cover data are used to modify runoff response while groundwater storage receives water through percolation (recharge) and releases water from this store as subsurface flow. For lateral flow between cells (both surface and subsurface), the G2G uses a kinematic wave formulation.

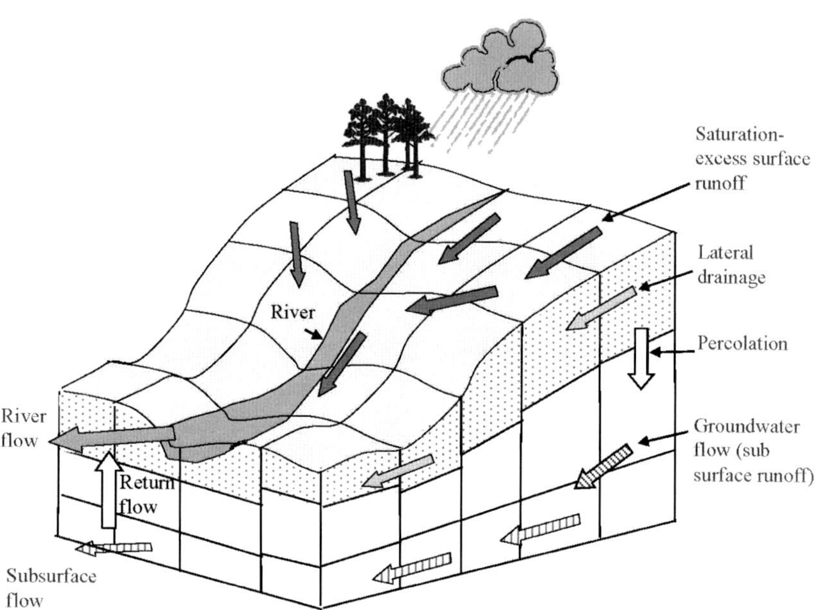

Fig. 4 Schematic of the G2G model showing configuration across a landscape.

In operational mode, the G2G employs three forms of error correction, namely state updating, flow insertion, and Autoregressive Moving Average (ARMA) error correction.

The G2G offers a number of advantages over other modeling configurations/types which make it suitable for large-scale, countrywide application (Cole & Moore, 2009; Cranston et al., 2012; Environment Agency, 2010; Price, Hudson, et al., 2012). The key benefits include the following:

- The use of a small set of regional parameters, supported by pre-existing digital data sets for model configuration and parameterization, making these tasks achievable at the national scale. This contrasts with lumped catchment models which require a model for each catchment/gauging station, each with parameters calibrated to observed flow at a catchment gauging station.
- The models distributed nature, at a fine spatial resolution, which allows better representation of the spatial variation in rainfall input and catchment response. This contrasts to lumped catchment models, which use catchment averaged data.
- A national calibration which allows for a manageable process undertaken at selected (reliable) river gauge locations. This contrasts with calibration across all modeled catchments individually.
- Forecasts for each grid cell, meaning that the model effectively forecasts "everywhere" within a defined domain. This contrasts with forecasts for single catchment outlet locations.
- It can be used to forecast for ungauged locations, highlighting its potential usefulness for flood forecasting across all locations.
- Only one model configuration is required for countrywide application, simplifying support and development.
- The model is computationally efficient and fast to run for nationwide real-time flood forecasting on a 1 km^2 model grid. This infers real advantages for operational model use, especially when running ensemble datasets.

However, results from the G2G model are not generally expected to be as good those produced by locally calibrated models. This is because the model is broadly calibrated across England and Wales with fewer parameters than most locally applied models (Environment Agency, 2015). The G2G model also uses a simple routing scheme which is unable to accommodate complex channel/floodplain flow processes, river control structures, or the fluvial/tidal boundary; thus performance in controlled or tidally influenced rivers can be compromised.

The choice of the G2G model is a compromise between the advantages and limitations listed above. However, it represents an appropriate choice of methodology given the requirements of the FFC and SEPA, which is to provide forecasts at extended lead times using ensembles of rainfall forecasts at a country level (Price, Hudson, et al., 2012).

In its present configuration, the G2G model receives observations and forecast data from a number of sources. Observed river level and flow data are available from the Environment Agency and Natural Resources Wales telemetry systems. Observed rainfall data is available from both the Environment Agency and Natural Resources Wales telemetry systems and the Met Office's radar network. Forecast rainfall data are available from a number of configurations of the Met Office Unified Model (UM). This includes a 5-day medium-range deterministic forecast at 4 km resolution, a high-resolution short-range ensemble forecast (MOGREPS-UK: Golding et al., 2014), and a nowcast ensemble forecast (STEPS: Seed, Pierce, & Norman, 2013). Data sets are also available from the Met Office UM covering snow data (to allow for modeling snowmelt) and potential evaporation.

G2G outputs river flow and this can be displayed either spatially as flow forecasts for each 1 km^2 grid cell for each timestep or by way of traditional hydrographs for specific gauged and ungauged locations (Fig. 5). The forecasts of flow are of limited use alone at the countrywide scale. Fundamental to forecasting flood risk at the FFC is the ability to make a link between fluvial flow and impact, where impact is measured in terms of flood damage to lives and livelihoods. While forecasting at the local (catchment) scale, the Environment Agency is able to link levels/flows to impact, whereas this is not possible at the national scale for all possible "at risk" locations. A useful approximation of impact is flow return period or rarity and although this does not give a measure of actual impact, it gives an indicative measure. Using this, assumption maps of return period for 1:2, 1:5, 1:10, 1:50, and 1:100 years were generated for all forecast points in the G2G model across England and Wales. Forecast flows are then compared to these static return period maps for all forecast points at each timestep during a model run, and levels of return period exceedance can be identified (Price, Hudson, et al., 2012).

To assess potential flood risk across the country over a 5-day forecast period, these data are presented as a single threshold exceedance for each

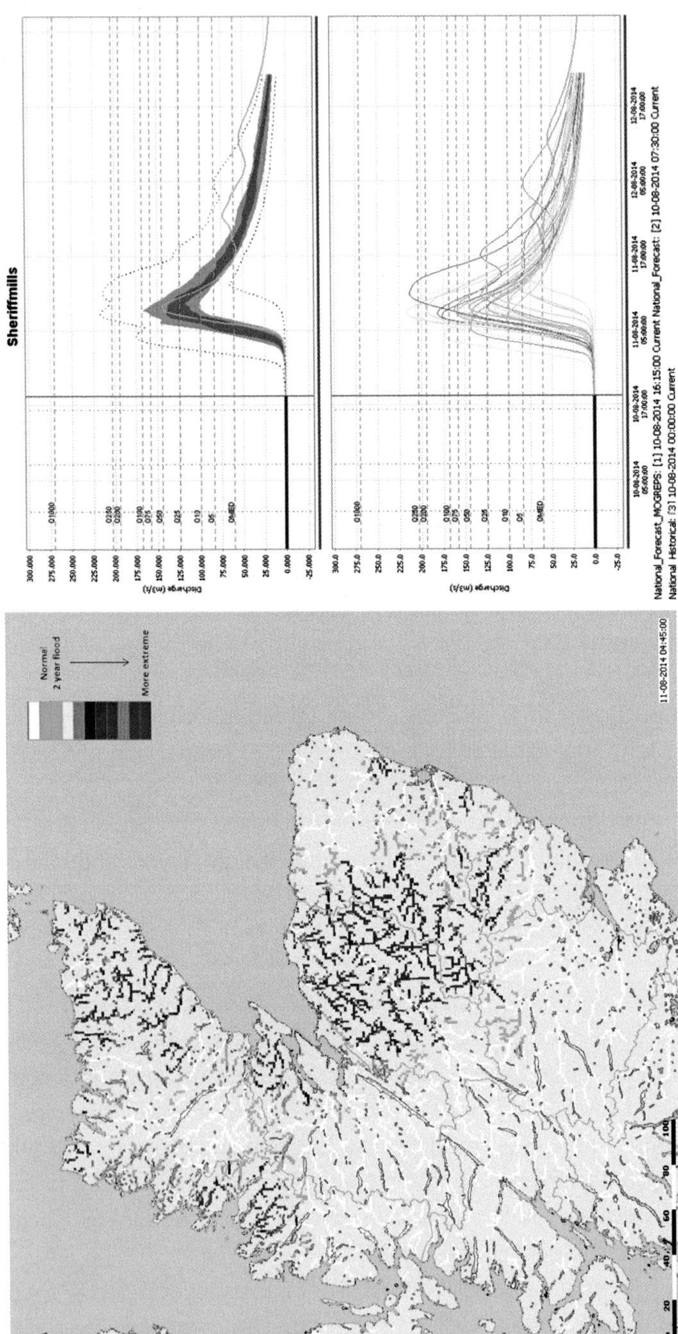

Fig. 5 Output from Grid-to-Grid for the Ex-Tropical Storm Bertha. Forecasts for countrywide 1 km gridded hydrological response compared with *Q(t)* grid to highlight the potential severity of flooding *(left)* and hydrological ensemble-based forecasts for the River Lossie at Sheriffmills in Moray *(right)*.

grid cell at each timestep, rendered spatially as a gridded map across the whole of England. For the nowcast and short-range ensemble forecasts, these are presented either as "spaghetti" plots for individual modeled locations (usually river gauging stations) in which the output from each individual run is shown or color-coded to illustrate the number of ensemble members above a given return period threshold and presented spatially as a gridded map across England and Wales. Displays are configured within the Delft-FEWS software package (Werner, van Dijk, & Schellekens, 2004), the platform used by the FFC for displaying and summarizing the results from the G2G model. A very similar approach is used by the SFFS, as illustrated in Fig. 5.

2.3 Countrywide Surface Water Flood Forecasting

The responsibility for surface water flooding across Britain varies by country. In England and Wales, the responsibility for providing forecasts for surface water flooding currently lies with Lead Local Flood Authorities (LLFAs), and not the Environment Agency or Natural Resources Wales, respectively. In Scotland, SEPA retains this responsibility but given the science capabilities, this remains in its infancy. In both cases the forecasting centers (FFC and SFFS) do include an assessment of flood risk from surface water in the Flood Guidance Statement (FGS).

The typical modeling approach to surface water flooding stems from initial work to develop a surface water flood forecasting (and warning) service, namely the "Extreme Rainfall Alert" (ERA) service (Dale, Davies, & Harrison, 2012; Halcrow, 2008). This service was developed jointly by the Environment Agency and the Met Office, and was introduced as a pilot in 2009. The service was based on an assessment of the likelihood of exceeding specified depth duration rainfall amounts and was linked to urban drainage design criteria — essentially the 30-year return period storm event. This service was delivered to a limited number of Category 1 and 2 responders. While a beneficial service, it was limited as it considered only rainfall intensity and took no account of surface/subsurface processes or vulnerability, which also contribute to determining surface water impacts.

In 2010, a more targeted and objective model, the Surface Water Decision Support Tool (SWFDST: Halcrow, 2011), was introduced operationally into the FFC. In its current form, the tool imports data on the likelihood of exceeding a number of pre-defined return period rainfall events derived from each run of the high-resolution ensemble MOGREPS-UK model. The tool then calculates a flood-impact weighted score for all of the 109 county and unitary authorities across England and Wales, taking into account urbanization and soil moisture status (as soil moisture deficit). For each county and unitary authority, a score of the maximum surface water flood risk is presented. The assessment of surface water flood impact is thus not based on storm intensity alone (unlike its predecessor, the ERA service), but takes into account the level of urbanization in an area and the prevailing antecedent soil moisture status. The tool has been used operationally in the FFC since 2010.

While it delivered benefits, the methodology employed in the SWFDST is limited both by the use of static vulnerability data (assessment of urbanization) and likelihood which is determined by rainfall intensity alone.

The FFC is in the early stages of further improving its surface water flood forecasting capability through the Natural Hazards Partnership (Met Office, 2015). The surface water hazard impact model (SWHIM: Cole et al., 2013) benefits from using the G2G hydrological model (Moore et al., 2006) to compute a surface water flood hazard footprint in response to a given rainfall event. This is then used with reference to vulnerability and exposure data provided by the Health and Safety Laboratory (HSL) to define overall flood risk. This methodology has been demonstrated at the proof of concept stage with work planned to develop a fully operational surface water flood (risk) forecasting system in 2016.

In Scotland, this approach was piloted by the SFFS for Glasgow during the 2014 Commonwealth Games with the development of an operational surface water flood risk forecast with a 24-h lead time. The pilot system delivered a novel method for forecasting the impacts of flooding in real-time and increased knowledge on communicating uncertainties in flood risk. The research application used the G2G model for surface water flood modeling across a domain of 10 by 10 km grid encompassing the East End of Glasgow. Forecast rainfall inputs used both the ensemble STEPS nowcast system and the blended precipitation ensemble forecast, combining the 2-km STEPS radar extrapolation forecast with the MOGREPS-UK 2.2 km forecast (Moore et al., 2015). The novel approach to impact assessment linked surface runoff to the severity of flooding impacts to people, property, and transport using information from SEPA's pluvial flood hazard maps (Fig. 6; Moore et al., 2015).

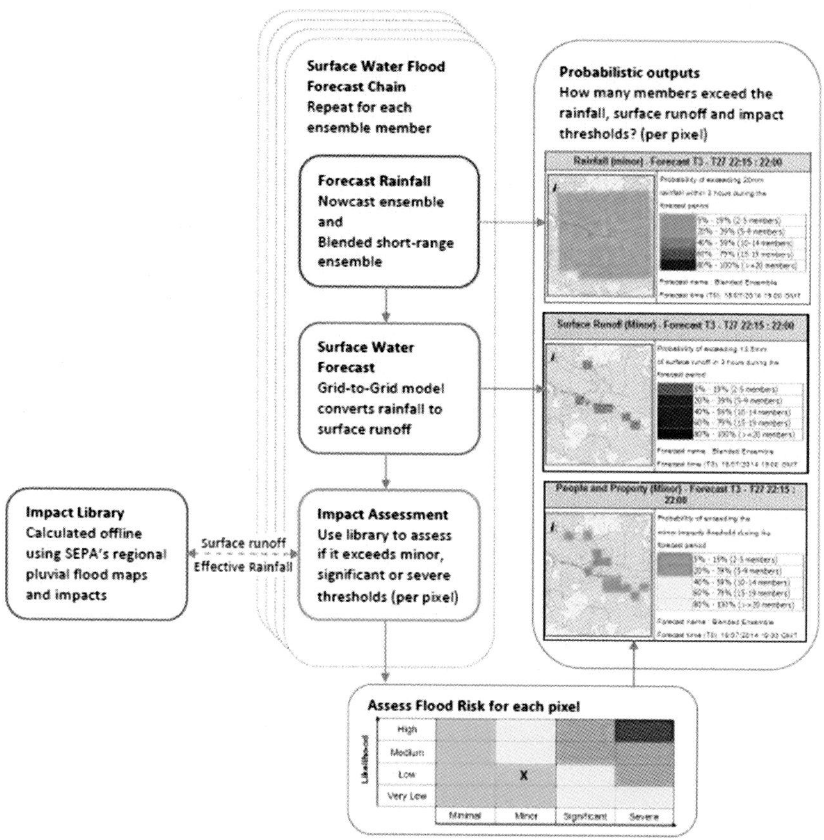

Fig. 6 Overview of the surface water flood forecasting chain, impact assessments and examples of the probabilistic outputs (Moore et al., 2015).

Case Study 2: Flash Flood Forecasting in Scotland

The problem of flash flooding is particularly challenging for flood forecasting professionals. The Scottish Highland village of Comrie is one such example which highlights the difficulties of providing timely and accurate flood forecasts. More than 100 homes were evacuated in storms in Aug. 2012 and once again that year in Nov. Flooding from intense rainfall across catchments like the Ruchill Water is typically driven by orographically enhanced rainfall over the upland areas. Short-range detection of this rainfall is often difficult given the under representation of radar and nowcast during certain synoptic conditions that result in a strong orographic component and has traditionally limited the role of flood warning (Cranston & Black, 2006).

However, some flood risk management measures are now in place for the Perthshire community, including the provision of flood warning; however, given the hydrological response of the river, these alerts could give just 30 min warning (Geldart, Speight, Tavendale, Maxey, & Cranston, 2013). The publication of new strategies for flood warning now provide specific targets for improving the science, including the development of methods for forecasting in rapid response catchments in Scotland (SEPA, 2012). This is supported through recent science development for applying short-term rainfall ensembles to gridded hydrological modeling. When applied to the Nov. 2012 storm, the predictions suggest a strong signal for the extreme runoff at the 18–21 h lead time using a 24-member T+24 of UKV blended ensembles. However, the predictions decrease in skill much closer to the peak using the T+7 Blended Ensemble of the STEPS nowcast. The decrease in performance is potentially related to how STEPS handles rainfall generated by orography, yet the high-resolution UKV aims to explicitly model the enhancement (Cole et al., 2013) (Fig. 7).

The study by Cole et al. (2013) concluded that:

- the G2G modeling approach has utility for forecasting in rapidly responding catchments, but methods of presenting probabilistic flood forecasts should be explored to make best use of the ensemble output;
- for rapid response catchments, further work needs to be done on the G2G model setup, including improvements to flow routing, runoff response, and data assimilation; and
- there was a need to maximize the utility of the hydrological predictions through the use of higher-resolution rainfall ensembles, for example, making the full use of MOGREPS-UK at 2 km resolution.

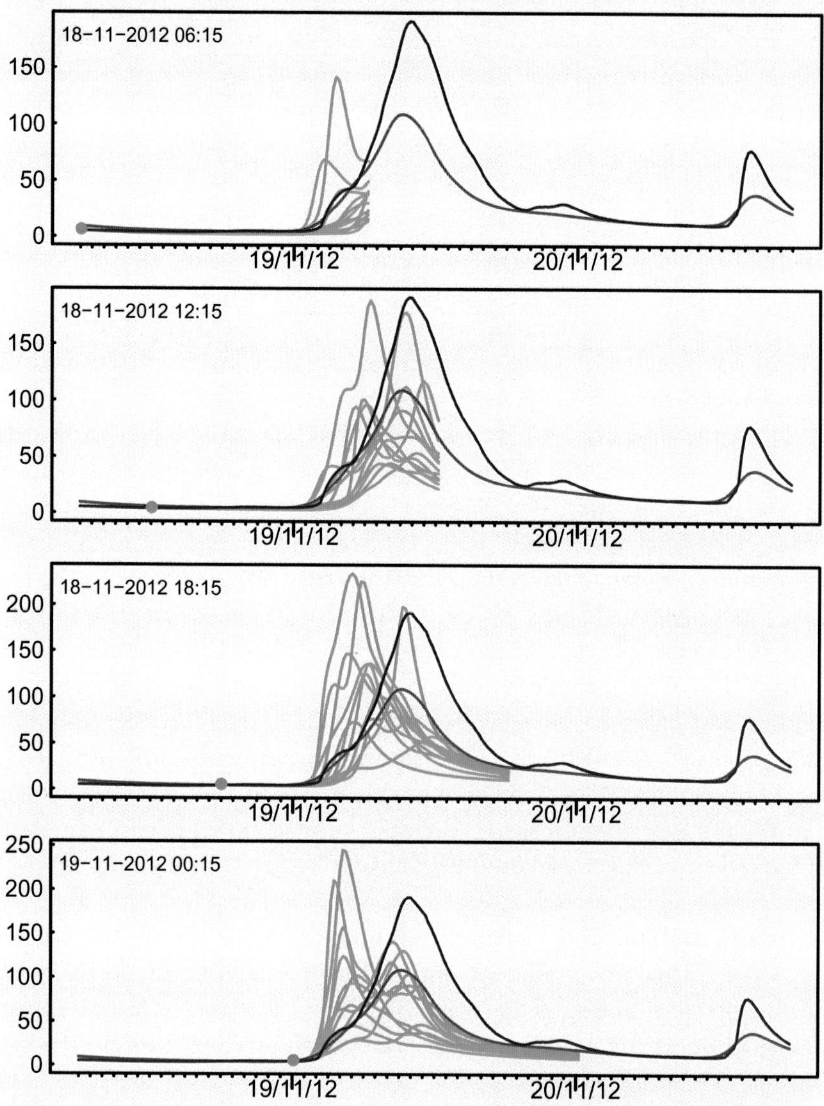

Fig. 7 Sequence of Blended Ensemble T+24 G2G flow forecasts for the Ruchill at Cultybraggan (*Blue* — observed, *Green* — G2G ensemble, *Red* — G2G modeled using raingauge input) (Cole et al., 2013).

2.4 Countrywide Coastal Flood Forecasting

The FFC provides national forecasting services to England and Wales, and — working with SFFS in Scotland — provides a full coastal capability as part of the United Kingdom Coastal Monitoring Forecasting Service (UKCMF). This complements existing public flood warning arrangements (Environment Agency, 2009c) delivered at a local level.

The FFC provides forecasts using observed and modeled data including; recent forecast verification, mean sea level pressure forecasts, surge and wave data, NWP wind forecasts, and astronomical tide level data. These data, along with discussions with monitoring and forecasting duty officers at local Environment Agency forecasting centers, are used to determine the overall coastal flood risk.

At shorter lead times, up to 7 days ahead of forecast flooding, surge data is routinely produced for specific reference ports around the coasts of the United Kingdom using the National Oceanographic Centre's Extended Area Continental Shelf Model (CS3X) operational surge forecasting model (Flowerdew, Mylne, Jones, & Titley, 2013) The CS3X consists of a 12-km shelf model with increased resolution (1 km^2) and one-dimensional fluvial models to represent the complexities of the Bristol Channel and Severn Estuary. The CS3X model is run in real time using Met Office deterministic and ensemble meteorological data sets, providing a deterministic surge forecast to T+48 h and a 12-member surge ensemble to T+162 (time lagged to generate a 24-member ensemble to 144 h), both four times a day. Fig. 8 illustrates a typical surge ensemble output, shown here for the port of Sheerness on the east coast of England. Alternative presentations of surge data from the same modeling system are illustrated in Case Study 3.

Short- and medium-range deterministic wave forecasts are supplemented by wave ensemble data out to 7 days, including significant wave height, direction, and swell. Forecast wind data is also used in the assessment of coastal flood risk, with empirical evidence showing that onshore winds can increase overtopping of coastal defences, exacerbating the impacts of coastal flooding. Forecasts of wind speed and direction are available to complement the surge and wave data and represent important components in the assessment of coastal flood risk.

At longer lead times, generally 7–15 days ahead of forecast flooding, medium-range ensemble weather forecasts from the European Centre for Medium-Range Weather Forecasts (ECMWF) are assessed to identify synoptic weather patterns that could lead to coastal flooding impacts. Historically this “longer range” analysis has been done subjectively by

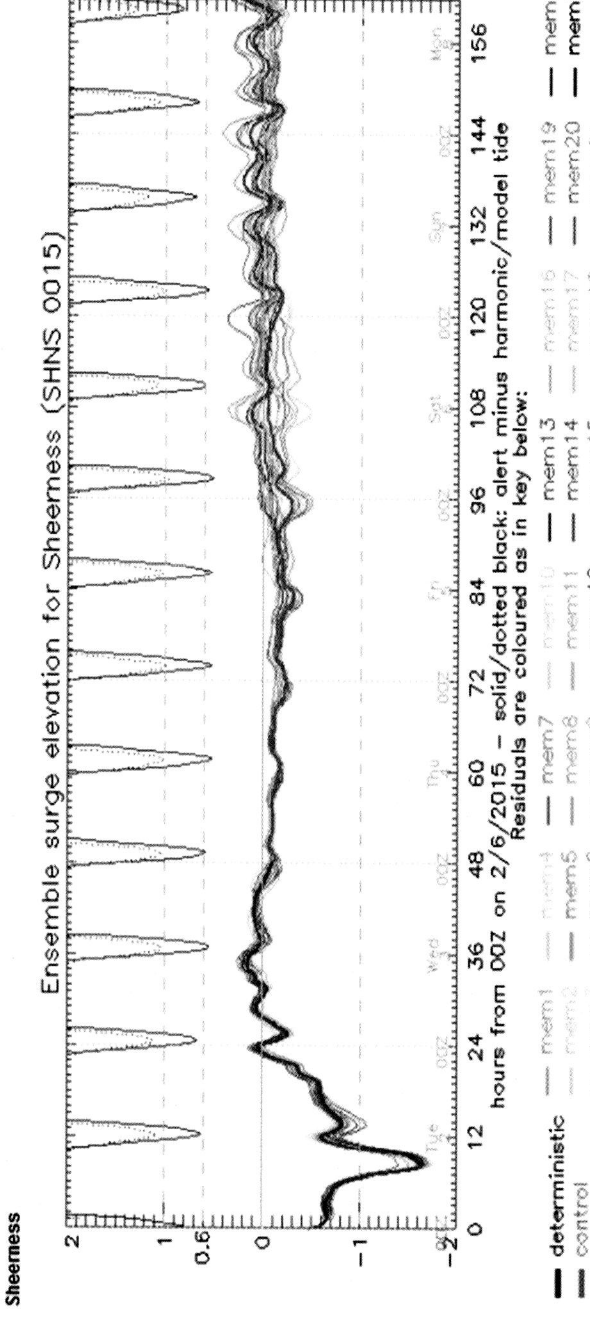

Fig. 8 Surge ensemble output for the port of Sheerness on the east coast of England, showing both the short-range deterministic and the medium-range ensemble forecasts.

"eye-balling" NWP "postage" stamp displays of ensemble data beyond day 7 (the range of the current forecast surge data) to identify synoptic patterns that might lead to an increased coastal flood risk. This process has been replaced recently by an objective measure derived from the Met Office's "Decider" tool, configured to objectively identify the risk of a significant coastal flood event in the 7–15 days period along the UK coastline.

This approach analyzes the 15-day ECMWF ensemble forecast to provide objective probabilities of a synoptic weather pattern occurring over the United Kingdom, the North Atlantic, and Western Europe, that may lead to a significant surge, and assesses this together with the astronomical tidal cycle.

The tool presents the probability of these high-risk regimes affecting Britain and provides guidance of the risk of coastal flooding in the medium-range period (7–15 days). Fig. 9 is an example of the output from the tool, based on the 12 GMT run of the ECMWF model on Oct. 26, 2014.

Probability of coastal flooding
related regimes affecting England and Wales
Based on the weather regime forecast tool
ECMWF 12Z run on Sat 4 Jul 2015

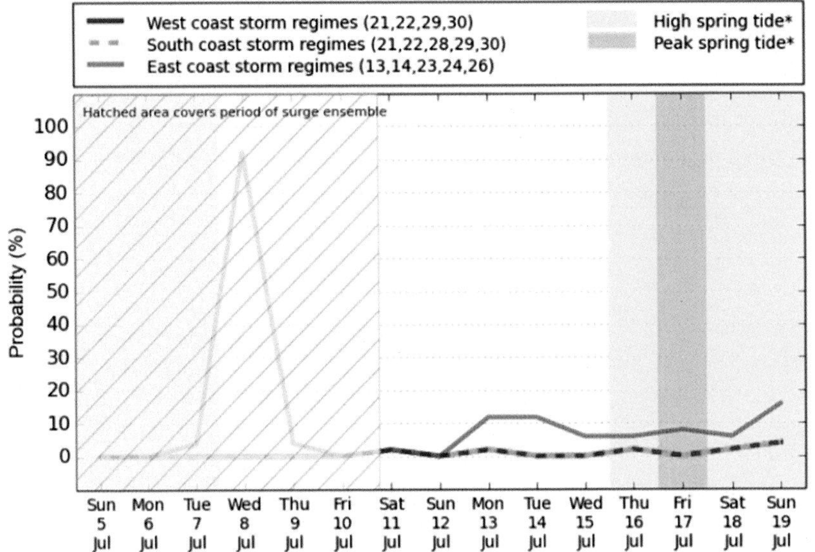

Fig. 9 Example output from the Weather Regime Analysis tool. Plot shows some increased probability of coastal flooding associated regimes for the east coast during the forecast period. Note that spring tide periods are highlighted in *yellow*.

Case Study 3: Coastal Flood Forecasting

Background

In 1953, a significant storm surge caused devastating floods along the east coast of England and Scotland. This, coupled with an inability to warn the public effectively at that time, led to what has been described as the worst national peacetime disaster to hit the United Kingdom, with 19 fatalities in Scotland and 307 fatalities in England. As a result of the 1953 floods, a network of tide gauges was set up around the United Kingdom to monitor tidal surges better, and the Storm Tide Warning Service was set up within the Met Office to provide a service for coastal forecasting.

Over time, the Storm Tide Warning Service has evolved to form the UKCMF service, providing both observed and forecast coastal data to the Environment Agency, Natural Resources Wales and SEPA. Among the services provided by UKCMF (including forecast tide and surge data) is the management of both the national tide gauge network (operated and maintained by the National Oceanography Centre (NOC)), and a network of wave buoys operated and maintained by the Centre for Environment, Fisheries and Aquaculture Science (Cefas).

There have been significant advances in the years since the 1953 floods. Forecasters have access to instantaneous sea level, wind, and wave observations, as well as gridded datasets showing the spatial variation of forecast surge, wind, and wave data. This then feeds smaller-scale, finer-resolution forecasting models which predict localized conditions at key forecast locations around the coastline. This data is provided to duty officers within the Environment Agency, Natural Resources Wales, and SEPA, who use the forecast information to inform the issue of Flood Alerts and Flood Warnings to members of the public and professional partners.

Coastal Flooding Case Study

A succession of rapidly deepening low pressure systems during the winter of 2013/14 presented significant coastal flood risk around the United Kingdom. Two such systems, with associated storm surge and large waves, produced coastal flooding around many areas of the United Kingdom, noticeably the north Wales coastline and areas of eastern England in Dec. 2013, and south and west coasts of Wales, along with southwest England, in Jan. 2014.

The event in Dec. 2013 occurred during a relatively minor spring tide period, with astronomical tides alone some way below flood alert levels. However, in the week leading up to the peak in the spring tide cycle, ensemble surge forecasts gave an indication of the potential of a large storm surge coinciding with the highest tide. This early understanding of risk was only possible due to the recent advances in probabilistic forecasting techniques, run on the Met Office High Performance Computer (HPC) and utilized by the FFC, Environment Agency and Natural Resources Wales, and by the SFFS and SEPA. This enabled discussions to take place between the FFC, Natural Resources Wales, and the Environment Agency on the potential impacts several days ahead of the largest tide. There followed an escalation in risk within the FGS and this prompted early conversations with professional partners at a local and national level, providing exactly the kind of early notification envisaged in the Pitt Review.

Fig. 10 illustrates the benefit gained by longer lead-time probabilistic forecasting. A tight banding of the ensemble members shows the agreement in marginal or slightly negative surge forecasts for the first 4 days. Beyond this there is much greater uncertainty, and the potential for a storm surge in excess of 1 m.

As the event progressed, the focus moved from long-range probabilistic modeling, giving an understanding of the potential risk, to shorter-range deterministic forecasting, using finer resolution Natural Resources Wales models, driven with data from the Met Office wind, wave, and surge models.

The nature of the conversations between the FFC and Natural Resources Wales also progressed from broad discussions around large-scale weather systems, to the development of storm surge at key locations. A good example of this is the primary tide gauge at Liverpool. As can be seen in Fig. 11, the surge was forecast to peak at 1.4 m during high tide, potentially causing a significant increase in sea level. However, in the hours before high tide, the observed residual suggested that the storm surge had peaked lower and earlier than forecast. Consultation with the FFC was able to establish that the meteorological conditions forecast to cause the storm surge were still present and this information was disseminated, reinforcing the decision to issue severe flood warnings and evacuate at specific risk areas. The plots in Fig. 11 show the resultant increase in storm surge, culminating at high tide and causing the highest recorded water level above Ordnance Datum (6.2 m AOD) since the gauge had been established in 1991. This was 0.3 m higher than the previous highest recorded level.

While there are lessons to be learned, and ongoing improvements will continue to be implemented, the management of this event would not have been possible prior to the formation of the FFC. In addition, the early notification and understanding of risk would not have been possible without long-range probabilistic forecasting.

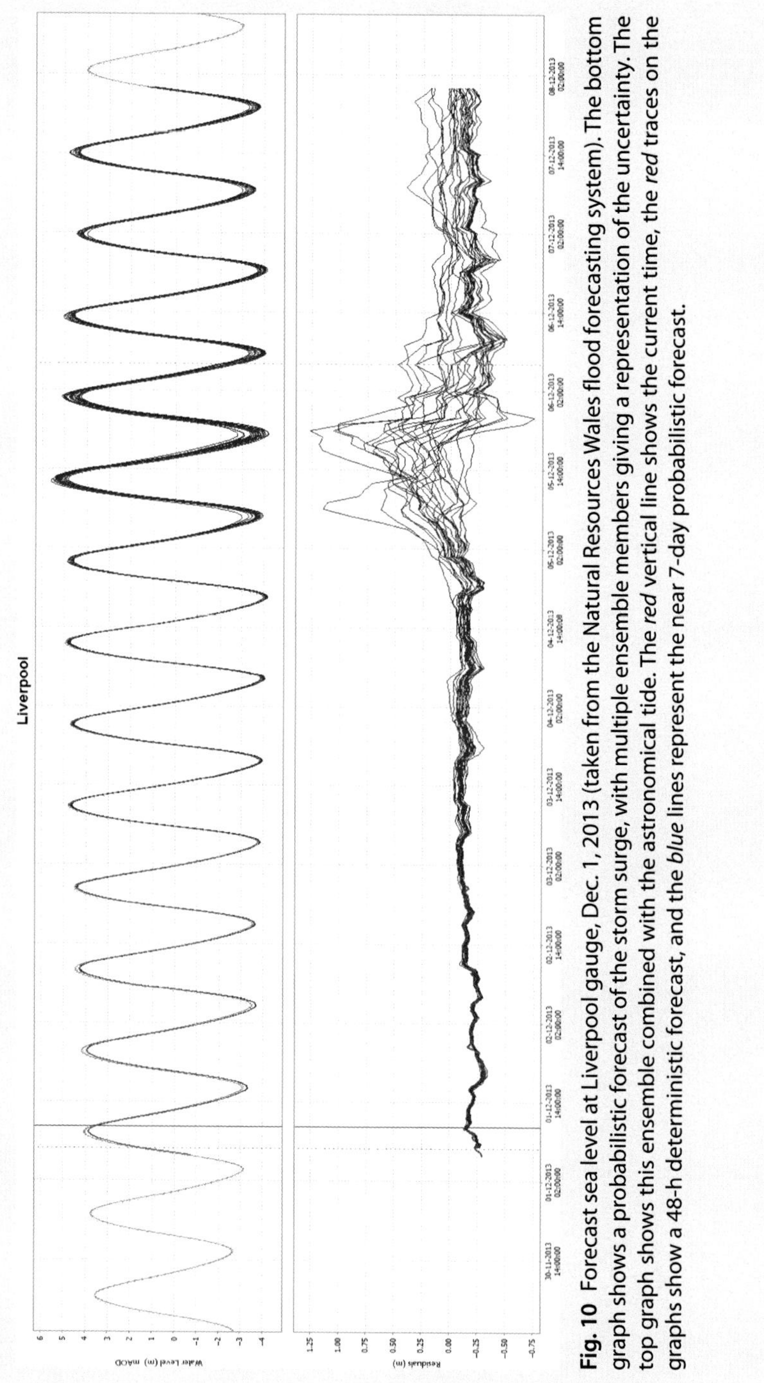

Fig. 10 Forecast sea level at Liverpool gauge, Dec. 1, 2013 (taken from the Natural Resources Wales flood forecasting system). The bottom graph shows a probabilistic forecast of the storm surge, with multiple ensemble members giving a representation of the uncertainty. The top graph shows this ensemble combined with the astronomical tide. The *red* vertical line shows the current time, the *red* traces on the graphs show a 48-h deterministic forecast, and the *blue* lines represent the near 7-day probabilistic forecast.

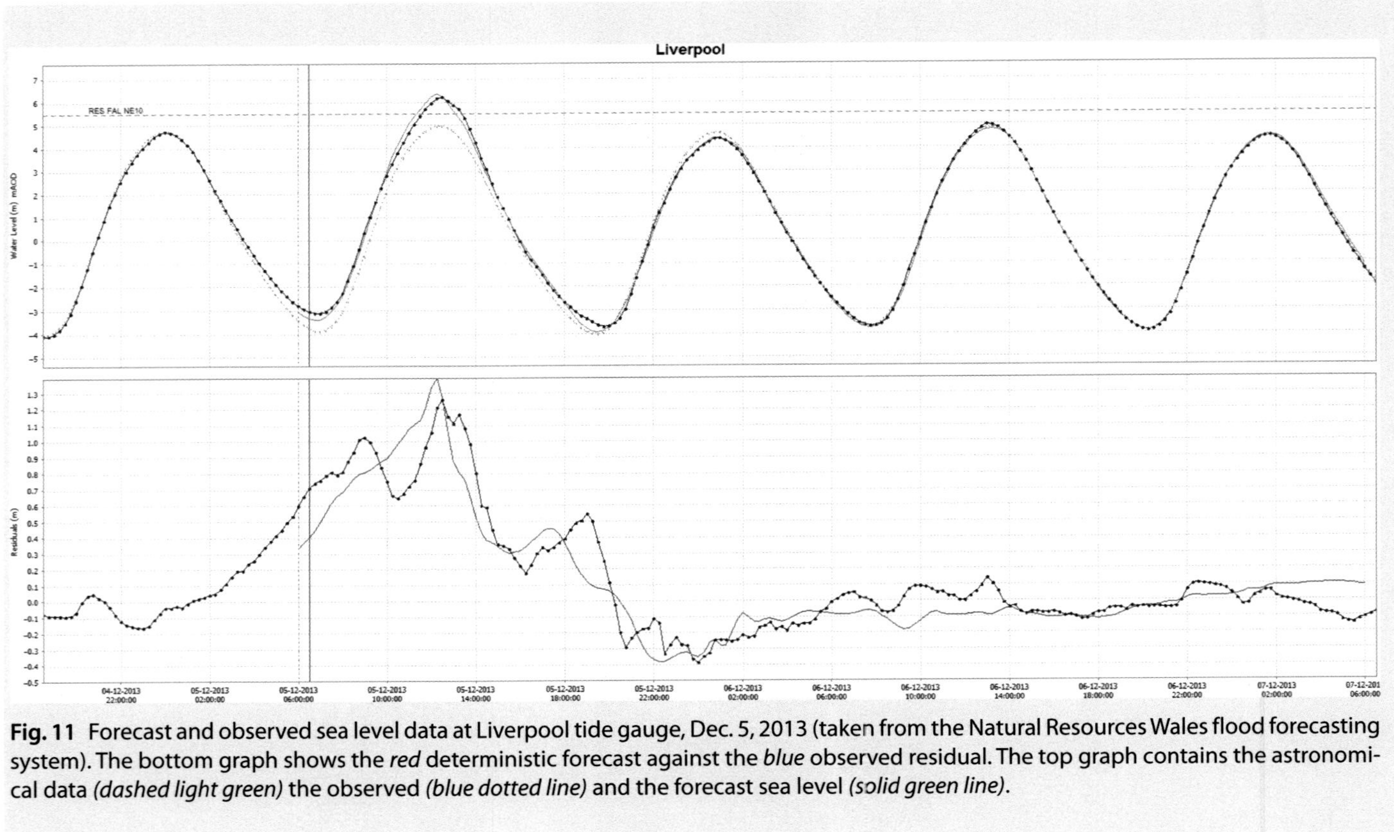

Fig. 11 Forecast and observed sea level data at Liverpool tide gauge, Dec. 5, 2013 (taken from the Natural Resources Wales flood forecasting system). The bottom graph shows the *red* deterministic forecast against the *blue* observed residual. The top graph contains the astronomical data *(dashed light green)* the observed *(blue dotted line)* and the forecast sea level *(solid green line)*.

2.5 Countrywide Groundwater Flood Forecasting

As with surface water flood risk, the responsibility for forecasting groundwater flood risk across Britain varies by country. In England the responsibility for providing forecasts for groundwater flooding currently lies with LLFAs. In Scotland this responsibility lies with SEPA.

Forecasts of groundwater flooding are currently only delivered for a limited number of locations, where this type of flooding has historically presented a risk. These locations are predominantly across southern England, but also include parts of the southeast and northeast, and are linked to the underlying chalk/limestone geology. Assessment of flood risk in these areas is currently achieved by the Environment Agency's monitoring of key boreholes, with local thresholds used for warnings. The FFC is in the early stages of exploring an approach that would provide objective, spatial forecasts of groundwater flood risk at longer lead times.

3 FORECAST DISSEMINATION PROTOCOLS AND PRODUCTS

3.1 Flood Forecasting and Warning Dissemination

There are well-established arrangements for the warning of severe weather and flooding across England, Scotland, and Wales. The Met Office is responsible for the dissemination of severe weather warnings through the National Severe Weather Warning Service (NSWWS), and Floodline (operated by the Environment Agency, SEPA and Natural Resources Wales) is the primary method of disseminating warnings of river and coastal flooding (Fig. 12). Strategic approaches for flood planning and response are then coordinated under frameworks such as the National Flood Emergency Framework for England (Defra, 2014). Such approaches provide clear thresholds and guidance for emergency response arrangements, and ensure a multiagency approach to managing flood events.

However, as previously described, the drivers for the establishment of the national flood forecasting centers were to fill a strategic level gap in flood forecasting, provide longer lead times for warning of flood risk, and to model, forecast, and warn for all sources of flooding. As such, recent developments in the dissemination of flooding information include countrywide scale guidance on flood risk, and products and services to cover a range of sources of flooding.

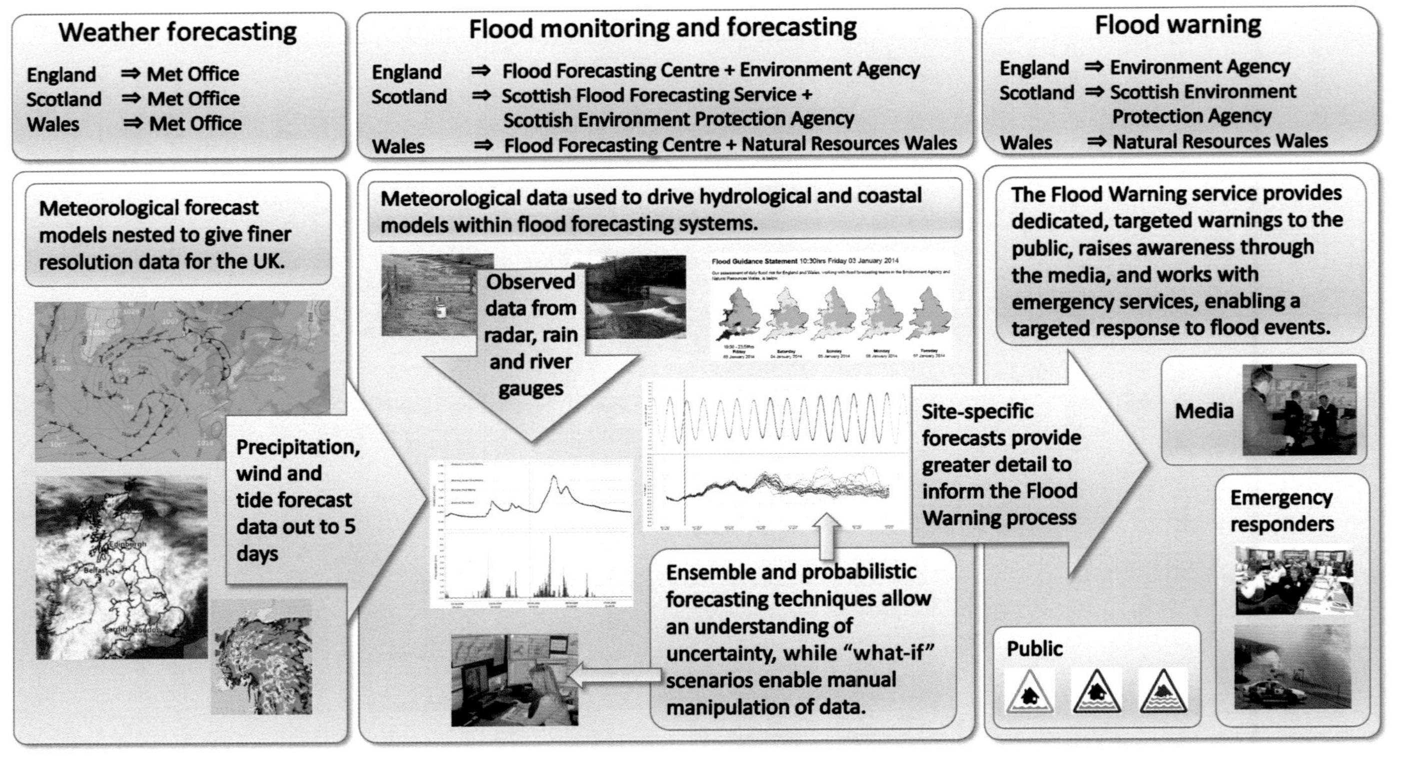

Fig. 12 Organizational arrangements for weather and flood monitoring, forecasting, warning, and response across England, Scotland, and Wales.

3.2 Flood Forecast Products and Services

Emergency responders, or Category 1 and 2 responders as defined by the Civil Contingencies Act, receive a range of flood forecasting and guidance products in the United Kingdom. These are produced by the FFC and the SFFS across a range of services. Customers include the Environment Agency, Natural Resources Wales, SEPA, national government, and the emergency response community such as the fire and rescue services and infrastructure operators. Products and services have been developed to meet customers' strategic needs and include the following:

- *Flood Guidance Statement (FGS)*: This is the main product for emergency responders and was introduced in Apr. 2009. It summarizes the forecast flood risk from rivers, the coast, and surface water for the next 5 days in England, Scotland, and Wales (groundwater flooding was included in 2012 in England and Wales). It includes a mix of maps and detailed commentary. The spatial detail is at a relatively broad scale, covering counties and unitary authorities, although more local detail is provided when appropriate. This is produced in collaboration with the flood warning authorities, Environment Agency, Natural Resources Wales, and SEPA.

  The FGS is issued daily and more frequently at times of higher risk. It is disseminated by email and is also available on a dedicated web portal for the responder community's Hazard Manager. It is currently emailed to over 3000 government and emergency responders. This service is supported by a telephone consultancy service, and during flood events regular advisory telephone conferences will take place between key organizations.

  At the heart of the FGS is the flood risk matrix. The risk is determined by assessing the potential impact and the likelihood of this, as illustrated in Fig. 13. The FGS addresses the recommendations in the Pitt Review by providing longer lead-time warnings and ensures that the responder community are aware of low-likelihood but potentially high-impact events.

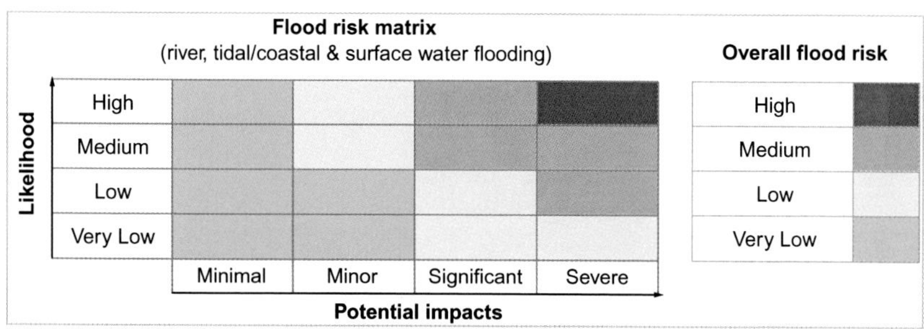

Fig. 13 The flood risk matrix that is used across England, Scotland, and Wales to communicate the likelihood and potential impacts of flooding.

- *Three-day flood risk forecast*: The target audience for this product is members of the public. It is an abridged version of the FGS with maps of flood risk for the day on which it is issued and the subsequent 2 days with a short commentary. It is updated in conjunction with the FGS and published on the web by the Environment Agency and Natural Resources Wales.

  The FGS and the 3-day flood risk forecast are generally aligned with the Met Office's NSWWS. A joint decision-making framework for flood and weather warnings ensures consistent messaging during flood events, a single authoritative voice, and clarity for customers.
- *Hydromet service*: This service is aimed at Environment Agency and Natural Resources Wales staff across England and Wales and was introduced in 2011. The Hydromet Service provides hydrometeorological guidance on a daily basis and includes bespoke rainfall, surge, wind, and wave forecasts as well as guidance on the performance of Met Office NWP models (deterministic and probabilistic) and more recently the performance of G2G. The FFC hydrometerologists also provide a 24/7 consultancy service to Environment Agency and Natural Resources Wales flood hydrologists.
- *UK Coastal Monitoring and Forecasting (UKCMF) service*: The FFC provides coastal flood forecasting guidance and assessment of the storm surge potential across the whole of UK coastal waters. Working in partnership this output supports the provision of coastal flood warning.
- *Internet-published flood forecasts*: Responders in Scotland have access to published hydrological and coastal flood forecast conditions made available through the FEWS Scotland system (Cranston et al., 2007). The reports present the latest deterministic forecast available from the forecasting systems which are a useful source of flood risk information during incidents. However, they do not capture any potential uncertainty in the predictions, and consequently SEPA and the SFFS work with responders to educate them in the use and interpretation of the available predictions.
- *Surface water flood forecasts*: In Scotland, a recent trial for alerting and communicating surface water flood hazards was piloted during the 2014 Commonwealth Games in Glasgow. A new daily surface water flood forecast was designed and produced based on operational requirements and emergency responder feedback, which linked the severity of flooding impacts on people, property, and transport to a probabilistic surface water forecast (Moore et al., 2015) (Fig. 14).

4 MEASURING PERFORMANCE OF THE FLOOD FORECASTING AND WARNING SERVICE

4.1 Introduction

It is a major challenge to measure the overall performance of a flood forecasting and warning service. This is because there are many interdependent components, all of which need to perform in order to ensure that recipients receive a timely and accurate warning. The quality of the service is dependent on the performance of:

- the detection elements (raingauges, radar observation products, telemetry system, local river gauges);
- the forecasting elements (rain forecast products from Met Office, guidance from FFC, forecasting platform architecture, local agency forecasting models, and added value by the Duty Officer who interprets the forecast); and
- the warning elements (has the forecast been properly communicated to the Warning Duty Officer, have they made a timely decision, has the Flood Warning system functioned correctly and sent a message to the public, and is the content of the message good).

The issue is made more complex with the Environment Agency being a distributed organization that delivers services at the local level. Historically, there has been local variation in the setting and operation of flood warning thresholds, making it difficult to measure lead-time performance for the onset of flooding in a consistent and meaningful way.

For these reasons, there have been challenges in developing a formal measure of the performance of the overall flood forecasting and warning service. This section focuses only on England and describes how the Environment Agency is building up a picture of the performance of the overall service.

Internal verification and performance measure work has focused on some of the individual elements (Met Office assessing performance of rainfall forecasting products, FFC assessing performance of the FGS, and the Environment Agency assessing the performance of their local catchment and river forecasting models). Recently, the Environment Agency has tried to measure the value added by the Environment Agency's monitoring and forecasting duty officers. However, experience has shown that during a major and prolonged event, such as the winter flooding of 2013/14, automation of the verification and performance monitoring process is required.

Glasgow Surface Water Flood Forecast
Issued at 17:00hrs Sunday 03 August 2014

Our assessment of surface water flood risk in Central Glasgow for the next 24 hours is below. This statement is valid at the time of issue.

Headline surface water flood risk

The overall surface water flood risk for Glasgow for Sunday evening is LOW. The flood risk on Monday is VERY LOW.

Minor disruption is possible including flooding and disruption to infrastructure and transport links. The risk is highest in the north and centre of Glasgow.

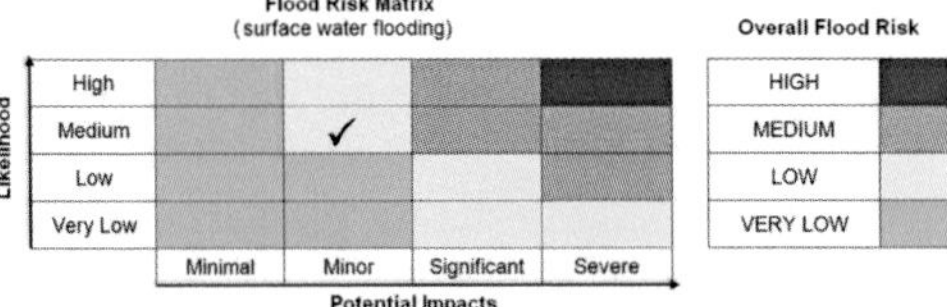

Weather Situation

Further rain this afternoon, evening and overnight, occasionally heavy but becoming patchier and unlikely to be as heavy as this afternoon. Rain gradually clearing north during Monday morning.

General overview of surface water flood risk in Glasgow

The surface water flood risk for Glasgow for Sunday evening is LOW and on Monday is VERY LOW.

Following heavy rain witnessed on Sunday afternoon and further lighter rain forecast throughout the evening, there is an increase in the flood risk to the north and centre of Glasgow, primarily to main transport routes in these areas.

Typical impacts will include standing water and spray on the road network. Transport links to and from event venues, especially the city centre, may be impacted. Rail links and infrastructure may also be at risk. Residual impacts may continue into Monday however these are expected to be very isolated and minor, if at all.

Potential flooding impacts may include:

Transport

- City centre roads
- M8 J14-18

Surface water flood risk heightened from: 15:00hrs 3rd August 2014
Expected time of highest risk: 17:00hrs 3rd August 2014
Expected end of heightened risk: 18:00hrs 3rd August 2014

General overview of other sources of flood risk in West Central Scotland (See the Scottish Flood Forecasting Service Flood Guidance Statement issued daily at 10:30hrs for details)

River Flood Risk: Very Low
Coastal Flood Risk: Very Low

No further Glasgow Surface Water Flood Forecast daily assessments are scheduled, unless the situation changes significantly.

Fig. 14 Example of the Glasgow daily surface water flood forecast issued at 17:00 Aug. 3, 2015 (Moore et al., 2015).

The following sections describe how local river model performance has been measured, why there has been a need to develop this into a national picture of how models are performing, how the national picture has been achieved, and how the Environment Agency is now looking to develop a formal measure of the overall performance of the Flood Forecasting and Warning service.

4.2 Assessing Local Model Performance

During the last 5 years, the Environment Agency has started to measure the performance of some of its local river models. The models have generally been assessed in three modes:

- *Perfect rainfall* — this assumes perfect fore-knowledge of rainfall, and uses observations of rainfall as input to the model. This is most useful to assess how good the model is without being impacted by errors in the rainfall forecast delivered by a different organization.
- *Forecast rainfall* — this uses the rain forecast available in real time and so better reflects the actual operational performance of the model. This is called "simulation" mode.
- A third *forecast mode* analyzes a model driven by forecast rainfall, as above, and/or an updating model capability — if the real-time system has this capability.

Assessment has tended to focus on an analysis of skill score performance using probability of detection (POD) and false alarm rate (FAR) to predict the crossing of flood warning thresholds (see Table 1). These measures have been used because they directly measure the ability of the models to meet forecast user requirements and the challenge of forecasting the high flow range accurately, unlike measures such as the Nash-Sutcliffe Efficiency (NSE or efficiency-R^2), which measure the overall fit of the model to the observations across the whole flow regime.

Table 1 Contingency table and performance measures derived by applying the contingency criteria

	Threshold observed	
Threshold forecast	Yes	No
Yes	a	b
No	c	d

False alarm rate: FAR $= b/(a+b)$
Probability of detection: POD $= a/(a+c)$
Critical success index: CSI $= a/(a+b+c)$

Although the POD and FAR metrics are good in principle, in practice they are of limited use due to the rarity of observed threshold crossings (ie, because most places don't flood in most years). For this reason the analysis can be applied to lower thresholds, for example, the level which has been exceeded 10 times in the historical record.

For each of these studies, decisions have to be made about:

- threshold value to use (as described above);
- whether or not a magnitude tolerance is applied to the threshold (usually 0.2 m);
- what "window" is applied to identify which forecasts are assessed for each lead time; and
- what "threshold crossing time window" to use to identify if a forecast can be counted as a hit (or near miss/close false alarm).

The above parameters have been varied, according to local need, and assessment done in one, two, or sometimes all three of the forecast modes mentioned above.

The benefits of these studies at the local level include:

- informing duty officers so they are able to make better operational decisions;
- identifying deficiencies in individual model/data components to inform investment decisions; and
- providing confidence to flood warning staff that forecasts are good enough to inform operational decisions.

However, after 10 years operating a wide range of real-time catchment and river model types, the Environment Agency has decided that it now needs to converge on the most effective subset of model types.

4.3 National Assessment of Local Model, and National Model Performance

To assess the merits or otherwise of the various model types and to develop a national picture of model performance, it should be possible to bring together all the skill score information provided by the many local model performance studies. However, there are important regional differences in how these studies have been undertaken. So it would be difficult to form a fair opinion of the merits of each model type applied across different parts of the country.

The Environment Agency has recently commissioned CEH Wallingford to paint this national model performance picture. The approach taken has been to gather from previous local performance studies the raw data: river flow observations, flow forecasts and historical simulation of flows. While there are important regional differences in how these data have been gathered, collation of the underlying project datasets has allowed standardization of the statistical methods (Environment Agency, 2015). This then enables a fair assessment of how each model type is performing across different parts of the country.

In addition, this report provides an assessment of how the national G2G model is performing across England and Wales. Where there are both local model and G2G performance data, then a comparison of how the G2G model is performing relative to the locally calibrated lumped model is also provided.

This is the first time such a countrywide picture of model performance has been compiled across England. It considers regional and model-type differences and presents an overview of the current forecasting capability of systems in current operational use (Environment Agency, 2015). The report presents detailed performance information (Table 2) in a template summary for each site.

Table 2 Performance information presented for each model

Performance metric	Further information
1. Thresholds and tolerances	Uses QMED* and QMED/2 with 20% tolerance
2. Historical simulation results	Uses Nash-Sutcliffe Efficiency
3. Flood hydrographs	For 10 largest events — includes simulation
4. Table of skill scores	POD, FAR, and CSI at a range of lead times
5. Additional statistic	Likelihood of actual crossing if forecast crossing occurs
6. Overall performance statistic	Weights all the metrics and combines to provide a single overall performance value

*The QMED at a gauging station is the median of the set of annual maximum flow data. (This equates to the 50% Annual Exceedance Probability)

An interesting feature of this work is that an "Overall performance statistic" is developed and trialed. Such a statistic will make it easier to compare model performance at a site or between sites, and to present an easy-to-take-in assessment of model performance by duty officers (Environment Agency, 2015).

4.4 Formal National Measure of Overall Performance of the Forecasting and Warning Service

For some years, the UK government have wanted a national performance measure of the overall flood forecasting and warning service. Such a measure needs to capture the outcome of the overall service to members of the public and businesses who are at risk of flooding, and to the response organizations that are responsible under the Civil Contingencies Act 2004.

It is considered that a simple and sustainable way to measure the outcome of the service is to measure the lead time of an accurate warning that reaches the recipient prior to the onset of flooding. Until recently, local variation in the setting of flood warning thresholds has made it impossible to measure lead time accurately in a consistent national way. However, on-going work to define the impact threshold — that is, the water level at which the onset of flooding occurs to the first property or infrastructure (such as a road) in a community — for every at-flood-risk community in England should be completed in 2016.

The lead time of a flood warning may then be defined consistently as the time from when a warning is sent to the recipient, to the time when the impact threshold is exceeded, as recorded by telemetry. The measured lead time may be compared against the Target Lead Time as published in the Local Flood Warning Plan. It is proposed that this information could form a core part of a new Quality measure reported to government from 2016/17, and work is ongoing to develop, refine, and base-line this measure.

5 FUTURE FORECASTING CHALLENGES

5.1 Recent Experience and Key Research Areas

With both the FFC and SFFS acting as catalysts, working relationships between the Met Office, Environment Agency, Natural Resources Wales, and SEPA have flourished, connecting relevant teams and joining up communications. This has resulted in a more efficient use of resources. The successful partnership between key public sector agencies has seen the FFC and SFFS grow quickly to become trusted advisors in improving resilience and preparedness to flood hazards.

Flood forecasting in Britain has undergone a number of challenging tests in recent years. In Nov. 2009, a slow moving weather front and orographically enhanced rainfall brought severe flooding to Cumbria, NW England, with over 5000 homes and businesses affected (eg, Sibley, 2010). During the summer of 2012, successive low pressure systems across England, Wales, and southern Scotland brought widespread flooding (eg, Marsh, Parry, Kendon, & Hannaford, 2013). Subsequently, winter 2013–14 was exceptional for the duration of flooding and was the wettest winter in the UK's observational records that extend back almost 250 years in England and Wales (Kendon & McCarthy, 2015). It was also exceptionally stormy, with a succession of coastal floods as documented in Sibley, Cox, and Titley (2015). Indeed, the substantial impacts from flooding during the winter of 2013–14 led to the UK's crisis response committee (COBR) convening on 35 occasions, a far greater number than during any previous period of flooding.

These events have highlighted and endorsed the benefits from the improvements already implemented in our forecasting capability, products, and services. They have also helped to distil our challenges that are likely to result in step changes over the following decade. The improvements that are likely to lead to these step changes, *as* identified from recent experience as well as recommendations following the winter 2013–14 floods (eg, Natural Resources Wales, 2014), the 2012 floods (eg, Environment Agency, 2013) and part of SEPA strategic developments in flood forecasting (SEPA, 2012), are summarized in Table 3.

Table 3 Areas to target for improvements as identified from flood events 2012–14

1. *Nowcasting science* to forecast rapidly developing situations in near real time and flash flooding in urbanized areas and rapid response catchments. Key areas include rainfall intensity, pattern of rainfall profile (intense rainfall once catchment has wetted up), antecedent conditions, urban modeling, probabilistic approaches.
2. *Estuaries and other high-risk areas.* Coupling national river flow models with coastal and hydrodynamic estuary models, and better routing models to provide improved forecasting in high-risk areas.
3. *Groundwater.* Benefit from a national groundwater flood risk capability, susceptibility assessments, hazard impact modeling approach.
4. *Longer-range precipitation and fluvial ensembles* to forecast response in slower-responding rivers. For example, River Thames 2014, extend detailed forecasts to day 6. Explore multi-meteorological and multi-hydrological model ensembles and improve understanding of hydrological uncertainty with respect to catchment status (eg, soil moisture).
5. *Enhanced longer range coastal ensembles* to develop and fully exploit the full potential of total water levels from surge and waves, and translating this into more detailed impact assessments.
6. *Longer lead-time projections.* For all sources of flooding, including at greater lead times than 5 days. For example, during winter 2015, COBR were seeking forecast assessments 2–6 weeks ahead. A number of approaches are being explored, including, for example, 7–15 day combined surge, wind, wave forecast.
7. *Concurrent flood risk.* Improved understanding of multiple sources and joint probabilities.
8. *Contextual information.* Understanding of severity of floods and attribution in a non-stationary climate.
9. *Multi-hazard impact forecasting.* Improving forecasting, observation, and reporting of flood impacts as well as the understanding of customer response triggers to improve categorization and focus.
10. *Making science relevant.* Communicating science, for example, the risk matrix, drawing on social science.

Improvements in operational meteorology, both deterministic and ensemble (probabilistic), will translate into improved flood forecasting; and these are considered first. Other key influencers in improved flood forecasting capability include improvements to observations, hydrological and hydrodynamic models, computer power, improvements in modeling storm surges, more integrated modeling, such as fluvial and coastal interactions in estuaries, as well as the interaction of fluvial and groundwater flooding.

Detailed impact modeling is also showing promise. For example, developments in the SWHIM combine high-resolution atmospheric models with a hazard footprint and time-varying national impact datasets (Moore et al., 2015; Pilling et al., 2014). Improvements in coupling multi-meteorological and multi-hydrological models are expected to realize significant improvements.

Appetite for increased lead times and a greater understanding of global teleconnections continues to drive improvements in the monthly, seasonal, and longer-range forecast systems. In addition, there is increasing interest in contextualizing flood events and attribution to climate drivers and climate change. Intertwined among all of this is the continual development of people and skills, and communication of products and services.

5.2 Nowcasting Science and Short Lead Times

The most dangerous floods include those that result from rapidly developing convective systems, or the organization or alignment of storm cells that may result in flash flooding over large urbanized areas or rapid response catchments such as Boscastle (Golding, Clarke, & May, 2006) and the Comrie (Geldart et al., 2013). These are also the most challenging to predict.

Improvements in real-time observations, such as dual polarization and Doppler radar capability and improved orographic correction schemes, will all translate into better estimates of rain rates and storm tracking. These improved observations, combined with more rapid and frequent data assimilation (eg, hourly) enabled by more powerful HPC will translate into improvements in flash flooding tools and techniques. While a next-generation surface water model is being developed, the SWHIM, further enhancements could enable fully distributed inundation models. These would be based on high-resolution (eg, 2 m) LIDAR data and would run on powerful HPCs for use in real-time forecasting.

The benefits of using ensemble approaches to model convective systems are well documented (eg, Golding, 2000; Sun et al., 2014). The Met Office currently runs MOGREPS-UK, a 2.2-km ensemble configuration with 12 members run every 6 h which are time lagged with the previous run creating a 24-member ensemble. This is a data feed currently used to drive G2G to provide the short-range hydrological ensemble (Section 2.1). Planned improvements with the new Met Office HPC will enable extension of lead times, hourly data assimilation, and increased ensemble size. Improvements in meteorological model physics from ENDGAME (Wood et al., 2013), which was implemented through 2014 and 2015, to a new dynamical core named "GungHo" will allow more efficient scalability to higher resolutions in the future. Further improvements to flood forecasting are likely to be realized through improvements to post processing and understanding of perturbation to hydrological conditions.

5.3 Observations and Instrumentation

Observations are closely linked to data assimilation, and therefore improvements to forecasts. It is worth considering improvements in technology that will provide more accurate estimates of observations. Examples include blending raingauge and radar information to optimize the benefits from gauges (eg, accuracy of rainfall totals reaching the ground) and radar (eg, spatial information). Advances in new instrumentation such as weighing raingauges will allow more accurate and continuous estimates compared to tipping bucket gauges (eg, Sevruk, Ondras, & Chvila, 2009) and soil moisture monitoring technology such as COSMOS probes (eg, Fry, 2015). Such improvements in instrumentation and technology will contribute to improved data assimilation. However, some consideration needs to be provided to understanding the sources of uncertainties in any merged product, including those based on the hydrological end-use (Cranston, 2014).

5.4 Estuaries and Coastal Modeling

Benefits from improvements in ensemble surge forecasts combined with ensemble wave forecast will offer the advantage of ensemble total water levels to be modeled. The success and benefits of long lead-time guidance was demonstrated by Sibley and Cox (2014) in the storm surge event that affected England and Wales on Dec. 5–6, 2013. Furthermore, hindcast experiments conducted for this storm surge demonstrated the even greater lead time would have been available from a higher-resolution atmospheric model (MOGREPS-G, ~33 km; relative to the MOGREPS-15, ~60 km). Indeed, this has now been implemented to routinely drive the operational storm surge model using the general approach described by Flowerdew et al. (2013). Experimental, ensemble wave forecasts are currently being trialed and assessed by the Met Office and the FFC. The combination of surge and wave ensembles with astronomical tides is likely to provide a key step forward.

Improvements in local wave transformation models and inundation modeling approaches will enable a more localized assessment of impacts. In addition, a new surge model, NEMO (O'Dea et al., 2012), which fully couples the ocean and atmosphere, is being developed and is planned to be operational in 2017. This will provide a framework for improved resolutions and closer coupling of systems that will offer improvements to coastal modeling, and in turn improvements to modeling interaction of rivers and sea in the high-risk estuary environments. In time, coastal hazard impact modeling may in turn offer intelligence for coastal erosion, landslides, and coastal debris hazards from high-energy waves.

5.5 Integrated Modeling and Forecasting

It is powerful to consider concurrent fluvial and coastal flooding. A typical scenario could be a protracted stormy period bringing successive intense low pressure systems across Britain. This could result in a storm surge coinciding with high tides and large waves being driven onshore, concurrent with saturated river catchments and high flows and further heavy rainfall.

An integrated modeling framework that considers the optimal coupling of physical interactions between the atmosphere, waves, oceans, land surface, and hydrology resolution is being explored under the UK Environmental Prediction prototype (Lewis et al., 2015). The aim is to provide the best possible national environmental prediction system at a 1-km spatial resolution. This will provide optimum boundary conditions in higher-resolution models, and a test bed for coupled environmental prediction across scales (Lewis et al., 2015). Both a component of and extension from this research is the development of multi-meteorological and multi-hydrological ensembles.

Flooding from groundwater has received relatively little attention from modeling and forecasting at the national scale. This is despite the significant flooding from groundwater in Feb. 2014. In 2015, the UK Groundwater Forum hosted a workshop to share good practice, identify lessons, and contribute to key recommendations that will help inform the UK government. However, this currently remains an area where a relatively modest investment could potentially provide significant improvements in national modeling and forecasting.

While not a common player to major floods, snowmelt can provide a significant contribution to flood hazards in British catchments such as the Tay (eg, Black & Anderson, 1994). G2G does include a snow hydrology component; however, further work has recently been initiated to optimize the best approach for handling precipitation, snow accumulation and melting for the purposes of snowmelt flood forecasting using G2G.

Alongside these developments, advances in downscaled medium-range atmospheric ensembles are being investigated. An example specific to the FFC and SFFS is the downscaling of MOGREPS-G to a 2-km spatial resolution to drive G2G. A key benefit from this is the provision of high-resolution medium-range flow ensembles out to day 6. This will provide longer lead times for the densely populated, slower-responding rivers such as the Lower Thames. To complement this, and illustrate how forecast events compare with climatology, improved contextual information such as $Q(t)$ return periods (Section 2.2) and extreme forecast indices are being explored.

5.6 Longer Lead Time Forecasting: Day 6 to Seasonal

Between Jan. and Mar. 2014, the UK government was requesting, on a regular basis, information from the Met Office and FFC about how long the unsettled conditions would continue. Existing flood forecasting lead times were challenged with a desire for information beyond day 5, with requests for forecasts between 2 and 6 weeks ahead not untypical. Improved understanding of global teleconnections, such as the El Niño and Polar Vortex, coupled with advances in seasonal weather forecasting, are illustrating significant skill in predicting the North Atlantic Oscillation (NAO) and temperature and precipitation forecasts over the UK at particular times of the year (Scaife et al., 2014). There are potential benefits to be explored in forecasting flood risk over the monthly to seasonal timescales for inland flooding, and also coastal flooding in the case of a strong positive NAO signal.

Drawing on approaches illustrated in Section 2.4 and Fig. 9, historical records have shown that particular weather regimes have strong positive relationships with surge and large waves, and so in turn are related to the incidence of coastal flooding. A prototype currently being trialed considers the likelihood of weather regimes, and the connections that these regimes have with the incidence of coastal flooding 15 and 32 days in advance (Section 2.4).

5.7 People, Skills, Interpretation, and Engagement

In order to realize the full operational benefits of scientific developments, it is essential to continue to develop our people who can use, interpret, and add value to what will be increasing amounts of information. This includes the advancement of hydrometeorology as a skill supported by vocational and professional training (PAA\VQSET, 2013, 2014).

There is a need to improve data, systems, and visualization using new technology and to share knowledge between operation and science. Indeed, a working model that draws on expertise from science in exceptional events is something that can be developed further. The FFC and SFFS should draw on recent flood events to help hone research priorities that are steered or influenced by operations and are quickly and seamlessly incorporated into operational improvements or new systems. The key is providing information to end-users with the skills and training to make informed decisions and take appropriate action in a staged approach as confidence increases in significant or severe events.

DISCLAIMER

The views expressed in this chapter belong to those of the authors and do not necessarily represent the views of the authors' organisations.

DEDICATION

This chapter is dedicated to Nigel Outhwaite–a quintessentially English flood forecaster who was a fountain of knowledge on the River Thames. Missed by all.

REFERENCES

Andryszewski, A., Evans, K., Haggett, C., Mitchell, B., Whitfield, D., & Harrison, T. (2005). Levels of service approach to flood forecasting and warning. In *International conference on innovation, advances and implementation of flood forecasting technology, Tromso, Norway, 17–19 October*.

BBC News. (2014). *Scotland weather: Flooding risk 'priority' pledge issued [online]*. Available from: http://www.bbc.co.uk/news/uk-scotland-south-scotland-25629614.

Bell, V. A., Kay, A. L., Jones, R. G., Moore, R., & Reynard, N. S. (2009). Use of soil data in a grid-based hydrological model to estimate spatial variation in changing flood risk across the UK. *Journal of Hydrology, 377*, 335–350.

Black, A. R., & Anderson, J. L. (1994). The Great Tay Flood of 1993. *Hydrological Data UK*. Wallingford (Oxfordshire): Institute of Hydrology.

Bye, P., & Horner, M. (1998). *1998 Easter floods*. Available from: https://www.gov.uk/government/publications/easter-1998-floods-review.

CEH. (1983). http://nrfa.ceh.ac.uk/sites/default/files/Flow_Gauging_on_River_Thames_100_Years.pdf.

Cole, S., & Moore, R. J. (2009). Distributed hydrological modelling using weather radar in gauged and ungauged basins. *Advances in Water Resources, 32*(7), 1107–1120.

Cole, S. J., Moore, R. J., Robson, A. J., Mattingley, P. S., Black, K. B., & Kay, A. L. (2013). *Evaluating G2G for use in rapid response catchments: Final report.* Bristol: Environment Agency.

Cranston, M. (2014). *Merging radar and raingauge data: Opportunities for a future UK-wide strategy*. Available from: https://radarhydrology.wordpress.com/2014/05/30/merging-radar-and-raingauge-data-opportunities-for-a-future-uk-wide-strategy/.

Cranston, M., & Black, A. R. (2006). Flood warning and the use of weather radar in Scotland: A study of flood events in the Ruchill Water catchment. *Meteorological Applications, 13*(1), 43–52.

Cranston, M., Werner, M., Jannsen, A., et al. (2007). Flood early warning system (FEWS) Scotland: An example of real time system and forecasting model development and delivery best practice. In *Proceedings of Defra conference on flood and coastal management, York, UK.* London, UK: Defra. Paper 02-3.

Cranston, M., Maxey, R., Tavendale, A., Buchanan, P., Motion, A., Cole, S., et al. (2012). Countrywide flood forecasting in Scotland: Challenges for hydrometeorological model uncertainty and prediction. In *Weather radar and hydrology, proceedings of a symposium held in Exeter, UK, April 2011* (pp. 538–543). IAHS Publ. 351.

Cranston, M., & Tavendale, A. (2012). Advances in operational flood forecasting in Scotland. *Water Management, 165*(2), 79–87.

Dale, M., Davies, P., & Harrison, T. (2012). Review of recent advances in UK operational hydrometeorology. *Water Management, 165*(2), 55–64.

Defra (Department for Environment, Food & Rural Affairs). (2014). *The national flood emergency framework for England.* Available from: https://www.gov.uk/government/publications/the-national-flood-emergency-framework-for-england.

Environment Agency. (2001). *Lessons learned from the August 2000 floods.* Available from: https://www.gov.uk/government/publications/autumn-2000-floods-review.

Environment Agency. (2009a). *Thames catchment flood management plan.* Available from: https://www.gov.uk/government/publications/thames-catchment-flood-management-plan.

Environment Agency. (2009b). *Flooding in England: A national assessment of flood risk.* https://www.gov.uk/government/uploads/system/uploads/attachment_data/file/292928/geho0609bqds-e-e.pdf.

Environment Agency. (2009c). *UK coastal monitoring and forecasting (UKCMF) service — Strategy for 2009 to 2019.* Bristol, England, UK: Environment Agency (31 pp.).

Environment Agency. (2010). *Hydrological modelling using convective scale rainfall modelling. Science Report — SC060087.* Bristol, UK: Environment Agency (240 pp.).

Environment Agency. (2013). *Managing flood and coastal erosion risks in England.* Report published by the Environment Agency.

Environment Agency. (2015). Understanding the performance of flood forecasting models for investment and incident management. R&D Project Report SC130006/R2. Authors: Robson, A. J., Moore, R. J., Wells, S. C., Rudd, A., Cole, S. J., Mattingley, P. S. (CEH Wallingford), EA/Defra/Welsh Government/NRW Flood and Coastal Erosion Risk Management R&D Programme, Research Contractor: CEH Wallingford, Environment Agency, Bristol, UK.

Flowerdew, J., Mylne, K., Jones, C., & Titley, H. (2013). Extending the forecast range of the UK storm surge ensemble. *Quarterly Journal of the Royal Meteorological Society, 139*, 184–197.

Fry, M. J. (2015). The provision of data from the COSMOS_UK soil moisture monitoring network. In *HIC2014 Conference Proceedings.*

Geldart, R., Speight, L., Tavendale, A., Maxey, R., & Cranston, M. (2013). The uses of radar, nowcast and numerical weather prediction in rainfall runoff modelling in Scotland. In *International Conference on Flood Resilience, 5–7 September 2013.*

Golding, B. W. (2000). Quantitative precipitation forecasting in the UK. *Journal of Hydrology, 239*(1–4), 286–305.

Golding, B. W., Ballard, S. P., Mylne, K., Roberts, N., Saulter, A., Wilson, C., et al. (2014). Forecasting capabilities for the London 2012 Olympics. *Bulletin of the American Meteorological Society, 95*, 883–896.

Golding, B. W., Clarke, P., & May, B. (2006). The Boscastle flood: Meteorological analysis of the conditions leading to flooding on 16 August 2004. *Weather, 60*, 230–235.

Halcrow. (2008). *Proposed pluvial flooding trial service final report.* Exeter, UK: Halcrow Group Limited. Report produced for the UK Met Office.

Halcrow. (2011). *Developing alerting criteria for surface water flooding, Report NA096.* Bristol, UK: Environment Agency (25 pp. + appendices).

Institute of Hydrology. (1983). Flow gauging on the River Thames — The first 100 years. *Hydrological Data UK.* Wallingford: Institute of Hydrology (pp. 34–41).

Kendon, M., & McCarthy, M. (2015). The UK's wet and stormy winter of 2013/14. *Weather, 70*(2), 40–47.

Lewis, H., Mittermaier, M., Mylne, K., Norman, K., Scaife, A., Neal, R., et al. (2015). From months to minutes — Exploring the value of high resolution rainfall observation and prediction during the UK winter storms of 2013/2014. *Meteorological Applications, 22*, 90–104.

Lin, B., Wicks, J. M., Falconer, R. A., & Adams, K. (2006). Integrating 1D and 2D hydrodynamic models for flood simulation. *Water Management, 159*, 19–25.

Marsh, T. J., Parry, S., Kendon, M. C., & Hannaford, J. (2013). *The 2010–12 drought and subsequent extensive flooding.* Available from: http://www.ceh.ac.uk/data/nrfa/nhmp/other_reports/The-2010-12-drought-and-subsequent-extensive-flooding-transformation.pdf.

McKenzie, A., & Ward, R. S. (2015). *Estimating numbers of properties susceptible to groundwater flooding in England.* Available from: http://nora.nerc.ac.uk/510064/1/OR15016.pdf.

Met Office. (2013). *Climate: National Meteorological Library and archive fact sheet 4.* Available from: http://www.metoffice.gov.uk/media/pdf/b/d/MetLIB_13_001_Factsheet_4.pdf.

Met Office. (2015). *Natural hazards partnership.* Available online http://www.metoffice.gov.uk/nhp/hazard-impact-model.

Moore, R. J. (1985). The probability-distributed principle and runoff production at point and basin scales. *Hydrological Sciences Journal, 30*(2), 273–297.

Moore, R. J., Cole, S. J., Bell, V. A., & Jones, D. A. (2006). Issues in flood forecasting: ungauged basins, extreme floods and uncertainty. In I. Tchiguirinskaia, K. N. N. Thein, & P. Hubert (Eds.), *Frontiers in flood research. 8th Kovacs Colloquium* (pp. 103–122). Paris: UNESCO. IAHS Publication 305.

Moore, R. J. (2007). The PDM rainfall-runoff model. *Hydrology and Earth System Sciences Discussions, 11*(1), 483–499.

Moore, R. J., Bell, V. A., Cole, S. J., & Jones, D. A. (2007). *Rainfall-runoff and other modelling for ungauged/low-benefit locations. Science Report — SC030227/SR1, Research Contractor.* Wallingford: Environment Agency, CEH.

Moore, R. J., Cole, S. C., Dunn, S., Ghimire, S., Golding, B., Pierce, C., et al. (2015). *Surface water flood forecasting for urban communities. CREW report CRW2012_03.* Available from: http://www.crew.ac.uk/sites/www.crew.ac.uk/files/publications/CREW_Surface%20water%20flood%20forecasting%20for%20urban%20communities_full%20report.pdf.

Natural Resources Wales. (2014). *Wales coastal flooding review.* Available from: http://www.naturalresources.wales/our-evidence-and-reports/flooding/wales-coastal-flooding-review-phase-1/?lang=en.

O'Dea, E. J., Arnold, A. K., Edwards, K. P., Furner, R., Hyder, P., Martin, M. J., et al. (2012). An operational ocean forecast system incorporating NEMO and SST data assimilation for tidally driven European North-West shelf. *Journal of Operational Oceanography, 5*(1), 3–17. http://www.tandfonline.com/doi/abs/10.1080/1755876X.2012.11020128.

PAA\VQSET. (2013). *PAA\VQSET Level 5 diploma in operational hydrometeorology (QCF).* http://register.ofqual.gov.uk/Qualification/Details/601_0757_1.

PAA\VQSET. (2014). *PAA\VQSET Level 6 diploma in operational hydrometeorology and flood forecasting (QCF).* http://register.ofqual.gov.uk/Qualification/Details/601_5050_6.

Pilling, C., Price, D., Wynn, A., Lane, A., Cole, S., Moore, R., et al. (2014). From drought to floods in 2012: Operations and early warning service in the UK. In *Hydrology in a changing world: Environmental and human dimensions. Proceedings of FRIEND-Water 2014, Montpellier, France, October 2014.* IAHS Publ. 363.

Pitt, M. (2008). *The Pitt review: Learning lessons from the 2007 floods [online].* Available from: http://archive.cabinetoffice.gov.uk/pittreview/thepittreview/final_report.html Accessed 28.01.09.

Price, D., Hudson, K., Boyce, G., Schellekens, J., Moore, R. J., Clark, P., et al. (2012). Operational use of a grid-based model for flood forecasting. *Water Management, 165*(2), 65–77.

Price, D., Pilling, C., Robbins, G., Lane, A., Boyce, G., Fenwick, K., et al. (2012). Representing the spatial variability of rainfall for input to the G2G distributed flood forecasting model: Operational experience from the Flood Forecasting Centre. In R. J. Moore, S. J. Cole, & A. J. Illingworth (Eds.), *Weather radar and hydrology, Proc. Exeter symp., April 2011.* (pp. 532–537): International Association of Hydrological Sciences. IAHS Publ., 351.

Scaife, A., Arribas, A., Blockley, E., Brookshaw, A., Clark, R., Dunstone, N., et al. (2014). Skilful predictions of European and North American winters. *Geophysical Research Letters, 41*, 2514–2519.

Seed, A. W., Pierce, C. E., & Norman, K. (2013). Formulation and evaluation of a scale decomposition-based stochastic precipitation nowcast scheme. *Water Resources Research, 49*(10), 6624–6641.

SEPA. (2011). *The national flood risk assessment.* Stirling, Scotland: Scottish Environment Protection Agency (14 pp.).

SEPA. (2012). *Flood warning strategy 2012–16.* Stirling, Scotland: Scottish Environment Protection Agency (18 pp.).

Sevruk, B., Ondras, M., & Chvila, B. (2009). The WMO precipitation measurement intercomparisons. *Atmospheric Research, 92*(3), 376–380.

Sibley, A. (2010). Analysis of extreme rainfall and flooding in Cumbria 18–20 November 2009. *Weather, 65*, 287–292.

Sibley, A., & Cox, D. (2014). Flooding along English Channel coast due to long-period swell waves. *Weather, 69*, 59–66.

Sibley, A., Cox, D., & Titley, H. (2015). Coastal flooding in England and Wales from Atlantic and North Sea storms during 2013/2014 winter. *Weather, 70*, 62–70.

Sun, J., Xue, M., Wilson, J. W., Zawadzki, I., Ballard, S., Onvlee-Hooimeyer, J., et al. (2014). Use of NWP for nowcasting convective precipitation: Recent progress and challenges. *Bulletin of the American Meteorological Society, 95*, 409–426.

Tavendale, A. (2009). Grant-aided flood management strategies in Scotland and England 1994–2004: Drivers, policy and practice (PhD thesis). University of Dundee, UK.

Werner, M., Cranston, M., Harrison, T., Whitfield, D., & Schellekens, J. (2009). Recent developments in operational flood forecasting in England, Wales and Scotland. *Meteorological Applications, 16*(1), 13–22.

Werner, M., van Dijk, M., & Schellekens, J. (2004). DELFT-FEWS: An open shell flood forecasting system. In *Proceedings of the 6th international conference on hydroinformatics.* (pp. 1205–1212).

Wood, N., Staniforth, A., White, A., Allen, T., Diamantakis, M., Gross, M., et al. (2013). An inherently mass-conserving semi-implicit semi-Lagrangian discretization of the deep-atmosphere global non-hydrostatic equations. *Quarterly Journal of the Royal Meteorological Society, 140*, 1505–1520.

CHAPTER 10

Flood Forecasting in the United States NOAA/National Weather Service

T.E. Adams III
University Corporation for Atmospheric Research, Boulder, CO, United States

1 INTRODUCTION

The National Oceanic and Atmospheric Administration (NOAA), National Weather Service (NWS) has the responsibility for producing and disseminating all public forecasts within the United States (U.S.). These forecasts are freely available using the world wide web (www), public media, and other sources. All hydrologic and hydraulic modeling that underlies the forecast process takes place in the 13 NWS River Forecast Centers (RFCs) (see Fig. 1). However, NWS policy and procedures have developed in a manner such that NWS Weather Forecast Offices (WFOs) utilize RFC forecast guidance to issue official NWS river and flood products to the public.

Details of RFC real-time operations, forecast procedures, methodologies, and modeling systems are discussed in subsequent sections. Readers will see that the NWS modeling system and methodologies used for flood forecasting are robust in terms of the application of current hydrologic and hydraulic science, and utilization of modern technologies that meet the constraints of real-time hydrologic forecasting. The two primary constraints facing real-time hydrologic forecasting are:

1. the availability of observational and forecast data in real-time for use as model inputs;
2. the computational efficiency of the models used.

Consequently, one should view NWS hydrologic forecast achievements using the filter of historical context. Advancements should be seen as an evolutionary process, but punctuated with several significant advances, which are described below. Improvements in forecast accuracy and services were made possible with prior improvements in, and

Flood Forecasting
http://dx.doi.org/10.1016/B978-0-12-801884-2.00010-4

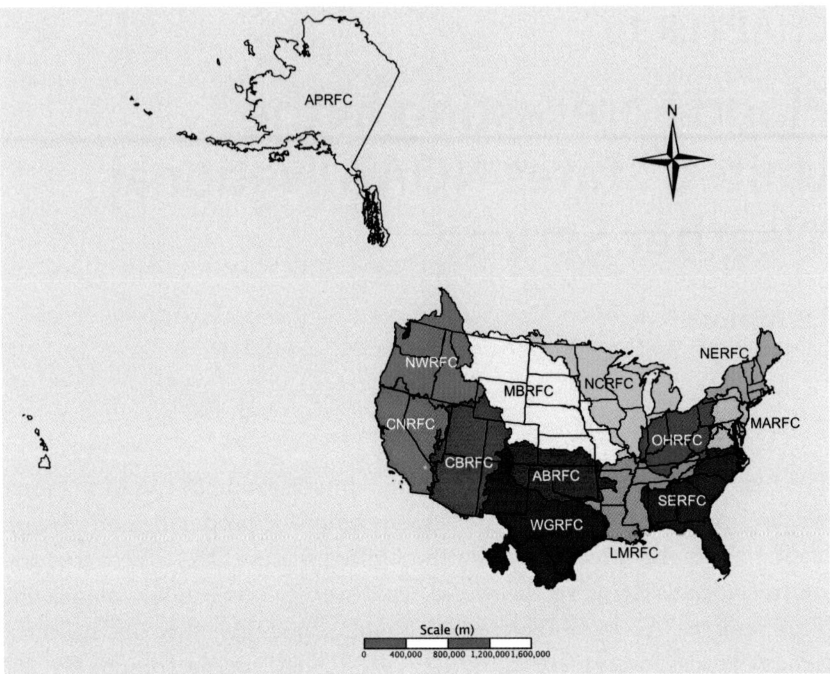

Fig. 1 Map showing the 13 NOAA/NWS River Forecast Centers; note that the RFC boundaries are defined on a river basin boundary basis and extend outside the USA into Canada and Mexico to cover complete river basin drainage. Also, note that five RFCs (ABRFC, LMRFC, MBRFC, NCRFC, and OHRFC) are used to support modeling and forecast responsibilities for the Mississippi River basin.

availability and utilization of, high-resolution datasets and the implementation of current science and technology. Gains in forecast accuracy exhibiting greater spatial and temporal detail during the 1990s and 2000s was achieved, in part, due to attention given to data visualization and direct hydrologic forecaster interaction with models in the NWS River Forecast System (NWSRFS) (U.S. Department of Commerce, 1972). The focus now is the implementation of more detailed, physically based, distributed hydrologic and hydraulic models and the quantification of forecast uncertainty, which is due to inherent model error and model forcings errors. An additional area of focus by the NWS is the recognition of how to best convey forecast uncertainty to the public and other users. These issues are addressed in Section 7, Future Developments.

2 HISTORY

The Mission Statement of the U.S. NWS is:[1]

> *The National Weather Service (NWS) provides weather, hydrologic, and climate forecasts and warnings for the United States, its territories, adjacent waters and ocean areas, for the protection of life and property and the enhancement of the national economy. NWS data and products form a national information database and infrastructure which can be used by other governmental agencies, the private sector, the public, and the global community.*

The NWS's River and Flood Program traces its origins to the start of the NWS. In 1870, Congress authorized the Army Signal Service Corps to create a river and stream gage program, as well as a weather observation and forecasting program. With the passage of the Organic Act of 1890 by the U.S. Congress, all weather and related river services were transferred into the Department of Agriculture in 1891, giving rise to the Weather Bureau. Provisions of the Organic Act of 1890 include:

> *15 USC 313. Duties. The Secretary of Commerce … shall have charge of the forecasting of the weather, the issuance of storm warnings, the display of weather and flood signals for the benefit of agriculture, commerce, and navigation, the gauging and reporting of rivers, the maintenance and operation of sea coast telegraph lines and the collection and transmission of marine intelligence for the benefit of commerce and navigation, the reporting of temperature and rainfall conditions for the cotton interests, the display of frost and cold-wave signals, the distribution of meteorological information in the interests of agriculture and commerce, and the taking of such meteorological observations as may be necessary to establish and record the climatic conditions of the United States, or as are essential for the proper execution of the foregoing duties.*

As the country grew, the need for expanded hydrologic services increased. Flooding events, such as the extensive flooding that struck the Kansas River in 1903,[2] raised public demands for improved flood forecasting and warnings. In response, Congress passed legislation making river and flood services a separate division within the Weather Bureau. With the passage of the Flood Control Act of 1938 (33 USC 706), the need for a more effective flood warning and control system was addressed. The Act

[1]From http://www.nws.noaa.gov/mission.php.

[2]See http://www.kshs.org/kansapedia/flood-of-1903/17221, http://www.kclibrary.org/blog/kc-unbound/devastation-and-controversy-history-floods-west-bottoms, and http://ks.water.usgs.gov/pubs/fact-sheets/fs.019-03.pdf.

authorized an allotment of funds from appropriations for flood control for the establishment, operation, and maintenance of an information service on precipitation, flood forecasts, and flood warnings. The Congressional bill reads:

> *33 USC 706. Current information, appropriations. There is authorized an expenditure as required, for any appropriations heretofore or hereafter made for flood control, rivers and harbors, and related purposes by the United States, for the establishment, operation, and maintenance by the Weather Bureau of a network of recording and non-recording precipitation stations, known as the Hydroclimatic Network, whenever in the opinion of the Chief of Engineers and the Secretary of Commerce such service is advisable in connection with either preliminary examinations and surveys or work improvement authorized by the law for flood control, rivers and harbors, and related purposes, and the Secretary of the Army upon the recommendation of the Chief of Engineers is authorized to allot the Department of Commerce funds for said expenditure.*

From 1891 to 1940, the Weather Bureau was part of the Department of Agriculture. In 1940, the Weather Bureau was moved to the Department of Commerce, and a river division within the U.S. Weather Bureau—the NWS Office of Hydrology—was formed. Hydrologic services expanded throughout the entire United States within the next few years. The river division was subsequently subdivided into river districts, each with an associated Weather Bureau office to focus on the hydrologic needs of that region. The Ohio River Forecast Center (Ohio RFC—OHRFC) was the first of the river divisions formed, opening in Sep. 1946, and was subsequently followed by the establishment of Missouri Basin RFC in Kansas City, Missouri.[3] The river districts were consolidated in the 1960s forming the 13 RFCs that exist today, as shown in Fig. 1 and Table 1. In 1970, the Weather Bureau name was changed to the NWS, and was relocated to the newly created NOAA within the Department of Commerce.[4]

Scientific and technological developments have been significant since the establishment of the 13 RFCs. Forecast methodologies began with the use of manually intensive graphical and hand-calculated procedures, to locally developed event-based hydrologic modeling systems in the 1970s and 1980s. This was followed by the emergence of the mainframe computer system-based NWSRFS in the 1970s and early 1980s, developed by the NWS Office of Hydrology and, significantly, it included the development

[3]See http://www.weather.gov/timeline.

[4]See https://commons.wikimedia.org/wiki/File:The_United_States_Weather_Service,_First_100_years_by_Bob_Glahn_part_1.pdf.

Table 1 Table providing the RFC identifier, RFC name, city and state location, and NWS Administrative Region of NOAA/NWS RFCs

RFC Identifier	Name	Location	Region
APRFC	Alaska-Pacific RFC	Anchorage, Alaska	Alaska-Pacific
ABRFC	Arkansas-Red Basin RFC	Tulsa, Oklahoma	Southern
CBRFC	Colorado Basin RFC	Salt Lake City, Utah	Western
CNRFC	Colorado-Nevada RFC	Sacramento, California	Western
LMRFC	Lower Mississippi RFC	Slidell, Louisiana	Southern
MARFC	Middle Atlantic RFC	State College, Pennsylvania	Eastern
MBRFC	Missouri Basin RFC	Pleasant Hill, Missouri	Central
NCRFC	North Central RFC	Chanhassen, Minnesota	Central
NWRFC	North West RFC	Portland, Oregon	Western
NERFC	Northeast RFC	Taunton, Massachusetts	Eastern
OHRFC	Ohio RFC	Wilmington, Ohio	Eastern
SERFC	Southeast RFC	Peach Tree City, Georgia	Southern
WGRFC	West Gulf RFC	Dallas-Fort Worth, Texas	Southern

of the Sacramento Soil Moisture Accounting (SAC-SMA) model. The creation of NWSRFS and the subsequent availability of UNIX based computer workstations made the development of interactive[5] hydrologic forecasting feasible in the early 1990s for use at RFCs for routine operations. This was a time of NWS modernization with deployment of the Advanced Weather Interactive Processing System (AWIPS) (see Section 5.1), Next Generation Radar (NEXRAD), and, within NWS hydrologic services, implementation of the Advanced Hydrologic Prediction Services (AHPS) (see Section 5.2). The modernization brought a significant infusion of hydrologic science and technology, including long lead-time probabilistic hydrologic forecasting, mainly for water resources applications, a www presence, widespread use of raingage-corrected NEXRAD radar based precipitation estimation,

[5]Interactive forecasting is a manual process of data assimilation and error correction that allows a hydrologic forecaster to make model state changes, correct hydrometeorological observation errors, and update newly acquired data through a series of trial model runs and adjustments until the forecaster is satisfied with the agreement between simulated and observed hydrographs. The justification of the iterative process is that model forecasts are best when agreement between model simulations and observations have been optimized.

expanded use of quantitative precipitation forecast (QPF), and use of gridded flash flood guidance (FFG). During the period of AHPS implementation most RFCs moved from the use of local Antecedent Precipitation Index models, which depend on basin-specific rainfall-runoff climatologies, to the conceptual, lumped-parameter SAC-SMA model.

With NWS reorganization in 2000, the Office of Hydrology (OH) was renamed the Office of Hydrologic Development (OHD) and the Hydrologic Research Laboratory (HRL), within OH, became simply the Hydrology Laboratory (HL). Further NWS reorganization in May 2015, by NWS Director Louis Uccellini, saw OHD evolve into the National Water Center (NWS) with Donald Cline named the inaugural Director (see Section 7.2).

3 A BRIEF HISTORY OF MODELS AND MODELING SYSTEMS USED BY THE NWS

Prior to NWS modernization with AWIPS, RFCs lacked a common hardware platform for implementation of NWSRFS (see Section 3.4). Operational hardware and modeling systems took many forms, ranging from, for example, implementation of NWSRFS on PRIME minicomputers[6] (at the MARFC), to hybrid platforms that utilized local personal computers in combination with batch remote job execution of NWSRFS rainfall-runoff models at the NOAA Central Computer Facility in Suitland, Maryland (at the OHRFC),[7] and the use of a locally developed modeling system on the NWS Automation of Field Operations and Services (AFOS) computer systems (at the NERFC). Installed in the early to mid-1980s, the AFOS computer system linked NWS offices for the transmission and display of weather data.

Major advancements in US hydrologic forecasting since the 1970s followed improvements in the quantity, quality, and increased resolution of new datasets, both spatially and temporally for observed and forecast

[6]See https://en.wikipedia.org/wiki/Prime_Computer.

[7]The locally developed hybrid forecast system at the OHRFC was called the Interactive Forecast Model (IFM), which relied on batch processing controlled by Job Control Language (JCL) to produce by runoff generated from the Antecedent Precipitation Index (API) Cincinnati, API-CIN and SNOW-17, snow accumulation and ablation models running within NWSRFS at the NOAA Central Computer Facility. Runoff values for individual basins (on the order of 300) were transmitted to the OHRFC, where forecasters used the IFM to convert the runoff to flows and route downstream within the interactive environment.

hydrometeorological variables. Through the 1970s, 1980s and into the 1990s, utilization of daily and 6-hourly subbasin mean quantities was necessary, but currently, 1-hourly, 4-km or less, datasets are utilized for model inputs. The availability of higher resolution model forcing data to RFCs has been significant, especially in terms of making increased use of detailed distributed hydrologic and hydraulic models possible. Equally important has been the availability of high-resolution (30 m horizontal resolution or less) Digital Terrain (or Elevation) Model, landuse/landcover, soils, vegetation data, etc, for use in Geographic Information Systems (GIS) for producing higher-quality watershed boundaries, stream networks, and improved model parameter estimates, and for aiding in hydrologic model calibrations.

3.1 NWS River Forecasting Paradigm

Since their formation, NWS RFCs have operated largely independently from each other and NWS National Headquarters. RFCs receive direction from the NWS Regional Administrative Headquarters[8] (see Table 1). Consequently, considerable differences in operational procedures exist between RFCs. However, these differences have been reduced measurably through the implementation of AWIPS (Section 5.1), AHPS (Section 5.2), and CHPS (Section 4.1) which, collectively, form the basis for current NWS RFC operations. The reduction in operational differences between RFCs can be largely attributed to the use of a common hardware and software platform for operations. A strength of AWIPS, AHPS, and CHPS is the flexibility of the systems to be customized to meet local needs in addressing local problems. In the interest of space and time, no attempt will be made to cover exhaustively the differences in operations or implementations of AWIPS, AHPS, and CHPS between the RFCs, although some significant operational concerns will be discussed in Section 5. In general, RFCs have the following common operational needs:

- data archiving—needed for near real-time forecast evaluation and verification, future model (re)calibrations, and training;
- model selection—there is flexibility within the NWS modeling/forecast systems to use model components somewhat interchangeably with different rainfall-runoff, hydrologic routing, and reservoir simulation models, depending on local conditions and need;
- initial model parameter value estimation—with more physically based models, model parameter values can be estimated directly from geophysical

[8]There are five NWS regions: Eastern, Southern, Central, Western, and Alaska-Pacific.

quantities, such as soil texture and depth, landuse/landcover type, etc; for reservoir simulation models, the storage-elevation relationship and operational rules are requirements; for hydraulic models, channel cross-section geometries and Manning roughness coefficients are needed, as well as design details of hydraulic structures such as levees (dikes), diversion structures, locks, and dams, etc;

- preprocessing of historical precipitation and temperature inputs for model calibration—data biases must be removed and erroneous data must be corrected or removed; data must be stored in a usable format for the modeling system;
- model calibration;[9]
- implementation of the modeling/forecast system, such as NWSRFS (see Section 3.4) or CHPS (see Section 4.1);
- operational data ingest;
- operational data quality control of:
  1. river water level (stage) data;
  2. station temperature and precipitation data;
  3. NEXRAD radar-based precipitation data;
  4. snow water equivalent (SWE) data;
  5. reservoir release data from the United States Army Corps of Engineers (USACE), Bureau of Reclamation, and Natural Resource Conservation Service (NRCS);
- forecast product generation—see, for example, Section 5.5 and Fig. 14;
- river stage and flow forecast verification.

The issue here is the manner in which each of these needs is addressed by the individual RFCs. What is apparent, given the flexibility of nationally supported NWS software, is that the approach taken by individual RFCs to utilize the software and hardware can vary widely. Where there is no NWS-wide solution to local needs, such as dealing with ice jams, stream withdrawals for irrigation, utilization of irrigation, de-watering practices, and other water resources applications, local *ad hoc* techniques must be developed in order to meet end-user and public needs in a timely way.

[9]Model calibration (Anderson (2002)) is procedure by which optimal model parameter values are determined through some combination of automatic optimization and manual human adjustments based on a trial-and-error iteration, where simulated values are compared to observed values. This should not be confused with a data assimilation process, whether automated using a variational assimilation scheme or using a Kalman filter, or through a manual adjustment process by a hydrologic forecaster to optimize model states immediately prior to real-time operations.

The process of model calibration and the setup and configuration of the forecast system is done well in advance of operational use. A substantial period of months of operational testing is needed to assure that the flow of data is stable, models are producing suitable results, and all operational procedures are fully in place. Best practices include the establishment of a Concept of Operations (CONOPS) or Station Duty Manual, as is known in the NWS, that details all aspects of forecast operations, the operational system, forecaster duties and responsibilities, hours of operation, staffing requirements during routine operational periods and exceptional flooding events, the physical setting, emergency contacts and procedures, etc. Also, use of operational checklists that detail routine and emergency or critical procedures is commonplace.

3.2 Antecedent Precipitation Index (API) Model

An important advancement in NWS hydrologic modeling during the early stages NWS operational hydrologic forecasting was the development of the Antecedent Precipitation Index (API) model in the 1940s by (Kohler, 1944). The model was developed in response to the need for tractable techniques that could simplify the relationships between rainfall and runoff for hydrologic prediction. At that time, many techniques to conceptualize soil characteristics through the application of infiltration theory and other models proved to be too complex, especially when trying to apply them to large basins. A major consideration in forecasting at that time was the computational time needed to produce forecast results using manual methods. Without the benefit of computers, less time-consuming methods were needed (Kohler & Richards, 1962; Betson, Tucker, & Haller, 1969). The relative unavailability of model forcing data and inability to estimate parameter values for more physically based models was another consideration in model selection. Generally, however, storm characteristics could be determined from an adequate network of precipitation stations but knowledge of soil moisture conditions throughout the basin was problematic.[10] Variations in soil and surface characteristics, vegetation differences, and land use added to the complexity of hydrologic modeling through the 1980s. Many factors have been used to index the moisture conditions, such as:

- days since last rain;
- discharge at the beginning of the storm;
- antecedent precipitation.

[10]Soil moisture estimation continues to be a significant impediment in operational forecasting due to extremely sparse soil moisture networks and the relative immature measurements from satellite-based remote sensing.

The first index is insensitive because it only accounts for the duration of periods without precipitation and does not account for the amount of recharge to the basin at the time of the last event. The second factor is seasonally sensitive and does not reflect changes by previous rains. The use of antecedent precipitation generally provides good results, assuming it is properly derived and uses a seasonal index or utilizes relationships that incorporate air temperature.

The variable API is a rough representation of initial soil-moisture conditions which can be readily determined by utilizing accumulated precipitation and by taking into account evaporation and infiltration. By using API, week of the year, and storm precipitation and duration as parameters, a relationship between storm runoff and precipitation using a graphical method of coaxial relations can be produced, as in Fig. 2. The assumption is that if any important factor is omitted from a relation, the scatter of points in a

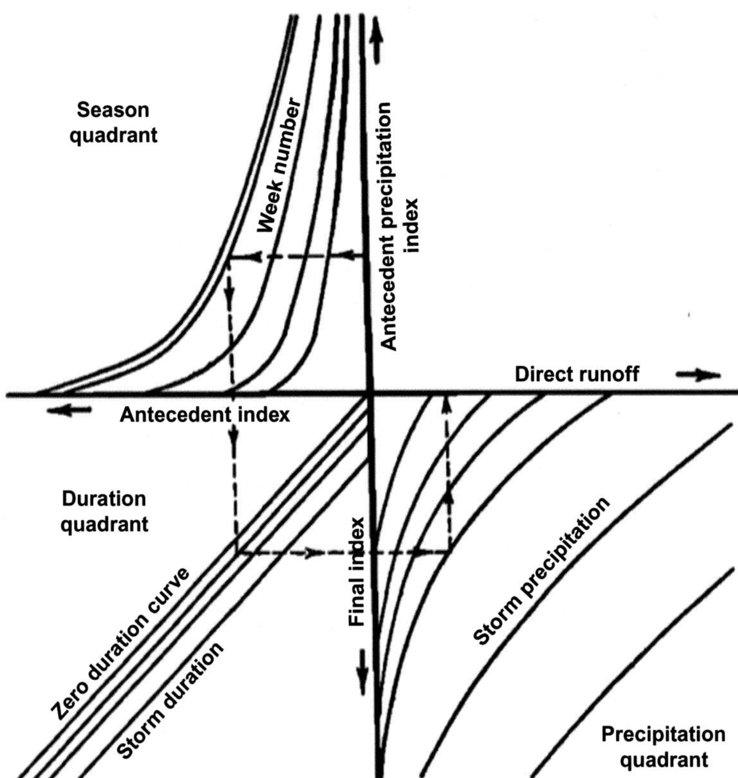

Fig. 2 Illustration showing the Antecedent Precipitation Index (API) relationship (from U.S. Department of Commerce, NOAA/NWS). Please see http://www.nws.noaa.gov/oh/hrl/nwsrfs/users_manual/part2/_pdf/23apislc.pdf.

plot of observed values of the dependent variable versus those computed by the relation will be at least partially explained (Kohler, Linsley, & Paulhus, 1958). The API procedure is a graphical set of three-variable relations arranged with common axes to facilitate manual computation. It was readily implemented in digital form within the NWSRFS. API models were developed for many of the RFC regions to account for regional rainfall-runoff relationship differences and inclusion of other factors.

Despite its significant utility through the 1990s, notable shortcomings can be seen; namely, the API model:

1. is a blackbox model, lacking model parameters that are either directly physically measurable or can be estimated from directly measurable physical quantities;
2. is outperformed by conceptual, continuous simulation models, such as the SAC-SMA and Stanford Watershed Model IV, particularly during and following extended dry periods; conceptual models provided enough memory to respond properly to conditions where relatively large amounts of rainfall produced little or no streamflow response (Schaake, 1976);
3. is climatologically based, with the implicit assumption that underlying rainfall-runoff relationships only change seasonally;
4. has difficulty accounting for conditions that differ significantly from the norm;
5. is an event-based model as it was initially formulated,[11] which was not suitable for use for long lead-time hydrologic simulations as discussed in Section 5.2;
6. state initializations can be a somewhat subjective process.

The API model was replaced during the implementation of AHPS (see Section 5.2) in the 2000s by most RFCs in favor of the conceptual, continuous simulation Sacramento Soil Moisture Accounting (SAC-SMA) model.

3.3 Data Acquisition and Processing

Data needs for RFC operations include:

- real-stream flow and stage (water level) observations, primarily from the U.S. Geological Survey;
- real-time raingage based precipitation and temperature measurements from Data Collection Platforms (DCPs) and the Cooperative Observer Program (discussed in Section 3.3.2);

[11]The event-based formulation was address with the development of the Continuous API (API-CONT) model by Eric Anderson, NOAA/NWS Office of Hydrologic Development, Hydrology Laboratory—see http://www.nws.noaa.gov/oh/hrl/nwsrfs/users_manual/part2/_pdf/23apicont.pdf.

- NWS NEXRAD-based precipitation data;
- SNOTEL snow water equivalent observations from the U.S. Army Corps of Engineers, U.S. Natural Resource Conservation Service, U.S. Bureau of Reclamation, and other sources;
- observed and projected reservoir flow releases from the U.S. Army Corps of Engineers, U.S. Natural Resource Conservation Service, U.S. Bureau of Reclamation, and other sources.

3.3.1 U.S. Geological Survey Stream Gaging Program

U.S. Geological Survey streamgage data are critical to RFC operations. As part of the National Streamflow Information Program, the U.S. Geological Survey (USGS) operates more than 7400 streamgages nationwide to provide streamflow information for a wide variety of uses.[12] USGS streamflow data are used in flood prediction, management and allocation of water resources, design and operation of engineering structures, scientific research, operation of locks and dams, and recreation. These streamgages are operated by the USGS in partnerships with more than 800 other federal, state, tribal, and local cooperating agencies. In 2007, nearly 90% of streamgages record and transmit streamflow information electronically so that streamflow information is available on the www in real time (http://waterdata.usgs.gov). Most streamgages transmit information via satellite, but a few use telephone and radio telemetry.

Streamgages transmit all stage information recorded since the last transmission to a Geostationary Operational Environmental Satellite (GOES), typically, on a preset schedule, every 1–4 h. Many streamgages have predetermined stage thresholds, which when exceeded, the time between transmissions to the satellite will decrease from 1–4 h to every 15 min, to provide more timely data during flooding or other emergency situations.

Transmissions from the streamgage are sent from the GOES satellite to the NOAA Command and Data Acquisition (CDA) facility at Wallops Island, Virginia. The received information is then immediately rebroadcast at much higher power to a domestic communications satellite (DOMSAT). This allows the information to be received at smaller Local Readout Ground Stations (LRGS) throughout the United States. The USGS maintains 21 LRGS systems to provide redundancy, in the event that one or more systems are not operating. Upon receipt at the LRGS, stage information is transferred almost immediately into the USGS National Water Information System (NWIS). NWIS consists of a network of more than 50 computers

[12]Taken from the USGS publication: http://pubs.usgs.gov/fs/2007/3043/FS2007-3043.pdf.

that collectively process all USGS water-resources data and store more than 100 years of streamflow, water-quality, and ground-water data from hundreds of thousands of sites across the country.

Most USGS streamgages operate by measuring the elevation of the water in the river or stream and then convert the water elevation/level (or stage) to a streamflow (discharge) by using a rating curve, which relates the water level elevation to a set of actual discharge measurements. The USGS currently utilizes rating curves because the technology is not available to measure water flow accurately enough directly.[13] The USGS standard is to measure river stage to 0.01 in. This is accomplished by the use of floats inside a stilling well, or with pressure transducers that measure how much pressure is required to a push a gas bubble through a tube (related to the depth of water), or with radar.

3.3.2 Cooperative Observer Program

The NWS Cooperative Observer Program (Coop)[14] is a nationwide weather and climate observing network comprised of more than 8700 volunteers. The data are gathered, generally, where individuals work and live. The NWS Cooperative Observer Program has two main goals, namely, to provide observational meteorological data:

1. of daily maximum and minimum temperatures, snowfall, and 24-h precipitation totals, required to define the climate of the United States and to help measure long-term climate changes;
2. in near real-time to support forecast, warning and other public service programs of the NWS.

The Cooperative Observer Program responsibilities include:

- selecting observation locations;
- recruiting, appointing and training observers;
- installing and maintaining equipment;
- maintaining station documentation;
- data collection and delivery to users;
- maintaining data quality control;
- managing fiscal and human resources required to accomplish program objectives.

Observers make temperature and 24-h precipitation observations and send observation reports electronically, daily, to the NWS and the National

[13]Please see http://water.usgs.gov/nsip/definition9.html#how_it_works.

[14]Please refer to http://www.nws.noaa.gov/om/coop/what-is-coop.html for details on the NWS Cooperative Observer Program, which are cited herein.

Climatic Data Center (NCDC). Many cooperative observers provide additional hydrological or meteorological data, such as evaporation or soil temperatures. Data is transmitted via telephone, computer, and, in some instances, by postal service. Equipment used at NWS cooperative stations must meet NWS equipment standards and may be owned by the NWS, the observer, or by a company or other government agency.

Coop data are used by RFCs for three principle purposes, including:

- quality control and bias correction of NEXRAD radar based precipitation estimates;
- supplementing hourly raingage precipitation observations, which requires use of disaggregation methods;
- RFC hydrologic model calibrations utilizing historical archives of Coop station data.

3.3.3 HADS

The Hydrometeorological Automated Data System (HADS) is a real-time data acquisition and data distribution system operated by the NWS Office of Dissemination.[15] The HADS system consists of data from more than 15,900 data point locations at which river and weather data are observed and subsequently processed. HADS processed data are transmitted through AWIPS (Section 5.1) automatically to RFCs and WFOs. Depending upon the requirements of the DCP owner and capabilities of the DCP, various sensors are interfaced and the data are collected and transmitted at different specified time intervals, primarily hourly. Agreements have been established between the NOAA National Environmental Satellite, Data and Information Service (NESDIS), and the owner of the DCP to allow NWS access to the data. HADS was designed to provide random and self-timed data to assist the forecaster in various program areas including Fire Weather, Agricultural Weather, and Hydrologic forecasting. HADS currently decodes a wide variety of hydrometeorological data types from reporting sites operated in all 50 US states, Canada, Mexico, Puerto Rico, the Virgin Islands, and several Central American countries.

The NESDIS GOES Data Collection System (DCS) is comprised of the DCPs, the NESDIS Command and Data Acquisition (CDA) System, and the Data Acquisition Processing System (DAPS). The components of the NESDIS systems are collectively referred to as the GOES DCS. This satellite-based system collects a variety of environmental data from locations

[15]Please see http://www.nws.noaa.gov/oh/hads.

in the Western Hemisphere. The system is a data relay network for more than 19,000 DCPs which transmit data to one of two GOES satellites (East and West). These data are relayed to the NESDIS CDA ground station located at Wallops Island, Virginia. The data are then relayed over dedicated telecommunication circuits to the NWS Telecommunications Gateway (NWSTG) in Silver Spring, Maryland, where the information is then routed to the HADS computer systems (Fig. 3).

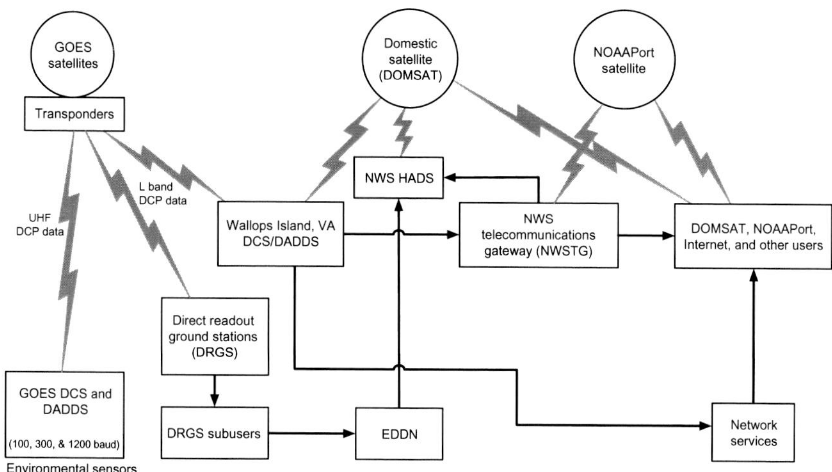

Fig. 3 Illustration showing the NESDIS DCS data flow path. From U.S. Department of Commerce, NOAA/NWS.

The NWSTG computers normally receive raw GOES DCP data via a telecommunication circuit within 2 minutes of the DCP transmission. This raw data is then transferred to another server in usually 1- to 2-min cycles. HADS then retrieves all new data files from this server and processes this data in a 3-min cycle.

Methods are in place specifically to filter out certain DCP messages that do not provide useful hydrometeorological data. This helps to increase the efficiency of HADS processing, since data from all of the active DCPs in the GOES DCS are forwarded to HADS by the NWSTG. An archive of 72 h of raw data is maintained within HADS. The DCS facility also maintains 72 h of raw data. When communication outages occur between the DCS at Wallops Island and the NWSTG in Silver Spring, the DCS is capable of holding and then forwarding all of the data not relayed during an outage. HADS has the capability of retrieving, processing, and, subsequently, distributing all back-logged data.

3.3.4 Community Collaborative Rain, Hail and Snow (COCORAHS) Network

The Community Collaborative Rain, Hail and Snow Network (CoCoRaHS) is an independent, non-profit, community-based network of volunteers who collaborate to measure and map precipitation (rain, hail, and snow). Members of the network emphasize the use of low-cost measurement tools, and stress training and education. Members of the CoCoRaHS network utilize interactive www-based technology to provide data for natural resource, education, and research applications. The network originated with the Colorado Climate Center at Colorado State University in 1998, largely in response to damaging flooding in Fort Collins, Colorado in 1997. The nationwide CoCoRaHS network includes thousands of volunteers and is independent of the NOAA/NWS-based Cooperative Observer Program discussed in Section 3.3.2. RFCs and WFOs use CoCoRaHS data to supplement the observational data received officially through HADS. CoCoRaHS data is not quality-controlled until reaching NWS RFCs and WFOs.

3.4 NWSRFS

The NWSRFS,[16] written predominantly in FORTRAN 77, was developed in the 1970s, with initial implementation in 1971 (Schaake, 1976). NWSRFS evolved into system consisting of four major modeling system components, which are discussed in subsequent sections. Also listed below is the Interactive Forecast Program (IFP). The major NWSRFS components are:

1. Operational Forecast System (OFS)—real-time operational software system used to generate short lead-time river and flood forecasts and maintain model state variables;
2. Calibration System (CS)—used to generate hydrologic time series based on historical data and refine initial model parameter estimates by comparing simulation results to observations;
3. Extended Streamflow Prediction (ESP) System[17]—uses current model states, calibrated parameters and historical time series to generate probabilistic forecasts extending weeks or months into the future;

[16]The complete NWSRFS User Manual is currently available at http://www.nws.noaa.gov/oh/hrl/nwsrfs/users_manual/htm/xrfsdocpdf.php.

[17]The use of the term "Extended" was later changed in favor of "Ensemble," in keeping with changes in scientific thought and usage.

4. FFG System—uses current model states to compute FFG values; FFG is the quantity of precipitation needed to force small stream to exceed bankfull conditions;[18]
5. Interactive Forecast Program (IFP)—graphical user interface (GUI) to the OFS, written in C, X Window System, and Motif, which allows a hydrologic forecaster to make model adjustments to NWSRFS OFS interactively during real-time operations.

3.4.1 Operational Forecast System (OFS)

The NWSRFS Operational Forecast System is a continuous river forecasting system used to make predictions of river flow, and generate flood forecasts and other hydrologic products. The system stores observed and future point data (precipitation, temperature, stage, etc), performs various preprocessor functions on the observed data (mean areal precipitation, stage to discharge conversion, etc), and produces forecast products (eg plots of predicted, simulated and observed river stage at selected points). Parameter values are determined from the Calibration System and are transferred to the OFS through a manual preparation step to create what is referred to as segment definitions, an example of which is given in Listing 1. This manual preparation is needed to add the description of operations which are particular to forecasting a particular basin, including, for example, upstream and downstream basins for streamflow routing, input/output time series data, models used, blend routines from observations to simulations, different plot options, etc.

3.4.2 Interactive Forecast Program (IFP)

Interactive Forecast Program (IFP) is the graphical user interface (GUI) front-end to the OFS, written to permit interactive adjustments, called runtime modifications, by hydrologic forecasters in a desktop workstation environment. The IFP allows the forecaster to change model states, correct observational errors, and make other forecast adjustments interactively in a real-time setting (see Adams, 1991; Adams & Smith, 1993). The IFP was written and implemented for operational use in the early 1990s.

[18]Some computations of FFG values attempt to account for the recognition that disastrous stream conditions can be reached with much less than bankfull flow at low-water crossings and in very steep canyons areas for recreational hikers, where vehicles can become buoyant and float and people can be swept away by relatively shallow, swiftly flowing water.

Listing 1 Example NWSRFS OFS Text Output of a Segment Definition (Partial Listing) from the Ohio River Forecast Center (OHRFC) for the Cheat River near Parsons WV

Asterisks '*' are used to indicate that lines have been omitted in the interest of brevity.

```
IDENTIFIER   PSNW2            39.12      79.68
TITLE       NR PARSONS , WV
UPSTREAM     BWRW2       HENW2      BOWW2
DOWNSTREAM   ROWW2
DEF-TS
PRNW2       MAP     6             INPUT          FPDB      CARD INPUT
PRNW2       MAP
dummy.map
PRNW2       MAPX    6             INPUT          FPDB      CARD INPUT
PRNW2       MAPX
PRNW2/PRNW2.MAPX
PRNW2       MAT     6             INPUT          FPDB      GENR INPUT
PRNW2       MAT
BLEND-TS
 1.00 0.99    48    6
PRNW2/PRNW2.MAT
PRNW2       RAIM    6             OUTPUT         FPDB      ESP  OUTPUT
PRNW2       RAIM     39.12      79.68      RAIN+MELT
PRNW2              0
PRNW2       INFW    6             OUTPUT         FPDB      ESP  OUTPUT
PRNW2       INFW     39.12      79.68      RUNOFF
PRNW2              0
PRNW2       SNWE    24            INPUT          FPDB      MSNG INPUT
PRNW2       SNWE
PRNW2       SWE     6             OUTPUT         FPDB      ESP  OUTPUT
PRNW2       SWE      39.12      79.68      SWE
PRNW2              0
PRNW2LC     SQIN    6             INTERNAL
PRNW2       SQIN    6             INTERNAL
PRNW2       SASC    6             INTERNAL
PSNW2       MAP     6             INPUT          FPDB      CARD INPUT
PSNW2       MAP
dummy2.map
PSNW2       MAPX    6             INPUT          FPDB      CARD INPUT
PSNW2       MAPX
PSNW2/PSNW2.MAPX
PSNW2       MAT     6             INPUT          FPDB      GENR INPUT
PSNW2       MAT
BLEND-TS
 1.00 0.99    48    6
PSNW2/PSNW2.MAT
*
*
*
END
CLEAR-TS
  PRNW2     RAIM    6
MERGE-TS     PRNW2
    2  PRNW2     MAPX     6    1
PRNW2       MAPX
PRNW2       MAP
SNOW-17      PRNW2
SHAVERS FK@PARSONS  1021. 38.6        YES  YES SUMS
    6  PRNW2     MAPX      1.000        PRNW2     RAIM
```

Listing 1 Example NWSRFS OFS Text Output of a Segment Definition (Partial Listing) from the Ohio River Forecast Center (OHRFC) for the Cheat River near Parsons WV—Cont'd

```
  PRNW2     MAT      6
  PRNW2     SNWE    24                                  0                  0
  PRNW2     SWE      6          PRNW2     SASC      6                  0
 1.20 1.10 0.550.150   80.      0
 0.15 0.50 0.00 0.00 0.10 0.10 0.00 0.00
 0.18 0.28 0.37 0.45 0.51 0.59 0.69 0.84 0.94
SAC-SMA      PRNW2
SHAVERS FORK@PARSONS          6  PRNW2     RAIM        PRNW2     INFW
  PRNW2     SASC      6  PRNW2      PRNW2     SUMS          24   24
                  1.0001.000  23.0  16.00.2450.0000.0000.060       00.100
                   70.0  2.24   48.  24.0   76.0.098.00700.1100.300  0.00
                   0.4  0.5  0.6  1.0  2.1  4.2  4.3  4.0  3.1  2.1  0.6  0.4
                   10.6   0.0   27.   0.6   37.   36.     1
 .000 .000 .000 .000 .000 .000 .000        0.00 0.00
FFG          PRNW2
PRNW2     SHAVERS FK@PARSONS        2 0.10 5.00
PRNW2     SAC-SMA    PRNW2      SNOW-17    PRNW2
CLEAR-TS
  PRNW2LC   SQIN     6
UNIT-HG      PRNW2
SHAVERS FORK@PARSONS         61.8     7       CARRY ENGL      0.000
  PRNW2     INFW     6  PRNW2LC   SQIN     6
    1581.7      3729.7      1388.6         0.0         0.0         0.0         0.0
    0.0028      0.0027      0.0027      0.0026      0.0026      0.0026
CLEAR-TS
  BOWW2R    SQIN     6
LAG/K        BOWW2
BOWW2     QINE   6 BOWW2R    SQIN   6     0      0 ENGL 0.00          0.0
          12.000
           0.000
           4
          17.568             6.000             16.849             12.000 X
           0.000             0.000              0.000              0.000
          18.315
CLEAR-TS
*
*
*
PLOT-TUL     PSNW2PLT
    1    1   51    0  100    6    6   14    8    I    -    0    0    0
    0   70    F    U    R  PSNW2    0    0    0
  MAP     RAIM    RO  BASE     REC  STG  QINE   SSTG
PRNW2     SNWE LIST       F7.2,      24
PSNW2     SNWE LIST       F7.2,      24
PSNW2     MAPX LIST       F7.2,       6
PSNW2     RAIM LIST       F7.2,       6
PSNW2     QIN  PLOT O                 1
PRNW2LC   SQIN BOTH M     F7.2,       6
PSNW2LC   SQIN BOTH L     F7.2,       6
HENW2     QINE PLOT 1                 6
BWRW2R    SQIN PLOT 2                 6
BOWW2R    SQIN PLOT 3                 6
PSNW2     STG  LIST       F7.2,       1
PSNW2     QINE PLOT *     F7.0,       6
PSNW2     STGE LIST       F7.2,       6
PSNW2     SQIN PLOT +                 6
END
STOP
```

3.4.3 Calibration System (CS)

The Calibration System is used to investigate the performance of various hydrologic techniques using historical data for calibrating hydrologic models (estimating parameters) used in the Operational Forecast System and for pure research purposes. Historical data access programs are available to inventory archived hydrometeorological data and convert it from the archive format to a standard data file format. Calibration preprocessor programs are available to compute mean areal values of precipitation (MAP), temperature (MAT), etc, from point values. The programs used to calibrate the hydrologic models are the Manual Calibration Program (MCP) and the Automatic Parameter Optimization Program (OPT). MCP operates by applying hydrologic models with user specified values for all parameters. OPT operates in a similar manner, but it includes a procedure to adjust parameter values automatically to improve the streamflow simulation, utilizing the Shuffle Complex Evolution (SCE) method (see Duan, Sorooshian, & Gupta, 1992, 1994, Duan, Gupta, & Sorooshian, 1993). The calibration software is intended for use on just one segment of a river system at a time. A major river system is calibrated one area at a time, with a typical calibration run spanning several years of historical data. In contrast, the OFS operates on an entire river system but for comparatively short time periods, on the order of 1 week.

3.4.4 Extended Streamflow Prediction (ESP) System

The Extended Streamflow Prediction (ESP) System, now known as Ensemble Streamflow Prediction, produces long-range probabilistic forecasts of hydrologic variables. ESP utilizes historical precipitation and temperature data as analogs for the future, with the assumption that historical meteorological data are equally likely to occur in the future. ESP accesses files in the Operational Forecast System for an estimate of the current hydrologic state and uses historical meteorological data to create many equally likely sequences of future hydrologic conditions each starting with current conditions. For ESP, each year of historical meteorological data produces one hydrologic ensemble streamflow trace; 50 years of historical meteorological data produce an ensemble of 50 streamflow members. The ESP system is designed to run a large river system and many years of historical data. The most typical use of ESP is for water supply forecasting, although the ESP logic can be used for a variety of long-range forecast needs (eg beyond 30 days in the future) (Day, 1985).

Advances to the initial development and implementation of ESP included the use of techniques to relax the assumption of equal weighting of past historical meteorological data, recognizing that skillful, near-term weather and climate outlooks could be useful in either weighting meteorological inputs with pre-processing techniques or the resulting hydrologic streamflow ensemble outputs with post-processing. Another technique is the adjustment of historical precipitation and temperature time-series data utilizing near-term weather and climate outlooks to reflect dryer or wetter and/or warmer or cooler atmospheric conditions. Consequently, the Ensemble Streamflow Prediction Analysis and Display Program (ESPADP) and associated software, described in http://www.nws.noaa.gov/oh/hrl/nwsrfs/esp/ESPADP_Manual/espadp_index.php, was developed for pre- and post-processing of the meteorological inputs and hydrologic ensemble outputs and to make time-series adjustments of the meteorological inputs.

Fig. 4 illustrates example output from ESPADP showing three analyses for the probability of exceedance for the period 03/11/2007 to

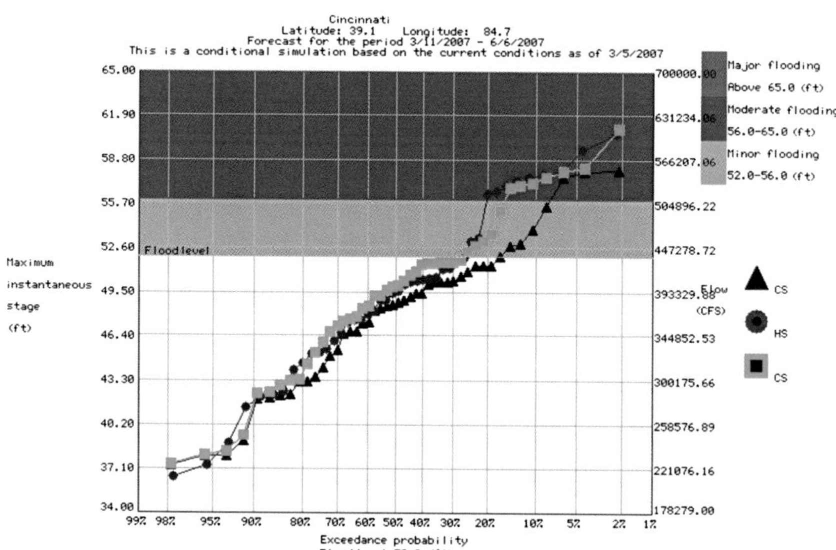

Fig. 4 Example AHPS 90-day ensemble, probabilistic hydrologic forecast for Cincinnati, Ohio (CCNO1) on the Ohio River for the period 03/11/2007–06/06/2007. The *orange region* designates minor flood category and the *red region* shows moderate flooding. The *black curve* is the operational conditional forecast, which includes NOAA/NWS Climate Prediction Center (CPC) and NOAA/NWS Weather Prediction Center (WPC) precipitation and temperature adjustments; *blue* is the historical simulation, and *green* is the conditional forecast without CPC and WPC adjustments.

06/06/2007 for the Ohio River at Cincinnati, OH, within the OHRFC forecast area. The *blue line*, labeled HS, shows the probability of exceedance for the historical simulation covering the 03/11/2007 through 06/06/2007 analysis period. The *black line* labeled CS represents the probability of exceedance for the operational 90-day ESP conditional simulation—that is, the simulation depends on the current basin soil moisture conditions. This simulation also utilizes time-series adjustments taken from extended 5-day NWS Weather Prediction Center QPF (http://www.wpc.ncep.noaa.gov/qpf/qpf2.shtml) and NWS Climate Prediction Center (CPC) 6- to 10-day and monthly climate outlooks (http://www.cpc.ncep.noaa.gov). The *green line* labeled CS shows the probability of exceedance for the 90-day ESP conditional simulation, reflecting current basin soil moisture conditions, but does not include the WPC and CPC historical meteorological time-series adjustments. A clearly discernible difference is apparent where the operational, climate adjusted time-series based conditional simulation shows a decreased likelihood of reaching a minor flood level compared to either the historical simulation or the conditional simulation without the climate-adjusted meteorological time-series forcings.

3.4.5 Flash Flood Guidance (FFG) System

The FFG System includes techniques and programs for computing FFG values. FFG is the amount of rainfall needed over a prescribed area, for a given duration to bring small streams to bankfull, flood conditions (see Sweeney, 1992; Sweeney & Baumgardner, 1999). The NWSRFS FFG operation used to compute FFG is a computationally efficient process that iteratively runs the operational hydrologic model, including snow accumulation and melt, over a basin until the precipitation needed to reach bankfull flow (or runoff) is reached. This process requires that prior hydrologic analyses be made to determine the peak streamflow (or normalized as peak runoff with respect to basin averaged peak flow) that corresponds to bankfull conditions—this index value is known as threshold runoff. For FFG purposes, a specific amount of rain is needed to produce a given amount of runoff based on estimates of current soil moisture conditions as maintained by soil moisture accounting models. The approach used to estimate FFG varies considerably between RFCs. A comprehensive review of the efficacy of NWS FFG systems, utilizing NWS Storm Data (http://www.ncdc.noaa.gov/IPS/sd/sd.html) for verification of FFG values, appears in Clark, Gourley, Flamig, Yang, and Clark (2014).

3.5 Models

NWSRFS is designed to be very flexible in its use of the models made available within the system through its modular design and architecture. The models used within NWSRFS include:

- hydrologic rainfall-runoff models;
- snow accumulation and melt models;
- reservoir simulation models;
- lumped parameter hydrologic routing models;
- miscellaneous models designed to handle unit hydrographs, handle glaciers, consumptive use in water resources, channel loss, baseflow, etc.

A complete list of available models within NWSRFS is given in Table 2.

3.6 Methods

Real-time forecasting with NWSRFS, as with most operational forecast systems, begins with the acquisition and processing of observations of real-time point station data for precipitation, temperature, and streamgage stage (river level) data. Observed stage data is converted to streamflow using rating curves obtained from the U.S. Geological Survey. Preprocessing utilities within NWSRFS convert observed point station data to mean areal quantities in the form of observed time-series data. Several temperature and precipitation data spatial interpolation preprocessing schemes are available within NWSRFS, including Thiessen weighting (Thiessen, 1911), Inverse Distance Weighting (IDW), and a method employing predetermined weights, which attempts to correct for orographic or terrain enhancement effects. Details can be found in the NOAA/NWS NWSRFS User Manual (http://www.nws.noaa.gov/oh/hrl/nwsrfs/users_manual/htm/xrfsdocpdf.php).

By necessity, NWSRFS was based on the use of conceptual, lumped parameter hydrologic modeling, which facilitated computationally efficient methods, algorithms, and models to meet the needs of a real-time operational hydrologic forecast system. NWSRFS encourages subdividing a large river basin into major tributary components, known as forecast groups, that can be modeled independently of each other for upstream basin areas. These larger subbasins (forecast groups) are further subdivided, first into segments and then, possibly, into multiple, smaller subbasin areas, which are known as mean areal precipitation (MAP) areas. Fig. 5 shows the OHRFC forecast area Forecast Groups and Mean Average Precipitation (MAP) basin areas. The MAP areas serve as the smallest hydrologic modeling area with RFC operations where the SAC-SMA, SNOW-17, or other models are applied. Typically, a segment has only a single MAP area, but the number is often

Table 2 Table providing NWSRFS model (operation) identifiers, a brief description and whether the model is used in the Operational Forecast System (OFS), forecast system component (FCST) or Calibration System (CS) (CALB), or both (BOTH)

Operation identifier	Description	Type
API-CIN	OHRFC API rainfall-runoff model	BOTH
API-CONT	Continuous API model	BOTH
API-HAR	MARFC API rainfall-runoff model	BOTH
API-HAR2	MARFC API rainfall-runoff model	BOTH
API-HFD	NERFC API rainfall-runoff model	BOTH
API-MKC	MBRFC API rainfall-runoff model	BOTH
API-SLC	CBRFC API rainfall-runoff model	BOTH
BASEFLOW	Baseflow simulation	BOTH
CHANLOSS	Conceptual model channel loss	BOTH
CONS-USE	Consumptive use model	BOTH
DHM-OP	Distributed hydrologic model	BOTH
DWOPER	Dynamic wave routing	BOTH
FFG	Flash flood guidance	FCST
FLDWAV	Generalized flood wave routing	BOTH
GLACIER	Glacier routing model	BOTH
LAG/K	Lag and K routing	BOTH
LAY-COEF	Layered coefficient routing	BOTH
MUSKROUT	Muskingum routing	BOTH
REDO-UHG	Reduced order unit hydrograph	BOTH
RES-J	Joint reservoir regulation model	BOTH
RES-SNGL	Single reservoir regulation model	BOTH
RSNWELEV	Rain-snow elevation computation	BOTH
SAC-SMA	Sacramento soil moisture accounting model	BOTH
SARROUTE	SSARR channel routing	BOTH
SNOW-17	Snow accumulation and ablation model	BOTH
SNOW-43	State-space snow accumulation and ablation model	BOTH
SS-SAC	State-space Sacramento model	FCST
SSARRESV	SSARR reservoir regulation	FCST
SWB-NILE	Simple water balance model	BOTH
TATUM	Tatum routing	BOTH
UNIT-HG	Unit hydrograph	BOTH
XIN-SMA	Xinanjiang soil-moisture accounting model	BOTH

See http://www.nws.noaa.gov/oh/hrl/nwsrfs/users_manual/part5/_pdf/532opers.pdf for details.

Fig. 5 Map showing the OHRFC forecast area with 31 Forecast Groups (*bold black outlines*) and ~700 Mean Average Precipitation (MAP) areas (*red outlines*). The Forecast Groups colored *green* are the local basin areas adjacent to the Ohio River mainstem. The *dark blue* areas show a portion of the Great Lakes. The light gray lines are U.S. State boundaries. The Ohio River extends approximately 1580 km from Pittsburgh, PA (most eastern) to the Mississippi River (most western).

greater. This requires, of course, that lumped parametric hydrologic routing must be included in the operational workflow for streamflow routing from upstream MAP areas and segments to downstream MAP areas and segments for the entire forecast area. An example of the text encoding of the NWSRFS workflow, called a segment definition, is given in Listing 1. The MAP area is the smallest region over which mean areal precipitation and temperature time series are calculated for both observed and forecasted quantities. RFCs have maintained the same nomenclature with CHPS (see Section 4.1) referring to modeling and operational workflow organization.

3.6.1 Quantitative Precipitation Estimation (QPE)

RFC precipitation estimation from observed data sources includes the use of point station, radar, and satellite measurements and the assimilation of these data sources into a single unbiased estimate. Estimates can take the form of either mean areal time series or gridded spatial fields. Specific NWSRFS techniques for time series estimation can be found at http://www.nws.noaa.gov/oh/hrl/nwsrfs/users_manual/part2/_pdf/26ofs_map.pdf and http://www.nws.noaa.gov/oh/hrl/nwsrfs/users_manual/part2/_pdf/26ofs_mapx.pdf.

Observed gridded precipitation estimation is covered, subsequently, in Section 5.3 covering the NWS Multisensor Precipitation Estimator (MPE).

3.6.2 Quantitative Precipitation Forecast (QPF)

NWS RFCs utilize Quantitative Precipitation Forecast (QPF)—that is, forecasted precipitation as one of the hydrologic model forcings. Through AWIPS (Section 5.1), wide area network (WAN) QPF guidance is received from the NWS Weather Prediction Center (WPC). RFCs analyze the 6-hourly WPC QPF and utilize custom AWIPS software to modify the 6-hourly gridded fields as changing conditions warrant. Typically, the WPC QPF is produced several hours, and possibly a full synoptic period, prior to the beginning of its valid period. So, rapidly changing meteorological conditions that were not evident in early numerical weather prediction model runs can occur, necessitating QPF changes from the WPC guidance.

3.6.3 Temperature Estimation

Mean areal temperature (MAT) estimation from point data collected at meteorological observation stations is described in the NWSRFS User Manual, found at http://www.nws.noaa.gov/oh/hrl/nwsrfs/users_manual/part2/_pdf/27ofs_mat.pdf. Many of the techniques described are outdated because of the availability of newer gridded data sets and newer techniques for improved temperature estimation in mountainous regions from point station data utilizing optimal techniques which have been incorporated into AWIPS software in the AWIPS Common AWIPS Visualization Environment (CAVE).

4 CURRENT MODELS AND MODELING SYSTEM

4.1 Community Hydrologic Prediction System (CHPS)

The Community Hydrologic Prediction System (CHPS) was developed based on the Flood Early Warning System (FEWS) from Deltares (the Netherlands) to replace the NWSRFS. FEWS consists of a Java-based client-server architecture, where data is passed from the core of FEWS to external models, such as the NWSRFS models listed in Table 3, through an XML read/write interface. External models are wrapped in model adapter coding to provide the needed XML read and write capability to communicate with the FEWS core. Essential features of the NWS CHPS implementation of FEWS are:

- the migration of all currently used models from NWSRFS to FEWS as well as many key operational modules for processing and manipulation of data;
- interactive capability very similar to NWSRFS-IFP was added to FEWS that had not previously existed.

Table 3 Table providing models available in the NWS Community Hydrologic Prediction System (CHPS)

Model description
BASEFLOW Simulation Model
Channel loss
Consumptive use
Continuous incremental API
Glacier routing
Gridded snow-17
Joint reservoir regulation (RES-J)
Lag and K routing
Layered coefficient routing
Muskingum routing
Rain-snow elevation
Sacramento Soil Moisture Accounting (SAC-SMA)
Sacramento Soil Moisture Accounting with Heat Transfer (SACHT)
Single reservoir regulation (RES-SNGL)
SNOW-17
SSARR reservoir regulation
SSAR channel routing
Tatum coefficient routing
Unit hydrograph
HEC-RAS – USACE Hydrologic Engineering Center – River Analysis System
HEC-ResSim – USACE Hydrologic Engineering Center—Reservoir Simulation System

See http://www.nws.noaa.gov/oh/hrl/general/indexdoc.htm for details.

Operational implementation of FEWS occurred in two phases; first, transition from NWSRFS to CHPS by four RFCs (NWRFC, CNRFC, ABRFC, and NERFC), called the CHPS Acceleration Team (CAT). Following the transition from NWSRFS to CHPS and real-time operational testing, the CAT RFCs aided the remaining RFCs with their transitions from NWSRFS to CHPS and real-time operational testing. Full RFC operational implementation of CHPS was completed in 2013; however, some use of NWSRFS remained through 2015 until the transition by RFCs to CHPS-based model calibration and long-lead time ESP hydrologic modeling was completed.

Significant advantages were gained by the adoption of CHPS, including much greater capability of use of external gridded data sources, customization of FEWS for data and model output visualization, and clearly improved capability for adding new models for operations than was possible with NWSRFS. Examples of this include additions of the U.S. Army Corps of Engineers (USACE) HEC-RAS 1-D, dynamic, unsteady flow routing model and HEC-ResSim reservoir simulation model to CHPS.

4.2 Models

Many of the models listed in Table 2 from NWSRFS remain available in CHPS-FEWS, with the addition of U.S. Army Corps of Engineers (HEC-RAS)1-D, unsteady flow routing and HEC-ResSim reservoir simulation models (Table 3). The HEC-RAS model replaces the NWS DWOPER and FLDWAV models for unsteady flow simulation.

4.2.1 Research Distributed Hydrologic Model (RDHM) Model

A major area of emphasis in NWS hydrologic services in recent years has been development and implementation of distributed hydrologic modeling for real-time operational forecasting. Koren, Reed, Smith, Zhang, and Seo (2004) report on the development of the NOAA, NWS Hydrology Laboratory Research Modeling System (HL-RMS), subsequently referred to as the Research Distributed Hydrologic Model (RDHM) (Smith, Koren, et al., 2004; Koren et al., 2004). The RDHM is comprised of several modeling components, including:

- SAC-SMA—Sacramento Soil Moisture Accounting Model;
- SNOW17—snow accumulation and ablation model;
- FRZ—frozen ground component;
- RUTPIX7—channel routing component;
- RUTPIX9—channel routing component;
- SACHTET—SAC-SMA model with (ground) heat transfer and with evapotranspiration.

The underlying modeling basis for the RDHM (OHD, 2013) is the Sacramento Soil Moisture Accounting Model (SAC-SMA) (Burnash, 1995), depicted schematically in Fig. 6, which is derived from the Stanford Watershed Model, developed in the early 1960s (Crawford & Linsley, 1966; Burnash, Ferral, & McGuire, 1973; Crawford & Burges, 2004). The SAC-SMA is a lumped-parameter, conceptual hydrologic model, but within the framework of the RDHM, the SAC-SMA is applied on a gridded basis and includes significant process enhancements to the SAC-SMA, including a frozen ground component, the ability to model soil moisture in a layered soil profile structure, and ability to model evapotranspiration (see Koren, Smith, Wang, & Zhang, 2000, 2003, 2004, 2010, Koren, Smith, Cui, & Cosgrove, 2007, Koren et al., 2004; 2010, Koren, Smith, & Duan, 2003; Smith, Seo, et al., 2004; Koren, 2006, 2011;

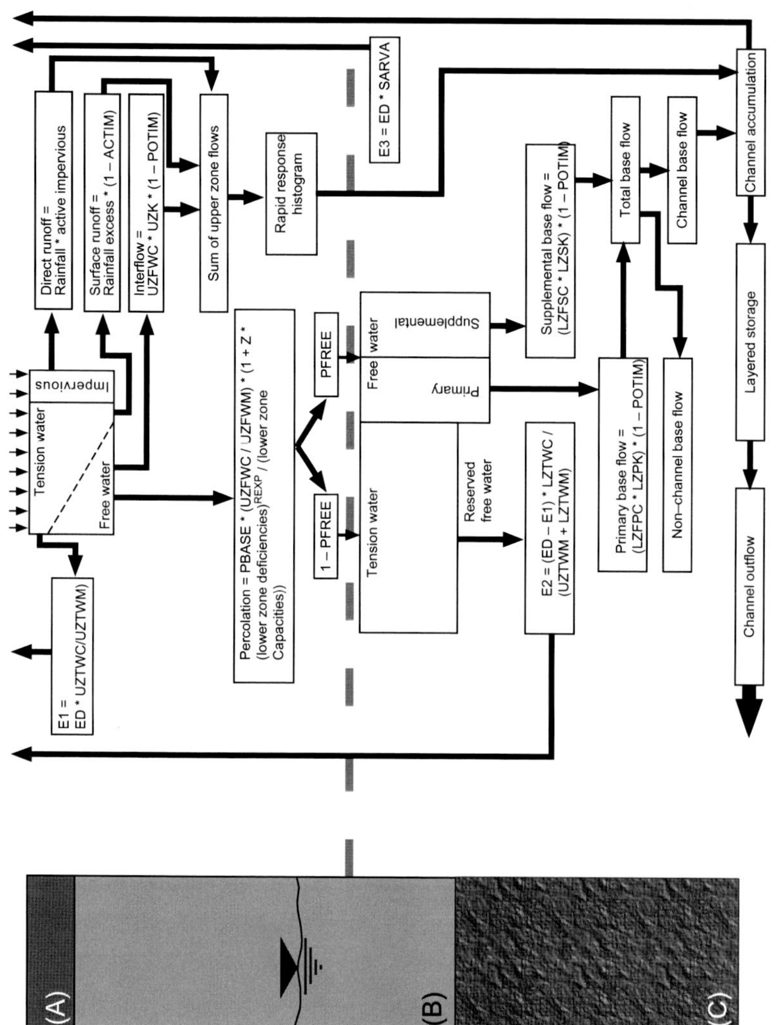

Fig. 6 Schematic of the conceptual SAC-SMA hydrologic model, showing the relation to an idealization of a soil profile and soil horizons, with saturated and unsaturated soil regions.

Cosgrove et al., 2012). With these enhancements, the RDHM becomes a robust, physically based distributed hydrologic model. Most NWS RFCs are actively pursuing operational implementation of the RDHM for flood forecasting within the CHPS-FEWS environment and aid in the production of real-time FFG. A significant research effort to evaluate strengths and weaknesses of distributed models in use around the globe was sponsored by the NOAA/NWS Office of Hydrologic Development. Research results from two separate evaluation studies called the Distributed Model Intercomparison Project, Phase-1 and Phase-2 were published (Smith, Seo, et al., 2004; Smith et al., 2012).

Additionally, the RDHM has threshold frequency based FFG generation capability (Cosgrove et al., 2012). The significance of this is that historical simulations are made for the period of observed record to generate the simulated historical distribution of flows at each modeled grid point (pixel); from this, operationally, the relative frequency of the simulated predicted flows can be determined. This methodology makes it possible to identify forecasted simulated peak flows as an *n*-year event (such as the 20-year flood), which gives us some sense of the relative significance of the forecasted flood event compared to past floods.

Since the RDHM is a physically based distributed hydrologic model, it is necessary to estimate model parameter values at every grid point location (pixel) within the domain being modeled. The NWS has developed methodologies to estimate initial, *a priori*, RDHM parameter values. Best/optimal simulation results are obtained utilizing a combination of manual calibration and automatic optimization within the RDHM. The RDHM utilizes a local search technique, referred to as the Stepwise Line Search (SLS), which employs a technique of successive minimization along coordinate directions (Press, Flannery, Teukolsky, & Vetterling, 1986), but with a fixed step size along each coordinate and one-step propagation at a time. The SLS algorithm consists of the following steps:

1. Start with the *a priori* estimates of the hydrologic model parameters.
2. With the all parameter estimates fixed to the *a priori* parameter values, increase or decrease the value of the first parameter by one step in the direction of decreasing objective function value.
3. With all other parameters now set to the new (or old, if the objective function value did not decrease) value, decrease or increase the value of the second parameter by one step in the direction of decreasing objective function value.

4. Repeat Step 3 until the objective function is minimized with respect to each of all remaining parameters.
5. Repeat Steps 2–4 until no further reduction in the objective function is realized.

RDHM SLS optimization is depicted graphically in Fig. 7. Hydrologic model parameter optimization for RFC hydrologic models has a long tradition (Duan et al., 1992, 1993, 1994).

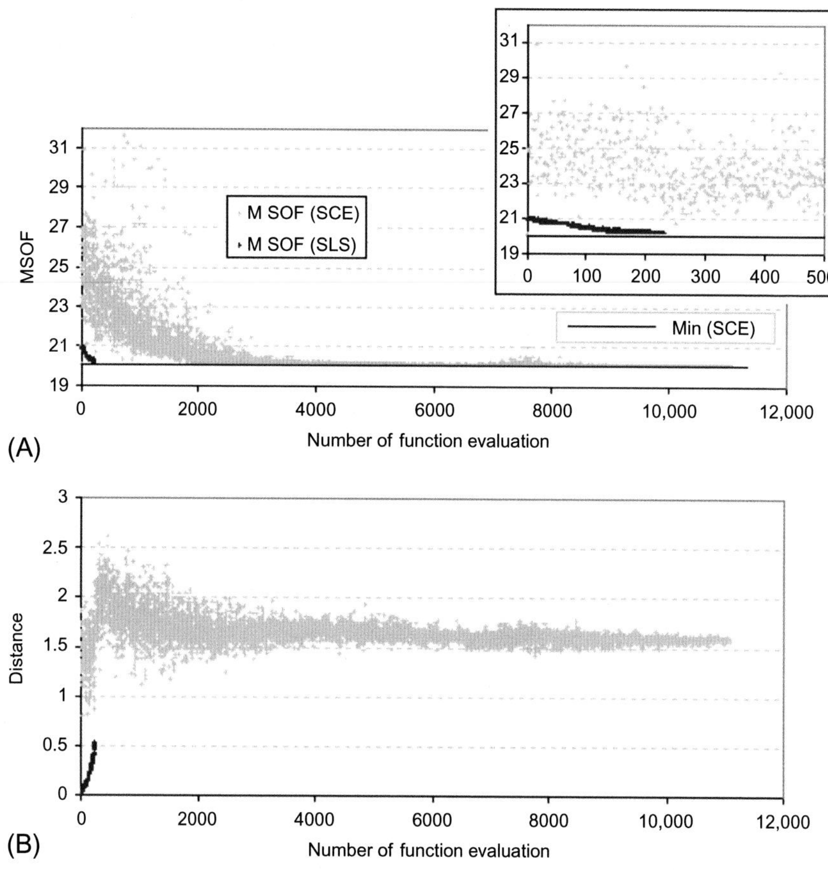

Fig. 7 An example of RDHM SLS parameter convergence. (A) Multi-scale objective function vs. the number of function evaluations in the optimization sequence. The upper-right plot is a zoom-up near the origin. (B) Distance from the *a priori* parameters vs. the number of function evaluations. The light and dark markers denote the multi-scale objective function value and the distance associated with the Shuffle Complex Evolution (SCE)- and SLS-optimized parameters, respectively. From OHD. (2013). Hydrology laboratory-research distributed hydrologic model (HL-RDHM) user manual. Technical report. NOAA/NWS, Office of Hydrologic Development, Silver Spring, MD, March.

While hydrologic model calibration is desired and highly recommended, with a scientifically well-formed model and with good parameter estimation, very good uncalibrated RDHM simulations are possible, as Fig. 8 demonstrates for the Potomac River at Little Falls Pump Station, near Washington, DC, covering the period Oct. 2009 through Jun. 2010 from a longer simulation period spanning Oct. 2007 to Dec. 2013. Under simulations during Dec. 2009 are probably due to either suboptimal SNOW-17 parameter estimates affecting snowmelt rates or due to low air temperature estimates, which would treat precipitation as snowfall rather than rainfall. In Jan. 2010, too much snow was available for melt, which gave rise to an over-simulation. The early Mar. 2010 simulation is too high and indicates snowmelt when there was probably little to no snow available, pointing to precipitation that was typed as snowfall in early Feb., that should have been typed as rainfall—leading to an under-simulation of the peak flow in early Feb. 2010.

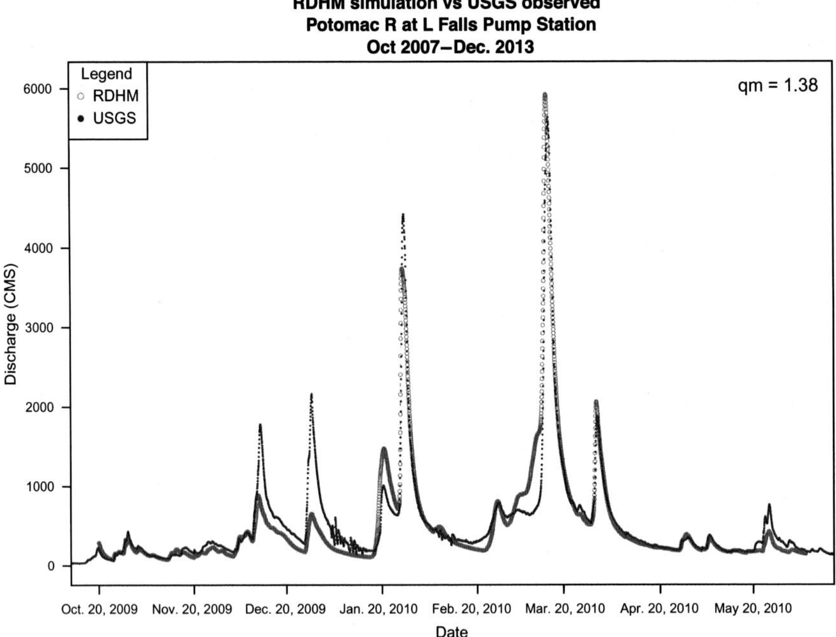

Fig. 8 Uncalibrated RDHM based simulation utilizing NEXRAD derived, xmrg-based precipitation forcings (*red circles*) compared against USGS measured streamflow (*blue dots*) for the Potomac River at Little Falls Pump Station, MD. The simulation includes the RDHM sac, snow17, frz, and rutpix9 model components.

Fig. 9 illustrates an instantaneous view of the gridded RDHM streamflow for the Potomac River basin at Little Falls Pump Station, near Washington, DC at 06/27/2006 19:00:00 UTC at the HRAP (~4 km) grid resolution, utilizing NEXRAD derived, xmrg-based precipitation and hourly North American Land Data Assimilation System (NLDAS) temperature forcings. Animations of RDHM streamflow simulations can be found at https://vimeo.com/113855426.

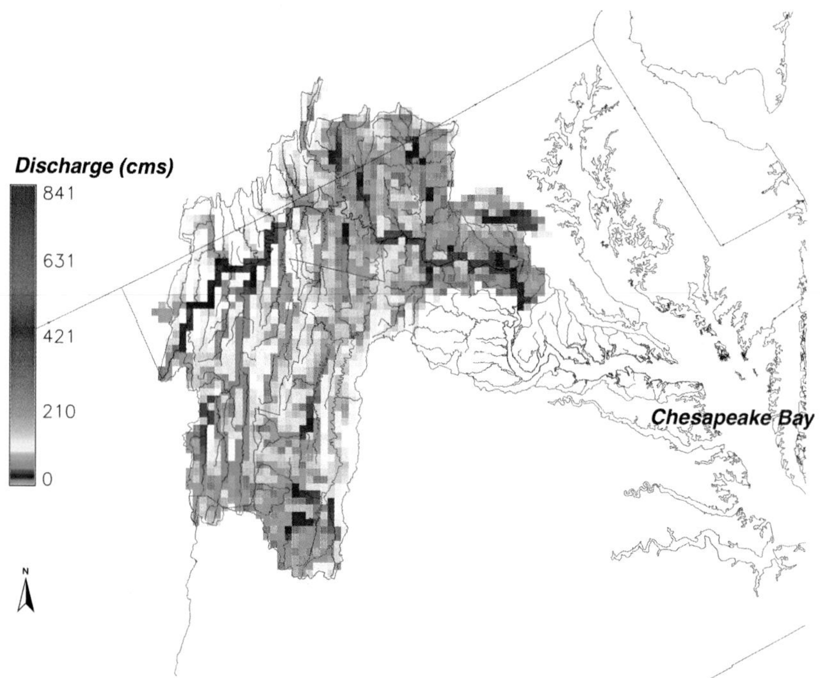

Fig. 9 Map showing uncalibrated RDHM discharge at 06/27/2006 19:00:00 UTC for the Potomac River basin upstream of Washington, DC, USA at the HRAP (~4 km) grid resolution, utilizing NEXRAD derived, xmrg-based precipitation and hourly NLDAS temperature forcings. The simulation includes the RDHM sac, snow17, frz, and rutpix9 model components. Discharge is shown in units of cubic meters per second (cms).

5 CURRENT OPERATIONS

Within the structure of NWS operations, RFCs provide hydrologic forecast guidance to NWS Weather Forecast Offices (WFOs) (see Fig. 10). There are currently 122 WFOs operating within the US; each of the 13 RFCs is co-located with one of the WFOs, occupying the same building space, with separate, but contiguous operations areas (see Fig. 11). There have been few

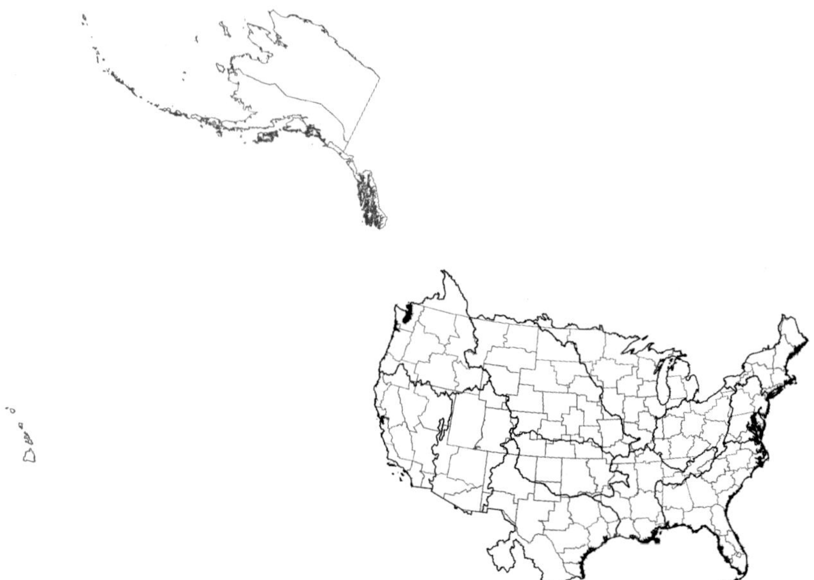

Fig. 10 Map showing 122 U.S. NWS Weather Forecast Office (WFO) areas of responsibility (*red*) overlain by 13 NOAA/NWS River Forecast Centers (*black*). Note that RFCs serve multiple WFOs and that the APRFC (Alaska) serves 4 WFOs, including Hawaii and Guam. The map is drawn to scale using an Albers Equal Area map projection.

Fig. 11 Photograph showing a partial view of the OHRFC operations area, c. 2008. The large screen display is used for operational briefings and for situational awareness during high-impact flooding events. The WFO Wilmington operations area is shown partially to the left of the OHRFC operations area.

substantial changes with the overall RFC operational workflow, shown in Fig. 12, with the transition from the NWSRFS based hydrologic forecasting to the CHPS-FEWS (see Section 4.1) based modeling framework. The most substantial changes found with the adoption of FEWS-based CHPS is the ease with which gridded hydrometeorological datasets can be utilized as hydrologic model inputs. Gridded datasets that are now readily usable within RFC workflows include output from numerical weather prediction (NWP) models using the European Centre for Medium-Range Weather Forecasts (ECMWF) gridded binary format (grib or grib2) (https://software.ecmwf.int/wiki/display/GRIB/Home) and UCAR/Unidata Network Common Data Format (NetCDF) (https://www.unidata.ucar.edu/software/netcdf/). Also, the relative ease at which new models and operational procedures can be added to CHPS-FEWS has allowed new models, not available in NWSRFS, to be added to RFC operational workflows. Both NWSRFS and CHPS-FEWS operate seamlessly within the NOAA/NWS AWIPS, which is discussed in Section 5.1.

All RFC modeling occurs within the AWIPS and CHPS environment. Nearly all hydrologic models within CHPS run at a 6-hourly time-step, but most RFCs are pursuing configuration changes that will permit hourly model time steps. Shorter time steps are necessary in order to model adequately smaller, fast responding basins that are subject to flash flooding. Unsteady flow hydraulic modeling is done with a version of the U.S. Army Corps of Engineers HEC-RAS model (http://www.hec.usace.army.mil/software/hec-ras/) which runs on Linux within CHPS. The internal 1-dimensional, unsteady flow HEC-RAS model time steps are considerably shorter than 1 h to maintain computational stability.

RFCs operate 365 days/year, 16 h/day, except when there is imminent or ongoing flooding within an RFC's area of responsibility. If there is a significant threat of flooding, RFCs extend operational coverage to 24 h/day. RFCs will also maintain 24-h/day operations in support of WFOs experiencing flash flooding or the threat of flash flooding within their hydrologic forecast areas. WFOs have the responsibility for issuing flash flood watches and warnings (see http://www.weather.gov/lwx/WarningsDefined)[19] as defined by NWS policy, which exists largely due to WFO daily 24-h operations schedule and overall responsibility to issue all public forecasts. Routine

[19]A Flash Flood Watch is issued when conditions are favorable for flash flooding. It does not mean that flash flooding will occur, but it is possible. A Flash Flood Warning is issued when flash flooding is imminent or occurring.

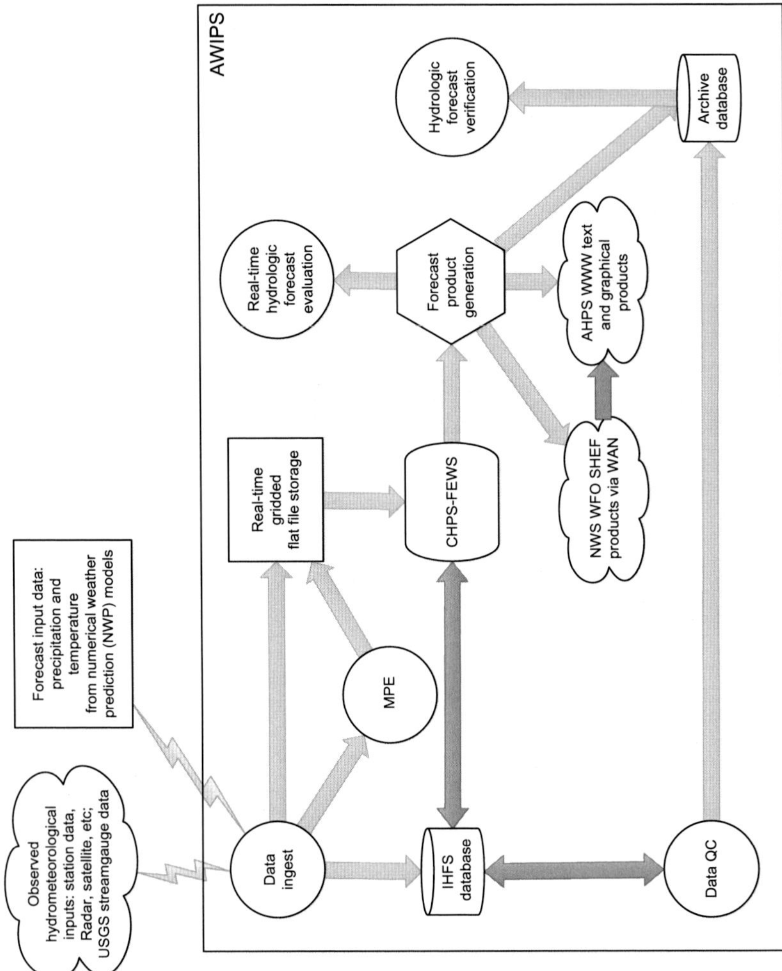

Fig. 12 Schematic of generalized RFC operational workflow, including data ingest and quality control (QC) within AWIPS, and CHPS-FEWS hydrologic and hydraulic modeling within AWIPS.

RFC operations are typically tied to meteorological synoptic times (00, 06, 12, and 18 UTC) due to the availability of meteorological forcing data from numerical weather prediction models. Consequently, routine daily forecasts in the U.S. tend to begin following 12 UTC and, for many RFCs, evening forecast updates begin at 00 UTC using interactive hydrologic forecasting techniques which serve as a manual data assimilation process (Adams & Smith, 1993). RFC hydrologic forecasts are updated as needed, depending on how newly acquired observed river stage/level values compare to forecasted values, utilizing real-time hydrograph visualization capabilities within the CHPS-FEWS environment. Fig. 12 shows a generalized operational hydrologic forecasting workflow for RFCs, critical steps of which are data quality control (QC) and radar derived precipitation estimation bias correction with the Multisensor Precipitation Estimator (MPE), described in Section 5.3.1 (Fig. 13).

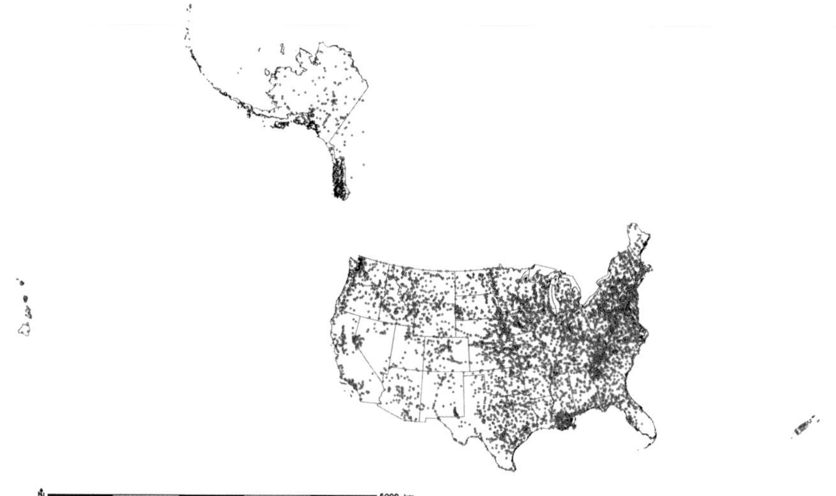

Fig. 13 Map showing the Continental United States (CONUS), Alaska, Hawaii, and Puerto Rico with 7280 NWS hydrologic forecast point locations, including both daily and flood-only forecast points. The map is drawn to scale using a Polar Stereographic map projection. (see http://water.weather.gov/ahps/download.php).

All NWS RFC and WFO operational applications and processes run within the AWIPS described in Section 5.1, including CHPS-FEWS at RFCs. RFC deterministic hydrologic forecast guidance is transmitted as Standard Hydrometeorological Exchange Format (SHEF) (see Listing 2 and Listing 3) to WFOs through the AWIPS wide area network (WAN),

Listing 2 Partial listing of the Ohio River Forecast Center (OHRFC) RVFSTG SHEF Encoded Forecast Product

```
RVFSTG
River Stage Forecasts
National Weather Service, Ohio River Forecast Center, Wilmington, Ohio
1152 AM EDT Wednesday, June 17, 2015

This is an INTERNAL guidance product issued by the River Forecast Center.
Official forecasts/warnings are issued by local NWS offices.

The following OHRFC guidance forecasts include QPF through at least 24 hours...
All forecasts are given in feet.
--------------------------------------------------------------------------------
Ohio River                  Latest                                  Crest       A/Abv FS
Forecast Point     FS       Obsv    Jun 18   Jun 19   Jun 20    Stage Da/Time   B/Blo FS
Pittsburgh         25.0     16.6     16.6     16.6     16.6
Dashields Lock     25.0     17.0     16.6     16.4     16.2
Montgomery Lock    33.0     16.3     16.3     15.5     14.9
New Cumberland     36.0     20.8     18.9     17.8     17.1
Wellsburg          36.0      M       26.9     26.5     26.3
Pike Island L&D    37.0      1.0     19.2     18.2     17.5
Wheeling           36.0     22.8     19.9     19.3     18.8
Hannibal Dam       35.0     20.1     16.7     15.6     15.0
Willow Island L    37.0     23.7     19.6     18.0     17.1
Marietta Pump H    35.0     23.1     21.4     20.2     19.5
Parkersburg        36.0     25.3     23.8     23.2     22.1
Belleville L&D     35.0     22.3     19.8     19.0     18.3
Racine Lock & D    41.0     23.5     21.1     19.9     18.9
Point Pleasant     40.0     26.2     25.6     25.6     25.4
R C Byrd Dam       50.0     25.5     24.5     24.0     22.8
Huntington         50.0     29.1     28.2     27.4     26.9
Ashland            52.0     35.3     35.2     35.0     34.9
Grayson            21.0      5.9      7.5      3.8      3.8
Lloyd Greenup D    54.0     26.1     27.2     25.2     24.5
Portsmouth         50.0     25.2     26.8     25.2     24.9
Maysville          50.0     34.9     35.7     35.4     35.2
Meldahl Dam        51.0     23.6     27.4     26.0     25.2
Cincinnati         52.0     30.5     32.9     33.0     32.1
Markland Dam       51.0     22.3     25.0     26.9     26.2
McAlpine Upper     23.0     12.8     12.9     12.9     12.9
McAlpine Lower     55.0     19.8     23.0     26.8     26.8
Cannelton Dam      42.0     14.8     16.8     19.1     21.2
Newburgh Lock &    38.0     16.7     19.1     21.9     24.7
Evansville         42.0     15.7     17.8     19.7     21.7
J T Myers Dam      37.0     17.6     21.1     22.9     25.1
Shawneetown        33.0     17.7     20.0     22.0     23.6
Golconda           40.0     30.0     29.9     30.9     31.6
--------------------------------------------------------------------------------
Allegheny River             Latest                                  Crest       A/Abv FS
Forecast Point     FS       Obsv    Jun 18   Jun 19   Jun 20    Stage Da/Time   B/Blo FS
Eldred             23.0      5.6      5.7      5.6      5.3
Olean              10.0      2.5      2.2      1.9      1.5
Salamanca          12.0      4.9      4.7      4.6      4.3
West Hickory       14.0      7.3      7.1      6.2      6.1
Meadville          14.0      8.2      7.3      6.0      5.1
Franklin           17.0      8.4      8.3      7.9      6.9
Parker             20.0      7.9      7.7      7.2      6.2
St Charles         17.0      5.5      5.1      4.7      4.3
 *
 *
 *
```

Listing 3 Example SHEF Encoded RVFLOM Forecast Product from the Lower Mississippi River Forecast Center (LMRFC)

```
295
FGUS54 KORN 281543
RVFLOM
River Forecast
NWS Lower Mississippi River Forecast Center, Slidell, LA
10:42am CDT Sunday, June 28, 2015

This is a NWS guidance product from the Lower Mississippi River Forecast Center.
Public forecasts and warnings are issued by NWS Weather Forecast Offices.

: Reserve - Mississippi River - LOM
: Zero Datum 0.01 ft (NGVD 1929 (1983 ADJ))  St. John The Baptist County, LA
: Flood Stage 22.0 ft    Action Stage 21.0 ft
: Moderate Stage 24.0 ft    Major Stage 26.0 ft
: Latest Stage 17.1 ft at 900am CDT on Sun, Jun 28
:
: Forecasts are in 6-hour increments.

: Forecasts include 24 hours of QPF.

.AR RRVL1 20150717 Z DC201506281542/DUE/DH12/HGIFFX :crest: 20.0
:  Obsv Valid  /   00Z /   06Z /   12Z /   18Z
:     Jun 26                           /  16.9
:     Jun 27   /  16.9 /  16.9 /  16.9 /  17.0
:     Jun 28   /  17.0 /  16.9 /  17.1
:
.ER RRVL1 20150628 Z DC201506281542/DUE/DH18/HGIFF/DIH6
:
:  Fcst Valid  /   00Z /   06Z /   12Z /   18Z
.E1 : Jun 28 :                         /  17.1
.E2 : Jun 29 : /  17.2 /  17.2 /  17.2 /  17.3
.E3 : Jun 30 : /  17.3 /  17.4 /  17.4 /  17.4
.E4 : Jul 01 : /  17.5 /  17.5 /  17.5 /  17.6
.E5 : Jul 02 : /  17.7 /  17.7 /  17.7 /  17.8
.E6 : Jul 03 : /  17.8 /  17.9 /  17.9
:______________________________________________________________________

$$
:
:...END of MESSAGE...
```

which is reviewed by WFO forecasters to catch potential problems. Often WFO forecasters will either call over telephone landlines or communicate via specialized internet chat software to discuss possible hydrologic forecast adjustments, which can involve either RFC hydrologic model re-runs or RFC forecast adjustments that are re-transmitted over the AWIPS WAN. Alternatively, an agreement can be made that the WFO forecaster will manually alter the hydrologic forecast locally. In all cases, the NWS has software as part of the AWIPS environment to decode SHEF messages, which can be displayed graphically for editing and, then, transmitted to NWS www servers for public access (see http://water.weather.gov/ahps). An example AHPS hydrograph is shown in Fig. 14.

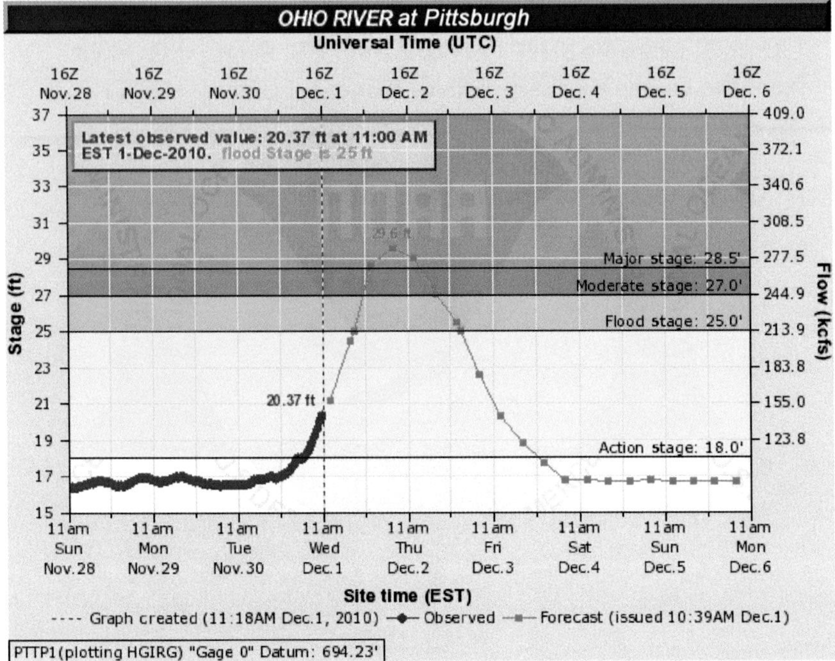

Fig. 14 Example AHPS webpage stage hydrograph for Pittsburgh, PA, showing observed 1-hourly river levels in *blue* and 6-hourly forecast in *green*, beginning at 11:00 AM Eastern Standard Time (EST) on Dec. 1, 2010. The forecast hydrograph is showing passing through minor, moderate, and major flood levels (stages). Courtesy U.S. Department of Commerce, NOAA/NWS.

5.1 AWIPS

The technological foundation for hydrometeorological forecasting in the NWS is the AWIPS, which includes telecommunications, data acquisition, database systems, computer systems for data and model display and analysis, and data archiving. Thus, AWIPS is a critical component of NWS operations. AWIPS hardware and software was deployed to Weather Forecast Offices (WFOs), RFCs, and other NWS sites throughout the United States from 1996 to 1999. The system has been in its Operations and Maintenance (O&M) phase of its lifecycle from 1999 through 2014. AWIPS O&M funding during this period was approximately $39.0 million per year. The NWS is currently deploying the second generation of AWIPS, AWIPS2, which includes restructuring of the AWIPS software into a Service Oriented Architecture (SOA). AWIPS technology infusion consists of separate projects that address three different areas of the AWIPS infrastructure: hardware, communications, and software. The hardware enhancements included migration from proprietary Unix hardware and software to Linux based hardware from 2003 through 2009 to provide increased processing and mass storage capacity. The hardware upgrades included the desktop workstations and local server systems.[20] AWIPS hardware is refreshed routinely on a cyclical basis. Communications enhancements have increased satellite network bandwidth and the software enhancements with AWIPS2 re-engineered AWIPS software into a standard SOA, making it easier and less expensive to integrate improved scientific methodologies and algorithms into AWIPS, while reducing software O&M costs.

The AWIPS Product Improvement (API) strategy was designed to increase system performance while reducing maintenance costs and processing latency. AWIPS collects, communicates, processes, displays, and analyzes hydro-meteorological data that is fundamental to the conduct of the NWS mission. Technology infusion is essential for the future of AWIPS, allowing AWIPS to accommodate the high volume, fine-scale data that are available from advanced satellite sensors (eg GOES-R, NPOESS[21]), new radars (eg Dual Polarity, which has been deployed operationally[22] and phased array radar in the future), and other ground-based automated observing systems, and advanced numerical weather and hydrologic prediction models.

5.2 AHPS

The AHPS was initiated by the NWS hydrologic services program in the early 1990s to advance the state of hydrologic and hydraulic modeling at higher resolution spatial and temporal scales and to deliver forecast products with improved detail, such as gridded, rather than county average, FFG. Greater emphasis was placed on coordination with local, state, and federal agencies, and with providing long lead-time probabilistic hydrologic forecasts for flood outlooks and water resources applications, especially directed at water supply issues for external decision makers (see McEnery, Ingram, Duan, Adams, & Anderson, 2005 and National Research Council, 2006b). The greatest water resources and water supply impacts can be found in

[20]Please refer to the NWS technical document found at http://www.nws.noaa.gov/ops2/ops24/documents/LX%20WS%20Replacement%20OAT%20Plan%20Final.pdf.

[21]Where we have Geostationary Operational Environmental Satellite—R Series (GOES-R) and National Polar-orbiting Operational Environmental Satellite System (NPOESS).

[22]As of the Spring of 2013, all NWS NEXRAD WSR-88Ds (Weather Surveillance Radar—1988 Doppler) received dual-polarization technology software and hardware upgrade, which greatly enhanced the radars by providing the ability to collect data on the horizontal and vertical properties of weather (eg, rain, hail) and non-weather (eg, insect, ground clutter) targets.

western states, but also the southeastern USA. The AHPS public interface includes a web-based suite of forecast products to display the magnitude and uncertainty of the occurrence of floods or droughts, from hours to days and months. See, for example, http://water.weather.gov/ahps and http://water.weather.gov/ahps/long_range.php. World-wide-web-based graphical products are aimed at providing information and planning tools for floodplain and emergency management purposes. The products enable government agencies, private institutions, and individuals to make more informed decisions about risk-based policies and actions to mitigate the dangers posed by floods and droughts. Graphical products most widely used by the general public, emergency managers, and others are the single-valued deterministic hydrologic forecasts, an example of which is shown in Fig. 14.

A major component of AHPS is the use of ESP (discussed in Section 3.4.4) to produce probabilistic flood, reservoir inflow, and streamflow outlooks extending 90 days into the future. A major concern is forecast uncertainty due to the use of QPF (Damrath et al., 2000). An example www-based product is depicted in Fig. 15, which shows a map displaying the 50% probability of exceedance for minor to major flood categories for NOAA/NWS RFC forecast point locations in the Mississippi River Valley, based on AHPS 90-day ensemble hydrologic forecasts. The actual web-based experimental graphics can be found at http://water.weather.gov/ahps/long_range.php.

5.3 Quantitative Precipitation Estimation (QPE)

In general, raingage networks are not capable of detecting high-intensity precipitation at the resolution needed for accurate flood forecasting applications for small watersheds due to inadequate gage densities. Significant research has shown that raingage network-based precipitation estimation suffers from degraded levels of accuracy with increased rainfall intensities during storms where convective processes are significant. Huff (1970), for instance, showed that correlations of gage rainfall accumulation within a dense raingage network diminish rapidly, both with increased distances between gages and with shorter accumulation periods. Using a dense raingage network over small watersheds, Fogel (1969) and Dawdy and Bergmann (1969) show the sensitivity of runoff prediction to accurate precipitation estimates. Thus, it is apparent that quantitative estimation of precipitation with radar is required as input to hydrologic models for flood forecasting of small basins, particularly at spatial scales where flash flooding is problematic (Smith, Seo, Baeck, & Hudlow, 1996; Johnson, Smith, Koren, & Finnerty, 1999). A major component of AWIPS and AHPS was the development and deployment of the NEXRAD system. Nominal NEXRAD radar coverage is shown in Fig. 16.

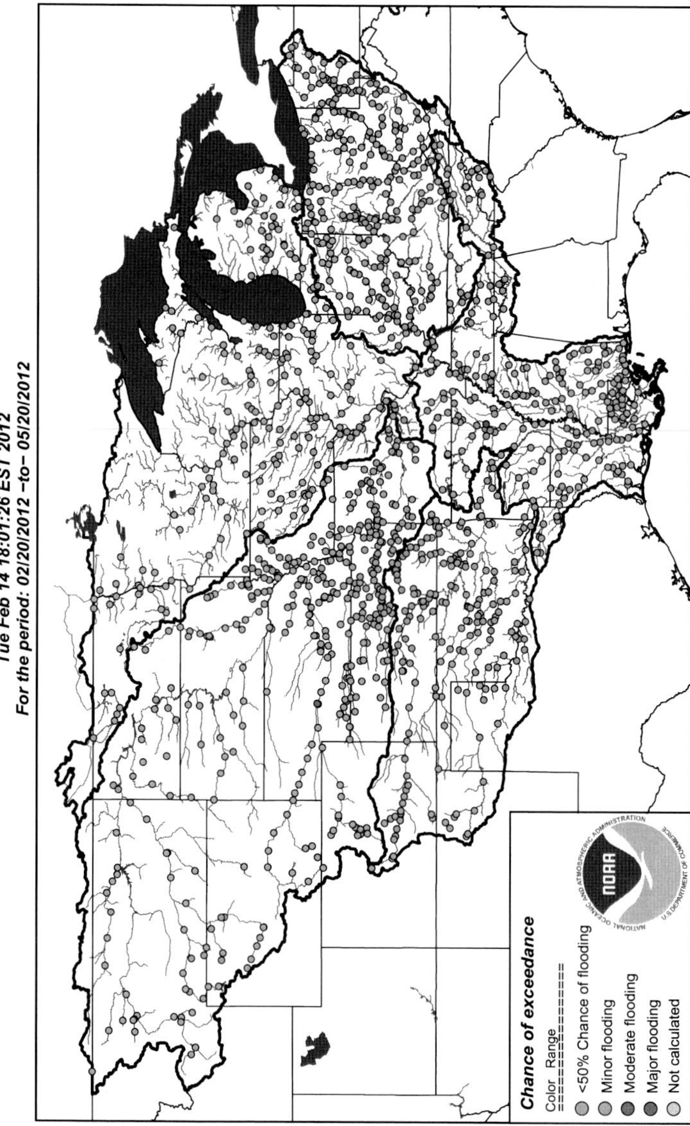

Fig. 15 Map showing the 50% probability of exceedance for minor to major flood categories for NOAA/NWS RFC forecast point locations in the Mississippi River Valley, based on AHPS 90-day ensemble hydrologic forecasts for the period Feb. 20–May 20, 2012. Web-based experimental graphics can be found at http://water.weather.gov/ahps/long_range.php. RFC boundaries for the MBRFC, NCRFC, OHRFC, LMRFC, and ABRFC are shown in *heavy black*.

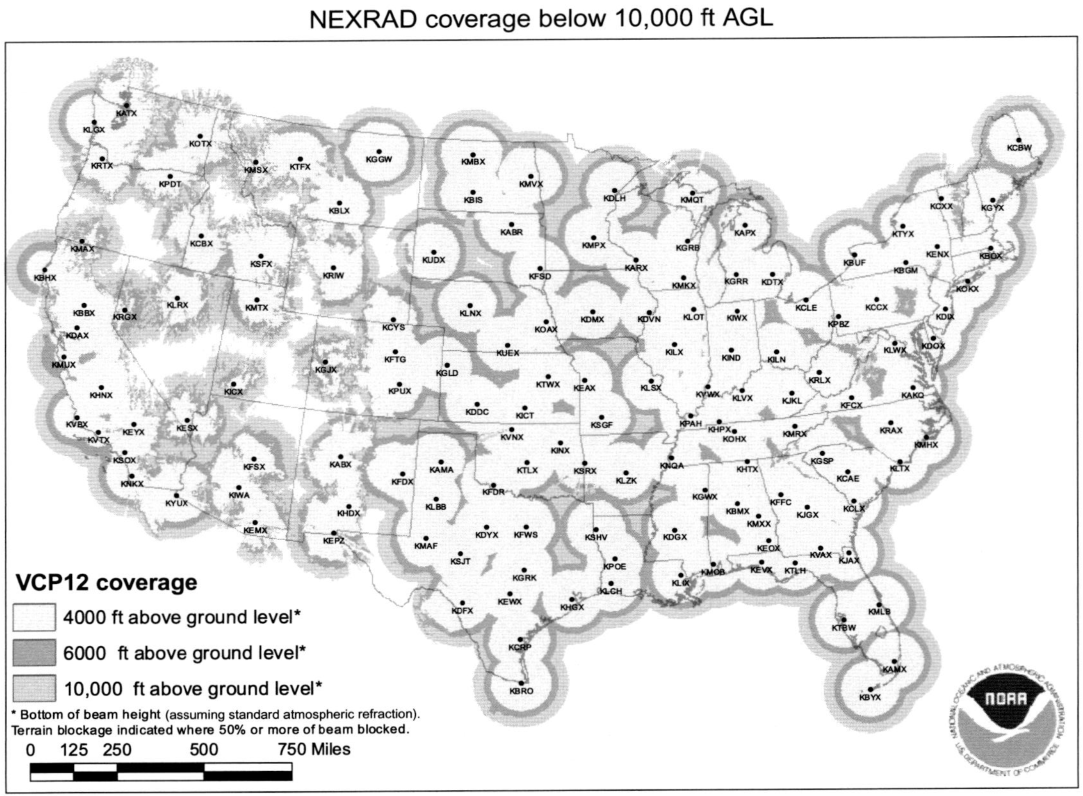

Fig. 16 Map showing the location of NWS NEXRAD radar sites and radar coverage below 10,000 ft above ground level (AGL). Note the areas in the western USA where there is no NEXRAD radar coverage (*white*). Courtesy U.S. Department of Commerce, NOAA.

Unfortunately, radar-based precipitation estimation can be problematic due to variations in the vertical profile reflectivity (VPR) and without sufficient quality control, bias correction, radar calibration, and use of appropriate Z-R relationships. Rainfall overestimation can be caused by a number of factors, including the presence of hail, large raindrops, melting ice, anomalous beam propagation, and passage of the radar beam through the atmospheric freezing layer, ground clutter and ground clutter suppression; underestimation can occur due to small raindrop size, dry ice, beam attenuation, beam over-shooting of heavy precipitation cores at lower atmospheric levels, partial beam filling by heavy precipitation cores (range effect), beam blockage, etc (Seo, 1998; Seo, Breidenbach, & Johnson, 1999).

Consequently, software processing of radar precipitation estimates must include bias correction utilizing raingage observations. As described by Hudlow et al., 1983, Anagnostou, Krajewski, Seo, & Johnson, 1998, Smith & Krajewski, 1991, Briedenbach, Seo, & Fulton, 1998, Fulton, Breidenbach, Seo, & Miller, 1998, and Fulton, 2002, the Stage III precipitation processing system was created specifically for the NWS RFCs which need rainfall estimates over a much larger area than that covered by an individual radar. The Stage III precipitation processing system mosaics together Stage II estimates from multiple radars onto a subset of the national HRAP[23] grid (see Fulton, 1998) covering the RFC area of responsibility. The resulting data format is a binary gridded file in what is known as the xmrg data format (refer to http://www.nws.noaa.gov/oh/hrl/misc/xmrg.pdf). Subsequent advances by the NWS Office of Hydrologic Development, Hydrology Laboratory led to the development of the Multisensor Precipitation Estimator (MPE) (Section 5.3.1) and the Multi-Radar, Multi-Sensor (MRMS) System by the National Severe Storms Laboratory (Section 5.3.2).

5.3.1 Multisensor Precipitation Estimator (MPE)

The precursor to MPE was the Stage III Precipitation Processing System described by Briedenbach et al. (1998). Fulton, Ding, and Miller (2003) report an algorithmic error, referred to as a truncation error, in low-level computations within the NEXRAD Radar Product Generator that produced a systematic low-bias for over a 10-year period through 2003. Corrections to

[23]Please refer to http://www.nws.noaa.gov/oh/hrl/nwsrfs/users_manual/part2/_pdf/21hrapgrid.pdf and http://www.nws.noaa.gov/oh/hrl/papers/wsr88d/hrapmap.pdf for details.

the truncation error were included NEXRAD software builds in the Open RPG (OPRG) through late 2003. The Multisensor Precipitation Estimator (MPE) was developed as a replacement to the Stage III precipitation processing system. The enhancements include (see www.nws.noaa.gov/hrl/papers/ffw_mpe_djs.pdf):

- delineation of effective radar coverage—limits the quantitative use of radar data to those areas where radar can detect precipitation consistently, based on a multi-year climatology of Digital Precipitation Array (DPA) data;
- mosaicking based on radar sampling geometry—in areas of radar coverage overlap, use the radar rainfall estimate from the lowest unobstructed sampling volume;
- RFC area-wide precipitation analysis;
- improved mean-field bias correction—based on near real-time raingage data; equivalent to adjusting the multiplicative constant, $A(t)$, in the Z–R relationship for each radar, that is, $Z = A(t)R^b$;
- local bias correction—radar bin-by-bin application of the mean field bias algorithm reduces systematic errors over small areas; equivalent to changing the multiplicative constant, $A(x,y,t)$, in the Z–R relationship for each radar bin, namely, $Z = A(x, y, t)R^b$; more effective in areas of high gage density.

Further enhancements included integration of additional data sources such as satellite precipitation estimates and using a post-analysis capability, including daily raingage observations, which are disaggregated into hourly and other multi-hour values with bias updating in the rerun mode. Fig. 17 shows an example OHRFC MPE 72-h precipitation accumulation for the period ending 06/08/2008 12:00:00 UTC, the majority of which fell within a 6-hour period. Precipitation accumulations for this example are shown in inches, with maximum amounts ranging from 9.0 to 12.0 in. (228 to 305 mm). Fig. 18 shows results from a bias analysis for annual Stage3/MPE accumulations compared to Parameter-elevation Regressions on Independent Slopes Model (PRISM) (Daly, Neilson, & Phillips, 1994; Daly, Taylor, & Gibson, 1997)[24] precipitation estimates covering the OHRFC area for the period 1997–2013. A substantial reduction of Stage3/MPE annual bias, where $bias = MPE / PRISM$, compared to PRISM estimates is evident from the analysis. Bias reductions can be traced directly to the

[24]For complete details, go to www.prism.oregonstate.edu

Fig. 17 Map showing the 72-h precipitation accumulation (inches) for the OHRFC area ending at 06/08/2008 12:00:00 UTC. *White lines* show major subbasin boundaries; state boundaries are colored *yellow*.

correction of the truncation error and move from the PRG and Stage III precipitation processing system to the OPRG and MPE.

The overall flow of hydrometeorological station and radar data is presented in Fig. 19, which gives a generalized depiction for RFC data acquisition, precipitation processing, and QPE use as input for CHPS-FEWS based hydrologic models. RFCs located in highly mountainous regions of the USA rely predominantly on gaged-based precipitation estimates due to beam blockage and inadequate spatial coverage by the NEXRAD system (see Fig. 16).

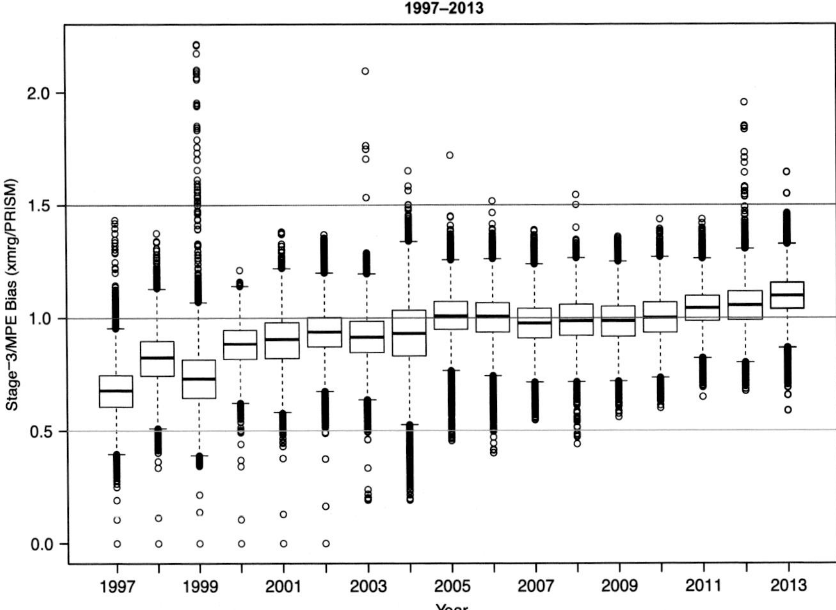

Fig. 18 Boxplot showing annual MPE/Stage-3 bias with respect to PRISM precipitation estimates at the ~4km (HRAP) grid scale for the OHRFC area, 1997–2013. *Horizontal green* and *blue lines* for biases equal to 0.5 and 1.5, respectively, are displayed for reference purposes only.

5.3.2 Multi-Radar, Multi-Sensor (MRMS) System

The Multi-Radar, Multi-Sensor (MRMS) System,[25] which is described by Zhang et al. (2015) and Cocks, Martinaitis, Kaney, Zhang, and Howard (2015), was developed by the NOAA/NWS National Severe Storms Laboratory (NSSL). Results reported by Kitzmiller et al. (2011) show consistent improved hydrologic model performance utilizing input derived from the MRMS system over MPE. Several RFCs utilize MRMS hourly gridded precipitation fields in place of initial mosaicked MPE fields as the starting point for manual quality QPE control within AWIPS MPE software. Experience at the OHRFC, for instance, has shown that this practice reduces boundary artifacts created by an MPE's local radar bias correction methodology and has helped to reduce the overall bias variance. However, referring to Figs. 18 and 20, there is an apparent upward drift of QPE bias in the OHRFC area that coincides with the use of MRMS estimates with MPE beginning in about 2010.

[25]The Multi-Radar, Multi-Sensor (MRMS) System was previously referred to as the National Mosaic and QPE algorithm package (NMQ).

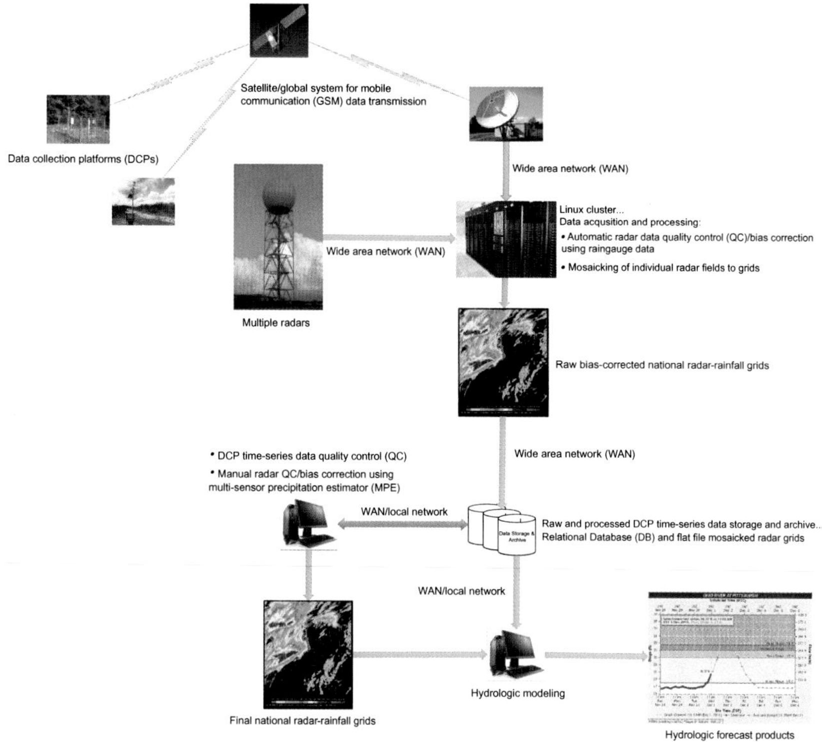

Fig. 19 Flow chart showing the flow of data from hydrometeorological data collection platforms (DCPs) and NEXRAD radar locations.

5.4 Hydrologic Ensemble Forecast System (HEFS)

The Hydrologic Ensemble Forecast System (HEFS) (Demargne et al., 2014) was developed as an extension of ESP (Section 3.4.4) methodologies and AHPS (Section 5.2) efforts to seamlessly span hourly to annual time-scales with ensemble, probabilistic hydrologic forecasts. Operational implementation of HEFS by NOAA/NWS RFCs is currently underway. The most visible use of HEFS provides hydrologic ensemble predictions of reservoir inflows to the New York City Department of Environmental Protection Public Affairs to support the Operations Support Tool to support reservoir operational decision making for water supply and water quality purposes. HEFS addresses recommendations made by National Research Council (2006a), and Schaake, Hamill, & Buizza (2007) and the Hydrological Ensemble Prediction Experiment (HEPEX; www.hepex.org). HEPEX is an international initiative established in 2004 to enable collaboration between the atmospheric and hydrologic research and forecast

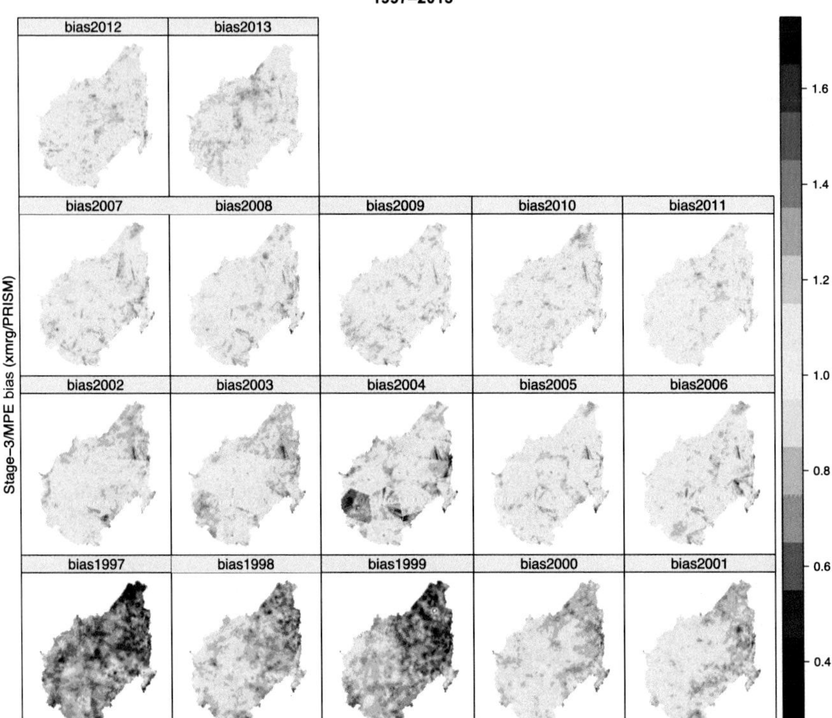

Fig. 20 Graphic showing annual MPE/Stage-3 spatial bias pattern with respect to PRISM precipitation estimates at the ~4 km (HRAP) grid scale for the OHRFC area, 1997–2013.

communities aimed at improving ensemble forecasts and to demonstrate their utility in decision-making in flood forecasting and water management. The Meteorological Model-based Ensemble Forecast System (MMEFS) (Adams & Ostrowski, 2010), similar to HEFS, was implemented operationally in 2013 by the MARFC, NERFC, OHRFC, and SERFC for short lead-time ensemble flood forecasting.

5.5 Forecast Products

Forecast products issued by RFCs take many forms, depending on the intended purpose, audience, regional needs, and historical context. Forecast products are transmitted electronically to the NWS WFOs using the Standard Hydrometeorological Exchange Format (SHEF) Code Manual[26] (NWS/HSD, 2012) through AWIPS (see Section 5.1).

[26]Please refer to http://www.nws.noaa.gov/om/water/resources/SHEF_CodeManual_5July2012.pdf.

5.6 Forecast Evaluation

Real-time visualization of forecasts is necessary to evaluate the overall quality of forecasts and to assess potential errors and biases that will assist the forecaster in making operational adjustments. Perhaps QPF was over estimated or model basin states are either too wet or dry. Adjustments to certain model states can be made by forecasters and QPF can be altered, if needed, following subsequent analysis of numerical weather prediction (NWP) model runs by RFC meteorologists. What is apparent in Fig. 21 is that initially with the first forecast (line labeled *a*), the QPF was too small. Subsequent forecasts demonstrate a methodological issue with the use of a limited duration of forecast periods, which is 8 6-h periods (48 h) of QPF. Any precipitation following after 48-hours is not reflected in the 10-day hydrologic forecast. With subsequent forecasts, beginning on Jul. 3 (line *d*), we see the initiation of a more substantial hydrograph rise. A confounding and complicating feature is the time delay apparent with the routing of upstream flows to the Evansville, IN (EVVI3) forecast point that contribute to the forecast increases in lines labeled *f* and *g*.

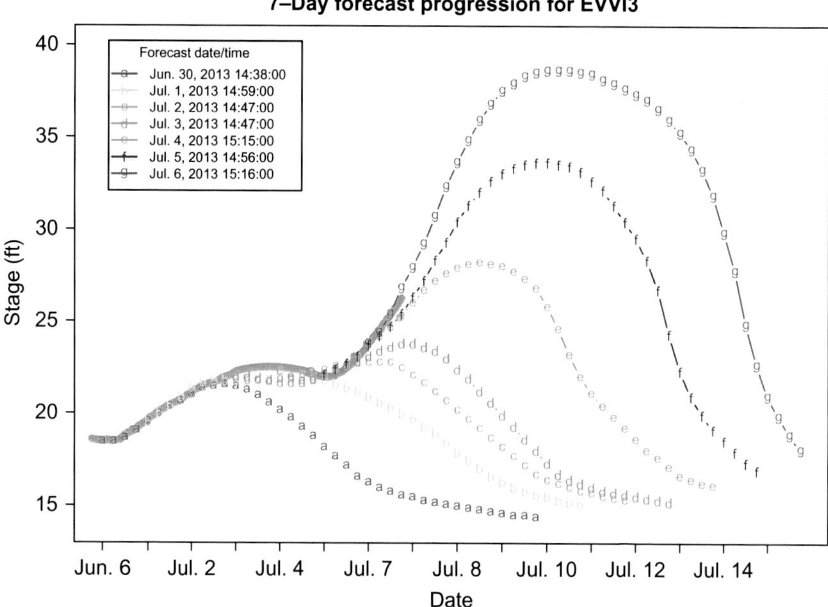

Fig. 21 Example 10-day forecast evaluation graphic, showing the Evansville, IN (EVVI3) forecast point on the Ohio River, used by some RFCs to make forecast comparisons (*colored and lettered lines*, beginning with *a*, the oldest, to *g*, the most recent, for the period Jun. 30–Jul. 16, 2013). Observed river levels (stages) are shown in *gray*, ending on Jul. 7, 2013.

5.6.1 Forecast Verification

The need for verification of single-valued, deterministic, and probabilistic hydrologic forecasts is well-documented (eg, OHD, 2000; Ebert & McBride, 2000; Ebert, 2001; Ebert, Damrath, Wergen, & McBride, 2003; Franz, Hartmann, Sorooshian, & Bales, 2003; Welles, Sorooshian, Carter, & Olsen, 2007; Demargne et al., 2009). Hydrologic forecast verification was formally initiated by the NWS for real-time, operational river stage (level) forecasts in 2003. RFCs utilize the Interactive Verification Program (IVP) to analyze a variety of statistics to measure accuracy and discrimination of forecasts. One of the first steps for computing most of the verification statistics in IVP is to define intervals for partitioning the data into categories based on stage. RFC river stage forecast verification using the IVP is run every month for all forecast point locations, the results of which are archived in a national verification database.[27] Verification of ensemble hydrologic forecasts is carried out using the Ensemble Verification System (EVS) (Brown, Demargne, Seo, & Liu, 2010; Demargne et al., 2010) found at http://amazon.nws.noaa.gov/ohd/evs/evs.html.

5.6.2 Service Assessment

As stated by the NWS, Service Assessments[28] are made to evaluate NWS performance following the occurrence of significant hydrometeorological, oceanographic, or geological events. Assessments may be initiated when one or more of the following criteria are met:

- major economic impact on a large area or population;
- multiple fatalities or numerous serious injuries;
- extensive national public interest or media coverage;
- unusual level of attention to NWS performance.

Typically, assessment teams are composed of experts from both within and outside the NWS. The stated purpose is to evaluate activities before, during, and after events to determine the usefulness of NWS products and services. Comprehensive reports produced by the assessment team serve as tools to identify and describe best practices in operations and procedures, and to identify and address NWS service deficiencies during

[27]For details, please refer to the NWS River Forecast Verification Plan at http://www.nws.noaa.gov/oh/rfcdev/docs/Final_Verification_Report.pdf.

[28]NWS Service Assessments, dating from 1957, can be found at http://www.nws.noaa.gov/om/assessments/index.shtml.

the event. Ultimately, the goal of the activity is to aid the NWS in improving its services to the nation.

6 ADDITIONAL TOPICS

NWS policies and directives have RFCs modeling hydrologically and hydraulically over their forecast regions, shown in Fig. 1, and providing forecast guidance to NWS Weather Forecast Offices (WFOs), which have the responsibility for flash forecasting. This is primarily due to 3 reasons, as WFOs:

1. maintain 24-h operations every day of the year (365 days);
2. have NWS official watch and warning forecast responsibilities;
3. are mandated to establish close relationships with and outreach to local communities.

The WFO hydrologic services program serves a critical role with both the flash flood watch and warning program, and by disseminating official flood and river stage forecast products that are derived from RFC flood forecast guidance, to the public and other local and state agencies. Additionally, WFOs, through community outreach, educate the public on safety measures to be taken during times of flooding. WFOs also work with localities and their emergency management offices to develop and implement Flood Warning Systems (FWS), which are observationally based, depending on critical observed rainfall rates/ accumulations or observed stream levels, rather than forecast-based.

6.1 Flash Flood Watches and Warnings

With RFC support, WFOs utilize a number of tools to meet the needs of the Flash Flood Watch and Warning Program. These include utilization of:

- flash flood guidance (FFG), discussed in Section 3.4.5;
- site-specific hydrologic modeling—AWIPS (see Section 5.1) based hydrologic modeling tool for small, locally important watersheds that are prone to flash flooding;
- headwater guidance—AWIPS, basin-scale, rainfall index based product, similar to FFG, to assess flooding potential;
- Flash Flood Monitoring and Prediction (FFMP)—AWIPS based tool that utilizes FFG values and observed and projected NEXRAD radar precipitation estimates to assess flash flooding potential, used in the process to issue flash flood watches and warnings (refer to http://www.nws.noaa.gov/mdl/ffmp).

6.2 AFWS

The NWS and the emergency management community recognize the importance of flood warning systems (FWS) for protecting lives and property. Communities across the country have created Automated FWS (AFWS) to warn areas of flood danger, and to provide NWS with critical rainfall, stream level, and other hydrometeorological data. Nearly all FWS are built and operated using gages and a communications system to collect and distribute information. AFWS, however, require a high level of ongoing commitment and support beyond one-time installation costs. Successful AFWS have committed local community participants, strong long-term operational funding, and a good coordination with the local NWS WFOs.[29]

The basic benefit of a local flood warning program is an increased lead time for watches and warnings at locations subject to flood risk. The information can be used to predict whether a flood is about to occur, when it will arrive, and how severe it will be. Organizations and individuals are given notice by the system so they can protect themselves and their property. The basic parts of a flood warning program are:

- the FWS, including equipment, people, and procedures for recognizing an impending flood and disseminating warnings;
- a prepared plan of action to be taken before and during the flood;
- arrangements for updating and maintenance of equipment and plans.

Local flood warning programs can be extremely effective. Those now in use have been credited with saving scores of lives and preventing millions of dollars of damage. They are most valuable where flooding occurs very quickly following heavy rains. Local flood warning programs also have been credited with preventing unnecessary evacuations and other overreactions in cases when floods threatened but did not occur. Thousands of communities that are threatened by floods lack the elementary protection of a flood warning program despite the many success stories, the relatively low cost for their development, and the simplicity of their operation. NWS can play an important role in promoting local flood warning systems by providing technical support, maintaining regular communications and feedback, and working with community officials on outreach and education programs about the risk from flooding (Fig. 22).

[29]Please refer to the NWS FWS manual at http://www.nws.noaa.gov/os/water/resources/Flood_Warning-Systems_Manual.pdf.

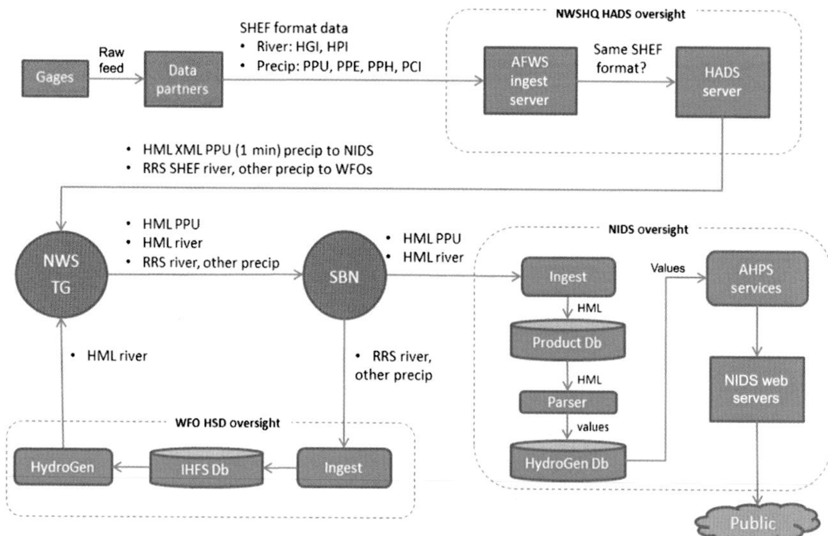

Fig. 22 Illustration showing the Automated Flood Warning System (AFWS). Courtesy NOAA/NWS.

7 FUTURE DEVELOPMENTS

7.1 Integrated Water Resources Science and Services (IWRSS)

The NOAA/NWS Integrated Water Resources Science and Services (IWRSS) initiative outlines a plan, in broad terms, to bring together US federal agencies to work cooperatively to address US national water resources problems. As stated in the *Integrated Water Resources Science and Services (IWRSS): An Integrated and Adaptive Roadmap for Operational Implementation*:[30]

> *[The]... objective of the IWRSS project is to demonstrate a broad integrative national water resources information system to serve as a reliable and authoritative basis for adaptive water-related planning, preparedness and response activities from national to local levels. The project seeks to make intersections between relevant systems more seamless, synthesize information better across systems to improve services and service delivery and improve the overall quality of information, and provide new information and services to better support the needs of water resources stakeholders.*

[30]The final plan can be found at http://www.nws.noaa.gov/oh/nwc/IWRSS_ROADMAP_FINAL.pdf.

The three main objectives stated in the IWRSS Roadmap are as follows:

- Integrate services and service delivery—improving interoperability between systems, exchanging data and information seamlessly between systems and consortium members, improving geospatial information accessibility, visualization, and interpretation within a transparent water resources information system.
- Increase accuracy and lead-time of river forecasts—achieved with improved data access and modeling capability, and strengthened collaboration between federal water agencies for a broad range of water resources activities, including river forecasting and management, flood forecasting, levee and dam failures, river ice, climate and drought mitigation, water supply, coastal environments, geo-intelligence, and research and development.
- Provide new "Summit-to-Sea" high-resolution water resources information and forecasts—this goal is concerned with the development and implementation of high-resolution models, interoperable tools and collaborative workflow from headwaters streams to coastal and estuarine areas, using consistent and seamless high-resolution GIS-ready geospatial data and information describing past, current, and future soil moisture, snowpack, evapotranspiration, groundwater, runoff, and flood inundation conditions, and the uncertainty associated with this information.

7.2 National Water Center

The NOAA/NWS National Water Center (NWC) opened on May 26, 2015 with an official ribbon cutting. The facility is located on the campus of the University of Alabama in Tuscaloosa. The main goals of the NWC are to implement the vision stated in the IWRSS roadmap (see Section 7.1), including the implementation of an operational national hydrologic modeling system that, when completed, meets the IWRSS goals. The NWS consists of four divisions, namely,

- *Interdisciplinary Science and Engineering Division*—this provides core science capacity, algorithm, and software component development and operational decision support for local, regional, national, and global scales. ISED has both developmental and operational functions. It develops: water-related core scientific knowledge; software applications; model components; new products and product improvements; skill evaluation techniques; and modeling and analytical methodologies and algorithms. It operates models, analytical tools, databases, and information systems. It produces: scientific publications; software applications; evaluations of algorithms, techniques, tools, products, and services; product documentation; incremental improvements in scientific or technical maturity of algorithms, models, and tools; evaluations of skill and performance of models, products, and services; new methodologies; and model parameterizations and calibrations. It maintains scientific expertise, product documentation, and developmental and operational software. ISED collaborates with partners and supports NWC and field operations, external partners, customers and stakeholders, and corporate knowledge management.

- *Analysis and Prediction Division*—this integrates science, software, and data components into operational water resources prediction systems for local, regional, national, and global implementation. APD has both developmental and operational functions. It develops water-related integrated information systems and infrastructure, integrated models, modeling and data assimilation tools and systems, and integrated model calibrations and parameterizations. It operates numerical models, analytical tools, databases, and information systems at national and global scales. It produces systems architecture and decision support products. It maintains systems, integrated models, databases, records, and documents. APD collaborates with partners and supports NWC and field operations, web data and product dissemination, external partners, customers and stakeholders, and corporate knowledge management.
- *Geospatial Intelligence Division*—this provides centralized and consistent data services, geospatial analyses, and cartographic expertise to support science and engineering development, systems implementation, and water resources operations at local, regional, national, and global scales. GID has both developmental and operational functions. It develops: water-related geospatial data; actionable intelligence derived from data; geospatial software applications; maps and graphics; new products and product improvements; spatial discretization techniques; and analytical methods. It operates: airborne survey systems for snow and soil moisture; geographic information systems; mapping and graphics software, systems, and tools; and databases, models, and geo-statistical analysis software. It produces and maintains enterprise geospatial datasets, maps, atlases, graphics, documentation, and geo-intelligence. GID collaborates with partners and supports NWC and field operations, external partners, customers and stakeholders, and corporate knowledge management (capturing, developing, sharing, and effectively using organizational knowledge).
- *Social Intelligence Division*—this provides geographic and socio-economic sector-specific water resources information, risk, impact, and economic assessments and decision support services. It marshals local, regional, and

national assets to ensure effective service delivery at all scales. SID has both developmental and operational functions. It develops: relationships with partners and stakeholders; decision support services; analyses of impacts and risks; requirements for improved information and services; and training and education programs. It operates socio-economic hazards and impacts databases, models, and information systems. It produces economic analyses, impact analyses, risk assessments, legal and policy assessments, outreach materials, and scientific publications. It maintains socio-economic databases and requirements databases. SID supports community resiliency, NWC and field operations, external partners, customers and stakeholders, training, and corporate knowledge management.

Ultimately, NWC modeling would produce detailed hydrologic forecasts at arbitrary distributed model grid point locations such as the example shown in Fig. 23, with results from the RDHM (refer to Section 4.2.1).

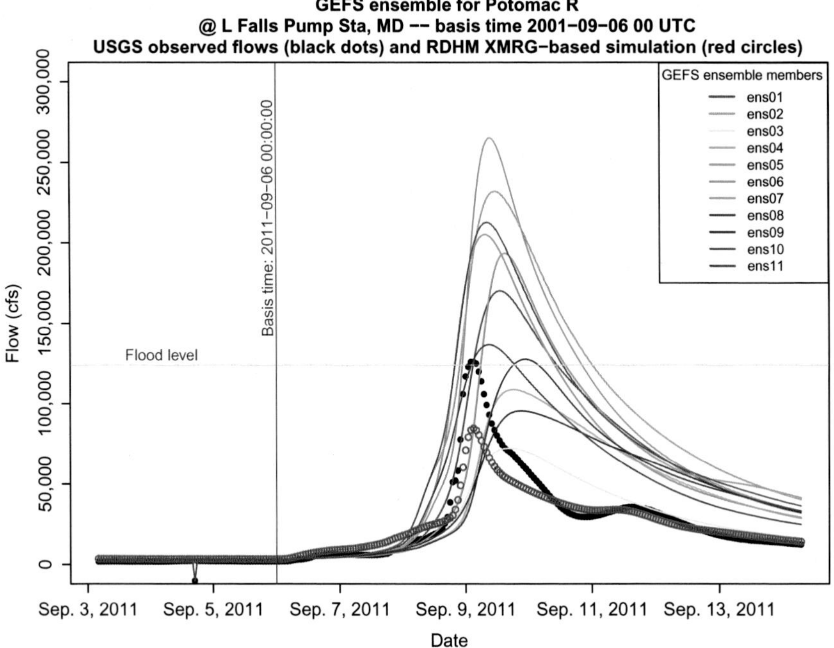

Fig. 23 RDHM based simulation utilizing NEXRAD derived, xmrg-based precipitation forcings (*red circles*) and ensemble hydrographs derived from NOAA/NWS/NCEP Global Ensemble Forecast System (GEFS) (*colored curves* identified as ens01, ens02…) numerical weather prediction (NWP) QPF and forecasted temperature fields, compared against USGS measured streamflow (*black dots*) for the Potomac River at Little Falls Pump Station, MD.

Hydrologic ensemble results utilizing output from the NOAA/NWS National Centers for Environmental Prediction (NCEP) Global Ensemble Forecast System (GEFS) are also shown to illustrate ensemble modeling goals of the NWC National Hydrologic Model. Hydrologic ensemble forecasts would be used to quantify flood forecast uncertainty, provide lateral inflows and boundary conditions for spatially detailed hydraulic models to identify flood inundation risks, reservoir inflow uncertainties, etc.

REFERENCES

Adams, T. E. (1991). Graphical user interface concepts for hydrologic forecasting in the modernized weather service. In *Seventh international conference on interactive information and processing systems for meteorology, oceanography, and hydrology, New Orleans, LA*: American Meteorological Society Preprints.

Adams, T., & Ostrowski, J. (2010). Short lead-time hydrologic ensemble forecasts from numerical weather prediction model ensembles. In *Proceedings of the world environmental and water resources congress 2010, Providence, RI*: EWRI.

Adams, T. E., & Smith, G. M. (1993). National Weather Service interactive river forecasting using state, parameter, and data modifications. In *Proceedings of the international symposium on engineering hydrology, San Francisco, CA*: American Society of Civil Engineers.

Anagnostou, E. N., Krajewski, W. F., Seo, D.-J., & Johnson, E. R. (1998). Mean-field radar rainfall bias studies for WSR-88D. *Journal of Engineering Hydrology (ASCE), 3*(3), 149–159.

Anderson, E. A. (2002). Calibration of conceptual hydrologic models for use in river forecasting. Technical report. U.S. National Weather Service, Office of Hydrology, Hydrology Laboratory.

Betson, R. P., Tucker, R. L., & Haller, F. M. (1969). Using analytical methods to develop a surface-runoff model. *Water Resources Research, 1*(2), 103–111.

Briedenbach, J., Seo, D. J., & Fulton, R. (1998). Stage II and III post processing of NEXRAD precipitation estimates in the modernized weather service. In *Proceedings of AMS 78th annual meeting, Phoenix, AZ, January*.

Brown, J. D., Demargne, J., Seo, D.-J., & Liu, Y. (2010). The Ensemble Verification System (EVS): A software tool for verifying ensemble forecasts of hydrometeorological and hydrologic variables at discrete locations. *Environmental Modelling and Software, 25*(7), 854–872.

Burnash, R. J. (1995). *The NWS River Forecast System — Catchment model* (1st ed.). Highlands Ranch, CO: Water Resources Publications.

Burnash, R. J., Ferral, R. L., & McGuire, R. A. (1973). A generalized streamflow simulation system: Conceptual modeling for digital computers. Technical report. U.S. Department of Commerce National Weather Service and State of California Department of Water Resources.

Clark, R. A., Gourley, J. J., Flamig, Z. L., Yang, H., & Clark, E. (2014). Conus-wide evaluation of national weather service flash flood guidance products. *Weather and Forecasting, 29*(2), 377–392.

Cocks, S. B., Martinaitis, S. M., Kaney, B., Zhang, J., & Howard, K. (2015). MRMS QPE *performance during the 2013–14 cool season. Journal of Hydrometeorology, 17*, 791–810.

Cosgrove, B. A., Clark, E., Reed, S., Koren, V., Zhang, Z., Cui, Z., & Smith, M. (2012). Overview and initial evaluation of the Distributed Hydrologic Model Threshold Frequency (DHM-TF) flash flood forecasting system. Technical report NOAA NWS Technical report NWS 54. Department of Commerce, NOAA/NWS, March.

Crawford, N. H., & Burges, S. J. (2004). History of the Stanford Watershed Model. *American Water Resources Association, Water Resources IMPACT, 6*(2), 3–5.

Crawford, N. H., & Linsley, R. K. (1966). Digital simulation in hydrology: Stanford watershed model IV. Technical report no. 39. Department of Civil Engineering, Stanford University, July.

Daly, C., Neilson, R. P., & Phillips, D. L. (1994). A statistical-topographic model for mapping climatological precipitation over mountainous terrain. *Journal of Applied Meteorology, 33*, 140–158.

Daly, C., Taylor, G., & Gibson, W. (1997). The PRISM approach to mapping precipitation and temperature. In *10th AMS conference on applied climatology, Reno, NV*: AMS.

Damrath, U., Doms, G., Fruehwald, D., Heise, E., Richter, B., & Steppeler, J. (2000). Operational quantitative precipitation forecasting at the German Weather Service. *Journal of Hydrology, 239*, 260–285.

Dawdy, D. R., & Bergmann, J. M. (1969). Effect of rainfall variability on streamflow simulation. *Water Resources Research, 5*(5), 140–158.

Day, G. N. (1985). Extended streamflow forecasting using NWSRFS. *Journal of Water Resources Planning and Management (ASCE), 3*, 157–170.

Demargne, J., Brown, J. D., Seo, D.-J., Wu, L., Toth, Z., & Zhu, Y. (2010). Diagnostic verification of hydrometeorological and hydrologic ensembles. *Atmospheric Science Letters, 11*(2), 114–122.

Demargne, J., Mulluski, M., Werner, K., Adams, T., Lindsey, S., Schwein, N., et al. (2009). Application of forecast verification science to operational river forecasting in the U.S. National Weather Service. *Bulletin of the American Meteorological Society, 90*(6), 779–784.

Demargne, J., Wu, L., Regonda, S. K., Brown, J. D., Lee, H., He, M., et al. (2014). The science of NOAA's operational hydrologic ensemble forecast service. *Bulletin of the American Meteorological Society, 95*, 79–98.

Duan, Q. A., Gupta, V. K., & Sorooshian, S. (1993). Shuffled complex evolution approach for effective and efficient global minimization. *Journal of Optimization Theory and Applications, 76*(3), 501–521.

Duan, Q., Sorooshian, S., & Gupta, V. K. (1992). Effective and efficient global optimization for conceptual rainfall-runoff models. *Water Resources Research, 28*(4), 1015–1031.

Duan, Q. A., Sorooshian, S., & Gupta, V. K. (1994). Optimal use of the SCE-UA global optimization method for calibrating watershed models. *Journal of Hydrology, 158*(5), 265–284.

Ebert, E. E. (2001). Ability of a poor man's ensemble to predict the probability and distribution of precipitation. *Monthly Weather Review, 129*, 2461–2480.

Ebert, E. E., Damrath, U., Wergen, W., & McBride, J. L. (2003). The WGNE assessment of short-term quantitative precipitation forecasts. *Bulletin of the American Meteorological Society, 84*(4), 481–492.

Ebert, E. E., & McBride, J. L. (2000). Verification of precipitation in weather systems: Determination of systematic errors. *Journal of Hydrology, 239*, 179–202.

Fogel, M. M. (1969). Effect of storm rainfall variability on runoff from small semiaird watersheds. *Transactions of the ASAE, 12*(6), 808–812.

Franz, K. J., Hartmann, H. C., Sorooshian, S., & Bales, R. (2003). Verification of National Weather Service ensemble streamflow predictions for water supply forecasting in the Colorado River Basin. *Journal of Hydrometeorology, 4*(12), 1105–1118.

Fulton, R. A. (1998). WSR-88D polar-to-HRAP mapping. Technical report. Hydrologic Research Laboratory, Office of Hydrology, National Weather Service, Silver Spring, MD, August.

Fulton, R. A. (2002). Activities to improve WSR-88D radar rainfall estimation in the National Weather Service. In *Proceedings of the second federal interagency hydrologic modeling conference, Las Vegas, NV, July*.

Fulton, R., Breidenbach, J., Seo, D.-J., & Miller, D. (1998). The WSR-88D rainfall algorithm. *Weather and Forecasting, 13*, 377–395.

Fulton, R. A., Ding, F., & Miller, D. A. (2003). Truncation errors in historical WSR-88D rainfall products. In *31st Conference on radar meteorology, August*. Seattle, WA: American Meteorological Society.

Hudlow, M. D., Greene, D. R., Ahnert, P. R., Krajewski, W. F., Sivaramakrishnan, T. R., Johnson, E. R., et al. (1983). Proposed off-site precipitation processing system for NEXRAD. In *Preprints, 21st conference on RADAR meteorology, Edmonton, AB, Canada, September*. (pp. 394–403).

Huff, F. A. (1970). Spatial distribution of rainfall rates. *Water Resources Research, 6*(1), 254–260.

Johnson, D., Smith, M., Koren, V., & Finnerty, B. (1999). Comparing mean areal precipitation estimates from NEXRAD and rain gauge networks. *Journal of Hydrologic Engineering, 4*(2), 117–124.

Kitzmiller, D., Van Cooten, S., Ding, F., Howard, K., Langston, C., Zhang, J., et al. (2011). Evolving multisensor precipitation estimation methods: Their impacts on flow prediction using a distributed hydrologic model. *Journal of Hydrometeorology, 12*, 1414–1431.

Kohler, M. A. (1944). The use of crest stage relations in forecasting the rise and fall of the flood hydrograph. Technical report (mimeo). U.S. Weather Bureau, September.

Kohler, M. A., Linsley, R. K., & Paulhus, J. L. (1958). *Hydrology for engineers*. New York: McGraw-Hill.

Kohler, M. A., & Richards, M. M. (1962). Multicapacity basin accounting for predicting runoff from storm precipitation. *Journal of Geophysical Research, 1*(12), 5187–5197.

Koren, V. I. (2006). Parameterization of frozen ground effects: sensitivity to soil properties. Predictions in ungauged basins: Promises and progress. In *Symposium S7, seventh IAHS scientific assembly, Foz do Iquacu, Brazil* (pp. 125–133): IAHS Publication 303.

Koren, V. I. (2011). Physically based modifications to the sac-sma, evapotranspiration component. Technical report NOAA NWS Technical report NWS 47. Department of Commerce, NOAA/NWS, July.

Koren, V., Reed, S., Smith, M., Zhang, Z., & Seo, D.-J. (2004). Hydrology laboratory research modeling system (HL-RMS) of the US National Weather Service. *Journal of Hydrology, 291*, 297–318.

Koren, V., Smith, M., Cui, Z., & Cosgrove, B. (2007). Physically-based modifications to the Sacramento soil moisture accounting model: Modeling the effects of frozen ground on the rainfall-runoff process. Technical report NOAA NWS Technical report NWS 52. Department of Commerce, NOAA/NWS, July.

Koren, V., Smith, M., Cui, Z., Cosgrove, B., Werner, K., & Zamora, R. (2010). Modification of Sacramento soil moisture accounting heat transfer component (SAC-HT) for enhanced evapotranspiration. Technical report NOAA NWS Technical report NWS 53. Department of Commerce, NOAA/NWS, October.

Koren, V., Smith, M., & Duan, Q. (2003). Use of a priori parameter estimates in the derivation of spatially consistent parameter sets of rainfall-runoff models. *Calibration of watershed models*. Washington, DC: American Geophysical Union.

Koren, V. I., Smith, M., Wang, D., & Zhang, Z. (2000). Use of soil property data in the derivation of conceptual rainfall-runoff model parameters. In *Conference on hydrology, Long Beach, CA*: AMS.

McEnery, J., Ingram, J., Duan, Q., Adams, T., & Anderson, L. (2005). NOAAÕs advanced hydrologic prediction service: Building pathways for better science in water forecasting. *Bulletin of the American Meteorological Society, 24*(3), 375–385.

National Research Council. (2006a). Completing the forecast: Characterizing and communicating uncertainty for better decisions using weather and climate forecasts. Technical report. Committee on Estimating and Communicating Uncertainty in Weather and Climate Forecasts, Washington, DC.

National Research Council. (2006b). Toward a new advanced hydrologic prediction service (AHPS). Technical report. Committee to Assess the National Weather Service Advanced Hydrologic Prediction Service Initiative, Water Science and Technology Board, Washington, DC.

NWS/HSD. (2012). Digital standard hydrometeorological exchange format (SHEF) code manual. Technical report. NOAA/National Weather Service, Office of Climate, Water, and Weather Services, Hydrologic Services Division, July.

OHD. (2000). National Weather Service verification software users' manual. Technical report. NOAA/NWS, Office of Hydrologic Development, Silver Spring, MD.

OHD. (2013). Hydrology laboratory-research distributed hydrologic model (HL-RDHM) user manual. Technical report. NOAA/NWS, Office of Hydrologic Development, Silver Spring, MD, March.

Press, W. H., Flannery, B. P., Teukolsky, S. A., & Vetterling, W. T. (1986). *Numerical recipes*. Cambridge: Cambridge University Press.

Schaake, J. C. (1976). Use of mathematical models for hydrologic forecasting in the National Weather Service. In *Environmental Protection Agency conference on modeling and simulation, Cincinnati, OH, April*: EPA.

Schaake, J., Hamill, T. M., & Buizza, R. (2007). Hepex: The hydrological ensemble prediction experiment. *Bulletin of the American Meteorological Society, 88*, 1541–1547.

Seo, D.-J. (1998). Real-time estimation of rainfall fields using radar rainfall and rain gauge data. *Journal of Hydrology, 208*, 37–52.

Seo, D.-J., Breidenbach, J., & Johnson, E. (1999). Real-time estimation of mean field bias in radar rainfall data. *Journal of Hydrology, 223*, 131–147.

Smith, M. B., Koren, V., Reed, S., Zhang, Z., Yu, Z., Moreda, F., et al. (2012). The distributed model intercomparison project—Phase 2—Motivation and design of the Oklahoma experiments. *Journal of Hydrology, 418*, 3–16.

Smith, M., Koren, V., Zhang, Z., Reed, S., Seo, D., Moreda, F., et al. (2004). NOAA NWS distributed hydrologic modeling research and development. Technical report NOAA NWS Technical report NWS 51. Department of Commerce, NOAA/NWS, April.

Smith, J. A., & Krajewski, W. F. (1991). Estimation of the mean field bias of radar rainfall estimates. *Journal of Applied Meteorology, 30*, 397–412.

Smith, J. A., Seo, D. J., Baeck, M. L., & Hudlow, M. D. (1996). An intercomparison study of NEXRAD precipitation estimates. *Water Resources Research, 32*(7), 2035–2045.

Smith, M. B., Seo, D.-J., Koren, V. I., Reed, S., Zhang, Z., Duan, Q.-Y., et al. (2004). The distributed model intercomparison project, DMIP—Motivation and experiment design. *Journal of Hydrology, 298*(1), 4–26.

Sweeney, T. (1992). Modernized areal flash flood guidance, noaa tech. rep. nws hydro 44. Technical report. NOAA/NWS/Hydrologic Research Laboratory, Silver Spring, MD.

Sweeney, T., & Baumgardner, T. (1999). Modernized flash flood guidance. rep. to nws hydrology laboratory. Technical report. NOAA/NWS/Hydrologic Research Laboratory, Silver Spring, MD.

Thiessen, A. H. (1911). Precipitation averages for large areas. *American Meteorological Society, Monthly Weather Review, 39*, 1082–1089.

U.S. Department of Commerce. (1972). National Weather Service River Forecast System (NWSRFS-Model). Technical report. NOAA Technical Memorandum NWS-Hydro-14, Washington, DC.

Welles, E., Sorooshian, S., Carter, G., & Olsen, B. (2007). Hydrologic verification: A call for action and collaboration. *Bulletin of the American Meteorological Society, 88*, 503–511.

Zhang, J., Howard, K., Langston, C., Kaney, B., Qi, Y., Lin, T., et al. (2015). Multi-Radar Multi-Sensor (MRMS) quantitative precipitation estimation: *Initial operating capabilities*. Bulletin of the American Meteorological Society. http://journals.ametsoc.org/doi/abs/10.1175/BAMS_D-14-00174.1.

PART 2

Continental Modeling and Monitoring — The Future?

CHAPTER 11

On the Operational Implementation of the European Flood Awareness System (EFAS)

P.J. Smith*, F. Pappenberger*, F. Wetterhall*, J. Thielen del Pozo†, B. Krzeminski†, P. Salamon†, D. Muraro†, M. Kalas†, C. Baugh*

*European Centre for Medium-Range Weather Forecasts, Reading, United Kingdom
†European Commission Joint Research Centre, Ispra, Italy

1 INTRODUCTION

In Europe, more than 40 rivers cross at least one border, with the most transnational river in Europe being the Danube, which is shared by 18 countries. In case of flooding, this means that different authorities involved in water resource management, civil protection and the organization of aid must communicate, share data and information and, ideally, take concerted actions to reduce the impact of the flooding along the course of the river. In these situations, flood risk management becomes challenging as inconsistent information, which may arise from incomplete communication between authorities, differing results from different forecasting models and subsequent assessment of the ongoing and forecasted flood event, or simply misunderstandings due to language barriers, can introduce uncertainties and errors in the assessment of the ongoing and upcoming situation. These errors could lead to incoherent and uncoordinated decision-making and actions across the chain of responsibilities, becoming counterproductive to reducing the impacts of the flood event (Demeritt et al., 2007).

In order to avoid discrepancies in information content, clear communication channels and agreed protocols for exchange of data and information are necessary, and many countries have agreed bilateral protocols accordingly. However, except for few examples such as the river Rhine, which needs to be managed across six different country borders and for which a single model is set up and information made available to all authorities concerned (Renner, Werner, Rademacher, & Sprokkereef, 2009), different

Flood Forecasting
http://dx.doi.org/10.1016/B978-0-12-801884-2.00011-6

models and forecasting systems exist for the different countries or even administrative units. This lack of reference information, which is consistent for all parties involved, can make an evaluation and assessment of the information complicated and difficult, in particular for those not covered by bilateral agreements with upstream countries or those responsible for the management of European aid.

Significant flooding across Europe at the start of this century highlighted the need for improvements in flood risk and crisis management. Post-event analysis lead the European Commission (EC) to initiate, among other important initiatives, the development of the European Flood Awareness System[1] (EFAS, Bartholmes, Thielen, Ramos, & Gentilini, 2009; Burek, del Pozo, Thiemig, Salamon, & de Roo, 2011; European Commission, 2002; Thielen, Bartholmes, Ramos, & de Roo, 2009), based on initial research activities (de Roo et al., 2003; Gouweleeuw, Thielen, Franchello, Roo, & Buizza, 2005; Pappenberger et al., 2005). The objectives of EFAS are to provide pan-European medium-range streamflow forecasts and early warning information in particular for large transnational river basins, in direct support to the national forecasting services, as well as harmonized information on possible high-impact flooding to the Emergency Response Coordination Centre[2] (ERCC) of the EC. In the case of major flood events, EFAS contributes to the better protection of the European Citizen, the environment, property, and cultural heritage.

From 2003 to 2012, EFAS was developed and tested at the Joint Research Centre (JRC), the EC's in-house science service, in close collaboration with various national hydrological and meteorological services across Europe, European Civil Protection through the ERCC and other research institutes (Buizza, Pappenberger, Salamon, Thielen, & de Roo, 2009; Cloke et al., 2009; Kalas, Ramos, Thielen, & Babiakova, 2008; Pappenberger, Bartholmes, et al., 2008; Pappenberger, Thielen, & Medico, 2011; Pappenberger, Stephens, et al., 2012; Ramos, Bartholmes, & del Pozo, 2007; Raynaud et al., 2014; Younis, Ramos, & Thielen, 2008). The EC's Communication "Towards Stronger European Union Disaster Response" adopted and endorsed by the Council in 2010 (European Commission, 2010) underpins the importance of strengthening concerted actions for natural disasters including floods, which are among the costliest natural disasters in the EU. Partially in response to this, EFAS became part of the Copernicus Emergency

[1]Previously the European Flood Alert System.

[2]Previously the Monitoring Information Centre (MIC).

Management Service (EMS) in 2011, and in 2012 it was transferred from research to operational service.

The importance of pan-European early warning systems in complementing national information systems was further highlighted in 2013 with the decision on a Union Civil Protection Mechanism, where it is stated that the EC "shall contribute to the development and better integration of transnational detection and early warning and alert systems of European interest in order to enable a rapid response, and to promote the interlinkage between national early warning and alert systems and their linkage to the ERCC and the CECIS [Common Emergency Communication and Information System]" (European Union, 2013). The Copernicus EMS, including early warning systems for better emergency management, was finally endorsed in Regulation (European Union, 2014). As a result, over the past 10 years EFAS has become increasingly integrated into national and European flood risk management.

Currently, more than 48 hydrological and civil protection services in Europe are part of the EFAS network. At this time, EFAS provides pan-European (Fig. 1) overview maps of riverine flooding hazards up to 10 days in advance, as well as post-processed forecasts at river gaging stations where the national services provide real-time data. In order to ensure that EFAS does not interfere in the one voice warning mandate postulated by the World Meteorological Organization, EFAS forecast products are not publicly available in real-time. Instead, national and EU authorities mandated to inform or act on ongoing or upcoming flood situations can get access to EFAS after having signed the condition of access, which regulates the dissemination of EFAS information for the EFAS operational centers and the partner organizations.

As a continental-scale trans-boundary forecasting system, EFAS offers forecast products that are complementary to national or region systems (see Alfieri, Salamon, Pappenberger, Wetterhall, & Thielen, 2012 for an overview) but does not attempt to resolve local-scale events for catchments below 2000 km^2, urban flooding or flash flood and debris flows like platforms such as FKIS-Hydro (Romang et al., 2010). In contrast to global flood forecasting initiatives such as GloFAS (Alfieri et al., 2013; Pappenberger, Dutra, Wetterhall, & Cloke, 2012), the significantly higher spatial resolution of EFAS allows for a more refined resolution of the hydrological processes. The wide range of products aiming to satisfy flood forecasters as well as civil protection and aid managers, along with the dissemination activities, means that EFAS is more than a software tool for aligning data and models in real-time and producing forecast products such as Delft-FEWS (Werner et al., 2013).

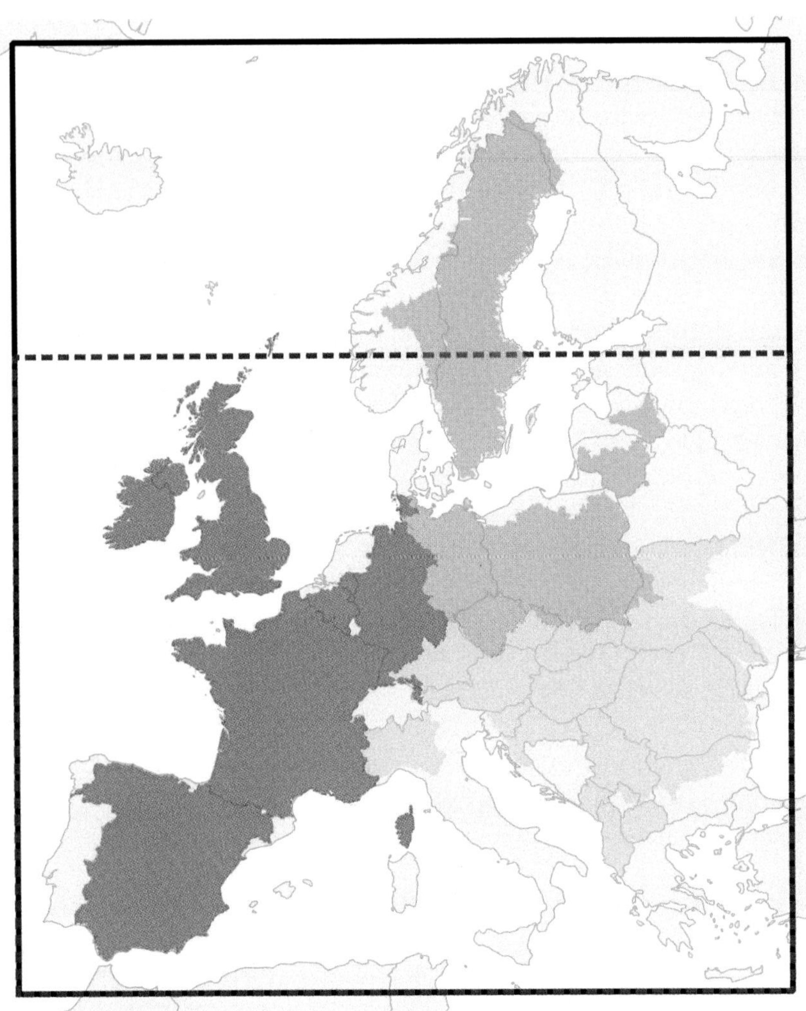

Fig. 1 Map showing EFAS (*black line*) and COSMO-LEPS (*red line*) domains. Shaded areas represent the river basins covered by partner authorities with the color indicating the corresponding dissemination center: Swedish Meteorological and Hydrological Institute (*orange*), the Slovak Hydrometeorological Institute (*blue*), or Rijkswaterstaat Waterdienst (*red*).

In this work, the status of the operational EFAS system as of Mar. 2015 is outlined. The background history and development of EFAS is not covered in detail. Readers are referred to Thielen, Bartholmes, et al. (2009) and the citations in this work. The description starts with an organizational overview (Section 3) before proceeding to outline the data acquisition (Section 4); the model components of the forecasting chain (Section 5), and the infrastructure utilized for generating forecasts (Section 6). Following this, the forecast products are described (Section 7) along with their dissemination (Section 8). The monitoring and operational performance of EFAS infrastructure are discussed in Section 9. Section 10 reviews the quality of the forecast and presents two case studies to illustrate the value of EFAS before conclusions are drawn.

2 EFAS STRUCTURE

EFAS follows many operational hydrometeorological systems in generating forecast products based on the output of a hydrological model forced by numerical weather predictions (NWPs). For each forecast the initial conditions of the hydrological model are derived using observed meteorological data. The forecast products are placed on a web platform available to the EFAS partners. These products are then analyzed and, if necessary, the awareness of responsible authorities to the potential for upcoming flood events is raised.

After the development phase, the operational EFAS has been outsourced to four centers while the overall management is continued by the EC. Following an open tendering process for contracts from 2012 to 2016, the following aspects of the system operations were issued:

1. *Hydrological data collection center*: a consortium of the Andalusian Environmental Information Network (REDIAM) and the Spanish ELIMCO Sistemas collects historic and real-time river discharge and water level data.
2. *Meteorological data collection center*: this runs onsite at the JRC. It collects historic and real-time observed meteorological data.
3. *Computational center*: the European Center for Medium-Range Weather Forecasts (ECMWF) collates NWPs, generates the forecast products and operates the EFAS Information System web platform.
4. *Dissemination center*: a consortium between the Swedish Meteorological and Hydrological Institute (SMHI), the Slovak Hydrometeorological Institute (SHMU), and the Rijkswaterstaat Waterdienst (RWS, the Netherlands) analyzes the results on a daily basis, assesses the situation, and disseminate information to the EFAS partners and to the EC.

The tendering for the next phase of EFAS operations was launched in Dec. 2014 for a duration of a further 6 years.

The division of work between four centers was designed to harvest the diverse skills within the European meteorological and hydrological communities by allowing institutions to focus on their areas of expertise.

Under the current contracts, the dissemination of EFAS results is performed by a consortium of national hydrometeorological services, ensuring that the distribution of EFAS information is executed by authorities which are experts in the field of flood forecasting as well as mandated to communicate with civil protection. This ensures the necessary competence to understand the complexity of legal issues associated with flood forecasting and civil protection within the countries. This is also necessary to build the trust between the different partners that the EU system at no point interferes with the national single voice warning principle.

The communication between the centers is ensured through a variety of standard means, including a dedicated communication platform within which video conferencing, electronic chat, document sharing, and issue tracking are implemented. Partner organizations can raise issues by contacting the centers or putting the issues in question on the agenda of the annual meeting.

3 DATA ACQUISITION

EFAS requires hydrological and meteorological data from in situ observations to calculate the initial hydrometeorological conditions and forecasting data to drive the flood forecasting system. Various meteorological and hydrological national services or river basin authorities provide real-time and historic data to EFAS. A complete list of data providers is provided at https://www.efas.eu/about-efas.html.

For EFAS, the meteorological and hydrological data collection centers are in charge of managing the existing network of providers of observed data. The centers can also contact potential providers and negotiate standard data license agreements between the provider and the COPERNICUS services. Data are collected on a 24/7 basis.

Hydrological data collection provides real-time and historic in situ hydrological observed data. Real-time data are used in the generation of post-processed forecast products while the historic data are also used in model calibration. Currently, data are collected for over 800 sites shown in Fig. 2. The meteorological data collection center collates several variables from gages including precipitation, temperature, and wind speed, though not all variables are collected from all stations. Fig. 3 indicates the coverage gages returning real-time precipitation data.

Alongside the in situ observations HSAF (http://hsaf.meteoam.it) satellite-derived soil moisture and snow coverage products are also

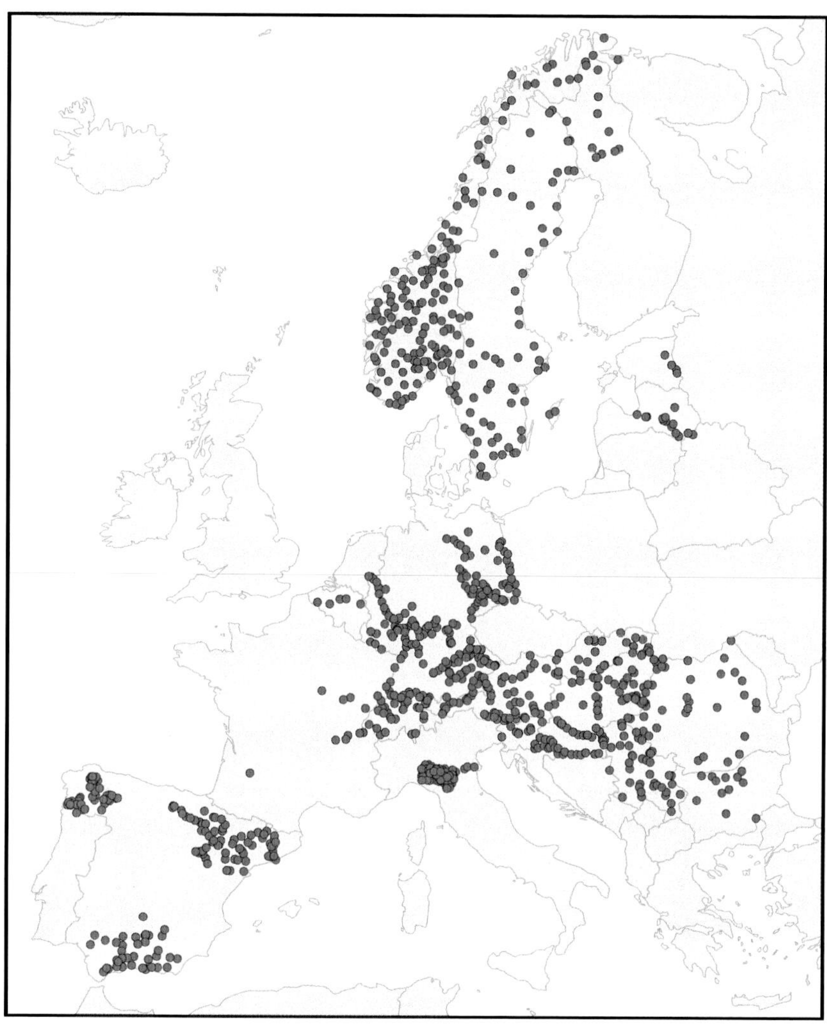

Fig. 2 Coverage of European Flood Awareness System (EFAS) real-time gaging stations.

collated for visualization purposes. Where the flood alerts issued by the national agencies are available, these are displayed in a common framework. For example, EFAS Information System (EFAS-IS) shows the warnings issued by the Swedish Hydrological Service to the public in the same way as it illustrates the warnings by the Bavarian water services. This provides a feedback loop from the officially issued warnings to the EFAS system.

Fig. 3 Coverage of EFAS real-time precipitation stations.

4 MODEL COMPONENTS

Within EFAS, hydrological forecasts are generated by cascading an ensemble of meteorological forecasts through a deterministic hydrological model. This section briefly outlines both the models that provide meteorological forcing and the hydrological model LISFLOOD.

4.1 Meteorological Models

In order to capture some of the uncertainty in the weather predictions, EFAS has been designed to operate with several NWP systems capable of providing the required forcings for the LISFLOOD hydrological model (see Section 5.2). Currently, EFAS makes use of four products (Table 1). Two are based on the ECMWF Integrated Forecasting System, of which the latest cycle (41r1) became operational on May 12, 2015. Details of older cycles can be found at http://www.ecmwf.int/en/forecasts/documentation-and-support/changes-ecmwf-model/cycle-41r1#versions. The ECMWF-HRES is a deterministic high-resolution run while the ECMWF-ENS is an ensemble forecast of lower resolution (Table 1).

Table 1 Summary details of the meteorological models used in generating EFAS forecasts

Product	Spatial resolution (km)	Vertical layers	Maximum lead time (days)	Number of members
ECMWF-HRES	T1279/16 km	137	10	1
ECMWF-ENS	T639/32 km for lead time 1–10 days, T319/64 km for lead time 11–15 days	91	15	51
German Weather Service	7 km up to day 3 then ~30 km	40	7	1
COSMO-LEPS	7 km	40	5.5	16

The German Weather Service provides a further deterministic forecast based on combining their global and limited-area models, essentially using the smaller-scale forecasts as a dynamical downscaling of the coarser, global model forecast (Schulz, 2005). The final meteorological product is the Limited-area Ensemble Prediction System (LEPS, version 5.1) of the Consortium for Small-scale Modeling (COSMO) (Montani, Cesari, Marsigli, & Paccagnella, 2011). This model, though of higher resolution, only covers the part of the domain (Fig. 1) with boundary conditions provided by ECMWF-HRES. All four models generate forecasts at 00:00 and 12:00 UTC.

4.2 LISFLOOD

LISFLOOD is a Geographic Information System (GIS) based spatially distributed hydrological rainfall-runoff model developed at the JRC (Bartholmes, Thielen, & Kalas, 2008; Knijff, Younis, & Roo, 2010). This model was developed at the JRC (JRC, EC) for operational flood forecasting

(Thielen, Bartholmes, et al., 2009) at pan-European scale. Driven by meteorological forcing data (precipitation, temperature, potential evapotranspiration, and evaporation rates for open water and bare soil surfaces), LISFLOOD calculates a complete water balance at a 6-hourly or daily time step and for every grid cell. Processes simulated for each grid cell include snowmelt, soil freezing, surface runoff, infiltration into the soil, preferential flow, redistribution of soil moisture within the soil profile, drainage of water to the groundwater system, groundwater storage, and groundwater base flow (see Fig. 4). Runoff produced for every grid cell is routed through the river network using a kinematic wave approach.

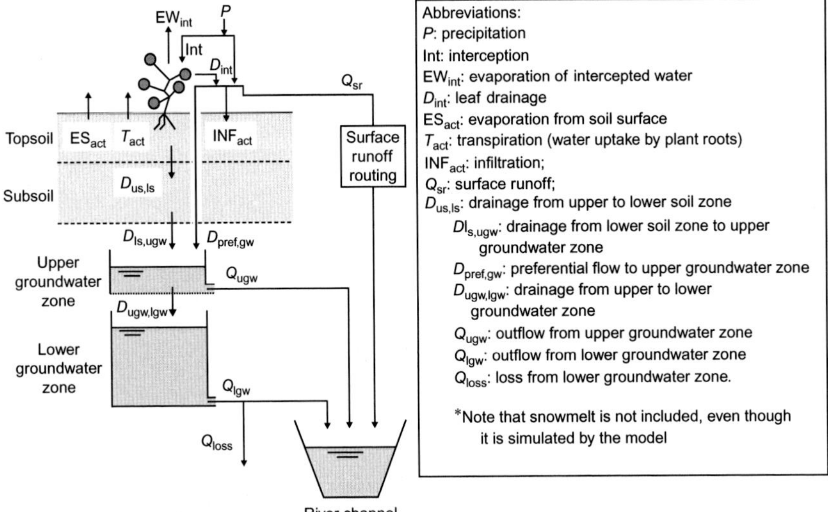

Fig. 4 Schematic description of the LISFLOOD model.

The pan-European set-up of LISFLOOD uses a 5 km grid on a Lambert Azimuthal Equal Area projection. Spatial data are obtained from various European databases with emphasis on having a homogeneous base for all over Europe. Data on soil properties are derived from the European Soil Geographical Database (King, Daroussin, & Tavernier, 1994). Vegetative properties (Leaf Area Index, LAI) were obtained from the GLOBCARBON project, based on monthly, 1 km resolution LAI data for the period of 1998–2007 (available at the SPOTIMAGE/VITO distribution site). The land cover dataset was created using the European Corine Land Cover 2000 (EEA, 2000) (CLC2000; 100 m — version 12/2009). The Global Land

Cover 2000 (GLC2000) database has been used for the missing areas of the European land cover database. Elevation data are obtained from the Shuttle Radar Topography Mission (SRTM) (Farr et al., 2007) and river properties were obtained from the Catchment Information System (Hiederer & de Roo, 2003). The meteorological data are extracted from the JRC MARS and the EU-FLOOD-GIS databases, which contain various data providers such as national institutions and continental-scale data providers, and are interpolated to the model grid using an inverse distance scheme (Ntegeka et al., 2013). All meteorological variables are interpolated on a 5×5 km grid using inverse distance weighting. Temperature variables are first corrected using the elevation. Observed river flow data at gaging stations from Europe taken from the Global Runoff Data Center (http://www.bafg.de/GRDC/EN/Home/homepage_node.html) as well as national/regional data providers were used during calibration.

A calibration exercise completed in 2013 (Zajac et al., 2013) produced Europe-wide parameter maps based on the estimation of parameter values for 693 catchments. Estimation was carried out using the Standard Particle Swarm 2011 (SPSO-2011) algorithm (Zambrano-Bigiarini & Rojas, 2013) and a root mean squared error criteria. For 659 of these, a set of nine parameters that control snowmelt, infiltration, preferential bypass flow through the soil matrix, percolation to the lower groundwater zone, percolation to deeper groundwater zones, residence times in the soil and subsurface reservoirs, and river routing were estimated by calibrating the model against historical records of river discharge. For the remaining 34 catchments, the option to represent reservoirs was used, requiring the calibration of four additional parameters related to reservoir operation, though neglecting the calibration of the deepest groundwater store resulted in 12 calibration parameters for these catchments.

Fig. 5 shows Nash-Sutcliffe efficiency (NSE) of the calibrated LISFLOOD model for the calibration (Jan. 1, 1994–Dec. 31, 2002) and validation (Jan. 1, 2003 to Dec. 31, 2012) time periods. In calibration, LISFLOOD is shown to have explanatory power for 90% of the catchments. For 32% of the catchments, LISFLOOD explains over three-quarters of the variance of the observed series. Visual and numeric comparisons of the calibration and validation periods show a broadly similar performance.

Notwithstanding the overall good agreement between the observed and simulated flow statistics, large discrepancies do occur at a small number of stations, particularly in the Iberian Peninsular and on the Baltic coasts.

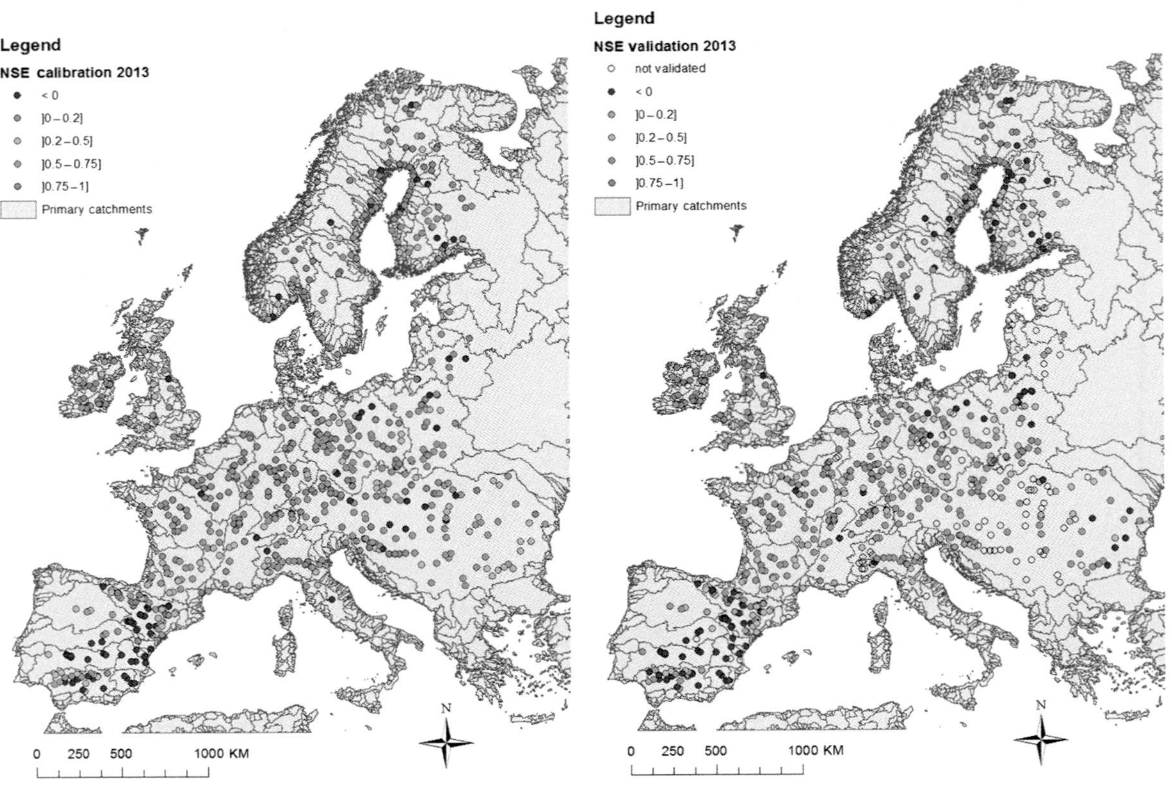

Fig. 5 The Nash-Sutcliffe efficiency (NSE) of LISFLOOD at the 693 sites for the calibration (left) and validation (right) periods.

Deviations from the observation-based statistics may be attributed to errors in meteorological forcing, the spatial interpolation of meteorological data, as well as to shortcomings in the hydrological model, its static input, and the calibration of its parameters. Some of the differences may also be due to those manmade modifications of flow regimes present in many catchments, but which are not fully accounted for in the hydrological model.

5 GENERATING FORECASTS

The generation of forecasts is the responsibility of the computational center. The task can be subdivided into three main components: (i) collating all the necessary forcing and input data; (ii) running LISFLOOD; and (iii) preparing results for visualization. Details of the scheduling and execution of these tasks are given in Section 6.2. Section 9 outlines the steps undertaken to monitor forecast generation. To give context to the later discussion, Section 6.1 provides an outline of the hardware used.

5.1 Infrastructure

The generation of the EFAS forecasts is executed on a dedicated "production" Linux cluster. The development and testing of the EFAS system is carried out on a separate general-purpose cluster with hardware specifications similar to the production cluster.

The production cluster comprises of eight nodes, each with two quad-core Xeon Intel processors and 128 GB of memory. The nodes run the same image of the operating system (SUSE Linux Enterprise server 13.1), but are configured differently such that one is an interactive node and the remaining seven are batch (or compute) nodes. A separate Management Workstation is used to provision, configure, and monitor the whole cluster. Portable Batch System (PBS) software is used to schedule, distribute, and manage jobs across the cluster. The cluster makes use of two storage units, each consisting of an I/O node (eight cores and 64 GB of memory) connected to via a double 8 GB/s Fiber Channel link to the IBM System Storage populated with an array of 300 GB SAS disks. All the hardware has redundant components in order to eliminate every single point of failure.

The production cluster and its storage have been installed in two racks, at different locations in the ECMWF computer hall. All areas of the computer hall are equipped with an inert gas fire suppression system. The ECMWF site is fed through two separate redundant power cables. The internal power distribution infrastructure is also redundant, with Diesel rotary UPS units (located in a separate building) providing emergency power.

5.2 Scheduling and Execution

EFAS forecasts are run through the Supervisor Monitoring Scheduler (SMS) software, a multi-threaded workflow package that enables users to run a large number of jobs (around 1100 in case of EFAS) with dependencies on each other and on time in a controlled environment. It provides for a reasonable tolerance to hardware and software failures, combined with good restart capabilities. It is used at ECMWF to run most of the operational suites across a range of platforms. The scheduling of tasks takes into account the dependencies between them as well as date and time dependencies. This makes SMS particularly suited for use in EFAS, where tasks require sequential evaluation yet must be performed simultaneously to ensure timely delivery of the forecasts — for example, running LISFLOOD to generate forecasts must occur after the initial conditions are determined, but each set of meteorological forecasts can be evaluated simultaneously.

The EFAS SMS suite is divided into "modules." A module is a collection of tasks sharing a common purpose and often the same work directory. Each module is further divided into a critical and noncritical stage. The critical stage performs operations which, if delayed, may result in a failure to generate some or all of the forecast products on time. The critical stage tasks are therefore closely monitored and supported on a 24/7 basis. Examples of critical stage tasks include preparing input data, running hydrological simulations, post-processing the results of these simulations, generating plots and tabular data, and publishing these products on the EFAS web interface. The noncritical stage includes storing of EFAS output in a tape archive and removing old output from work directories. Any delays or failures of these tasks require an investigation, but are not critical to the delivery of forecasts.

The remainder of this section describes the scheduling of the modules in EFAS which control the evaluation of LISFLOOD to generate an ensemble of hydrological forecasts. Other modules are evaluated after each forecast run to produce additional analysis for the forecast products outlined in Section 7.

EFAS hydrological forecasts are produced twice a day as part of the 00 and 12 cycles. The cycle names relate to the nominal time that the meteorological forecasts used as forcings. Each cycle runs four variations of hydrological forecasts. The variations arise due to forcing each one with a different meteorological forecast which improves forecast performance (Pappenberger, Bartholmes, et al., 2008; Ye et al., 2013). In the case of the ensemble meteorological forecasts, each ensemble is evaluated separately. Details of these variation and their evaluation are given in Table 2.

Table 2 Summary of hydrological forecast variations with approximate UTC run times

Forecast variation	NWP forcing	12 cycle Start	12 cycle End	00 cycle Start	00 cycle End	Max. number of simultaneous tasks
dwd	German met. service	17:50	18:00	06:20	06:30	2
eud	ECMWF-HRES	19:00	19:10	07:00	07:10	2
eue	ECVMWF-ENS	20:30	21:00	08:30	09:00	51
cos	COSMO-LEPS	21:15	21:25	09:15	09:25	16

Alongside the two forecast cycles, a water balance module is evaluated. This is a simulation run of LISFLOOD driven by inputs based on meteorological observations. On a given day, the model evaluates the 24h up to 06:00 UTC, starting at −42h and ending at −18h, relative to the nominal time of the subsequent "00" hydrological forecasts. As the final state of this simulation is valid at 18h prior to the start of the 00 hydrological forecast simulations, it cannot be used directly as initial conditions for these simulations. To fill this gap, a short 18-h LISFLOOD simulation is run, driven by either DWD or ECMWF deterministic forecasts (depending on the variation of the subsequent hydrological forecast). Similarly, 30-h long "fill-up" simulation is performed to create initial conditions for the 12 hydrological forecast run.

6 FORECAST PRODUCTS

From the ensemble of hydrological forecasts, a number of forecast products are derived. The form of these products are one of the most dynamic parts of the system, with their evolution being driven by user requests and comments (de Roo et al., 2011; Pappenberger, Stephens, et al., 2012; Ramos, Thielen, & de Roo, 2009; Wetterhall et al., 2013). In this section, four of the key EFAS products are outlined.

6.1 Flood Alerts

EFAS only provides information to the national hydrological services when there is a danger that critical flood levels might be exceeded. In EFAS, the critical thresholds are needed at every grid point and therefore cannot be derived from observations. Instead, based on observed meteorological data, long-term discharge time series are calculated at each grid with the same LISFLOOD model parameterization that is set-up in the forecasting system.

From these long-term simulations return periods are estimated — the 1-, 2-, 5-, and 20-year return periods. All flood forecasts are compared against these thresholds — at every pixel — and the threshold exceedance calculated. Only when critical thresholds are exceeded persistently over several forecasts, is information at these locations is produced, for example, in the form of color-coded overview maps or time series information at control points. The persistence criteria, currently three consecutive forecasts with greater than a 30% probability of exceeding a threshold based on the forecasts derived from the ECMWF-ENS forcing, has been introduced to reduce the number of false alerts and focus on large fluvial floods caused mainly by either widespread severe precipitation, combined rainfall and snowmelt, or prolonged rainfalls of medium intensity.

6.1.1 EFAS Thresholds and Return Periods

The EFAS thresholds are based on a 22-year model run using observed meteorological data as input, producing a surface re-analysis (similar to Balsamo et al., 2015). A Gumbel distribution; fitted using the L-moments procedure; is applied to each pixel in the LISFLOOD discharge output maps to obtain return periods. The return periods are then associated to EFAS alert levels as described in Table 3.

Table 3 EFAS thresholds and return periods

EFAS threshold	Description	Impact
Severe	Alert level corresponds to a simulated flood event with a return period of >20 years	Potentially severe flooding expected
High	Alert level corresponds to a simulated flood event with a return period >5 years and <20 years	Significant flooding is expected
Medium	Alert level corresponds to a simulated flood event with a return period >2 years and <5 years	Bankfull conditions or slightly higher expected. If flooding occurs, no significant damages are expected
Low	Alert level corresponds to a simulated flood event with a return period >1.5 years and <2 years	Water levels higher than normal or up to bankfull conditions, but no flooding is expected

6.2 Post-Processed Forecasts

At a given location, the forecasts can be post-processed not only to minimize errors in the timing, volume, and magnitude of the peak when compared to the observed, but also to derive more accurate calibrated probabilistic forecasts (Bogner & Kalas, 2008; Bogner & Pappenberger, 2011; Bogner, Pappenberger, & Cloke, 2012; Bogner, Meißner, Pappenberger, & Salamon, 2014). The approach used is a two-step process and is applicable at points along the river network where both historic and real-time observations of discharge are available.

The first step of the process is correction of each member of the ensemble of forecasts using the approach outlined in Bogner and Pappenberger (2011). The second step combines the forecast up to 10 days' lead time using Bayesian Model Averaging (BMA, Raftery, Gneiting, Balabdaoui, & Polakowski, 2005), the parameters of which are estimated for each lead time using a moving window of past forecasts. An example output is shown in Fig. 6. Both the forecast hydrograph and the probability of crossing thresholds derived from the historical observed data are shown.

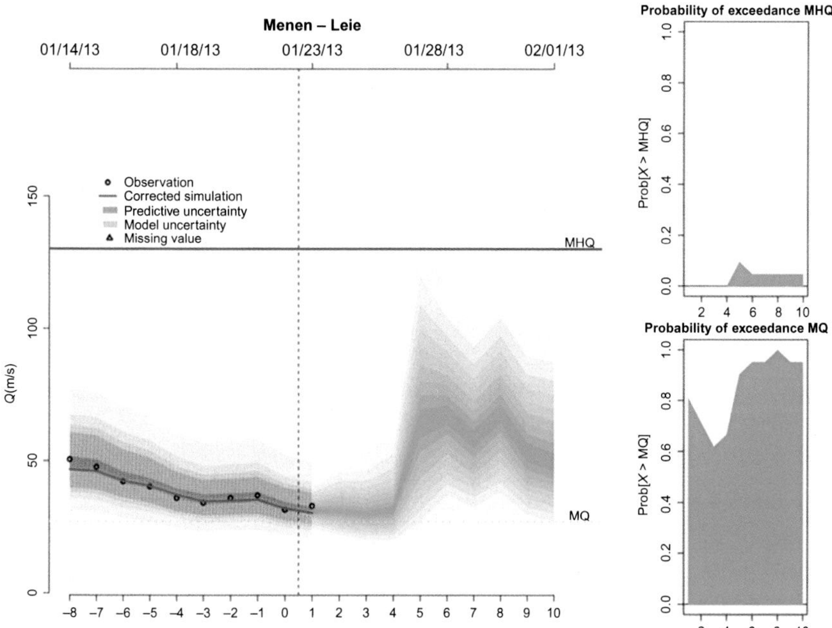

Fig. 6 Post-processed 10-day discharge forecast at Menen on the River Leie showing summaries of the post-processed forecast, along with the probability of exceeding the daily mean (MQ) and mean annual maxima (MHQ) observed discharges.

6.3 Flash Flood Alerts

Although designed for larger, riverine floods, the concepts and methodologies of EFAS have been shown to be also applicable for the detection of flash floods (Alfieri, Smith, del Pozo, & Beven, 2011; Alfieri, Velasco, & Thielen, 2011; Younis, Anquetin, & Thielen, 2008). The EFAS user community welcomed the inclusion of a flash flood indicator as a novel product. For EFAS, flash flood early warning is performed through the detection of rainstorms with extreme rainfall accumulations over short durations (6, 12, and 24h) and within small-size catchments (<5000 km^2) prone to flash flooding. The European Precipitation Index based on simulated Climatology (EPIC, Alfieri & Thielen, 2012) is used as an indicator of upcoming hazardous events. System results only depend on the Quantitative Precipitation Forecast (QPF) and on the modeled river network. EPIC is calculated twice per day with a probabilistic approach, using COSMO-LEPS forecasts and a grid resolution of 1 km. At those locations with significant probabilities of exceeding reference warning thresholds (ie, return periods of 2, 5, and 20 years), reporting points are created and geo-located in the web interface. For each point, a return period plot is produced, showing the uncertainty range of EPIC return periods over the 132-h forecast horizon, as described by Alfieri and Thielen (2012). An example plot is also shown in Fig. 7. An analysis of daily EPIC forecasts over 22 months ending in Sep. 2011 denoted a probability of detection of rainstorm events and flash floods of 90%, corresponding to 45 events correctly predicted, with average lead time of 32h (Alfieri, Thielen, & Pappenberger, 2012). A future development to this system will replace the EPIC method with the European Runoff Index based on Climatology (ERIC, Raynaud et al., 2014). This works in the same way as EPIC but is based on the surface runoff values calculated by the LISFLOOD hydrological model. Therefore it better represents antecedent catchment conditions which may exacerbate the flash flood severity. ERIC became operational in the summer of 2015.

6.4 Rainfall Animation

The rainfall animation based on the COSMO-LEPS ensemble and ECMWF deterministic models are available. Images can be shown for different timesteps as well as the continuous animation of the sequence over the forecast range. For the deterministic model, the visualization routine uses a standard approach, where rainfall rates are divided into 16 classes of variable size and shown with a rainbow-like color palette. For COSMO-LEPS ensemble forecast product, a novel visualization technique is implemented.

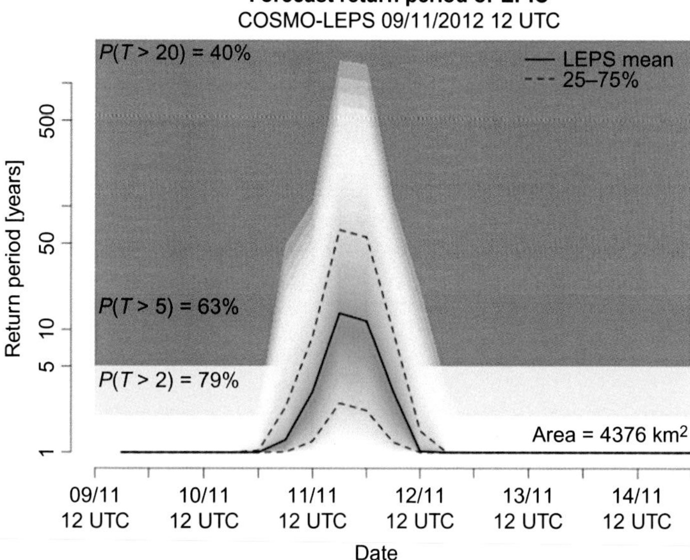

Fig. 7 Return period plot of probabilistic the European Precipitation Index based on simulated Climatology (EPIC) forecast for Sep. 11, 2012 12 UTC. Reporting point for the Piave river at the outlet, NE Italy.

The ensemble mean of rainfall rates is linked to a color palette, following the same approach as of the deterministic forecast. In addition, the ensemble spread is organized into three classes, according to the coefficient of variation (CV) of forecast values against the ensemble mean. Each class is then shown with different level of transparency, which increases with the CV. Fig. 8 shows an example of this type of image.

6.5 Soil Moisture and Snow Anomaly Maps

EFAS also displays some of the initial conditions of the hydrological model such as the soil moisture and the snow water equivalent. However, as in flood forecasting it is often more important how different the soil moisture and snow conditions are in comparison to the "normal" situation, so anomalies are also displayed. The anomalies are calculated by scaling the value using the mean and stand deviation of the values taken on that year day in the LISFLOOD long-term model run used to derive the warning thresholds.

In the case of the snow water equivalent, the simulated variable corresponds to a 10-day average. Therefore also the long-term average and the standard deviation for the snow water equivalent are derived using 10-day average values from the LISFLOOD long-term run.

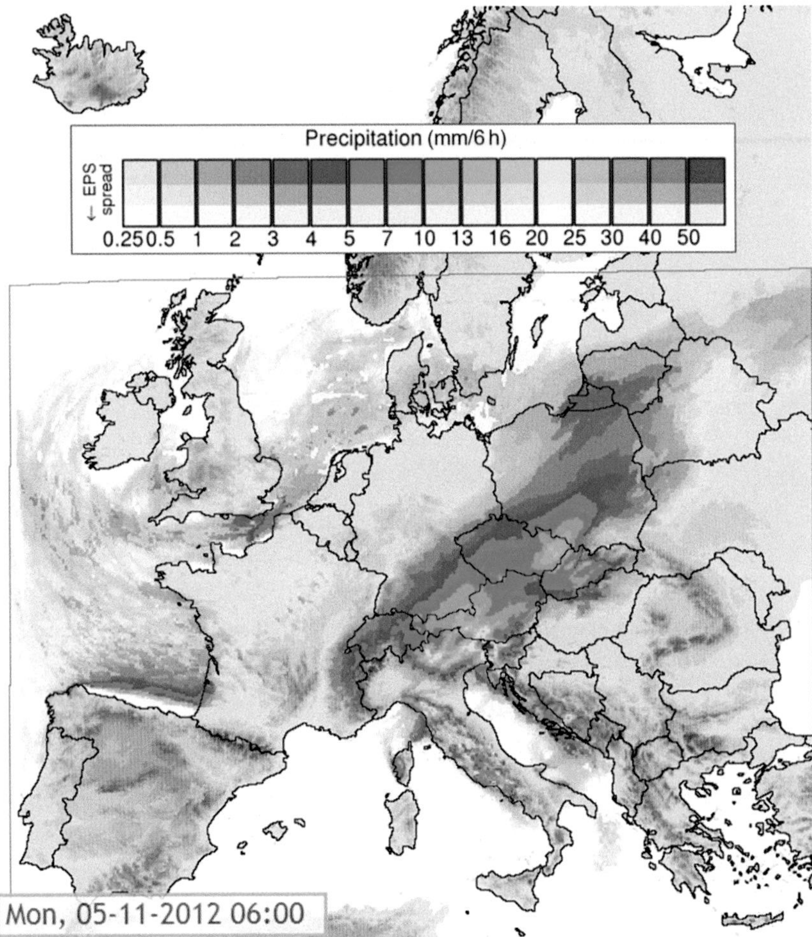

Fig. 8 Six-hour ensemble precipitation forecast from COSMO-LEPS run of Nov. 3, 2012 12 UTC. Forecast lead time of 36–42 h.

7 FORECAST DISSEMINATION

Dissemination of the forecasts to end user is carried out in two ways. The first is through the use of a password-protected web-based interface; the EFAS-IS accessible only to registered users. The second is for the dissemination center to proactively contact end users when alerts are issued within their domain. In the following sections, these two methods are introduced.

7.1 EFAS-IS

The EFAS-IS (https://www.efas.eu) is a Rich Internet Application (RIA, Fig. 9) providing the same level of interactivity and responsiveness as desktop

Fig. 9 Screen shot of the EFAS Information System (EFAS-IS) showing the tabbed layout and menu for selection of products to be visualized.

applications. It was carefully designed alongside the forecast products with the aims of end users in mind. The EFAS-IS allows control and management of the content within the web portal based on user specific roles and permits various workflows in a collaborative environment. It grants end users the ability to contribute to and share information and helps improve communication by allowing users to raise queries with the EFAS centers. Alongside restricted information for EFAS partners, public information, such as the bimonthly bulletins designed to review recent floods and inform about ongoing system improvements, are available on the web portal.

7.1.1 EFAS Web Services

Alongside EFAS-IS, two services are provided to partner organizations for downloading data from EFAS for further analysis and incorporation into their own systems. These services use Open Geospatial Consortium (OGC) standards to deliver either data about individual points; the EFAS SOS (Sensor Observation Service) services; or maps; the EFAS WMS-T (Web Map Service Time).

7.2 Email Alerts and Daily Overview

The EFAS dissemination center sends out warning emails to the corresponding EFAS partners in order to inform them of a possible upcoming flood event. The emails are, however, just a call for attention to the concerned EFAS partners. More details can then be found on the EFAS-IS. In this situation, three types of emails can be sent by an EFAS forecaster relating to three types of EFAS warning: EFAS Flood Alert, EFAS Flood Watch, and EFAS Flash Flood Watch. There are strict criteria on the activation, upgrading, and deactivation of these warnings; these are outlined in Table 4.

Table 4 Rules for activation, upgrading, and deactivation of EFAS flood alerts and watches

Action	EFAS flood alert	EFAS flood watch	EFAS flash flood watch
Activation	1. Catchment part of agreed list of catchments 2. Catchment area is larger than 4000 km^2 3. Event more than 48 h in advance with respect to the forecast date 4. Forecasts are persistent (three consecutive forecasts with more than 30% exceeding EFAS high threshold according to ECMWF-ENS) 5. At least one of the deterministic forecasts (ECMWF or DWD) exceeds also the EFAS high threshold	1. Catchment part of agreed list of catchments 2. Any of Flood Alert criteria is not met but the forecasters thinks the authorities should be informed 3. Any other doubt	1. Catchment part of agreed list of catchments 2. Probability of exceeding the flash flood high threshold is forecast to be greater than 60%

Table 4 Rules for activation, upgrading, and deactivation of EFAS flood alerts and watches—Cont'd

Action	EFAS flood alert	EFAS flood watch	EFAS flash flood watch
Upgrading	If an EFAS Flood Watch has been sent but in the following forecasts all the conditions for an EFAS Flood Alert are fulfilled, then the EFAS Flood Watch can be upgraded to an EFAS Flood Alert		
Deactivation	Observations reported by the national/regional hydrological service clearly indicate that the EFAS Flood Alert/Watch is a false alarm Observations reported by the national/regional hydrological service clearly indicate that discharges/water levels have decreased already to normal values; meanwhile, EFAS simulations still show that simulated discharge exceed the EFAS high threshold The simulated EFAS discharge at the reporting point(s) for which the EFAS Alert/Watch was issued falls below the EFAS high threshold		Probability of exceeding the flash flood high threshold is forecast falls to less than 60%

Alongside these, a daily overview is sent to the ERCC of the EC which contains information on ongoing floods in Europe as reported by the national services, EFAS Flood Alerts, EFAS Flood Watches, and EFAS Flash Flood Watches.

8 OPERATIONAL PERFORMANCE

8.1 Monitoring

The entire chain of EFAS computations as well as the underlying hardware and software infrastructure are monitored at all times to ensure uninterrupted availability and timely delivery of the EFAS products.

The core monitoring services and first-level support are provided by a dedicated team of operators who are available on the premises at all times. This core group follows established procedures to rectify issues themselves or forward the issue to the second-level support staff. The second-level support is provided by specialized teams of experts on 24/7/365 call-out duty with remote access ECMWF IT infrastructure. Finally, the third-level support is provided by in-house and third party technical and scientific experts.

Operators on duty have several mechanisms at their disposal to monitor the activity of the EFAS system. The first one is built into the SMS job scheduling software — its graphical user interface (Xcdp) visualizes the progress of computations and instantly alerts operators about any failures. The second mechanism is a dedicated subset of EFAS watchdog jobs which are executed at specified times and check if EFAS computations have reached their expected stage. Additionally, the state of the EFAS system and underlying infrastructure (computational cluster, web servers, and network) is monitored by the OpsView service (http://www.opsview.com), which is a monitoring and alerting tool for servers, switches, applications, and services. The acquisition of input data from external providers is monitored via the web interface built into the ECMWF Product Dissemination System (ECPDS) data acquisition system which, for example, raises an alarm if no new data has been received for a prolonged period of time.

In Table 5, various services are listed along with the action which is taken following a failure, the response time, and how the response is triggered. If the failures will result in late, incomplete, or incorrect products, the JRC and dissemination center are informed.

Table 5 Summary of monitoring procedures for the operational EFAS system

Service	Action	Response	Trigger
Data acquisition	Operator informs the data provider	24/7	Alarm raised in the ECPDS monitoring interface
EFAS computations	Operator follows a recovery procedure or calls an analyst	24/7	Failure/delay signaled by Xcdp or abnormal state detected by OpsView
EFAS web interface	Operator follows a recovery procedure or calls an analyst	24/7	Abnormal state detected by OpsView; an email or phone call from user

8.2 System Performance

The current deadline for delivering EFAS 12 UTC and 00 UTC forecasts is 02:00 UTC and 14:00 UTC the following day, respectively. Overall, the performance of the system is very high, with a greater-than-99% reliability

Table 6 Delays in EFAS products delivery for the period between 01.04.2014 and 01.04.2015

Date	Delay	Description
17.04.2014	2h	All products delayed due to network file-system issues
27.04.2014	1 day	Products based on COSMO-LEPS forecast delayed due to issues with the COSMO-LEPS model over the weekend
24.06.2014	2 days	Products based on DWD forecast delayed due to unexpected change of DWD data format
22.07.2014	2h	Products based on COSMO-LEPS forecast delayed due to late arrival of forecast data
27.09.2014	7h 30min	Products based on COSMO-LEPS forecast delayed due to late arrival of forecast data
30.09.2014	3h 30min	Products based on COSMO-LEPS forecast delayed due to late arrival of forecast data
26.11.2014	30min	All products delayed as a result of unusually high workload on the computational cluster
23.02.2015	12h	Products based on DWD forecast delayed — data not sent by the provider

due to strict quality assurance measures allowing the system to capture and promptly correct the majority of problems as they arise.

Table 6 lists incidents between Apr. 2014 and Apr. 2015 which led to delays and missed deadlines. In two of these incidents, all products were delayed. In the remaining six cases, the delay involved only some of the products. In each case, EFAS users were informed in a timely manner by email.

9 CASE STUDIES

9.1 General Performance

The forecast performance of EFAS is continually monitored in terms counting the number of watches, alerts, and flood watches sent out. As far as possible, the alerts are counted as hits or false alarms in comparison with observed floods, otherwise they are assigned as unknown. Fig. 10 shows a large inter-annual variation, and that 2013 and 2014 stand out as having a greater number of warnings. The main reasons for this are two major events: the central European floods of 2013 and the Balkan floods of 2014, which are discussed below.

The observed occurrence of a flood in Fig. 10 was extracted from the International Disaster Database (ref: http://www.emdat.be/database, accessed Jun. 10, 2015). There are clear trends in the data. It appears the

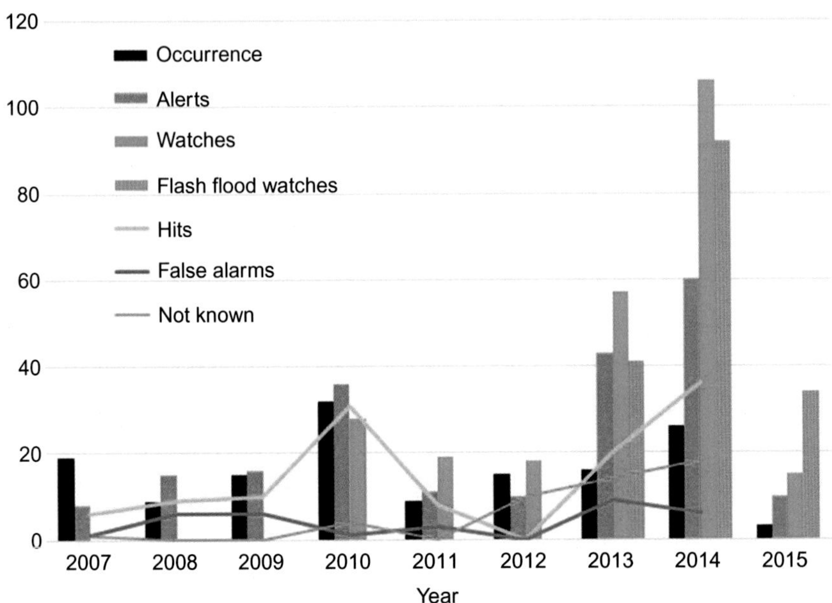

Fig. 10 Number of flood alerts, watches, and flash flood watches sent since 2007. The lines show the verified hits and false alarms over the first years of EFAS.

system has increased in activity over the years when comparing the number of reported events with warnings and alerts. This could be due to the fact that the number of EFAS members has grown over the years, or that due to changes in the criteria for issuing alerts and watches more are being issued. However, there is a much higher correlation between the number of affected people and the number of issued flood alerts (0.89), than with the number of events (0.65). This is an effect of how an event is classified in the database. The number of people affected is a better measure of the total extent of the flood, and this is what is reflected in an increase in the number of alerts.

The performance of EFAS is also continuously monitored in terms of skill scores such as bias, NSE, and continuous ranked probability scores (CRPS). Performance results are published in a bimonthly bulletin and the scientific literature (eg, Alfieri et al., 2014; Pappenberger, Thielen, et al., 2011), and new user focused scores are developed as needed (eg, Cloke & Pappenberger, 2008; Pappenberger, Bogner, et al., 2011, Pappenberger, Cloke, Persson, & Demeritt, 2011; Pappenberger, Scipal, & Buizza, 2008). The system is evaluated against its own climatology, which is the water balance run of LISFLOOD (Pappenberger, Ramos, et al., 2015). The performance of the model is steadily increasing; however, there are inter-annual

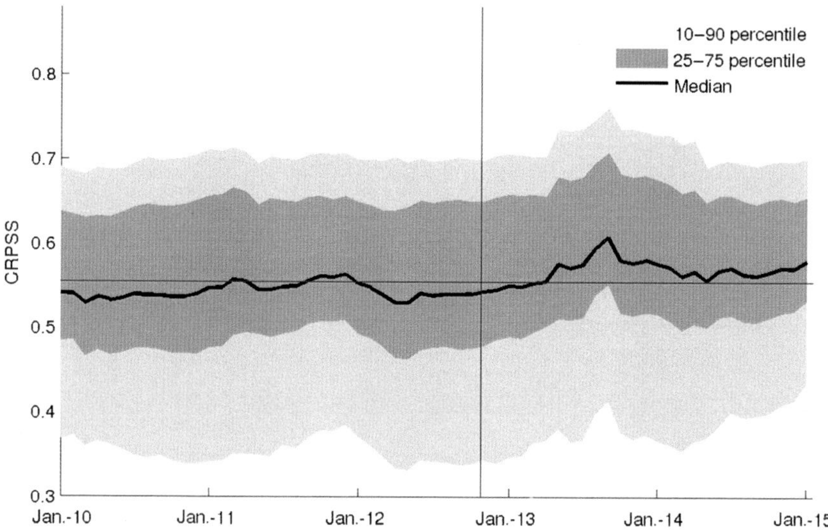

Fig. 11 CRPS of EFAS driven by ECMWF-ENS over the period Jan. 1, 2010–Apr. 30, 2015 as an annual running mean. The *lighter gray* areas show the 10–90th percentiles, and the *darker gray* 25–75th percentiles. The results are shown for all the areas larger than 4000 km^2.

variations (Fig. 11). The recent drop in performance is due to the poor performance of the winter of 2015, which was difficult to predict for all weather centers.

9.2 Central European Floods in Summer 2013

The Central European flood event of Jun. 2013 was the first large-scale crisis during which the operational EFAS was actively reporting to the ERCC. ECMWF as a current EFAS operational center published a detailed analysis of this event (Haiden et al., 2014; Pappenberger, Wetterhall, Albergel, et al., 2013). The Jun. 2013 flood event was a severe, large-scale event that affected several countries and led to the loss of lives as well as considerable damage in two major European catchments (Danube and Elbe). Over the last week of May 2013, EFAS forecasts showed a rapidly increasing probability of exceeding flood warning thresholds for wide areas in Central Europe including Germany, Poland, Austria, the Czech Republic, and Slovakia. Between 28 and 31 May, 14 EFAS flood warnings of different severity levels (both flood alerts and watches) were issued for some of the major rivers (eg, Elbe, Danube, Rhine, and Odra) up to eight days before the beginning of the extreme streamflow conditions (Fig. 12). Cities such as Wittenberg (Germany) were severely affected by the rising waters of the Elbe, where the record high of the "flood of the century" in the year 2002 was surpassed by more than half a meter on Jun. 9, 2013.

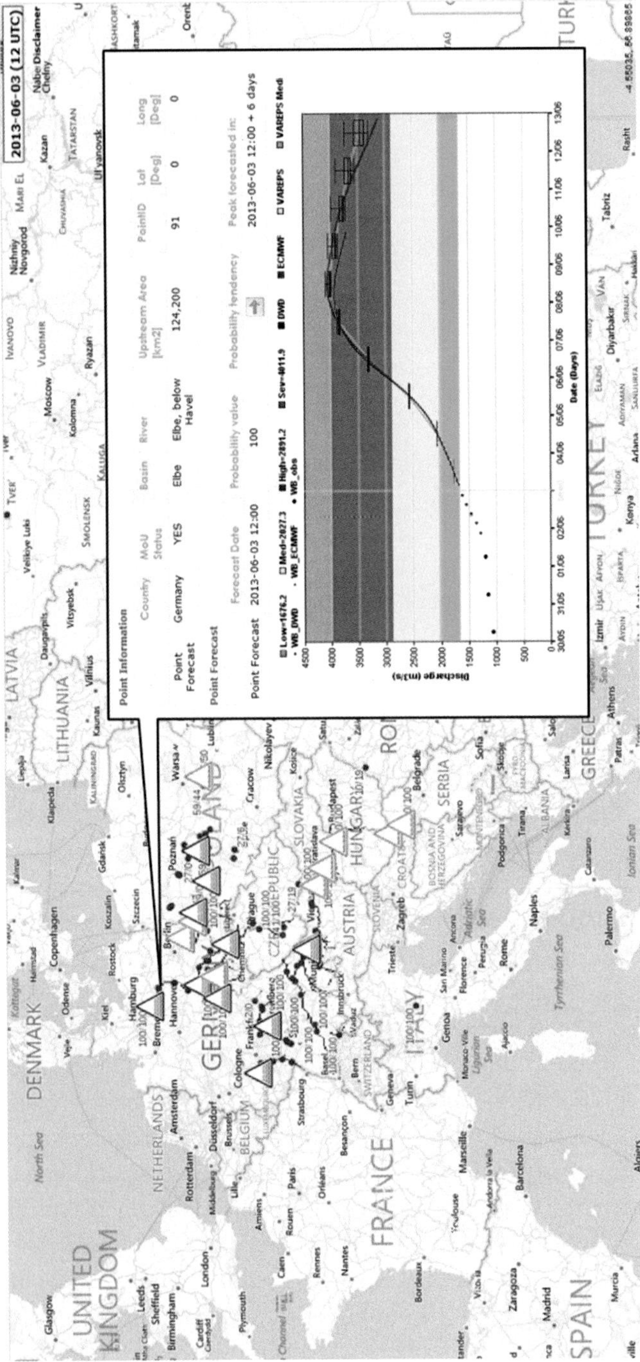

Fig. 12 European Flood Awareness System (EFAS) active alerts (*red*) and watches (*orange*) on Jun. 3, 2013, and multi-model streamflow prediction for the Elbe River at Wittenberg, Germany, based on 12:00 forecasts on Jun. 3, 2013 and valid for the next 10 days.

9.3 Floods on the Balkan in May 2014

Exceptionally intense rainfalls from May 13 onwards following weeks of wet conditions led to disastrous and widespread flooding in the Balkans, in particular Bosnia-Herzegovina and Serbia. Critical flooding was also reported in other countries including Southern Poland, Slovakia, and the Czech Republic. The events in Bosnia-Herzegovina and Serbia are reported to be the worst in more than 100 years, with 44 reported casualties (ECMWF, 2015). Also one person died in each of Croatia and the Czech Republic due to flooding. More than a million inhabitants are estimated to have been affected by this flood event. Both Bosnia-Herzegovina and Serbia activated the EU Community Mechanism for help in the afternoon of the May 15th and again on the 17th for further assistance for Bosnia-Herzegovina. EFAS started providing the relevant national authorities and the ERCC with EFAS notifications from May 11 onwards (Figs. 13 and 14).

Fig. 13 EFAS interface showing all EFAS notifications to national authorities based on 12:00 forecasts from May 12, 2014.

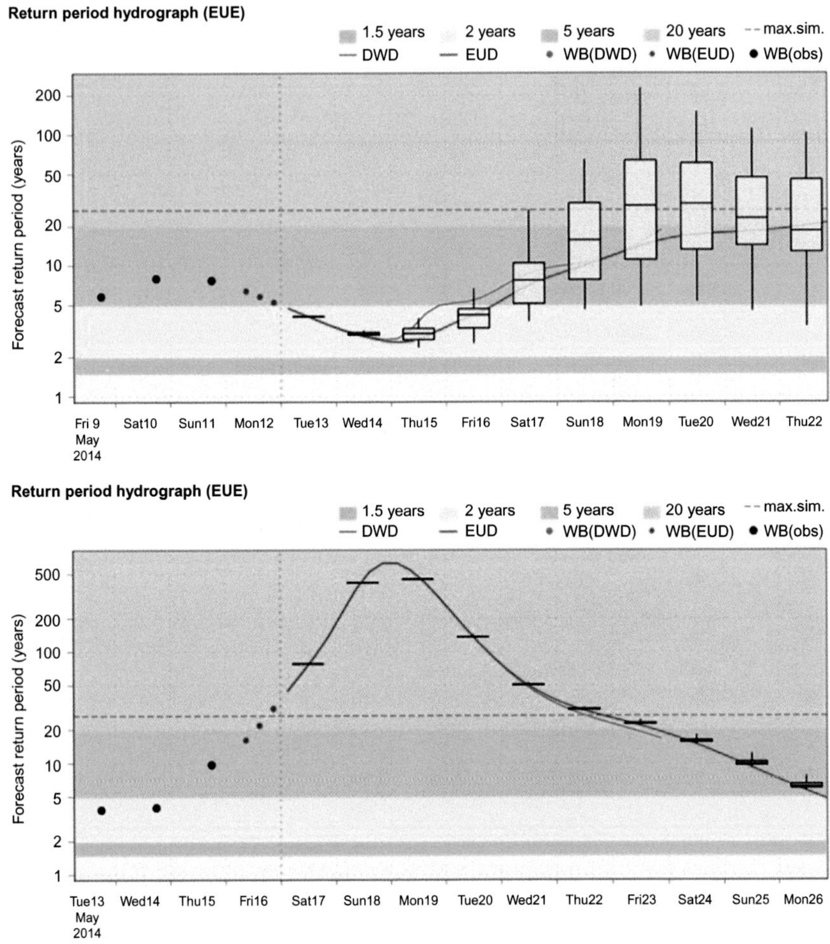

Fig. 14 Return period for the flow in river Sava for a point close to Belgrade. The forecasts area initialized May 9 (top) and May 13 (bottom).

10 CONCLUSIONS

Following a devastating, transnational flood event affecting several countries in Europe, the development of a pan-European flood forecasting system was launched to enhance the EU's capabilities for flood preparedness and coordination of aid. The EFAS was developed over a 10-year period from 2003 to 2012 before being transferred to a fully operational system under the Copernicus EMS providing early flood warnings across Europe. The system has grown from a research experiment used to provide forecast information on an ad hoc basis to a complex operational system in which the hydrological model forms a small part of a sophisticated forecasting chain. This chapter represents a snapshot of the current set-up forecasting system including the set-up of different operational centers and all the modeling components. The EFAS operational forecasting systems can be divided into six major function blocks:

(a) *Data acquisition*, which includes the acquisition of all static and dynamic data used to operate the EFAS system including weather forecasts, hydrological and meteorological observations or national warnings.

(b) *Model components*, which includes the hydrological model and general set-up of it.

(c) *Forecast infrastructure*, which relates to the underlying hardware infrastructure and the way the workflow of the forecasting system is managed.

(d) *Forecast products*, which includes all products produced as part of the forecast including flood alerts, post-processing and auxiliary information, such as rainfall animations or soil moisture/snow anomaly maps.

(e) *Forecast dissemination*, which deals with all aspects of disseminating products, and as such includes the web site and data distribution services.

(f) *Performance monitoring*, which includes the monitoring of the technical system performance and reliability as well as statistical skill of the forecasts and warnings.

This chapter demonstrates how these function blocks successfully performed in two major floods across Europe in 2013 and 2014.

The described systems are continuously evolving by, for example: adapting to new temporal and spatial resolutions from the forcings (Wetterhall, He, Cloke, & Pappenberger, 2011); adding and inventing new products to push the limits of predictability further in the future (Lavers, Pappenberger, & Zsoter, 2014; Thielen, Bogner, et al., 2009); probing new methods such as data assimilation (Neal et al., 2009); exploring new avenues for decision-making (Dale et al., 2012; Demeritt, Nobert, Cloke, & Pappenberger, 2012; Pappenberger & Brown, 2012; Ramos, van Andel, & Pappenberger, 2013); transferring experience to new geographical domains such as Africa or China (He et al., 2010; Thiemig et al., 2010; Thiemig, Bisselink, Pappenberger, & Thielen, 2014); and balancing the end user needs of a higher resolution information with scientific and operational demands (Beven & Cloke, 2012; Beven, Cloke, Pappenberger, Lamb, & Hunter, 2014; Wood et al., 2011).

The latter is also a challenge that is key for many operational flood forecasting systems and whose challenges were generalized by Pagano et al. (2014) as: making the most of available data; making accurate predictions; turning forecasts into effective warnings; and operating a reliable operational service.

It has been demonstrated that an early flood warning systems such as EFAS provides an immense monetary benefit (about 400 Euros for every 1 Euro invested) (Pappenberger, Cloke, et al., 2015). Nevertheless, there is a fundamental question of whether a reliable service that is only needed for a small fraction of time (during the times of floods) is sustainable financially and scientifically in the long term. It may require a merger of parts of the system components in a wider framework of, for example, a natural hazard warning service to pool resources and exploit synergies (Pappenberger, Wetterhall, Dutra, et al., 2013), which is an exciting future journey and opportunity.

ACKNOWLEDGMENTS

The authors wish to acknowledge that the ongoing operation and develop of EFAS has benefited from the contribution of many people, too numerous to name, and has received funding from multiple sources.

REFERENCES

Alfieri, L., Burek, P., Dutra, E., Krzeminski, B., Muraro, D., Thielen, J., et al. (2013). GloFAS — Global ensemble streamflow forecasting and flood early warning. *Hydrology and Earth System Sciences, 17*, 1161–1175. http://dx.doi.org/10.5194/hess-17-1161-2013.

Alfieri, L., Pappenberger, F., Wetterhall, F., Haiden, T., Richardson, D., & Salamon, P. (2014). Evaluation of ensemble streamflow predictions in Europe. *Journal of Hydrology, 517*, 913–922. http://dx.doi.org/10.1016/j.jhydrol.2014.06.035.

Alfieri, L., Salamon, P., Pappenberger, F., Wetterhall, F., & Thielen, J. (2012). Operational early warning systems for water-related hazards in Europe. *Environmental Science and Policy, 21*, 35–49. http://dx.doi.org/10.1016/j.envsci.2012.01.008.

Alfieri, L., Smith, P. J., del Pozo, J. T., & Beven, K. J. (2011). A staggered approach to flash flood forecasting — Case study in the Cévennes region. *Advances in Geosciences, 29*, 13–20. http://dx.doi.org/10.5194/adgeo-29-13-2011.

Alfieri, L., & Thielen, J. (2012). A European precipitation index for extreme rain-storm and flash flood early warning. *Meteorological Applications, 22*, 3–13. http://dx.doi.org/10.1002/met.1328.

Alfieri, L., Thielen, J., & Pappenberger, F. (2012). Ensemble hydro-meteorological simulation for flash flood early detection in southern Switzerland. *Journal of Hydrology, 424–425*, 143–153. http://dx.doi.org/10.1016/j.jhydrol.2011.12.038.

Alfieri, L., Velasco, D., & Thielen, J. (2011). Flash flood detection through a multi-stage probabilistic warning system for heavy precipitation events. *Advances in Geosciences, 29*, 69–75. http://dx.doi.org/10.5194/adgeo-29-69-2011.

Balsamo, G., Albergel, C., Beljaars, A., Boussetta, S., Brun, E., Cloke, H., et al. (2015). ERA-Interim/Land: A global land surface reanalysis data set. *Hydrology and Earth System Sciences, 19*, 389–407. http://dx.doi.org/10.5194/hess-19-389-2015.

Bartholmes, J., Thielen, J., & Kalas, M. (2008). Forecasting medium-range flood hazard on European scale. *Georisk: Assessment and Management of Risk for Engineered Systems and Geohazards, 2*, 181–186. http://dx.doi.org/10.1080/17499510802369132.

Bartholmes, J. C., Thielen, J., Ramos, M. H., & Gentilini, S. (2009). The European Flood Alert System EFAS — Part 2: Statistical skill assessment of probabilistic and deterministic operational forecasts. *Hydrology and Earth System Sciences, 13*, 141–153. http://dx.doi.org/10.5194/hess-13-141-2009.

Beven, K. J., & Cloke, H. L. (2012). Comment on "Hyperresolution global land surface modeling: Meeting a grand challenge for monitoring Earth's terrestrial water" by Eric F. Wood et al. *Water Resources Research, 48*. http://dx.doi.org/10.1029/2011wr010982.

Beven, K., Cloke, H., Pappenberger, F., Lamb, R., & Hunter, N. (2014). Hyperresolution information and hyperresolution ignorance in modelling the hydrology of the land surface. *Science China Earth Sciences, 58*, 25–35. http://dx.doi.org/10.1007/s11430-014-5003-4.

Bogner, K., & Kalas, M. (2008). Error-correction methods and evaluation of an ensemble based hydrological forecasting system for the Upper Danube catchment. *Atmospheric Science Letters, 9*, 95–102. http://dx.doi.org/10.1002/asl.180.

Bogner, K., Meißner, D., Pappenberger, F., & Salamon, P. (2014). Korrektur von Modell- und Vorhersagefehlern und Abschätzung der prädiktiven Unsicherheit in einem probabilistischen Hochwasservorhersagesystem. *Hydrologie und Wasserbewirtschaftung*. http://dx.doi.org/10.5675/HyWa_2014,2_2.

Bogner, K., & Pappenberger, F. (2011). Multiscale error analysis, correction, and predictive uncertainty estimation in a flood forecasting system. *Water Resources Research, 47*, n/a–n/a. http://dx.doi.org/10.1029/2010wr009137.

Bogner, K., Pappenberger, F., & Cloke, H. L. (2012). Technical note: The normal quantile transformation and its application in a flood forecasting system. *Hydrology and Earth System Sciences, 16*, 1085–1094. http://dx.doi.org/10.5194/hess-16-1085-2012.

Buizza, R., Pappenberger, F., Salamon, P., Thielen, J., & de Roo, A. P. (2009). EPS/EFAS probabilistic flood prediction for Northern Italy: The case of 30 April 2009. *ECMWF Newsletter, 120*, 10–16.

Burek, P., del Pozo, J. T., Thiemig, V., Salamon, P., & de Roo, A. P. (2011). Das Europäische Hochwasser-Frühwarnsystem (EFAS). *Korrespondenz Wasserwirtschaft, 2011*, 196–202. http://dx.doi.org/10.3243/kwe2011.04.001.

Cloke, H. L., & Pappenberger, F. (2008). Evaluating forecasts of extreme events for hydrological applications: An approach for screening unfamiliar performance measures. *Meteorological Applications, 15*, 181–197. http://dx.doi.org/10.1002/met.58.

Cloke, H., Thielen, J., Pappenberger, F., Nobert, S., Balint, G., Edlund, C., et al. (2009). Progress in the implementation of Hydrological Ensemble Prediction Systems (HEPS) in Europe for operational flood forecasting. *ECMWF Newsletter, 121*, 20–24.

Dale, M., Wicks, J., Mylne, K., Pappenberger, F., Laeger, S., & Taylor, S. (2012). Probabilistic flood forecasting and decision-making: An innovative risk-based approach. *Natural Hazards, 70*, 159–172. http://dx.doi.org/10.1007/s11069-012-0483-z.

de Roo, A. P., Gouweleeuw, B., Thielen, J., Bartholmes, J., Bongioannini-Cerlini, P., Todini, E., et al. (2003). Development of a European flood forecasting system. *International Journal of River Basin Management, 1*, 49–59. http://dx.doi.org/10.1080/15715124.2003.9635192.

de Roo, A. P., Thielen, J., Salamon, P., Bogner, K., Nobert, S., Cloke, H., et al. (2011). Quality control, validation and user feedback of the European Flood Alert System (EFAS). *International Journal of Digital Earth, 4*, 77–90. http://dx.doi.org/10.1080/17538947.2010.510302.

Demeritt, D., Cloke, H., Pappenberger, F., Thielen, J., Bartholmes, J., & Ramos, M. (2007). Ensemble predictions and perceptions of risk, uncertainty, and error in flood forecasting. *Environmental Hazards, 7*, 115–127. http://dx.doi.org/10.1016/j.envhaz.2007.05.001.

Demeritt, D., Nobert, S., Cloke, H. L., & Pappenberger, F. (2012). The European Flood Alert System and the communication, perception, and use of ensemble predictions for operational flood risk management. *Hydrological Processes, 27*, 147–157. http://dx.doi.org/10.1002/hyp.9419.

ECMWF. (2015). http://www.ecmwf.int/en/about/media-centre/news/2014/severe-floods-balkans Accessed 14.07.15.

EEA. (2000). Corine Land Cover raster database 2000.

European Commission. (2002). Communication from the commission to the European parliament and the council: The European Community response to the flooding in Austria, Germany and several applicant countries — A solidarity-based initiative. Brussels, 28.8.2002 COM(2002) 481. http://eur-lex.europa.eu/legal-content/EN/TXT/?uri=CELEX:52002DC0481.

European Commission. (2010). Communication from the commission to the European parliament and the council: Towards a stronger European disaster response: The role of civil protection and humanitarian assistance. Brussels, 26.10.2010 COM(2010) 600. http://eur-lex.europa.eu/legal-content/EN/TXT/?uri=CELEX:52010DC0600.

European Union. (2013). Decision no 1313/2013/EU of the European Parliament and of the council of 17 December 2013 on a union civil protection mechanism (1). *Official Journal of the European Union, 56*, 924–947. http://dx.doi.org/10.3000/19770677.L_2013.347.eng.

European Union. (2014). Regulation (EU) no 377/2014 of the European Parliament and of the council of 3 April 2014 establishing the Copernicus programme and repealing regulation (EU) No 911/2010 (1). *Official Journal of the European Union, 57*, 44–66.

Farr, T. G., Rosen, P. A., Caro, E., Crippen, R., Duren, R., Hensley, S., et al. (2007). The shuttle radar topography mission. *Reviews of Geophysics, 45*. http://dx.doi.org/10.1029/2005rg000183.

Gouweleeuw, B. T., Thielen, J., Franchello, G., Roo, A. P. J. D., & Buizza, R. (2005). Flood forecasting using medium-range probabilistic weather prediction. *Hydrology and Earth System Sciences, 9*, 365–380. http://dx.doi.org/10.5194/hess-9-365-2005.

Haiden, T., Magnussona, L., Tsonevsky, I., Wetterhall, F., Alfieri, L., Pappenberger, F., et al. (2014). ECMWF forecast performance during the June 2013 flood in Central Europe. *ECMWF Technical Memorandum, 723*.

He, Y., Wetterhall, F., Bao, H., Cloke, H., Li, Z., Pappenberger, F., et al. (2010). Ensemble forecasting using TIGGE for the July-September 2008 floods in the Upper Huai catchment: A case study. *Atmospheric Science Letters, 11*, 132–138. http://dx.doi.org/10.1002/asl.270.

Hiederer, R., & de Roo, A. (2003). *A European flow network and catchment data set*. Joint Research Centre, Commission of the European Communities. Ispra (VA), Italy.

Kalas, M., Ramos, M.-H., Thielen, J., & Babiakova, G. (2008). Evaluation of the medium-range European flood forecasts for the March-April 2006 flood in the Morava River. *Journal of Hydrology and Hydromechanics, 56*, 116–132.

King, D., Daroussin, J., & Tavernier, R. (1994). Development of a soil geographic database from the Soil Map of the European Communities. *Catena, 21*, 37–56. http://dx.doi.org/10.1016/0341-8162(94)90030-2.

Knijff, J. M. V. D., Younis, J., & Roo, A. P. J. D. (2010). LISFLOOD: A GIS-based distributed model for river basin scale water balance and flood simulation. *International Journal of Geographical Information Science, 24*, 189–212. http://dx.doi.org/10.1080/13658810802549154.

Lavers, D. A., Pappenberger, F., & Zsoter, E. (2014). Extending medium-range predictability of extreme hydrological events in Europe. *Nature Communications, 5*, 5382. http://dx.doi.org/10.1038/ncomms6382.

Montani, A., Cesari, D., Marsigli, C., & Paccagnella, T. (2011). Seven years of activity in the field of mesoscale ensemble forecasting by the COSMO-LEPS system: Main achievements and open challenges. *Tellus A*. http://dx.doi.org/10.3402/tellusa.v63i3.15816.

Neal, J., Schumann, G., Bates, P., Buytaert, W., Matgen, P., & Pappenberger, F. (2009). A data assimilation approach to discharge estimation from space. *Hydrological Processes, 23*, 3641–3649. http://dx.doi.org/10.1002/hyp.7518.

Ntegeka, V., Salamon, P., Gomes, G., Sint, H., Lorini, V., Zambrano-Bigiarini, M., et al. (2013). EFAS-Meteo: A European daily high-resolution gridded meteorological data set for 1990–2011. European Union, Scientific and technical research series (pp. 1831–9424) (26408).

Pagano, T. C., Wood, A. W., Ramos, M.-H., Cloke, H. L., Pappenberger, F., Clark, M. P., et al. (2014). Challenges of operational river forecasting. *Journal of Hydrometeorology, 15*, 1692–1707. http://dx.doi.org/10.1175/jhm-d-13-0188.1.

Pappenberger, F., Bartholmes, J., Thielen, J., Cloke, H. L., Buizza, R., & de Roo, A. (2008). New dimensions in early flood warning across the globe using grand-ensemble weather predictions. *Geophysical Research Letters, 35*. http://dx.doi.org/10.1029/2008gl033837.

Pappenberger, F., Beven, K. J., Hunter, N. M., Bates, P. D., Gouweleeuw, B. T., Thielen, J., et al. (2005). Cascading model uncertainty from medium range weather forecasts (10 days) through a rainfall-runoff model to flood inundation predictions within the European

Flood Forecasting System (EFFS). *Hydrology and Earth System Sciences, 9*, 381–393. http://dx.doi.org/10.5194/hess-9-381-2005.

Pappenberger, F., Bogner, K., Wetterhall, F., He, Y., Cloke, H. L., & Thielen, J. (2011). Forecast convergence score: A forecaster's approach to analysing hydro-meteorological forecast systems. *Advances in Geosciences, 29*, 27–32. http://dx.doi.org/10.5194/adgeo-29-27-2011.

Pappenberger, F., & Brown, J. D. (2012). HP today: On the pursuit of (im)perfection in flood forecasting. *Hydrological Processes, 27*, 162–163. http://dx.doi.org/10.1002/hyp.9465.

Pappenberger, F., Cloke, H. L., Parker, D. J., Wetterhall, F., Richardson, D. S., & Thielen, J. (2015). The monetary benefit of early flood warnings in Europe. *Environmental Science and Policy, 51*, 278–291. http://dx.doi.org/10.1016/j.envsci.2015.04.016.

Pappenberger, F., Cloke, H. L., Persson, A., & Demeritt, D. (2011). *HESS Opinions* "On forecast (in)consistency in a hydro-meteorological chain: Curse or blessing?". *Hydrology and Earth System Sciences, 15*, 2391–2400. http://dx.doi.org/10.5194/hess-15-2391-2011.

Pappenberger, F., Dutra, E., Wetterhall, F., & Cloke, H. L. (2012). Deriving global flood hazard maps of fluvial floods through a physical model cascade. *Hydrology and Earth System Sciences, 16*, 4143–4156. http://dx.doi.org/10.5194/hess-16-4143-2012.

Pappenberger, F., Ramos, M., Cloke, H., Wetterhall, F., Alfieri, L., Bogner, K., et al. (2015). How do I know if my forecasts are better? Using benchmarks in hydrological ensemble prediction. *Journal of Hydrology, 522*, 697–713. http://dx.doi.org/10.1016/j.jhydrol.2015.01.024.

Pappenberger, F., Scipal, K., & Buizza, R. (2008). Hydrological aspects of meteorological verification. *Atmospheric Science Letters, 9*, 43–52. http://dx.doi.org/10.1002/asl.171.

Pappenberger, F., Stephens, E., Thielen, J., Salamon, P., Demeritt, D., van Andel, S. J., et al. (2012). Visualizing probabilistic flood forecast information: Expert preferences and perceptions of best practice in uncertainty communication. *Hydrological Processes, 27*, 132–146. http://dx.doi.org/10.1002/hyp.9253.

Pappenberger, F., Thielen, J., & Medico, M. D. (2011). The impact of weather forecast improvements on large scale hydrology: Analysing a decade of forecasts of the European Flood Alert System. *Hydrological Processes, 25*, 1091–1113. http://dx.doi.org/10.1002/hyp.7772.

Pappenberger, F., Wetterhall, F., Albergel, C., Alfieri, L., Balsamo, G., Brogner, K., et al. (2013a). Floods in Central Europe in June 2013. *ECMWF Newsletter, 136*, 9–11.

Pappenberger, F., Wetterhall, F., Dutra, E., Giuseppe, F. D., Bogner, K., Alfieri, L., et al. (2013b). Seamless forecasting of extreme events on a global scale, climate and land surface changes in hydrology. In: *Proceedings of H01, IAHS-IAPSO-IASPEI Assembly, Gothenburg, Sweden, July 2013* (IAHS Publ. 359, 2013).

Raftery, A. E., Gneiting, T., Balabdaoui, F., & Polakowski, M. (2005). Using Bayesian model averaging to calibrate forecast ensembles. *Monthly Weather Review, 133*, 1155–1174. http://dx.doi.org/10.1175/mwr2906.1.

Ramos, M.-H., Bartholmes, J., & del Pozo, J. T. (2007). Development of decision support products based on ensemble forecasts in the European Flood Alert System. *Atmospheric Science Letters, 8*, 113–119. http://dx.doi.org/10.1002/asl.161.

Ramos, M.-H., Thielen, J., & de Roo, A. P. (2009). Prévision hydrologique d'ensemble et alerte avec le système européen d'alerte aux crues (EFAS): cas des crues du bassin du Danube en août 2005. In J.-M. Tanguy (Ed.), *Applications des modèles numériques en ingénierie 1*. Traité D'Hydraulique Environnementale 7, 69–86. Hermes - Lavoisier, Paris.

Ramos, M. H., van Andel, S. J., & Pappenberger, F. (2013). Do probabilistic forecasts lead to better decisions? *Hydrology and Earth System Sciences, 17*, 2219–2232. http://dx.doi.org/10.5194/hess-17-2219-2013.

Raynaud, D., Thielen, J., Salamon, P., Burek, P., Anquetin, S., & Alfieri, L. (2014). A dynamic runoff co-efficient to improve flash flood early warning in Europe: Evaluation on the 2013 central European floods in Germany. *Meteorological Applications*, n/a–n/a. http://dx.doi.org/10.1002/met.1469.

Renner, M., Werner, M., Rademacher, S., & Sprokkereef, E. (2009). Verification of ensemble flow forecasts for the River Rhine. *Journal of Hydrology, 376*, 463–475. http://dx.doi.org/10.1016/j.jhydrol.2009.07.059.

Romang, H., Zappa, M., Hilker, N., Gerber, M., Dufour, F., Frede, V., et al. (2010). IFKIS-Hydro: An early warning and information system for floods and debris flows. *Natural Hazards, 56*, 509–527. http://dx.doi.org/10.1007/s11069-010-9507-8.

Schulz, J.-P. (2005). Introducing the Lokal-Modell LME at the German Weather Service. *COSMO Newsletter, 5*, 158–159.

Thielen, J., Bartholmes, J., Ramos, M.-H., & de Roo, A. (2009). The European Flood Alert System — Part 1: Concept and development. *Hydrology and Earth System Sciences, 13*, 125–140. http://dx.doi.org/10.5194/hess-13-125-2009.

Thielen, J., Bogner, K., Pappenberger, F., Kalas, M., del Medico, M., & de Roo, A. (2009). Monthly-, medium-, and short-range flood warning: Testing the limits of predictability. *Meteorological Applications, 16*, 77–90. http://dx.doi.org/10.1002/met.140.

Thiemig, V., Bisselink, B., Pappenberger, F., & Thielen, J. (2014). A pan-African flood forecasting system. *Hydrology and Earth System Sciences Discussions, 11*, 5559–5597. http://dx.doi.org/10.5194/hessd-11-5559-2014.

Thiemig, V., Pappenberger, F., Thielen, J., Gadain, H., de Roo, A., Bodis, K., et al. (2010). Ensemble flood forecasting in Africa: A feasibility study in the Juba-Shabelle river basin. *Atmospheric Science Letters, 11*, 123–131. http://dx.doi.org/10.1002/asl.266.

Werner, M., Schellekens, J., Gijsbers, P., van Dijk, M., van den Akker, O., & Heynert, K. (2013). The Delft-FEWS flow forecasting system. *Environmental Modelling and Software, 40*, 65–77. http://dx.doi.org/10.1016/j.envsoft.2012.07.010.

Wetterhall, F., He, Y., Cloke, H., & Pappenberger, F. (2011). Effects of temporal resolution of input precipitation on the performance of hydrological forecasting. *Advances in Geosciences, 29*, 21–25. http://dx.doi.org/10.5194/adgeo-29-21-2011.

Wetterhall, F., Pappenberger, F., Alfieri, L., Cloke, H. L., del Pozo, J. T., Balabanova, S., et al. (2013). HESS Opinions "Forecaster priorities for improving probabilistic flood forecasts". *Hydrology and Earth System Sciences, 17*, 4389–4399. http://dx.doi.org/10.5194/hess-17-4389-2013.

Wood, E. F., Roundy, J. K., Troy, T. J., van Beek, L. P. H., Bierkens, M. F. P., Blyth, E., et al. (2011). Hyperresolution global land surface modeling: Meeting a grand challenge for monitoring Earth's terrestrial water. *Water Resources Research, 47*. http://dx.doi.org/10.1029/2010wr010090.

Ye, J., He, Y., Pappenberger, F., Cloke, H. L., Manful, D. Y., & Li, Z. (2013). Evaluation of ECMWF medium-range ensemble forecasts of precipitation for river basins. *The Quarterly Journal of the Royal Meteorological Society, 140*, 1615–1628. http://dx.doi.org/10.1002/qj.2243.

Younis, J., Anquetin, S., & Thielen, J. (2008). The benefit of high-resolution operational weather forecasts for flash flood warning. *Hydrology and Earth System Sciences, 12*, 1039–1051. http://dx.doi.org/10.5194/hess-12-1039-2008.

Younis, J., Ramos, M.-H., & Thielen, J. (2008). EFAS forecasts for the March-April 2006 flood in the Czech part of the Elbe River Basin — A case study. *Atmospheric Science Letters, 9*, 88–94. http://dx.doi.org/10.1002/asl.179.

Zajac, Z., Zambrano-Bigiarini, M., Salamon, P., Burek, P., Gentile, A., & Bianchi, A. (2013). Calibration of the lisflood hydrological model for europe - calibration round 2013. Technical report, Joint Research Centre, European Commission.

Zambrano-Bigiarini, M., & Rojas, R. (2013). A model-independent Particle Swarm Optimisation software for model calibration. *Environmental Modelling and Software, 43*, 5–25. http://dx.doi.org/10.1016/j.envsoft.2013.01.004.

CHAPTER 12

Developing Flood Forecasting Capabilities in Colombia (South America)

M. Werner*,†, J.C. Loaiza‡, M.C. Rosero Mesa‡, M. Faneca Sànchez†, O. de Keizer†, M.C. Sandoval§
*UNESCO-IHE, Delft, Netherlands
†Deltares, Delft, the Netherlands
‡IDEAM, Bogotá, Colombia
§CVC, Cali, Colombia

1 INTRODUCTION

Climate variability in Colombia is strongly influenced by the El Niño Southern Oscillation (Poveda, Álvarez, & Rueda, 2011), with the La Niña phase typically being associated to a stronger influx of moisture from the Choco current, resulting in an increase to the number of extreme precipitation events, saturation of soils (Grimm & Tedeschi, 2009), and consequently an increased incidence of floods. The second half of 2010 marked the onset of prolonged La Niña conditions, which prevailed through to the first quarter of 2012, despite a brief reestablishing of neutral conditions in the second part of 2011 (Fei, Lisha, & Jiang, 2015). These events led to widespread floods in Colombia as a result of long duration and intense precipitation during the 2010–11 boreal winter wet season, as well as during the latter part of 2011 and early 2012. The consequences of these floods were widespread. Over 2.7 million people were directly affected, and more than 400 lives were lost (Rodríguez-Gaviria & Botero-Fernández, 2013). The floods led to the internal displacement of 1.5–2.2 million people (Shultz et al., 2014). Particularly affected were the peripheral areas to the large cities, such as Cali and the capital Bogotá, with communities in extensive areas of informal development found in the urban fringe having a high vulnerability to being affected by floods (Rogelis, Werner, Obregón, & Wright, 2015). Many of these cities are located in the Andean mountains, and hazardous events due to prolonged and intense rainfall include debris flows and mudflows, with the lack of structured land cover related to these informal settlements exacerbating the susceptibility of such events occurring (Rogelis & Werner, 2013).

Flood Forecasting
http://dx.doi.org/10.1016/B978-0-12-801884-2.00012-8

The flood events prompted the Colombian government at both national and regional level to develop a new approach to dealing with flood risk. In 2010, a new policy on river basin planning had been adopted (MAVDT, 2010), though this did not extensively cover the aspect of flood risk. Additionally, although Colombia has a relatively well-developed disaster response network at local and regional level, preparedness to the flood events was generally low, with lack of coordination and insufficient capacity inhibiting an effective response (Brickle & Thomas, 2014). An added complexity due to the extensive geographic scale of the flood events was that responsibilities at the regional level are vested in regional water management authorities, whose jurisdictional areas follow administrative, rather than river basin, boundaries. In the aftermath of these events, several recommendations have been made to improve the management of (future) flood events. These called for: a more extensive inclusion of flood management in river basin planning; the development of a more integrated approach to floods, particularly at the basin level; and the strengthening of information and early warning systems.

In response to these recommendations, the national hydrometeorological agency in Colombia, the Instituto de Hidrología, Meteorología y Estudios Ambientales (IDEAM), who at the national level has the mandate for monitoring and warning of climate-related events, has initiated efforts to develop forecasting and warning capabilities at the national level. This includes upgrading the real-time network of hydrometric and meteorological stations, installation of weather radars, and the development of operational flood forecasting capabilities. The latter has been piloted in three basins, that of the Bogotá river, the Upper Cauca river, and a section of the Magdalena river, with an operational system being established at the national headquarters of IDEAM. The goal of this pilot was not only to establish a first step for a national hydrological forecasting system that could gradually be expanded to provide hydrological forecasts across the country, but also to pilot the organizational setup of forecasting and warning in the country.

This chapter first provides some background to the hydrological structure of Colombia, as well as more detailed information on the river basins included in the initial phase of the forecasting system. Some background to forecasting in Colombia is provided, followed by a description of the structure of the pilot forecasting system, as well as some current developments that are already expanding the original scope of the pilot. Section 6 provides some reflections on the progress that has been made in establishing what is to become a national forecasting system, as well as on some of the technical and organizational challenges that are faced in a country that is considered to be among the upper-middle-income economies (World Bank, 2015).

2 PHYSIOGRAPHY AND DEMOGRAPHICS

Colombia, in northern South America lies just above the equator in the tropics, with its climate strongly influenced by the intertropical convergence zone, or monsoon belt. In large parts of the country this results in a bimodal system, with two wet and two dry seasons annually. The first wet season is in the months of Apr. to May and the second from Oct. to Dec. In other parts of the country the regime is unimodal, with only a single wet season. Rainfall during these wet seasons is typically convective of nature (Daza & Umba, 2005). An important aspect of Colombian climatology is the elevation. The Andean mountain chain runs from south to north through the country. At the southern border the chain divides into three branches, the Eastern *Cordillera Oriental*; the Central *Cordillera Central*, and the Western *Cordillera Occidental*. Elevations in the river valleys between these chains range between 200 and 1000 m, and may contain extensive flat areas, while the chains themselves may reach elevations up to 5500 m, including a number of volcanic peaks with permanent snow caps and glaciers. These flat areas are often extensively developed, including urban and extensive agricultural areas. Above the (tropical) forest line at elevations of some 3500 m, areas covered with Páramo can be found. This is a high montane wetland ecosystem typical to the Andes, and these reach up to the snowline at elevations of some 5000 m. The nature of precipitation in these high elevation areas is quite different, with lower interseasonal variation than at lower elevations (Buytaert, Celleri, Willems, De Bièvre, & Wyseure, 2006). The southeastern part of the country is largely flat and sparsely inhabited, forming the Upper Orinoco and Northern Amazonian basins.

The country can be divided into five large hydrological regions: the Andean, Orinocian, Amazonian, Atlantic, and Pacific. The characteristics of these regions in terms of flooding vary significantly. The Andean regions are mountainous with steep slopes and rapid responses, leading to flash floods and depending on geomorphology and land cover also to mudflows and debris flows (Rogelis & Werner, 2013). In other regions, larger river systems can be found, meandering across alluvial plains. Floods in these rivers are typically slower to develop but generally of larger extent and longer duration. The most important basin in Colombia is the Magdalena-Cauca basin (see Fig. 1). The largest of the two, the Magdalena river, flows through the valley between the Cordillera Oriental and Central, while the Cauca River drains the valley between the Cordillera Central and Occidental. They join in a large inner delta in the Momposina depression, before flowing into the Atlantic Ocean at the city of Barranquilla. Part of this lower basin is referred

Fig. 1 Map of Colombia showing the basins initially considered in the forecasting system.

to as La Mojana, which is a poorly developed area that was inundated much more extensively during the 2010–11 events than is normal for the inner delta. The Magdalena-Cauca basin is the most important in the country primarily due to some 70% of the Colombian population living in the basin (Alfonso, He, Lobbrecht, & Price, 2012), as well as it being the focus of economic activities. All the largest cities, including Bogotá (the capital), Medellin, Cali, Barranquilla, and several others, are within the basin.

3 DEVELOPMENT OF A PILOT FOR THE NATIONAL HYDROLOGICAL FORECASTING SYSTEM

Prior to the current initiative to develop hydrological forecasting, very little hydrological forecasting existed in Colombia. At the national level, IDEAM has the national mandate for forecasting and warning, and operated a series of simple regressions between major gaging stations on the larger rivers. Additionally, some smaller scale systems were in operation; a community-based system had been setup for the Río Frío river, a right bank tributary to the Bogotá river (see Fig. 1). A small integrated forecasting system combining a HEC-HMS hydrological event model and a HEC-RAS hydrodynamic model in the Tunjuelo basin was in operation (Rogelis, 2006). The Tunjuelo river again forms a tributary (left bank) to the Bogotá river, with the lower reaches of the Tunjuelo river passing through the Bogotá urban area. Perhaps one of the earliest operational systems developed in Colombia is that for the Upper Cauca basin, though this was developed not as a flood forecasting scheme, but rather for the prediction of monthly to seasonal inflows to the Salvajina reservoir in the upper reaches of the basin, using a conceptual hydrological model (the Swedish HBV Model). An advanced operational system has recently been established for the metropolitan area of Medellin, integrating a dense network of meteorological and river gages, radar rainfall, and hydrological models. A complete overview of these small-scale operational systems in the country can be found in Domínguez-Calle and Lozano-Báez (2014). Although several of these have served their purpose well, providing level and flow forecasts for the locations they were established for, the coverage from the national perspective is limited. Additionally, a variety of different approaches has been taken, with some such as that for the Río Frío being essentially based only on observed levels, while that in Medellin is technically much more advanced, making use of radar rainfall and models. For providing forecasts at the national level, the intention of IDEAM was to develop a consistent approach to providing forecasts. This approach should be able to incorporate, if required and if desired, the models and data used in the existing forecasting schemes, as well as allowing the system to grow gradually, for example, as models are developed for basins where there are no current capabilities. To develop this national forecasting system, a pilot for the national forecasting system was initiated.

For the development of this pilot, a set of criteria was established for basins that could be incorporated in a first stage. These criteria included that these basins had been affected during the recent flood events, are of socio-economic relevance to the country, have a well-developed monitoring network with data available in real time, and, preferably have existing models available that could be readily incorporated into an operational forecasting system. The underlying reason for the latter two criteria were that the approach taken was to rapidly develop the pilot of the national forecasting system in the country and make this operational, without waiting for models and data to become available across the country. The operational system could then gradually be expanded as observational networks and new modeling capabilities become available. This resulted in a set of three basins being identified as pilot catchments: the basin of the Upper Cauca, the Bogotá river basin, and the middle reach of the Magdalena river (see Fig. 1).

3.1 Magdalena River

The Magdalena river has a length of some 1540 km from its source to the mouth, and a drainage area of 257,000 km^2, which includes its main tributary, the Cauca river. On average, the river discharges some 7100 m^3/s into the Caribbean Sea. The reach that has been included initially in the forecasting system is a short 250 km reach between the stations at Puerto Salgar in the Department of Cundinamarca and Barrancabermeja in the department of Santander. This reach was selected as there is an existing Mike11 1D hydraulic model available for the main stem of the river, with a HEC-HMS hydrological model available for the primary tributaries on the reach.

3.2 Upper Cauca Basin

The Cauca River is the main tributary of the Magdalena river, with a length of some 1350 km and drainage area of 63,300 km^2. At the mouth, the river discharges on average 2200 m^3/s into the Magdalena (Sandoval & Ramírez, 2007). The area considered initially, however, includes only the Upper Cauca basin between the Salvajina dam (including the basin and tributaries flowing into the dam) and the station at La Victoria, some 450 km downstream. The city of Cali, the main city in the area, is on the left bank floodplain about halfway down the reach. The city and surrounding industrial and agricultural areas were severely affected by the 2010–11 floods, as well as in the 2012 La Niña event. This reach falls within the jurisdiction of the regional water authority Corporación Autónoma Regional del Valle del Cauca (CVC), who manage a well-developed network of hydrometeorological stations, and have also developed several models, including a hydraulic model (Mike11) of the reach, and a conceptual hydrological (HBV) model of the main river and its tributaries.

3.3 Bogotá River

The third basin considered is the Bogotá river basin, which is the smallest of the three. The Bogotá forms a right bank tributary to the main Magdalena river, and so is actually already included in the reach of the Magdalena considered. However, given that the river passes the capital city of Bogotá, which with a population of some 8 million is the largest and most important city of Colombia, it is considered separately, and at a more detailed level. The part of the basin considered in detail is again only that of the upper basin, upstream of the city of Bogotá. The city of Bogotá lies on a high-altitude plain with an elevation of some 2600 m, with the Bogotá River arising in the *Cordillera Oriental*, flowing across the high-altitude plain, before tumbling over 2000 m down to the Magdalena river in its final reach. The basin falls entirely within the jurisdiction of the regional water authority Corporación Autónoma Regional de Cundinamarca (CAR), who equally operate a real-time network of hydrological and meteorological stations in the basin. The downstream end of the basin was selected to be at Puente La Virgen, a station close to the satellite town of Chía, which was severely affected during the 2010–11 floods (Melgarejo & Lakes, 2014).

4 DESIGN AND SETUP OF THE NATIONAL HYDROLOGICAL FORECASTING SYSTEM

Within Colombia, IDEAM has the mandate for the provision of meteorological and hydrological forecasts at the national level. Despite this, the regional water authorities (34 in total) also have responsibility within their own region, and would therefore need to work closely with IDEAM at the national level. To allow this, the Delft-FEWS operational forecasting platform (Werner et al., 2013) was selected as the core of the operational system. This system has been widely used both in single river basins such as the Rhine basin (Werner, Reggiani, De Roo, Bates, & Sprokkereef, 2005) as in national flood forecasting systems across the United Kingdom (Werner, Cranston, Harrison, Whitfield, & Schellekens, 2009), and the United States (Demargne et al., 2014), as well as in several other countries. This platform was designed as an open platform to support integration of data from different sources in real time, process these data, and run models in real time to provide operational hydrological products such as flood and seasonal forecasts.

4.1 Forecasting System Infrastructure

The setup of the system in Colombia using Delft-FEWS follows the approach commonly taken in many applications of the system. The system is composed of central servers, including a master controller server that orchestrates all operational processes, a database server, and computational units (forecasting shell servers) for executing model runs. In this case, IDEAM hosts the master controller server and the data server, while CVC and CAR have clients through which they can connect to these central servers from any location through the Internet to view and analyze forecasts, or to run additional forecast processes. This allows forecasting staff at the regional authorities collaborating in this effort to log in remotely and operate the system with the same functionality available to them as to the staff in the offices of IDEAM. The layout of the system is depicted in Fig. 2. Staff at the offices of CVC and CAR are on separate institutional networks, but the firewalls have been configured to allow the Java Messaging Service, which is the mechanism used to synchronize data between the local clients and the central database, to pass. Additionally, data from the telemetry systems at CVC and CAR is first transferred through a satellite link from the measurement stations to their respective offices and imported into their respective institutional databases. These databases are then polled at regular intervals and subsequently transferred through FTP to the servers at IDEAM for import into the Delft-FEWS database.

4.2 Monitoring Data

One of the main inputs in the forecasting system is formed by real-time and near-real-time data from hydrological and meteorological stations. IDEAM, as the national hydrometeorological institute, operates an extensive network of stations across Colombia. These include real-time stations that relay observations via the GOES satellite at regular intervals, as well as conventional stations where data are transmitted by observers through telephone and/or specially made web applications. This is made available on a daily basis, and

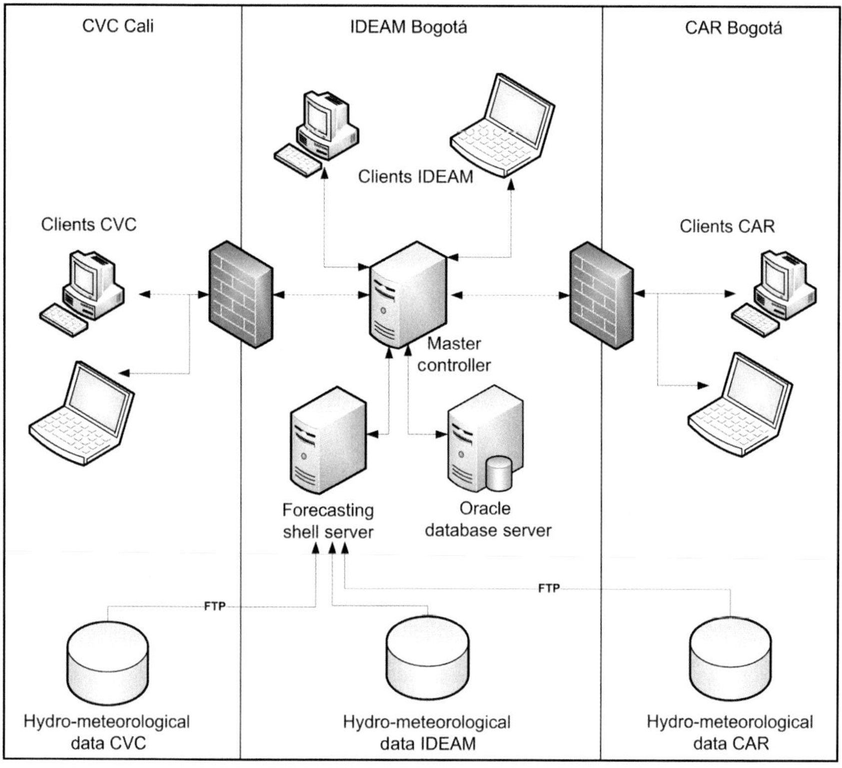

Fig. 2 Layout of the FEWS Colombia system.

is considered to be near real time. The two other regional water management authorities, CVC and CAR, operate similar networks in the Upper Cauca and Upper Bogotá basins, respectively. The distribution of stations across the country (see Fig. 3) is, however, uneven, with most stations in the more densely populated Andean region. The data networks of the agencies involved are to some extent overlapping, with a high density of stations in the socio-economically important basins of the Upper Cauca and Upper Bogotá. Unfortunately, of the 34 regional water authorities in Colombia, CVC and CAR have the most extensive monitoring networks. While there are a few other regional agencies that operate their own networks, many rely solely on the (in some basins sparse) national network operated by IDEAM. This may well be a constraint in expanding the forecasting system to cover other basins. It is expected, however, that this will improve, as in a parallel effort to the development of national forecasting capabilities, the coverage of the national network is being extended and many stations upgraded to provide data in real time (Table 1).

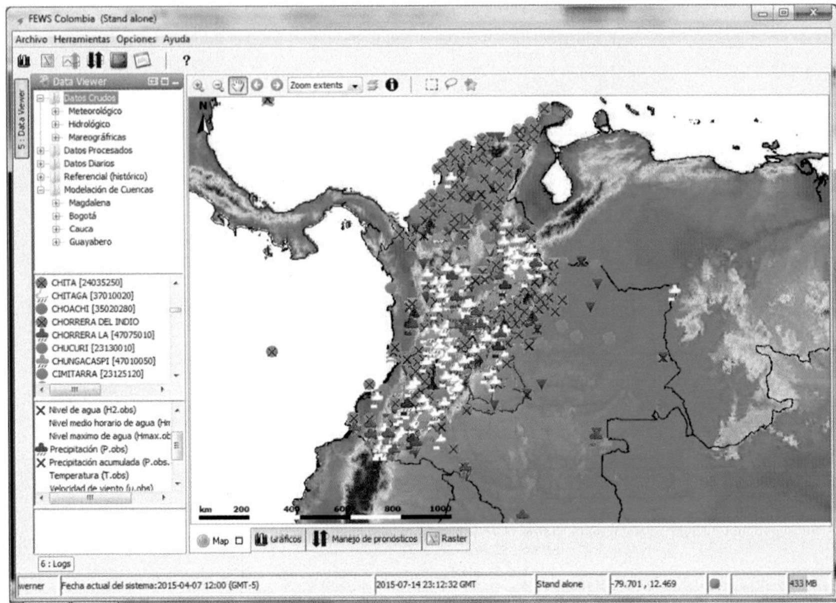

Fig. 3 Main map display of FEWS Colombia, showing the distribution of stations across the country. Triangles represent hydrological sites, while *green* dots represent meteorological stations. The icons indicate the status of data for these sites, with clouds indicating the intensity of precipitation. *Red* crosses at stations indicate that recent data at these is missing.

Table 1 Active hydrological and meteorological stations across the country that have been incorporated into the operational forecasting system

Agency	Extent (area in km^2)	Type of station: Satellite	Type of station: Conventional
IDEAM	National (1,141,748)	267	2454
CVC	Upper Cauca basin (16,254)	55	150
CAR	Upper Bogotá basin (2281)	61	327

4.3 Forecasting Processes and Models

Within the operational system, forecast processes have been defined for the three basins described. In these processes the raw import data is validated, transformed to equidistant time steps where required, interpolated spatially to calculate average rainfall over (sub) basins, and subsequently used as boundary conditions to a cascade of hydrological and hydraulic models. This reflects a common setup of such models in forecasting environment (Werner, Schellekens, & Kwadijk, 2005). The models used in each of the three basins included in the pilot were for the largest part existing models, although in most cases these had not been developed primarily for running as a part of an operational forecast process. Where required, amendments were then made to ensure these could run smoothly within the forecasting environment, and subsequently linked through the definition of a workflow process in Delft-FEWS. Fig. 4 shows, as an example, the model setup for the Magdalena river. In this case, the upstream boundary condition is defined as the outflow of the Betania reservoir, as observed at the Puente Santander station just downstream of the dam. A simple regression is then applied to translate this outflow to the station at Puerto Salgar, where it is then applied as the upstream boundary of the Mike11 hydrodynamic model, with the station at Barrancabermeja acting as the downstream boundary. Three major left bank tributaries have been developed using the HEC-HMS conceptual hydrological model. Other tributaries, where no existing models are

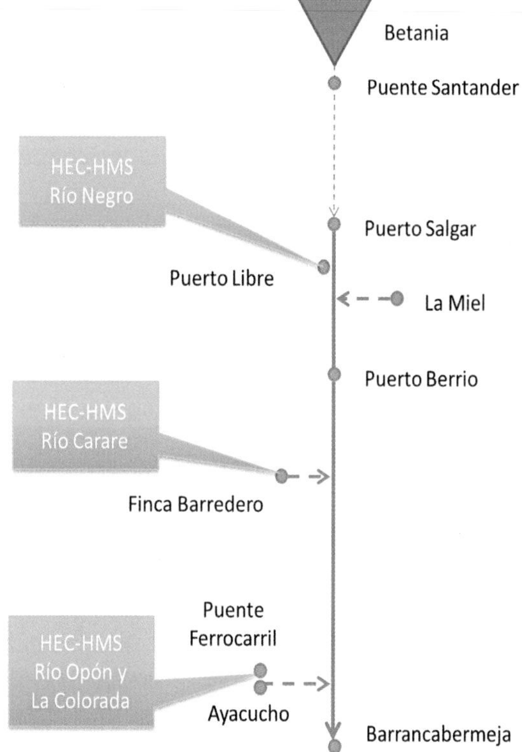

Fig. 4 Model schematization of the reach of the Magdalena river incorporated into FEWS. *Orange* circles represent hydrological stations that provide observed level and flow data in real time.

Table 2 Overview of hydrological and hydraulic models integrated into FEWS Colombia for providing hydrological forecast in the three pilot basins

	Type of model	
Basin	**Hydrologic**	**Hydrodynamic**
Bogotá	HEC-HMS (16 subbasins) Hourly time step	HEC-RAS Hourly time step
Cauca	HBV (62 subbasins) Daily time step	MIKE 11 (Salvajina — La Virginia) Daily time step
Magdalena	HEC-HMS (10 subbasins) Event model; daily time step	MIKE 11 (Puerto Salga — El Banco) Daily time step

available, are included using simple regression and extrapolation models. At all stations where a model result is available (such as at the outlets of the HEC-HMS model), an ARMA model is used as output correction to reduce forecast errors.

Table 2 provides an overview of the hydrological and hydraulic models that have been integrated to provide forecasts for the three pilot basins. These models are forced using observed data in the historic period. For the forecast period, a meteorological weather forecast provided by a weather research and forecasting (WRF) numerical weather model run by IDEAM is applied. This model is available at a resolution of 15 km for most of the country, while for the region of the capital, Bogotá, there are higher resolution models available at 5 and 1.67 km resolution (Arango & Ruiz, 2011). These provide precipitation and temperature forecasts with a lead time of 72 h. Results of these WRF models show that precipitation is, however, overestimated. This overestimation has also been found in other applications of WRF models simulating monsoonal rainfall (Yang, Zhang, Qian, Huang, & Yan, 2014), though this can be improved through choice of parameterization.

4.4 Forecasting Outputs and Products

The system provides several outputs that can be used in disseminating forecasts results to response agencies and the public. Within the system, displays are available showing graphs of water levels, discharges, and other relevant variables. An example of a typical forecast provided by the system at the station of Puente La Virgen is shown in Fig. 5. Other data can be viewed spatially, such as interpolated precipitation across the basins. These data are, however, internal to the system and available only to the forecasters working with it. To disseminate forecasts to a wider public, a web platform has

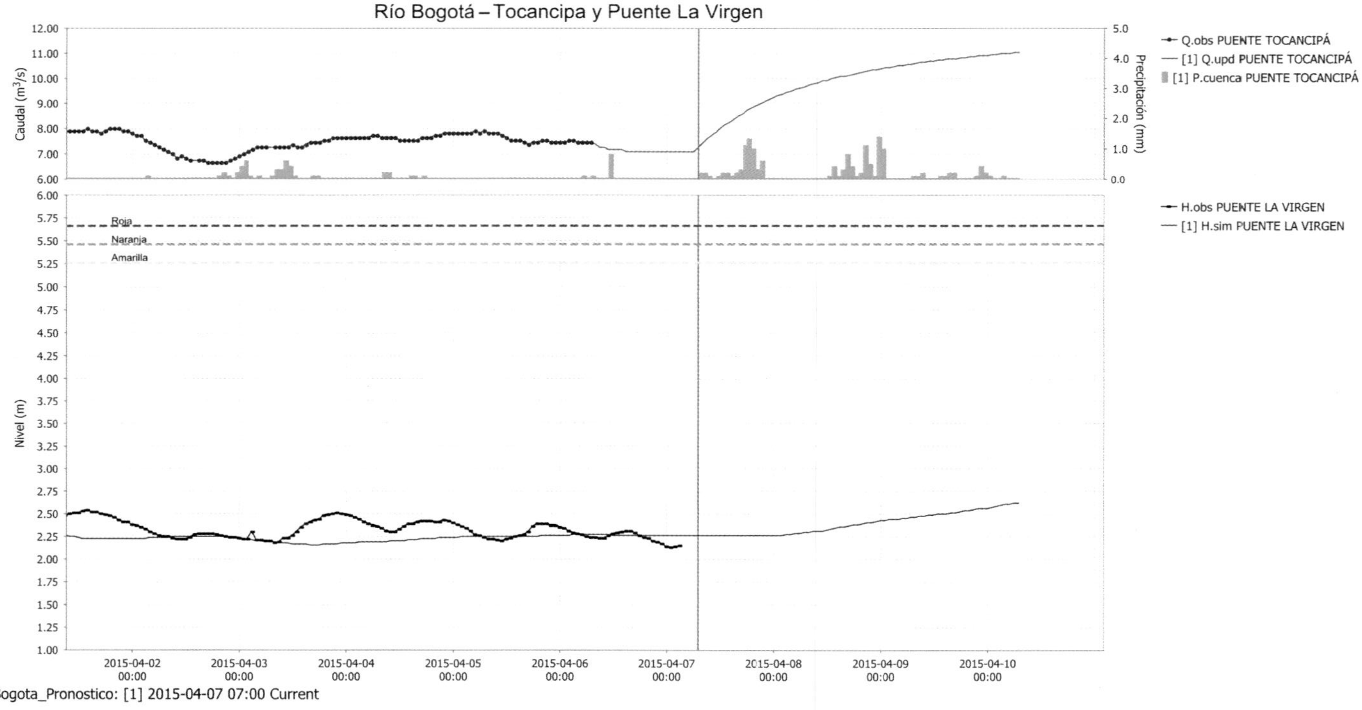

Fig. 5 Example of a short-term (3-day lead time) forecast for the Station at Puente la Virgen, just upstream of Bogotá. The upper graph shows catchment averaged rainfall, observed and simulated discharge at the station of Tocancipá, which is in the upper catchment, while the lower graph shows observed and forecast levels at Puente la Virgen.

been created that displays forecast results in the form of graphs and maps, exported at the end of each forecast run. This may be viewed by a wider audience within IDEAM, but is (as yet) not available to the general public. As the system matures, and forecast results become more reliable with the improvement of observational data coverage and better models, IDEAM has the intention to include graphs exported from the system in the hydrological forecast bulletin. This bulletin is issued to the general public on a daily basis by IDEAM, combining a narrative of the hydrological situation drafted by the duty forecaster, as well as graphs of observed (and in future forecasted) water levels and discharges. Currently these graphs are also developed on a daily basis by the duty forecaster, which is a laborious process. It is expected that with the introduction of FEWS Colombia, the effort required to produce the supporting figures will be minimized, thus allowing more time for interpretation.

5 EXPANSION OF THE PILOT SYSTEM

Since its initial installation early in 2014, the system is gradually being expanded to include other models, as well as broader applications such as water resources forecasting. Recently a reservoir optimization model was included for the operation of the Salvajina dam in the Upper Cauca basin. This uses a 6-month probabilistic forecast of inflows into the dam generated using extended streamflow forecasting (Day, 1985), that resamples historical precipitation and temperatures to force the HBV hydrological model of the Upper Cauca basin. An example of a 6-month forecast is shown in Fig. 6. This shows results for a forecast run made in 2014, using the precipitation and temperature data for the year of 1992, which was one of the most severe drought years in the region. The graph shows the inflow forecast to the reservoir (upper graph), the pool level in the reservoir (middle graph), and the releases from the dam (lower graph). These releases are optimized to maintain a flow target at the gage of Juanchito, some 200 km downstream. At this gage a minimum discharge is to be maintained, to guarantee the water supply intake to the city of Cali. The uncertainty of the water levels due to uncertainty in the reservoir inflow is also shown (middle graph). Inflows to the same reservoir are also predicted with a month lead time using a statistical multiple input linear regression model based on the southern oscillation index, as well as flow and precipitation of the previous month (Fig. 7).

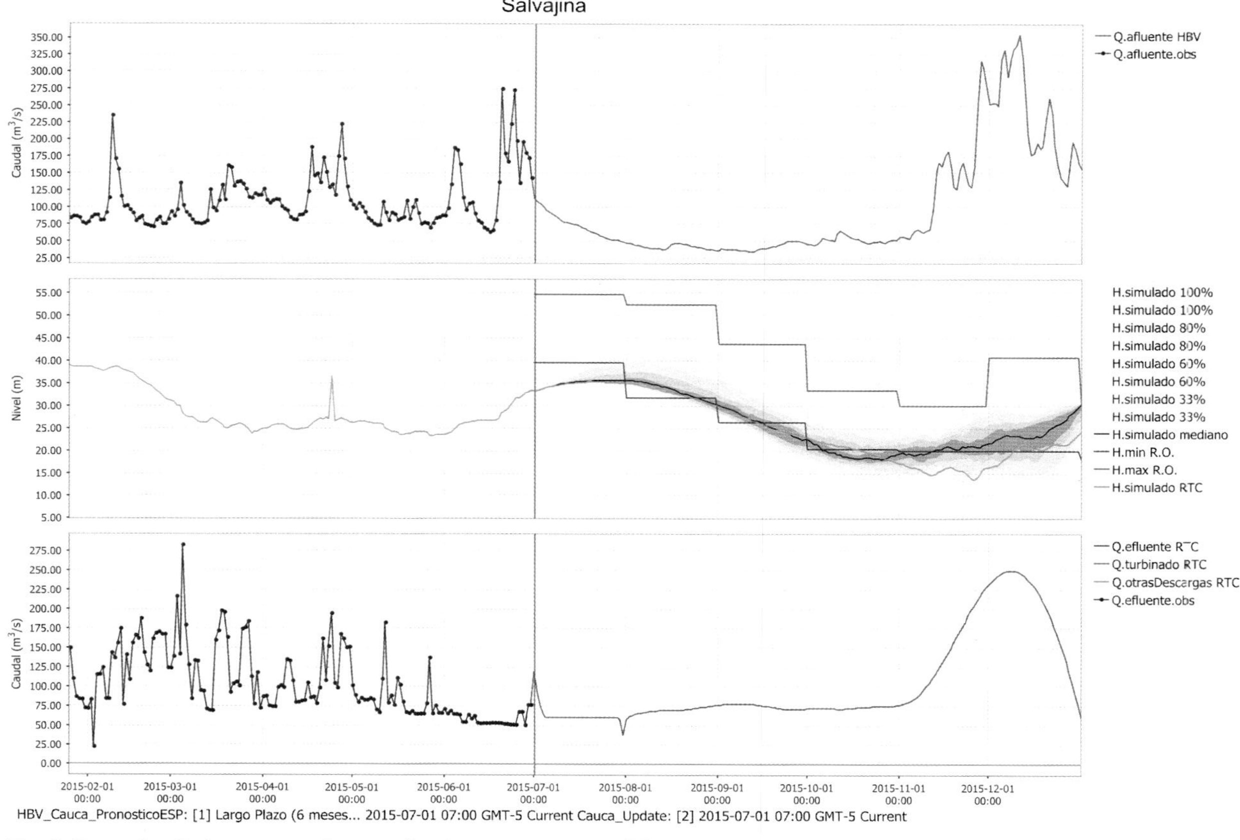

Fig. 6 Example of a long-range forecast for the optimization of the operation of Salvajina dam.

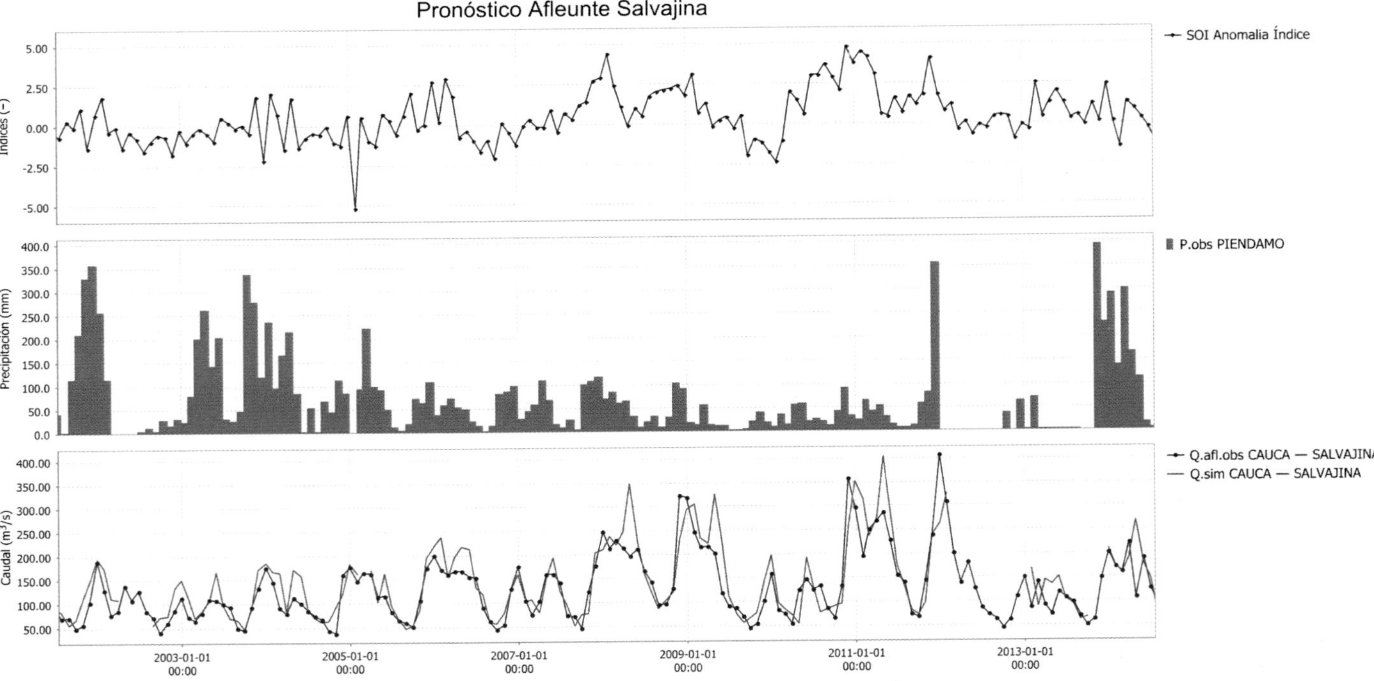

Fig. 7 Prediction of the inflow to Salvajina reservoir with a lead time of 1 month, based on a linear regression model. The upper two graphs show the southern oscillation index (SOI) anomaly index and the monthly precipitation at the station of Piendamo. The lowest graph shows the predicted inflow *(blue line)* compared to the observed inflow *(red line)*.

6 DISCUSSION AND OUTLOOK

The establishment of the pilot operational forecasting system operated by IDEAM is a first milestone in the development of operational hydrological forecasting capabilities at the national level in Colombia that can provide operational flood forecasts and, through providing timely information to response agencies and the public, contribute to a reduction of the impacts of floods to people and to the economy. Additionally, the development of the system has strengthened the capabilities of IDEAM as the national hydrometeorological institute in Colombia, complementing the other early warning systems for meteorological and forest fire events that were already operational at the institute.

The system now forms an integral part of the operational services of IDEAM, and although hydrological forecasting capabilities have initially been developed for three pilot basins only, the hydrological and meteorological data that has been integrated is from all stations in the country, as well as from the two regional authorities. This has been one of the main innovations of the system. While IDEAM was already operating an extensive network of real time and conventional stations, there was a lack of integrated systems that could easily combine these different sources and provide a comprehensive overview of the hydrological situation in the country. Such integration is often a challenge in a developing country setting, where data collection and modeling systems are often setup through independent projects using proprietary software, with data then often difficult to share between different systems. The flexibility of Delft-FEWS in bringing these sources together has at the outset proved to be of the most added-value to IDEAM as an operational tool.

The cooperation between different institutions in the implementation of the system is also seen as a major step. Both the national agency IDEAM, and the regional water management authorities, CVC and CAR, collaborated in the establishing of the system, with data being exchanged freely. This was one of the first examples of interinstitutional collaboration in the country between this type of national and regional institution, allowing data as well as knowledge to be exchanged. Still challenges remain in this context. Continued commitment of the different agencies to maintain both technical and human capacity are prerequisite for the sustainability of the system. The embedding of the system in the operational processes, such as it has been at IDEAM, will contribute to ensuring that sustainability, though continuing institutional commitment is required to further develop and maintain the system. Within the next years it is expected that the system will expand substantially. Currently it is used more as a monitoring than a forecasting tool, but as data processes are consolidated and better models are developed and incorporated, it is expected that this will gradually change. The introduction of seasonal forecasts to support operation of reservoirs has already widened the scope, while at IDEAM and CVC, new hydrological models for additional basins are being calibrated for inclusion into the system.

Organizationally there are equally several questions that remain to be answered. There are plans at IDEAM to establish regional forecasting centers in the main cities of Colombia, which, combined with the installation of seven new weather radar stations and expansion of the network of hydrological and meteorological stations, will considerably improve the available hydrological and meteorological data. However, how the institutional mandates are shared between the IDEAM as a national agency, the regional forecasting centers of IDEAM, and the existing regional water management agencies will need to be defined, to ensure that information and knowledge is shared between them to the best effect.

The current system, as a pilot, is a first step towards a fully operational forecasting and warning service at the national scale in Colombia. Currently the system is institutionally embedded within IDEAM, particularly to support monitoring based on real-time information. It is expected that as hydrological and hydraulic models come on board, and forecasting requirements, experience and institutional responsibilities become clearer, the system will start to fulfill its potential in contributing to the reduction of risk.

REFERENCES

Alfonso, L., He, L., Lobbrecht, A., & Price, R. (2012). Information theory applied to evaluate the discharge monitoring network of the Magdalena river. *Journal of Hydroinformatics, 15*(1), 1–18. http://dx.doi.org/10.2166/hydro.2012.066.

Arango, C., & Ruiz, J. F. (2011). *Implementación del modelo WRF para la sabana de Bogotá*. Bogotá, Colombia. Retrieved from, http://www.ideam.gov.co/documents/21021/21132/Modelo_WRF_Bogota.pdf/f1d34638-e9f8-4689-b5f4-31957c231c46.

Brickle, L., & Thomas, A. (2014). Rising waters, displaced lives. *Forced Migration Review, 45*, 34–35.

Buytaert, W., Celleri, R., Willems, P., De Bièvre, B., & Wyseure, G. (2006). Spatial and temporal rainfall variability in mountainous areas: A case study from the south Ecuadorian Andes. *Journal of Hydrology, 329*(3–4), 413–421. http://dx.doi.org/10.1016/j.jhydrol.2006.02.031.

Day, J. (1985). Extended streamflow forecasting using NWSRFS. *Journal of Water Resources Planning and Management, 111*, 157–170.

Daza, M. H., & Umba, H. A. S. (2005). *Atlas climatológico de Colombia*. Bogotá, D.C: IDEAM.

Demargne, J., Wu, L., Regonda, S. K., Brown, J. D., Lee, H., He, M., et al. (2014). The science of NOAA's operational hydrologic ensemble forecast service. *Bulletin of the American Meteorological Society, 95*(1), 79–98. http://dx.doi.org/10.1175/BAMS-D-12-00081.1.

Domínguez-calle, E., & Lozano-báez, S. (2014). Estado del arte de los sistemas de alerta temprana en Colombia. *Revista de la Academia Colombiana de Ciencias Exactas, Físicas y Naturales, 38*(148), 321–332.

Fei, Z., Lisha, F., & Jiang, Z. H. U. (2015). An Incursion of off-equatorial subsurface cold water and its role in triggering the "Double Dip" La Nina event of 2011. *Advances in Atmospheric Sciences, 32*, 731–742. http://dx.doi.org/10.1007/s00376-014-4080-9.1.

Grimm, A. M., & Tedeschi, R. G. (2009). ENSO and extreme rainfall events in South America. *Journal of Climate, 22*(7), 1589–1609. http://dx.doi.org/10.1175/2008JCLI2429.1.

MAVDT. (2010). *Politica Nacional para la gestión integral del recurso Hídrico — Documento final*. Bogotá, Colombia: Ministerio de Ambiente, Vivienda y Desarrollo Territorial (124 pp.).

Melgarejo, L. F., & Lakes, T. (2014). Urban adaptation planning and climate-related disasters: An integrated assessment of public infrastructure serving as temporary shelter during river floods in Colombia. *International Journal of Disaster Risk Reduction, 9*, 147–158. http://dx.doi.org/10.1016/j.ijdrr.2014.05.002.

Poveda, G., Álvarez, D. M., & Rueda, Ó.A. (2011). Hydro-climatic variability over the Andes of Colombia associated with ENSO: A review of climatic processes and their impact on one of the Earth's most important biodiversity hotspots. *Climate Dynamics, 36*(11–12), 2233–2249. http://dx.doi.org/10.1007/s00382-010-0931-y.

Rodríguez-Gaviria, E., & Botero-Fernández, V. (2013). Flood vulnerability assessment: A multiscale, multitemporal and multidisciplinary approach. *Journal of Earth Science and Engineering, 2*, 102–108.

Rogelis, M. C. (2006). *Sistema de alerta temprana del río Tunjuelo.* Bogotá, Colombia.

Rogelis, M. C., & Werner, M. (2013). Regional flood susceptibility analysis in mountainous areas through the use of morphometric and land cover indicators. *Natural Hazards and Earth System Sciences, 1*(6), 7549–7593. http://dx.doi.org/10.5194/nhessd-1-7549-2013 (discussions).

Rogelis, M. C., Werner, M., Obregón, N., & Wright, N. (2015). Regional prioritisation of flood risk in mountainous areas. *Natural Hazards and Earth System Sciences, 3*(7), 4265–4314. http://dx.doi.org/10.5194/nhessd-3-4265-2015 (discussions).

Sandoval, M. C., & Ramírez, C. (2007). El río Cauca en su valle alto. *Un aporte al conocimiento de uno de los ríos más importantes de Colombia*. Cali, Colombia: Corporación Autónoma Regional del Valle del Cauca, CVC-Programa Editorial Universidad del Valle.

Shultz, J. M., Milena, Á., Ceballos, G., Espinel, Z., Oliveros, S. R., Fonseca, M. F., et al. (2014). Internal displacement in Colombia Fifteen distinguishing features. *Disaster Health, 2*(1), 13–24. http://dx.doi.org/10.4161/dish.27885.

Werner, M., Cranston, M., Harrison, T., Whitfield, D., & Schellekens, J. (2009). Recent developments in operational flood forecasting in England, Wales and Scotland. *Meteorological Applications, 16*(1), 13–22.

Werner, M., Reggiani, P., De Roo, A., Bates, P., & Sprokkereef, E. (2005). Flood forecasting and warning at the river basin and at the European scale. *Natural Hazards, 36*(1–2), 25–42.

Werner, M., Schellekens, J., Gijsbers, P., van Dijk, M., van den Akker, O., & Heynert, K. (2013). The Delft-FEWS flow forecasting system. *Environmental Modelling & Software, 40*, 65–77. http://dx.doi.org/10.1016/j.envsoft.2012.07.010.

Werner, M., Schellekens, J., & Kwadijk, J. (2005). 23: Flood early warning systems for hydrological (sub) catchments. *Encyclopedia of hydrological sciences.* Hoboken, NJ: John Wiley and Sons.

World Bank. (2015). *World development indicators — Colombia*. Retrieved from, http://data.worldbank.org/country/colombia.

Yang, B., Zhang, Y., Qian, Y., Huang, A., & Yan, H. (2014). Calibration of a convective parameterization scheme in the WRF model and its impact on the simulation of East Asian summer monsoon precipitation. *Climate Dynamics, 44*, 1661–1684. http://dx.doi.org/10.1007/s00382-014-2118-4.

PART 3

Challenges Facing Flood Forecasting

CHAPTER 13

Streamflow Data

G.J. Wiche*, R.R. Holmes Jr.†
*U.S. Geological Survey, Bismarck, ND, United States
†U.S. Geological Survey, Rolla, MO, United States

1 INTRODUCTION

Streamflow is the volumetric discharge expressed in volume per unit time (typically cubic feet per second (ft^3/s) or cubic meters per second (m^3/s)) that takes place in a stream or channel and varies in time and space. Excess streamflow can create a flooding hazard, and although excess streamflow is a natural occurrence and healthy for the ecosystem, mitigation may be required when a stream is located near people or infrastructure. Lack of streamflow can be detrimental to humans (inadequate water supply) and ecosystems (inadequate environmental flows to sustain aquatic life). Knowledge of the quantity and temporal and spatial distribution of streamflow is crucial to the science of hydrology, emergency management, flood forecasting, and associated fields such as water-resources design, water-resource management, and environmental protection.

The U.S. Geological Survey (USGS) collects hydrologic information, which includes streamflow, used for flood forecasting, flood control reservoir operation, and flood fighting operations in the United States. The fact that the USGS does not design, build, or operate water projects and does not have any regulatory responsibility provides the basis for impartiality that is a hallmark of USGS hydrologic monitoring activities. The USGS has collected streamflow information for the nation's streams since 1889, and as of 2015, the USGS operated about 8100 streamgages nationwide (Fig. 1). In addition to flood forecasting, streamgages provide streamflow information for many other uses including water management and allocation, engineering design, operation of locks and dams, and recreational safety and enjoyment (Norris, 2009). These streamgages are operated by the USGS in partnerships with more than 800 federal, state, tribal, and local cooperating agencies. The following list includes the guiding principles of the streamgaging network:

Flood Forecasting
http://dx.doi.org/10.1016/B978-0-12-801884-2.00013-X
2016, Published by Elsevier Inc.

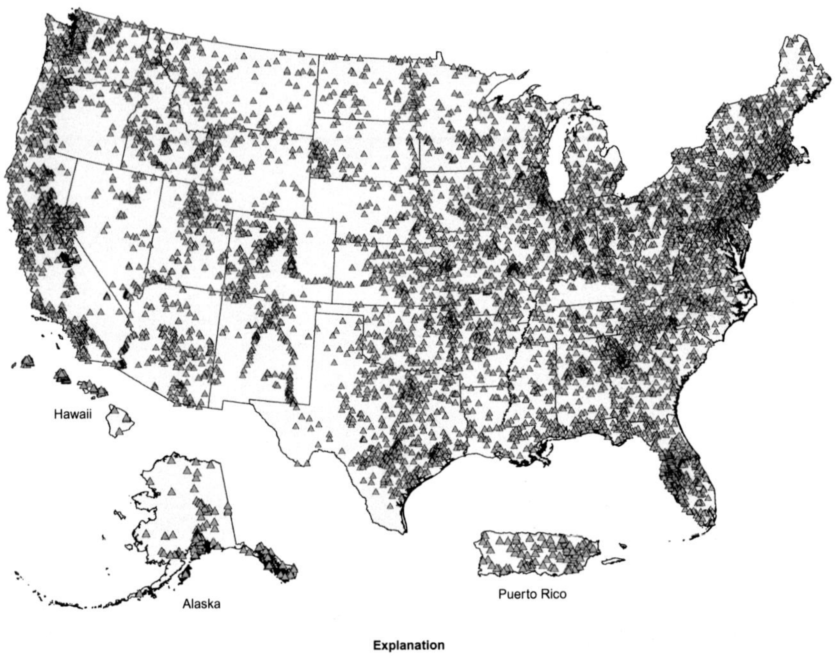

Fig. 1 Active U.S. Geological Survey streamgages in water year 2014 (a water year is the 12-month period Oct. 01 through Sep. 30, designated by the calendar year in which it ends).

- Streamgaging stations are funded by the USGS and partner agencies to meet the Federal mission goals of the USGS and the goals of the partner agencies.
- Data are free and available to the public.
- USGS operates the network on behalf of all partnering agencies using consistent methods; thus, the network achieves economies by eliminating multiple infrastructures for testing, providing training to staff, developing and maintaining the communications and database systems, and providing quality assurance.
- USGS has staff distributed throughout the United Sates, and staff can respond to catastrophic floods or regional droughts (U.S. Geological Survey, 1998).

The purpose of this chapter is to: (1) describe the history of streamgaging in the United States; (2) describe how the USGS obtains streamflow information; (3) describe how the streamflow information is delivered; and (4) describe the issues and opportunities facing streamflow-monitoring programs.

2 HISTORY OF STREAMGAGING IN THE UNITED STATES

Langbein and Hoyt (1959) provide a comprehensive description of the development of streamgaging in the United States and point out that the earliest hydrologic assessment programs began in the early 1800s; also, Langbein and Hoyt report that the streamgaging information provides for stream navigation and for water supply analyses in the northeastern United States. Most assessment programs were cursory and often inadequate for the intended purpose of water-resource planning (Matalas & Barnes, 1977). Also, in the 1840s the U.S. Army Corps of Engineers (USACE) completed hydraulic studies related to navigation on the Ohio and Mississippi Rivers. Although early water-resource assessments and studies could have benefited from streamgaging information, regular streamgaging did not begin until the 1860s in an effort to understand better the availability of water for large cities such as Boston and New York (Langbein & Hoyt, 1959).

In 1879, the United States Congress created the USGS and appointed Clarence King the first Director. The first duty of the USGS was to classify the more than 1.2 billion acres of public lands in the United States. After organizing the USGS, Clarence King resigned in 1881, and John Wesley Powell was selected as the second Director of the USGS (Rabbitt, 1989). During the 1870s and 1880s, large numbers of people settled in the western United States, and the need for dependable water supplies increased rapidly. In Oct. 1888, Congress authorized a survey to investigate the extent to which the arid region of the United States could be irrigated and to select sites for reservoirs and other hydraulic works for storage and utilization of water for irrigation and flood mitigation (Rabbitt, 1989). Powell now had the opportunity to determine the suitability of arid lands for settlement, and he advocated for the establishment of streamgages to provide the information needed to make water-resource decisions related to water storage and irrigation. Powell's vision was to create an organization capable of collecting reliable streamflow information. In an effort to meet his vision, 14 men were sent to a training camp in Embudo, New Mexico, in 1888 to train and develop streamflow-measuring techniques (Gunn, Matherne, & Mason, 2014). Fredrick H. Newell, a young Massachusetts Institute of Technology engineering graduate, was hired by Powell to lead the group, and Newell is credited with developing many of the first streamgaging methods and policies used by the USGS. In fact, Newell is considered "the Father of Systematic Streamgaging" (Follansbee, 1919). By the time Newell left the Chief Hydrographer's position in 1902 to head the Reclamation Service, about 158 streamgages existed in the United States.

In the early 1920s, Congress was keenly interested in hydroelectric development, and the Secretary of War directed the USACE to provide Congress an estimate of a river survey. Congress funded the USACE, under the Rivers and Harbors Act of 1927, to conduct the river surveys ("308" investigations), and streamflow information was needed to complete the "308" investigations. Thus, USACE funded the USGS to establish and maintain many new streamgaging stations, but the funding was discontinued after the "308" investigations were completed (Follansbee, 1939).

Although the impetus for streamgaging in the USGS was driven by the need to monitor the amount of water available for irrigation in the western United States, floods such as the great Mississippi River flood of 1927 and floods in New England in 1928 were followed by increased support for the streamgaging program. Severe flooding in several areas of the United States in 1935 and 1936 led to the passage of the Flood Control Act of 1936, which led to the investigation and construction of hundreds of flood control dams. At about the same time the Flood Control Act was passed, Congress increased the funding available to the USGS to match 50-50 with the States (Cooperative Water Program). The net result of the Flood Control Act and the Cooperative Water Program funding led to an almost tripling of streamgages from about 1800 streamgages in 1928 to 4160 streamgages in 1939 (Follansbee, 1928, 1939).

Streamgaging continued to increase during the war years (1939–47) from 4160 streamgages in 1939 to 5812 streamgages in 1947 (Follansbee, 1947). Much of the increase in streamgages was driven by the need for streamflow data for USACE flood control plans and for Bureau of Reclamation (Reclamation) irrigation studies in the western United States. Rapid industrial expansion during World War II also led to an increase in the number of streamgages funded through the Cooperative Water Program. Floods, droughts, sudden industrial expansion, water-development programs, and new water-resource commissions created the need for additional streamgages after World War II, and by 1957, about 6900 streamgages were in operation (Ferguson et al., 1990). The streamgages operated by the USGS continued to increase and peaked at about 8300 streamgages in 1968 (Fig. 2). The number of streamgages decreased slowly from 1968 until 1997, when 6790 streamgages were in operation.

In response to the decline in the number of streamgages, the U.S. Congress requested that the USGS evaluate the streamgage network and produce a report documenting the findings (U.S. Geological Survey, 1998). One of the conclusions in the report indicated that the need for streamgages

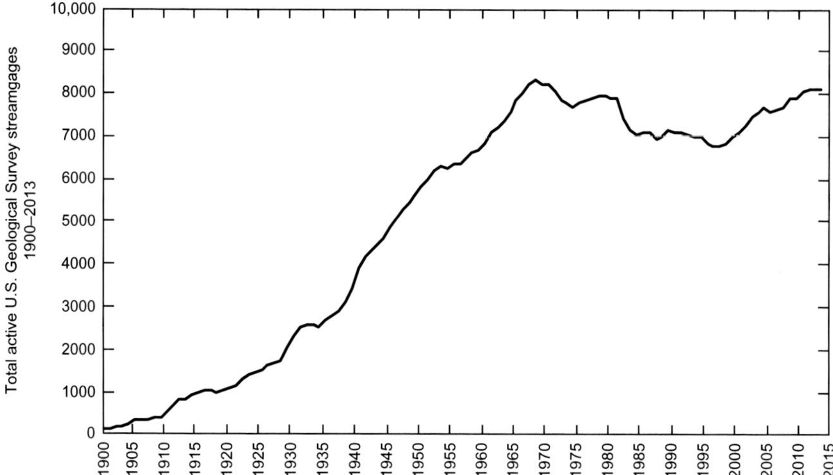

Fig. 2 The number of active U.S. Geological Survey streamgages from 1900 to 2015. Data from U.S. Geological Survey. (2015). USGS water data for the nation. U.S. Geological Survey National Water Information System. http://waterdata.usgs.gov Accessed 01.06.15.

used to prevent flood losses was increasing at the same time that funding from streamgages was decreasing (U.S. Geological Survey, 1998).

The National Streamflow Information Program (NSIP) was established in 2003 in response to Congressional and stakeholder concerns about the stability of the streamgage network (Bales et al., 2004). The concerns were caused by: (1) the decrease in the number of operating streamgages, including a disproportionate loss of streamgages with a long period of record; (2) the inability of the USGS to continue operating high-priority streamgages in an environment of reduced funding through partnerships; and (3) the increasing demand for streamflow information because of emerging resource-management issues and new data-delivery capabilities. The mission of the NSIP is to provide the streamflow information necessary for: federal, state, and local economic well-being; the protection of life and property; and efficient and effective water-resource management.

The NSIP's priority goal is to establish a stable "backbone" streamgage network to meet critical federal needs for streamflow information, and one of the needs is to operate streamgages to support flood and other streamflow forecasting by the National Weather Service (NWS) and other federal agencies across the United States. Additional NSIP information is documented at (http://water.usgs.gov/nsip/reports.html) and in Bales et al. (2004). A total of 4759 streamgage locations distributed across the country have been identified to meet federal needs, including 3631 of the locations being identified to support the federal need of flood and streamflow forecasting, most of which are NWS flood forecasting sites. More than 900 of the 4759 streamgages will meet multiple federal critical needs.

3 STREAMFLOW COLLECTION AND COMPUTATION

Knowledge of the amount of and temporal and spatial distribution of streamflow is crucial to the science of hydrology, emergency management, and associated fields such as water-resources design, water-resources management, and environmental protection. A direct determination of a continuous time series of streamflow is not economically feasible as of 2015. Streamflow is typically determined by using surrogate measures of variables such as water-surface elevation (stage), water-surface slope, rate of water-surface rise, or index velocity. Streamflow is computed by developing a relation between the surrogate variable and individual measurements of streamflow (discharge measurements). The relation is called a "rating" (Kennedy, 1984, chap. A10).

Streamflow data include stage and volumetric discharge (discharge) data that are collected at various intervals depending on the need. Often streamflow data are collected using permanent or semipermanent structures, which are termed streamgages, installed on or near the stream (Fig. 3); however,

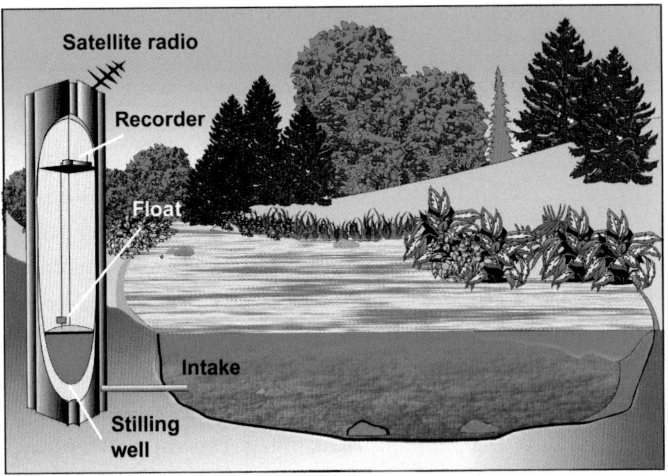

Fig. 3 Diagram of streamgage. Stream stage is the height of the water surface above a reference elevation or datum. Stage can be measured by a variety of methods, including a float, as shown in this illustration. Modified from Nielsen, J. P., & Norris, J. M. (2007). From the river to you: USGS real-time streamflow information. U.S. Geological Survey fact sheet 2007-3043 (4 pp.).

names such as gages, streamflow-gaging stations, streamflow-monitoring stations, and river gages also are used to refer to these installations. The streamgage contains a device to measure, store, and transmit the water level and other hydrologic variables in the river. The systematic collection of streamflow data at a streamgage can be classified into two categories: periodic and continuous. Periodic streamflow data are collected intermittently through time (sometimes with no set frequency). Continuous streamflow data are collected at a frequency sufficient to fully characterize the streamflow magnitude through time over the complete hydrograph. This chapter will report on continuous streamflow data. Typical frequency of continuous streamflow data collected at streamgages ranges from 1 to 5 min for small streams (drainage area typically less than 50 km^2) which are "flashy" to 60 min for large rivers (drainage area typically greater than 250,000 km^2), which rise and fall more gradually.

3.1 Measuring Stage

A fundamental part of any streamgage is the collection of water level or stage data, which is governed by guidelines and standards to ensure quality data (Sauer & Turnipseed, 2010, chap. A7; World Meteorological Organization, 2010). Much of the material in this chapter describing measuring stage is from Sauer and Turnipseed (2010). The stage of a stream is the height of the water-surface elevation above a particular local datum, often known as the gage datum. This gage datum is often established such that a zero value is below the channel bed to ensure nonnegative values at all stages. Gage datum is typically surveyed into known geodetic datum to relate stage to known elevations of adjoining physical features and enable comparison of stages between streamgages (Rydlund & Densmore, 2012, chap. D1).

A streamgage is a structure located beside a river that contains a device to measure, store, and transmit the stage data for the river. Stage, sometimes called gage height, is measured using a variety of methods. A stilling well placed in the riverbank (Fig. 3) or attached to the bridge pier are methods used to measure stage. The stilling well is connected to the river by a set of intake pipes and the water level in the well is the same as the level in the stream. Traditionally, stage has been measured inside the stilling well using a float. Also, stage can be measured using a submersible or nonsubmersible pressure transducer placed in the stream. Stage can be determined by measuring the pressure required to maintain a small flow of gas through a tube and bubbled out at a fixed location under the water in the stream (nonsubmersible pressure transducer) (Olson & Norris, 2007). The measured pressure is directly related to the height of water over the tube outlet in the stream, and as the depth of water increases, more pressure is required to push the gas bubbles through the tube. In the case of a submersible pressure transducer, an electronic cable is used to transmit the vertical position of the water (Sauer & Turnipseed, 2010, chap. A7). In recent years, the primary method of sensing the stage is to use either a nonsubmersible or submersible pressure transducer. Also in recent years, acoustic, radar, laser, and optical pulse sensors have grown in popularity because the instruments can be located on any fixed structure over the stream such as a bridge handrail (Fig. 4). The radar sensors have the advantage of not being in the water; thus, the sensors usually are safer to operate and do not get damaged as easily by ice and debris. However, under certain environmental conditions (such as wind and waves, floating debris, and ice cover) the sensors may not provide a valid stage.

Fig. 4 Rapid deployment streamgage with a radar-level sensor used to measure stage.

At each streamgage, a nonrecording gage is used as a reference gage to set the correct value and stage and to ensure the gage is reading correctly during periodic inspections. The types of nonrecording gages used are staff gage, wire weight gage, float tape gage, and electric tape gage (Sauer & Turnipseed, 2010, chap. A7).

Streamgages operated by the USGS provide stage measurements that are accurate to the nearest 0.01 ft or 0.2% of the effective stage, whichever is greater (Sauer & Turnipseed, 2010, chap. A7). The accuracy criteria apply to the total error for all the equipment necessary to sense, record, and retrieve the stage information. Sauer and Turnipseed (2010) fully describe individual sources of stage-measurement error. A permanent gage datum must be established and used as a reference elevation for the stage at the streamgage. The permanent gage datum should be established below the lowest expected stage, which will be at or below the elevation of zero flow on the control for all conditions. To maintain accuracy and to ensure that stage is being measured above a constant reference elevation, the elevations of streamgage structures and the associated stage measurement are routinely surveyed relative to the permanent elevation of benchmarks near the streamgage (Kenney, 2010).

3.2 Measuring Streamflow

Streamflow or discharge is the volume of water moving past a fixed location on a river in a unit of time, commonly expressed as cubic feet per second or cubic meters per second. A variety of methods following standard protocols (Benson & Dalrymple, 1967, chap. A1; Herschy, 2008; Rantz et al., 1982; Turnipseed & Sauer, 2010, chap. A8; World Meteorological Organization, 2010) are used to measure discharge, and the methods are grouped into two broad categories: direct and indirect methods. Direct measurements of discharge use instruments, sensors, and devices to "directly" measure the discharge on site when the streamflow of interest is taking place. Indirect methods are used primarily to measure the peak discharge from a flood event after the flood has passed and the stage has receded. Indirect methods use forensic surveys of the remnant high-water marks and the channel geometry in concert with the estimates of the boundary roughness to estimate the peak discharge from the principles of conservation of energy, momentum, and mass. A variety of indirect measurement techniques exist that can be used based on the stream hydraulics and presence or absence of structures (Benson & Dalrymple, 1967, chap. A1; Bodhaine, 1968, chap. A3; Dalrymple & Benson, 1967, chap. A2; Matthai, 1967, chap. A4).

Most of the discharge measurements made at streamgages are direct measurements based on the velocity-area method, whereby the cross section of the stream is subsectioned and the total discharge is the sum of the products of the area and mean velocity of each subsection (International Standards Organization, 2007a)

$$Q = \sum_{i=1}^{n} \left(a_i v_i \right), \quad (1)$$

where Q is the total discharge, n is the number of subsections, a_i is the area of subsection i, and v_i is the average velocity of subsection i. The midsection

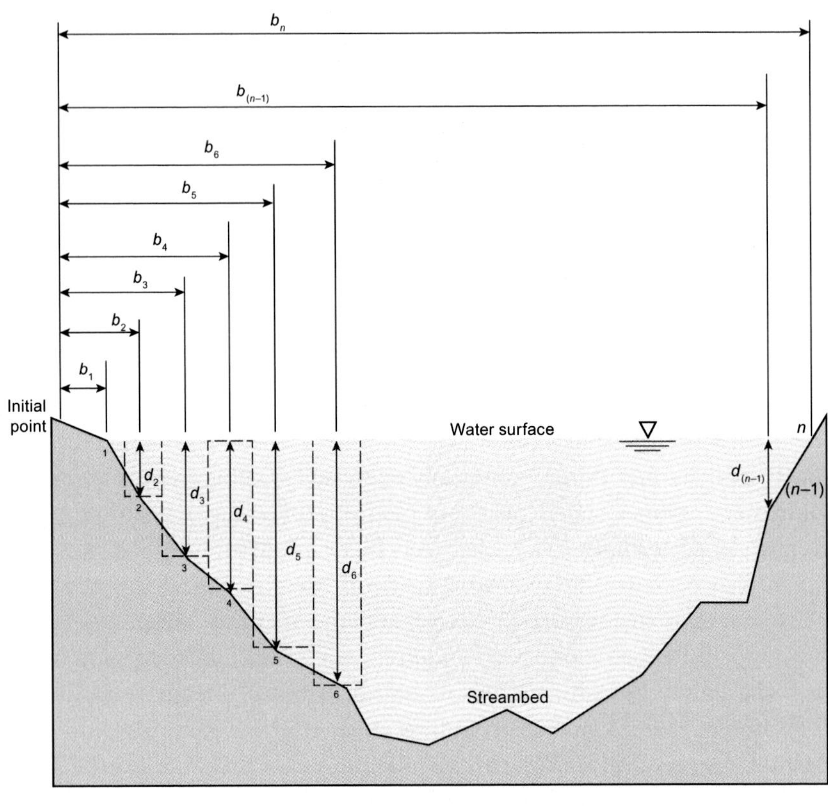

Fig. 5 Midsection area-velocity method to determine the total discharge. Modified from Turnipseed, D. P., & Sauer, V. B. (2010). Discharge measurements at gaging stations. U.S. Geological Survey techniques and methods (87 pp.). Arlington, VA: U.S. Geological Survey. Available at: http://pubs.usgs.gov/tm/tm3-a8 (Book 3, chap. A8).

velocity-area method (Fig. 5) is a cross-section discretization method commonly used to determine discharge in the United States (Turnipseed & Sauer, 2010, chap. A8). The total discharge is equal to the sum of the discharges computed for each horizontally discretized subsection. Each subsection discharge is the product of the area and mean velocity of the subsection determined from velocity and depth at the midpoint of each subsection. With the advent of acoustic Doppler current profilers (ADCPs) technology, efficient discretization of the channel cross section into small subsections (both horizontal and vertical discretization) has increased the number of data points collected in each measurement and reduced the measurement time. Thousands of velocity and depth observations are made as the ADCP travels approximately perpendicular to streamflow (Mueller, Wagner, Rehmel, Oberg, & Rainville, 2013, chap. A22), with the ADCP mounted rigidly either inside a tethered boat or near the bow of a manned boat.

Instruments to determine velocity include mechanical current meters (vertical axis and horizontal axis type meters), electromagnetic meters, ADCPs, and optical strobe velocity meters (Turnipseed & Sauer, 2010, chap. A8). Mechanical meters have been around for over 100 years. The predominant mechanical meter in use in the United States today is the AA Price mechanical current meter, which was first introduced in the 19th century (Frazier, 1974).

Continuous measurement of velocity, termed "index velocity," is used at selected sites to derive streamflow by the index-velocity method (Levesque & Oberg, 2012, chap. A23). The four types of metering systems that have typically been used to measure index velocity are standard mechanical current meters, deflection meters, electromagnetic meters, and acoustic meters (Rantz et al., 1982). Over time, the acoustic velocity meter (AVM) and acoustic Doppler velocity meter (ADVM) have become the predominant tools for index-velocity methods (Levesque & Oberg, 2012, chap. A23). The AVM uses two sensors positioned at an angle to the flow and uses the acoustic travel time technique (Laenen, 1985, chap. A17). ADVMs, introduced in 1997, consist of a single unit using the Doppler principle (Morlock, Nguyen, & Ross, 2002). In the United States, the relatively low-cost but robust ADVM has become more predominant than the AVM to measure index velocity (Levesque & Oberg, 2012, chap. A23).

3.3 Stage-Discharge Relation

In the absence of discrete measurements of streamflow (discharge measurements), streamflow is usually determined through surrogate measures of other variables such as stage, water-surface slope, rate of water-surface rise, or index velocity at a streamgage. Streamflow is derived from the surrogate variables using what is termed a "rating." The rating model is developed and calibrated using discharge measurements collected onsite by field staff employing standard methods (Turnipseed & Sauer, 2010, chap. A8).

Ratings can be divided into two broad categories: simple stage-discharge ratings (simple ratings) and complex ratings. The first step in determining

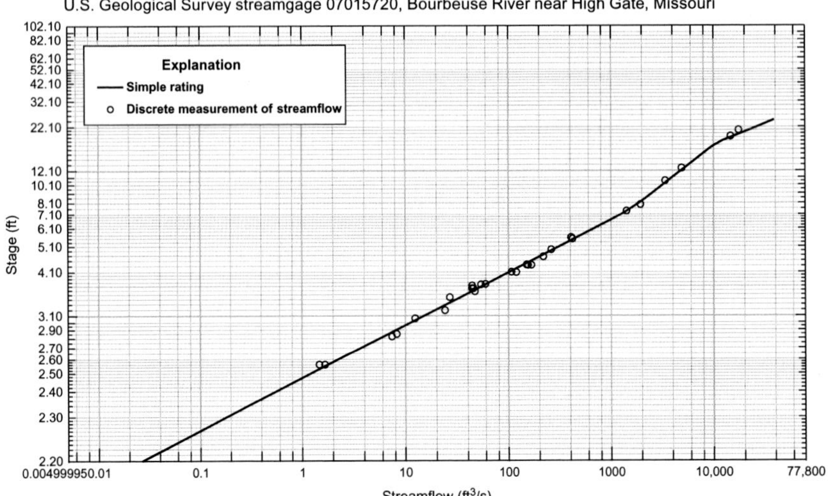

Fig. 6 Example of a simple rating for the Bourbeuse River near High Gate, Missouri (U.S. Geological Survey streamgage 07015720). Data from U.S. Geological Survey. (2014). USGS water data for the nation. U.S. Geological Survey National Water Information System. http://waterdata.usgs.gov Accessed 30.09.14.

the type of rating required is to plot the discharge measurements along with stage. A simple rating relates discharge to stage, assuming a unique one-to-one relation between stage and discharge (Fig. 6). Simple ratings often have multiple linear segments, and each segment represents a different control feature. USGS simple ratings are often developed using logarithmic plots. A straight-line relation on a logarithmic rating can be used as a guide when the rating needs to be extended (interpolated) above the maximum measured discharge; however, the rating should not be extended beyond the stage where the channel changes shape and a new control takes effect. Kennedy (1984) provides an extensive discussion about the development of ratings.

A variety of complex rating methods exist to deal with situations when simple ratings do not work well. A complex rating relates discharge to state and other variables because of lack of unique one-to-one relation between stage and streamflow. Complex ratings (Fig. 7) often are required for locations on low-gradient streams, streams with large amounts of channel or overbank storage, or streams with highly unsteady flow. Complex rating methods range from simply adding a second independent variable to the process of computing streamflow to computer models that solve conservation-of-momentum and conservation-of-mass partial differential

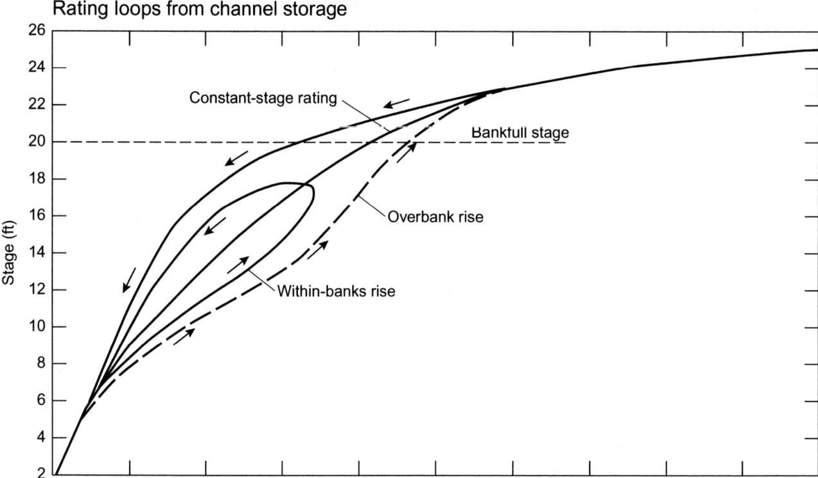

Fig. 7 Example of a hypothetical complex rating.

equations (Kennedy, 1984, chap. A10). More discharge measurements are required to characterize complex ratings than simple ratings.

Traditionally, the simplest form of rating model (stage-discharge) is tried first, with progressively more complex models developed if the simpler forms fail to provide an accurate relation. Rating development and verification constitute an ongoing process throughout the life of a streamgage, requiring systematic analysis and periodic evaluation to refine (or reconstruct) the rating in response to changes in stream hydraulics. Changes in stream hydraulics can be caused by changing channel conditions or the construction of flood protection works such as dykes. A thorough understanding of both the science of river hydraulics and the many assumptions made for various rating methods must be the foundation for rating development. The sites producing the best rating-derived streamflow data are the result of a systematic process that is put in place to regularly analyze available rating methods in response to new data, keeping in mind that alterations and corrections made based on new data must accurately reflect the underlying stream hydraulics; therefore, the alterations and corrections are rational and defensible. Discharge measurements are critical, especially during floods and droughts.

3.3.1 Rating Controls

For a simple rating, it is assumed that a one-to-one relation between stage and discharge is controlled by a section (section control) or reach of channel (channel control); both of these features are referred to as station control

(Kennedy, 1984, chap. A10). Corbett (1943) explains this as follows: "In order to have a definite and enduring stage-discharge relation, the channel must have certain physical features capable of regulating or stabilizing flow past the gage to such an extent that for a given stage of the water surface the discharge will always be the same." A station control essentially "controls" the relation between stage and discharge, and prevents hydraulic disturbances from translating past the control (and affecting the stage-discharge relation).

For section control, a single section of the channel is the controlling feature and, strictly defined, is configured (eg, by constriction, bump in channel bottom, gravel/rock riffle, artificial weir/flume, or some combination) such that the flow passes through the critical-flow state. For medium to high streamflow, the influence of the section control is lost (submerged) and channel control is typically in effect. Channel control consists of all the physical features of the downstream channel (such as geometry, bed slope, expansion/contraction, sinuosity, and bed/bank roughness) that determine the stage of the river for a particular discharge (Rantz et al., 1982). Given the assumption of steady, uniform flow, a simple stage-streamflow relation exists with channel control; however, steady, uniform flow rarely exists in natural streams because flow is seldom steady and channel roughness and geometry in the controlling cross section are usually changing. As such, the controls on the stage-streamflow rating are complex. The components affecting the channel control often produce a hysteresis effect most often called the loop rating (Kennedy, 1984, chap. A10), or loop-rating curve (Henderson, 1966). Kennedy (1984) describes the loop effects when hysteresis is tied to channel storage during the rise and fall of a hydrograph or when streams with large floodplains transition into and out of overbank flows.

Although all natural streams have hysteresis (Fig. 7), a complex rating is not required in all cases, because sometimes the degree of hysteresis is small and well within the uncertainty of discharge measurements made by the hydrographer.

3.3.2 Rating Extension

Discharge measurements over a range of stage form the basis for construction of accurate rating curves. Given that discharge measurements have associated errors, and adequate discharge measurements are sometimes not available to define the entire range of the rating, hydraulic theory helps guide accurate construction of the complete rating curve.

Slope-conveyance and step-backwater analyses are two hydraulic methods that can be used to provide quantitative estimates of the relation between stage and streamflow under channel control conditions (Rantz et al., 1982). These methods are useful to supplement discharge measurements in shaping the rating curve, as well as extrapolating the rating under certain constraints.

Both methods work well for rating development and shaping, particularly in ranges of stage ranges where discharge measurements are available to augment the methods. In the situations where ratings need to be extrapolated beyond the highest observed measurement, great care must be exercised. The USGS policy is not to extrapolate ratings to more than twice the highest discharge measured by discrete observation (Rantz et al., 1982) unless detailed hydraulic analysis (such as step backwater) is completed. The slope-conveyance method is not considered a "detailed hydraulic analysis" but, because the method uses discharge measurements and channel geometry, it can be used for extrapolation of ratings beyond twice the highest measurement. Note, however, that the method should only be used within the same control conditions as the highest discharge measurement. For example, if the highest discharge measurement was made below bankfull conditions, the slope-conveyance method should not be used to extrapolate the rating to above bankfull conditions. This prohibition is driven mainly by the fact that the relation of friction slope to stage at one control condition may not be the same as that for a different control condition.

Step-backwater studies are considered to be detailed hydraulic analyses, but even these studies should be completed using a calibrated step-backwater model. Calibration would consist of adjusting the model parameters (mainly roughness) so that the modeled water-surface profile matches (within tolerances) the observed water-surface profile for a known streamflow.

3.3.3 Temporary Changes in Control and the Shifting-Control Method for Simple Ratings

Natural streamgage controls may change temporarily because of a variety of reasons that include scour/fill of the stream channel, changes in aquatic vegetation, and accumulation of drift/debris. The changes in the control are detected in the plotting position of the discharge measurement on the simple rating. The need for discharge measurement to detect changes to the control increases during floods. An example of control changes that would only be detected by streamflow measurements is shown in Fig. 8. Also, Fig. 8 shows that two direct streamflow measurements made during flooding in 2009 (numbers 353 and 354) on the Platte River near Kearney, Nebraska (location not shown) resulted in a more than plus 1-ft correction of the rating curve at the upper end (above a stage of 5.9 ft) of the rating.

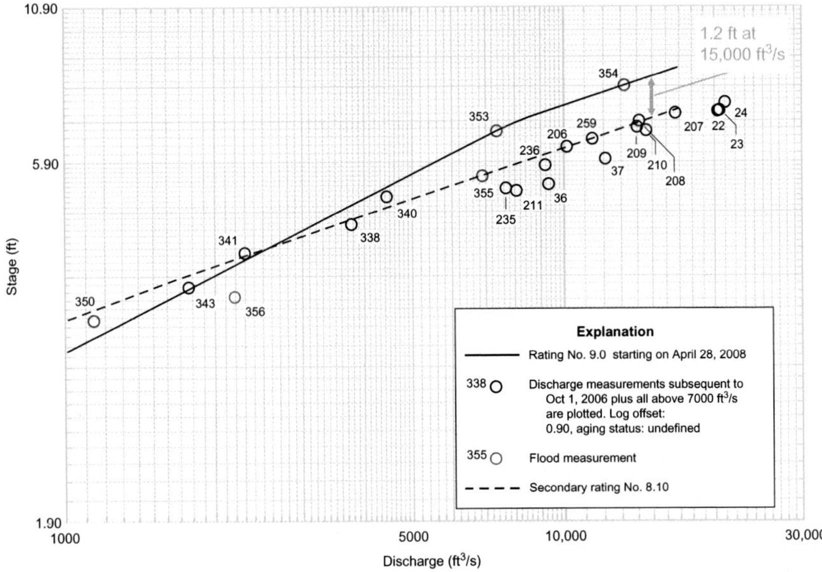

Fig. 8 Changes in the rating curve for the Platte River near Kearney, Nebraska (USGS streamgage 06770200). Data from Holmes, R. R., Jr., Koenig, T. A., & Karstensen, K. A. (2010). Flooding in the United States Midwest, 2008. U.S. Geological Survey professional paper 1775 (64 pp.).; U.S. Geological Survey. (2014). USGS water data for the nation. U.S. Geological Survey National Water Information System. http://waterdata.usgs.gov Accessed 30.09.14.

At a streamflow of 15,000 ft^3/s, the stage on the rating curve changed by plus 1.2 ft, from approximately 7.0 to 8.2 ft.

3.3.4 Index-Velocity Method

An increasingly common method to derive a continuous time series of streamflow, particularly at streamgages with rating complexity, is the index-velocity method (Levesque & Oberg, 2012, chap. A23). Streamflow can be computed at a channel cross section as the product of the mean velocity and the cross-sectional area.

The cross-sectional area can be readily determined through bathymetric and traditional land surveys. Using these data, the relation between stage and cross-sectional area (Fig. 9A) can be determined (stage-area rating). The frequency of re-survey and verification of the stage-area rating depends on the stability of the channel. The USGS determined that the channel be re-surveyed every year for the first 3 years, followed by a minimum of every 3 years so long as the channel remains stable (Levesque & Oberg, 2012, chap. A23).

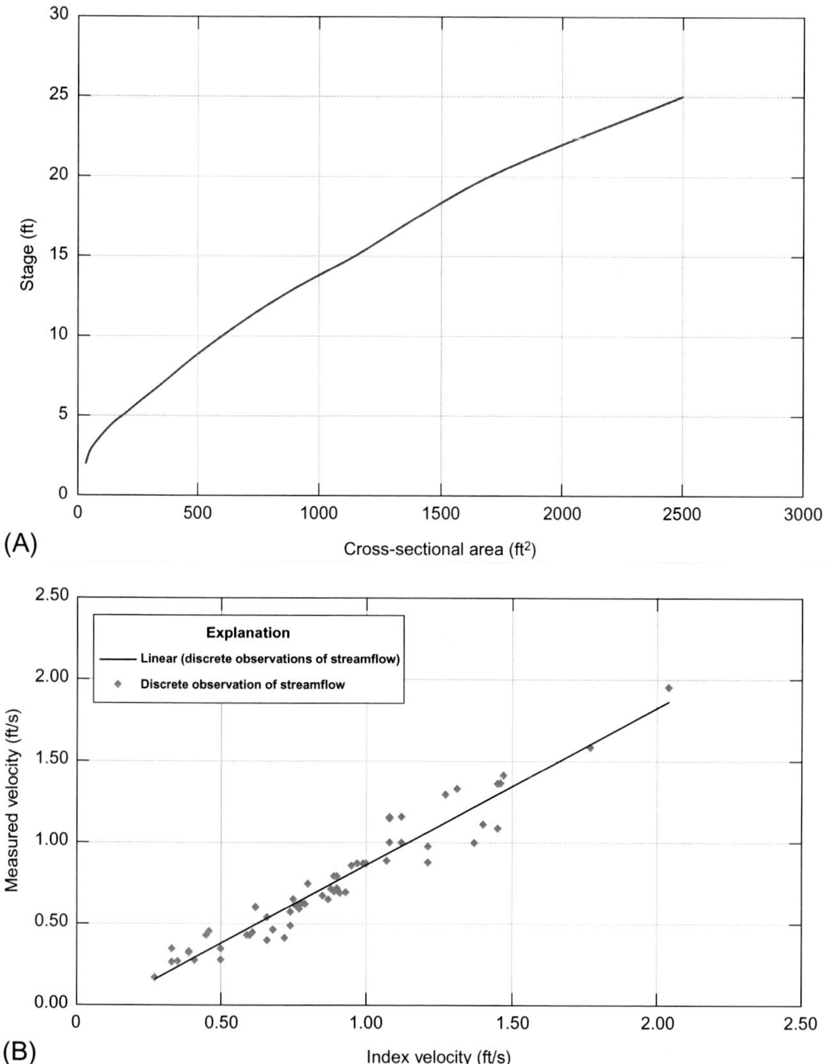

Fig. 9 (A) Stage-area rating and (B) index-velocity rating for Iroquois River at Foresman, Indiana (USGS streamgage 05524500). Data from U.S. Geological Survey. (2014). USGS water data for the nation. U.S. Geological Survey National Water Information System. http://waterdata.usgs.gov Accessed 30.09.14.

Autonomous, continuous-direct determination of mean-channel velocity is difficult and typically not cost-efficient; however, continuous measurement of velocity (index velocity) of a certain part of a river, stream, or canal can be made and used as an index to estimate the mean-channel velocity (Levesque & Oberg, 2012, chap. A23). Discharge measurements and concurrent readings of the index velocity are made at a range of expected mean velocities. Once sufficient discharge measurements have been made, a relation to predict mean velocity can be developed (index-velocity rating). Mean velocity can typically be a function of the stream-wise component of the measurement index velocity (Fig. 9B), but may also require additional independent variables such as stage. Accuracy of the index-velocity rating must be checked periodically by making additional discrete observations of streamflow and adjusted accordingly.

3.3.5 Uncertainty in Ratings

Over the past few decades, awareness has increased regarding the importance and subsequent need for estimates of the uncertainty of streamflow data. Much work has been done to derive methodologies for the uncertainty of discharge measurements (Cohn, Kiang, & Mason, 2013; Herschy, 1999; International Standards Organization, 2007b; Oberg & Mueller, 2007; Pelletier, 1988, 1989) and the uncertainty of streamflow ratings used to derive continuous streamflow data (Clarke, 1999; Masson, Ghio, Lallement, Parsy, & Philippe, 1987; Morlot, Perret, Favre, & Jalbert, 2014; Schmidt, 2002). Unfortunately, this awareness has not translated into an agreed-upon, systematic method to estimate the uncertainty of rating curves, and the subject remains a matter of research (Le Coz, 2012).

The following list includes factors that rating curve uncertainty depends on: (1) characteristics of hydraulic controls at the site; (2) stability and condition of hydraulic controls at the site; (3) frequency and distribution of discharge measurements (through time and range of streamflow); (4) accuracy of discharge measurements; and (5) accuracy of stage observations. Current research notwithstanding, the standard operational practice as of 2015 in the United States by the USGS was to categorize the final derived streamflow data as excellent, good, fair, or poor, based on the analyst's qualitative assessment of the above factors at the site for various data collection periods (Rantz et al., 1982).

4 DELIVERY OF STREAMFLOW INFORMATION

The daily streamflow record was the best-known and main hydrologic product produced by the USGS for about 80 years (Hirsch & Fisher, 2014). Until the mid-1980s, the stage record was collected and stored in analog format (Bales, 2014), and the process of converting water levels obtained from chart recorders or digital recorders to discharge daily took from a few months to a year to finalize. Starting in the mid-1980s, digital storage devices became a cost-effective method for storing hydrologic information collected at a streamgage; however, a site visit was required to download the information. The streamflow record, therefore, was useful for hydrologic investigations, calibration of NWS hydrologic models to historic floods, water planning, and resource assessments. However, the daily streamflow had little value for short-term water decisions such as flood forecasting, emergency response, and reservoir releases made by federal, state, and local agencies with flood protection and warning responsibilities.

The first systems used to transmit near real-time hydrologic information were landline telephone based (starting in 1921) and radio transmission based (1931). The early transmission systems were complex and designed for a specific agency, and the data were not available to the public (Hirsch & Fisher, 2014). During the 1960s and 1970s, experiments and research were conducted to use radio transmission communications via satellites as a component of the data collection systems (DCSs) to deliver hydrologic information (note: DCS that have the ability to transmit remotely the data are termed data collection platforms). By the mid-1970s, satellite-equipped DCS were advanced enough that the USGS began investigations to identify the costs and benefits of real-time hydrologic data collection. The investigations indicated that flood forecasts, flood warnings, and reservoir management, in addition to other water-resource uses, would benefit from installation and use of satellite DCS (Paulson & Shope, 1984). Paulson and Shope (1984) provide an excellent discussion of the evolution and ultimate use of satellite DCS to deliver hydrologic information. The number of streamgages instrumented with real-time telemetry grew from 120 stations in 1978 to about 5100 in 1999 and 9600 in 2014 (Hirsch & Fisher, 2014).

Streamflow information collected by a DCS is most commonly relayed to the USGS Water Science Centers by communication through the Geostationary Operational Environmental Satellite (GOES) system. Data transmitted through the GOES telemetry are relayed through one or more satellites to a ground receiving station and then on to USGS servers for processing and dissemination via the Internet through the National Water Information System (NWIS, http://waterdata.usgs.gov/nwis). In some cases, landline and cellular telephone technology is used to interrogate the DCS and deliver the streamflow information using the same USGS servers, processing, and delivery as the satellite system. The various telemetry options allow the USGS to select the best system for the environmental conditions at the streamgage to deliver real-time streamflow information.

4.1 Streamflow Forecast

The NWS's 13 regional River Forecast Centers (RFCs) provide hydrologic forecasts for approximately 4000 locations in the United States using calibrated hydrologic and hydraulic models. The hydrologic forecast contains information for the time and stage, the time when the river is expected to exceed flood stage, and the time when the river is expected to drop below flood stage if within the time period covered on the graph (http://www.nws.noaa.gov/os/water/ahps/resources/Guide_to_Hydrologic_Information_Brochure.pdf). The operation and maintenance of data networks supply critical hydrometeorologic information, such as precipitation, real-time streamflow, and reservoir data used by the RFC to develop the hydrologic forecasts. The USACE, USGS, and Reclamation operate networks that supply much of the information needed by the RFC's.

The USGS supplies much of the streamflow information used by the RFC and, under full implementation of the NSIP, the USGS goal is to operate 3631 streamgages at NWS flood forecast locations. As of 2015, about 2465 streamgages were in operation at forecast locations. The streamgages operated at NWS forecast locations are some of the most highly visible and receive the most usage of any type of streamgage operated by the USGS. Typically, streamflow is output from the NWS hydrologic forecast models, and a rating is used to convert the flow to stage. As the severity of a flood increases, the need for a verified and accurate rating increases. The frequency of discharge measurements and the number of times the shifted rating changes is dependent on the stability of the control, the magnitude of hysteresis, and the severity of the flood.

In 2004, the USGS developed a national depot to serve the most up-to-date shift-adjusted rating curves for all USGS streamgages displayed on the National Water Information System Web (NWISWeb, http://waterdata.usgs.gov/nwis/sw). For each streamgage serving stage and discharge over the NWISWeb, the current stage-discharge rating, including any applicable shift, is retrieved daily from the USGS Water Science Center that operates the streamgage and the information is transferred to the depot. Users need to check the depot frequently to ensure the current shifted rating is used for current data analyses. The USACE and the NWS have automatic procedures for obtaining the most recent shifted rating from the depot. In parts of the United States, the USGS, NWS, and USACE hold frequent flood information teleconferences during large regional floods and often discuss and agree on the need for discharge measurements to confirm a rating at problem streamgage locations.

NWSChat is an Instant Messaging program used by the NWS operational staff, the emergency response agencies, and the media since 2010 to share critical weather and hydrologic information during hazardous events (https://nwschat.weather.gov). In the case of floods, NWSChat has been used extensively by federal, state, and local agencies involved in all levels of flood forecasting and disaster warning and response events in river basins monitored by many RFCs. NWSChat provides an efficient method to deliver hydrologic information from one user to many users, eliminating the need to email or call interested parties. During floods, many USGS staff use NWSChat to post results of discharge measurements immediately on completion and to document streamgage operational status.

The calibration and use of the hydrologic and hydraulic models used by the RFC for flood forecasting is an ongoing process, because even a well-calibrated model will not always be static. Changes in river basin characteristics, such as land-use changes, urbanization, drainage improvements, channel modifications, and levee construction, can change the timing and magnitude of runoff and change the stage-streamflow rating in use at a streamgage. Thus, a continuous cycle of model input updates, recalibration, and rating adjustments must be part of a sound flood forecasting program.

5 OPPORTUNITIES FOR THE STREAMGAGE PROGRAM

Even with about 8100 streamgages in the USGS network nationwide, stage and streamflow data are needed during floods at locations that are ungaged. Gages at these additional locations are needed to provide such things as flood forecasting, flood control reservoir operation, flood fighting operations, and emergency management. To meet these needs, the USGS developed, built, and maintains a cache of rapid deployment gages (RDGs) that can be installed and report near real-time data within hours to days of a recognized need.

An example of the need for RDGs was during the spring of 2009 when record flooding took place across the Northern Great Plains. The flooding devastated many communities along rivers and streams in the Missouri River and Red River of the North Basins in North Dakota, western Minnesota, and southeastern South Dakota (Macek-Rowland & Gross, 2011). Although the USGS streamgaging network provided information for flood warning and reservoir operations in the James River (tributary to the Missouri River) before, during, and after the flood, state and federal agencies needed additional flood monitoring at many ungaged locations.

In the days leading up to the flooding on the James River and many of its tributaries (Fig. 10), 14 RDGs with satellite telemetry were installed in the James River Basin to supplement the existing streamgage network. The RDGs were used by USACE and Reclamation staff to monitor the timing of the flood wave into Pipestem and Jamestown Reservoirs and to balance releases from Pipestem and Jamestown Reservoirs against downstream flooding at communities along the James River. As indicated in Table 1, the RDGs were installed for several different purposes, but as soon as the data

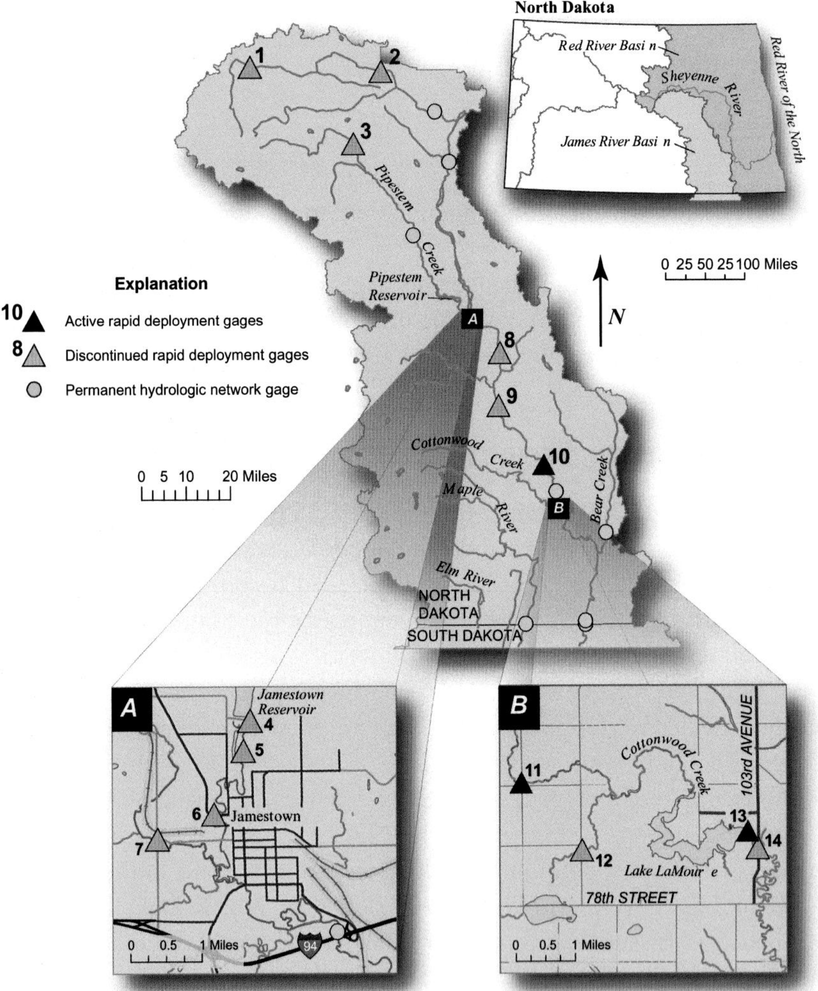

Fig. 10 Location of rapid deployment gages in the James River Basin, North Dakota, during 2009 spring flooding in North Dakota.

Table 1 Purpose of rapid deployable gage installations

Sites	Original purpose of installation
1–4	Monitor time of flood wave into reservoirs
5–7	Used to maximize reservoir releases without overtopping levees
8–10	Used to manage reservoir releases in an effort to protect life and property
11–14	Hydrologic monitoring at reservoir in jeopardy of emergency spillway failure

were provided electronically over the Internet through the USGS NWIS (http://waterdata.usgs.gov/nwis/rt), the uses of the data expanded rapidly. As an example, record-breaking inflow to Lake LaMoure (Fig. 10) created a potentially life-threatening situation when outflow caused major scouring in the emergency spillway (Fig. 11). Prior to the flood, no streamflow gages were located on Cottonwood Creek. Therefore, two RDGs were installed to monitor inflow, one RDG was installed to monitor the Lake LaMoure water level, and one RDG was installed to monitor outflow. The RDGs were used in combination with several streamflow measurements to make decisions about evacuations and potential infrastructure damage downstream.

Although the RDGs helped the USACE and Reclamation manage the reservoirs in the James River Basin, many emergency managers, law enforcement personnel, and the general public also used the uninterrupted water information to save lives and protect property. Generally within 48 h

Fig. 11 Lake LaMoure emergency spillway during 2009 spring flooding in North Dakota. Photo courtesy of the North Dakota State Water Commission.

of requesting the RDGs from the USGS Hydrologic Instrumentation Facility, Stennis Space Center, Mississippi, the RDGs were delivered to the North Dakota Water Science Center in Bismarck, North Dakota, the RDGs were installed, the data transmitted from the RDGs were delivered via the Internet, and a link to access the data was added to the James River Flood Information Website (http://nd.water.usgs.gov/floodinfo/james.html).

During Hurricane Rita in 2005, the USGS first developed and implemented rapidly deployable storm-surge sensors (SSSs) to observe and document hurricane-induced storm-surge interaction with coastal areas (http://ga.water.usgs.gov/flood/hurricane). The SSSs (47 during Hurricane Rita and 230 during Hurricane Sandy in 2013) are strapped to bridge piers, power and light poles, and other structures along and inland of the coast in the days just before the hurricane or tropical storm makes landfall. Water level and barometric pressure are recorded and stored on the SSS every 30 s. Although the SSSs do not transmit real-time information, scientists, engineers, and emergency managers are using the information to design and calibrate complex hydrodynamic models in an effort to understand storm-surge processes better.

Dissemination of real-time stage and streamflow data is a crucial requirement of the NWS to support flood forecasts and other forecasting done by the NWS. Two changes in streamflow collection technology in the last 100 years revolutionized the collection and delivery of streamflow information. In the 1990s, the development of data loggers and satellite radio transmission (first technological change) allowed the USGS to store large amounts of hydrologic information digitally and transmit the hydrologic information in near real-time. During the last 30 years, latency time (time between collection and dissemination to the end user) has decreased dramatically.

The second major technological change was the development of the ADCP for measuring streamflow. The use of ADCPs has reduced the time needed to make discharge measurements, along with allowing hydrographers to make discharge measurements in some flood conditions that were not possible using conventional discharge equipment. Thus, the hydrographer's ability to make more and better flood measurements has improved the upper end of the stage-streamflow rating at many locations. Typically, the hydrographer will relay the ADCP measurement, via cell phone, to the USGS office and the upper end of the rating will be defined, checked, and sent to the ratings depot the same day the measurement was made.

Federal and state agencies that have a flood warning and emergency management mission were the primary users of stage and streamflow in the early years of real-time delivery of stage and streamflow information; however, the transition to real-time delivery of streamflow information led to outcomes the USGS had not anticipated (Hirsch & Fisher, 2014). AS of 2015, the USGS delivered real-time stage and streamflow data in tabular or graphic form for each streamgage through NWISWeb (http://waterdata.usgs.gov/nwis) for every streamgage equipped for real-time transmission. Availability of the streamflow data in NWISWeb allows the general public, such as local government officials, business owners, and farmers to use the information and make decisions regarding the protection of life and property. Historic streamflow data are also available on NWISWeb, and are important to enable contextual understanding of streamflow probabilities, rainfall-runoff characteristics, and stream hydraulics.

Value-added synthesis of historic and real-time streamflow information, such as USGS WaterWatch (http://waterwatch.usgs.gov), provide powerful hydrologic situational awareness through displays of maps, graphs, and tables describing real-time, recent, and past streamflow conditions for the United States. The real-time information is generally updated on an hourly basis. WaterWatch provides streamgage-based maps that: show the location of more than 3000 long-term (30 years or more) USGS streamgages; use colors to represent streamflow conditions compared to historical streamflow; feature a point-and-click interface allowing users to retrieve graphs of stream stage and streamflow; and highlight locations where extreme hydrologic events, such as floods and droughts, are occurring. The real-time streamflow maps highlight flood and high-flow conditions.

WaterWatch summarizes streamflow conditions in a region (state or hydrologic unit) in terms of the long-term typical condition at streamgages in the region. Summary tables are provided along with time-series plots that depict variations through time. WaterWatch also includes tables of current streamflow information and locations of flooding.

REFERENCES

Bales, J. D. (2014). Progress in data collection and dissemination in water resources, 1974–2014. *Water Resources Impact, 16*(3), 18–22.

Bales, J. D., Costa, J. E., Holtschlag, D. J., Lanfear, K. J., Lipscomb, S., Milly, P. C. D., et al. (2004). *Design of a National Streamflow Information Program, report with recommendations of a committee.* U.S. Geological Survey open-file report 2004-1263 (42 pp.).

Benson, M. A., & Dalrymple, T. (1967). General field and office procedures for indirect discharge measurements. *U.S. Geological Survey techniques of water-resources investigations.* Arlington, VA: U.S. Geological Survey. Book 3 (30 pp.).

Bodhaine, G. L. (1968). Measurement of peak discharge at culverts by indirect methods. *U.S. Geological Survey techniques of water-resources investigations.* Arlington, VA: U.S. Geological Survey. Book 3 (60 pp.).

Clarke, R. T. (1999). Uncertainty in the estimation of mean annual flood due to rating-curve indefinition. *Journal of Hydrology, 222*, 1–4.

Cohn, T. A., Kiang, J. E., Mason, R. R., Jr. (2013). Estimating discharge measurement uncertainty using the interpolated variance estimator. *ASCE Journal of Hydraulic Engineering, 139*, 502–510.

Corbett, D. M. (1943). Stream gaging procedure. *U.S. Geological Survey water-supply paper 888* (245 pp.).

Dalrymple, T., & Benson, M. A. (1967). Measurement of peak discharge by the slope-area method. *U.S. Geological Survey techniques of water-resources investigations.* Arlington, VA: U.S. Geological Survey. Book 3 (12 pp.).

Ferguson, G. F., et al. (1990). *A history of the Water Resource Division, U.S. Geological Survey—Volume V, July 1, 1947, to April 30, 1957.* Washington, DC: United States Government Printing Office (309 pp.).

Follansbee, R. (1919). *A history of the Water Resources Branch of the United States Geological Survey — Volume 1.* Administrative document (458 pp.).

Follansbee, R. (1928). *A history of the Water Resources Branch of the United States Geological Survey — Volume 2, years of increasing cooperation, July 1, 1919, to June 30, 1928.* Unpublished administrative document (202 pp.).

Follansbee, R. (1939). *A history of the Water Resources Branch of the United States Geological Survey—Volume 3, years of 50-50 cooperation, July 1, 1928, to June 30, 1939.* Unpublished administrative document (386 pp.).

Follansbee, R. (1947). *A history of the Water Resources Branch of the United States Geological Survey—Volume 4, years of World War II, July 1, 1939, to June 30, 1947.* Unpublished administrative document (398 pp.).

Frazier, A. H. (1974). Water current meters — In the Smithsonian Collections of the National Museum of History and Technology. *Smithsonian studies in history and technology.* (Vol. 28). Washington, DC: Smithsonian Institution Press (95 pp.).

Gunn, M. A., Matherne, A. M., Mason, R. R., Jr. (2014). The USGS at Embudo, New Mexico — 125 years of systematic streamgaging in the United States. *U.S. Geological Survey fact sheet 2014-3034.* http://dx.doi.org/10.3133/fs20143034 (4 pp.).

Henderson, F. M. (1966). *Open channel flow.* New York, NY: McMillan Company (522 pp.).

Herschy, R. W. (1999). Uncertainties in hydrometric measurements. *Hydrometry* (2nd ed., pp. 355–370). New York, NY: John Wiley.

Herschy, R. W. (2008). *Streamflow measurement* (3rd ed.). Abingdon: Taylor and Francis Group (507 pp.).

Hirsch, R. M., & Fisher, G. T. (2014). Past, present, and future of water data delivery from the U.S. Geological Survey. *Journal of Contemporary Water Research and Education, 153*, 4–15.

International Standards Organization. (2007a). *Hydrometry — Measurement of liquid flow in open channels using current meters or floats.* ISO 748 (46 pp.).

International Standards Organization. (2007b). *Hydrometry — Velocity-area method using current-meters — Collection and processing of data for determination of uncertainties in flow measurement.* ISO 1088 (48 pp.).

Kennedy, E. J. (1984). Discharge ratings at gaging stations. *U.S. Geological Survey techniques of water-resources investigations.* Arlington, VA: U.S. Geological Survey. Book 3 (59 pp.), http://pubs.usgs.gov/twri/twri3-a10/.

Kenney, T. A. (2010). *Levels at gaging stations: U.S. Geological Survey techniques and methods 3-A19.* (60 pp.).

Laenen, A. (1985). Acoustic velocity meter systems. *U.S. Geological Survey techniques of water-resources investigations.* Arlington, VA: U.S. Geological Survey. Book 3 (38 pp.).

Langbein, W. B., & Hoyt, W. G. (1959). *Facts for the nation's future.* New York, NY: The Ronald Press Company (288 pp.).

Le Coz, J. (2012). A literature review of methods for estimating the uncertainty associated with stage-discharge ratings. *Technical report of the WMO initiative on assessment of the performance of flow measurement instruments and techniques.* (21 pp.).

Levesque, V. A., & Oberg, K. A. (2012). Computing discharge using the index velocity method. *U.S. Geological Survey techniques and methods.* Arlington, VA: U.S. Geological Survey. Book 3 (148 pp.).

Macek-Rowland, K. M., & Gross, T. A. (2011). 2009 Spring floods in North Dakota, western Minnesota, and northeastern South Dakota. *U.S. Geological Survey scientific investigations report 2010-5225.* (41 pp.).

Masson, J., Ghio, M., Lallement, C., Parsy, C., & Philippe, J. (1987). Debitmetrie: Precision des stations de jaugeage. *La Houille Blanche,* 333–338.

Matalas, N. C., Barnes, H. H., Jr. (1977). A bicentennial reflection: Streamgauging in the United States. *Hydrological Sciences Bulletin, 22*(1), 41–59. Available at http://www.tandfonline.com/loi/thsj19.

Matthai, H. F. (1967). Measurement of peak discharge at width contractions by indirect methods. *U.S. Geological Survey techniques of water-resources investigations.* Arlington, VA: U.S. Geological Survey. Book 3 (44 pp.).

Morlock, S. E., Nguyen, H. T., & Ross, J. H. (2002). Feasibility of acoustic Doppler velocity meters for production of discharge records from U.S. Geological Survey streamflow-gaging stations. *U.S. Geological Survey water resources investigations report 01-4157.* (56 pp.).

Morlot, T., Perret, C., Favre, A.-C., & Jalbert, J. (2014). Dynamic rating curve assessment for hydrometric stations and computation of the associated uncertainties: Quality and station management indicators. *Journal of Hydrology, 517,* 173–186.

Mueller, D. S., Wagner, C. R., Rehmel, M. S., Oberg, K. A., & Rainville, F. (2013). Measuring discharge with acoustic Doppler current profilers from a moving boat (ver. 2.0, December 2013). *U.S. Geological Survey techniques and methods.* Arlington, VA: U.S. Geological Survey. http://dx.doi.org/10.3133/tm3A22 Book 3 (95 pp.).

Norris, J. M. (2009). National Streamflow Information Program, implementation status report. *U.S. Geological Survey fact sheet 2009-3020.* (6 pp.).

Oberg, K. A., & Mueller, D. S. (2007). Validation of streamflow measurements made with acoustic Doppler current profilers. *ASCE Journal of Hydraulic Engineering, 133,* 1421–1432.

Olson, S. A., & Norris, J. M. (2007). U.S. Geological Survey streamgaging. *U.S. Geological Survey fact sheet 2005-3131.* (4 pp.).

Paulson, R. W., Shope, W. G., Jr. (1984). Development of earth satellite technology for the telemetry of hydrologic data. *Water Resources Bulletin, 20*(4), 611–618.

Pelletier, P. M. (1988). Uncertainties in the single determination of river discharge: A literature review. *Canadian Journal of Civil Engineering, 15*(5), 834–850.

Pelletier, P. M. (1989). Uncertainties in streamflow measurement under winter ice conditions, a case study: The Red River at Emerson, Manitoba, Canada. *Water Resources Research, 25*(8), 1857–1868.

Rabbitt, M. C. (1989). The United States Geological Survey: 1879–1989. *U.S. Geological Survey circular 1050.* (52 pp.).

Rantz, S. E., et al. (1982). Measurement and computation of streamflow. *U.S. Geological Survey water-supply paper 2175.* (631 pp.).

Rydlund, P. H., Jr., & Densmore, B. K. (2012). Methods of practice and guidelines for using survey-grade global navigation satellite systems (GNSS) to establish vertical datum in the United States Geological Survey. *U.S. Geological Survey techniques and methods.* Arlington, VA: U.S. Geological Survey. Book 11 (102 pp. with appendixes).

Sauer, V. B., & Turnipseed, D. P. (2010). Stage measurement at gaging stations. *U.S. Geological Survey techniques and methods.* Arlington, VA: U.S. Geological Survey. Book 3 (45 pp.). Available at http://pubs.usgs.gov/tm/tm3-a7/.

Schmidt, A. R. (2002). *Analysis of stage-discharge relations for open-channel flows and their associated uncertainties.* (Ph.D. thesis) Urbana-Champaign: University of Illinois (329 pp.).

Turnipseed, D. P., & Sauer, V. B. (2010). Discharge measurements at gaging stations. *U.S. Geological Survey techniques and methods.* Arlington, VA: U.S. Geological Survey. Book 3 (87 pp.). Available at http://pubs.usgs.gov/tm/tm3-a8/.

U.S. Geological Survey. (1998). *A new evaluation of the USGS streamgaging network: A report to congress.* (20 pp.).

World Meteorological Organization. (2010). *Manual on stream gauging; Volume I: Field work, World Meteorological Organization Number 1044-V1.* (252 pp.).

CHAPTER 14

A Simple Streamflow Forecasting Scheme for the Ganges Basin

Y. Jiang*, W. Palash*, A.S. Akanda†, D.L. Small*, S. Islam*
*Tufts University, Medford, MA, United States
†University of Rhode Island, Kingston, RI, United States

1 INTRODUCTION

1.1 Floods in Bangladesh and the Role of Forecasting

The Ganges-Brahmaputra-Meghna (GBM) river system, with a combined drainage area of approximately $1500 \times 10^3\,km^2$, is the third largest freshwater outlet to the ocean in the world (Chowdhury & Ward, 2004). The Ganges, with a basin area of about $907 \times 10^3\,km^2$, the largest of three in terms of land area, originates in the southern foothills of the Himalayan Mountains in India and flows 2000 km before entering Bangladesh. The river continues to flow southeast further and joins the Brahmaputra and the Meghna before flowing into the ocean (Rasid & Paul, 1987). Being the most downstream riparian country of the GBM system, a massive discharge flows into the Bay of Bengal through Bangladesh (Fig. 1), particularly during the monsoon season. As a result, flooding is an annual recurring event during monsoon, with approximately one-fifth of its area inundated by flood waters every year and as much as two-thirds inundated during extreme events (Monirul Qader Mirza, Warrick, Ericksen, & Kenny, 2001).

The upstream region of the GBM basins receives highly seasonal rainfall (Rasid & Paul, 1987) and the river flow within Bangladesh is primarily generated by rainfall over the GBM catchment area outside the country (Chowdhury & Ward, 2007). Because Bangladesh is very flat, with an elevation gradient from the coast to northward is about 1 m per 50 km, once the water level exceeds flood stage, floodwaters could quickly spread out across the extensive floodplain (Webster et al., 2010). The flatness nature of the country also contributes to the stagnation of floodwaters, impedes the recession of excess water during high floods, and contributes to water-borne disease epidemics (Akanda et al., 2013).

Flood Forecasting
http://dx.doi.org/10.1016/B978-0-12-801884-2.00015-3

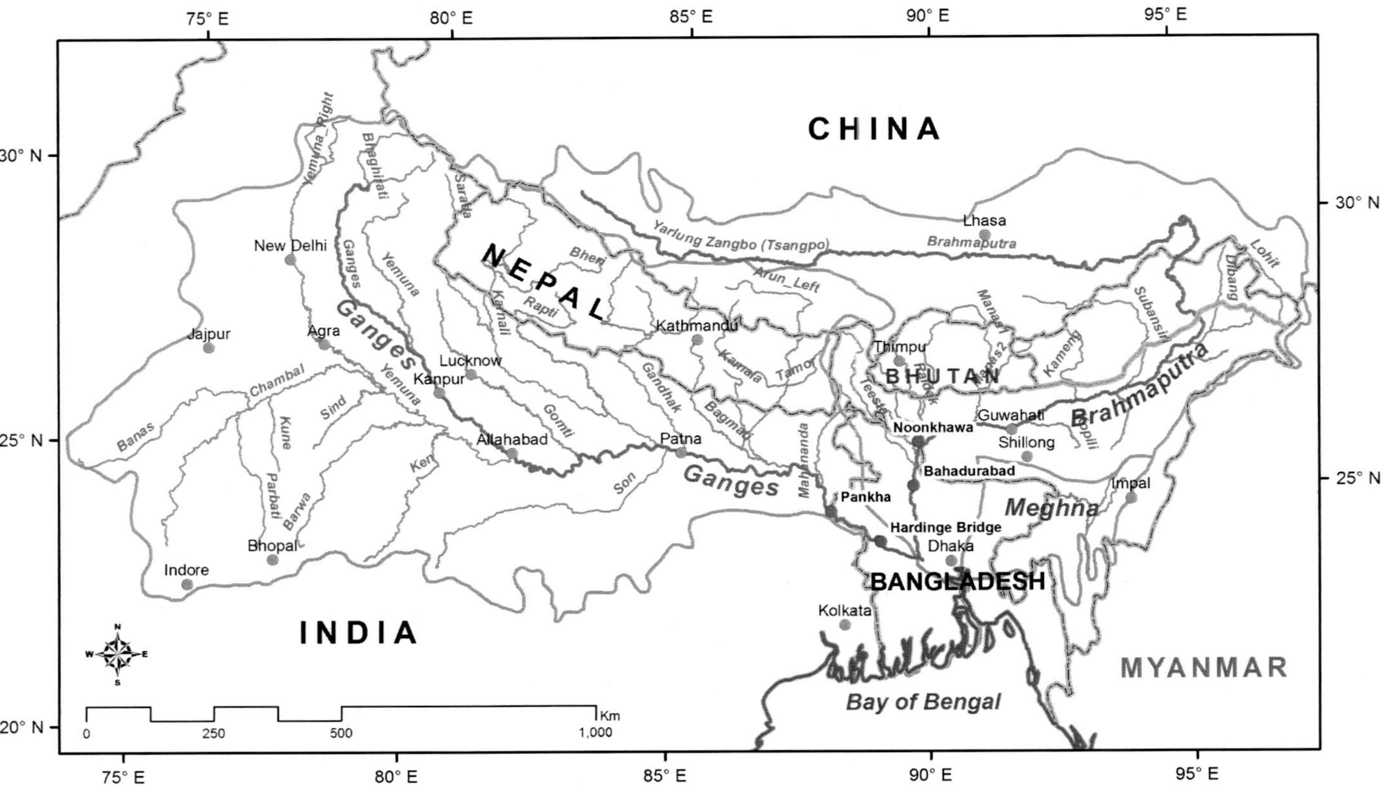

Fig. 1 Geographical setting of the Ganges-Brahmaputra-Meghna river system and riparian countries.

The flood-prone areas in Bangladesh make up 42% of the country's geographical area, and the loss caused by floods is tremendous (Monirul Qader Mirza et al., 2001). Rasid and Paul (1987) suggested that the most direct way to prevent flood damage would be to enforce floodplain regulation and sufficient delineation of flood hazard areas. However, due to limited land resources and high population density throughout the country, relocating the floodplain population to safer areas is not a feasible alternative (Chowdhury, 2000). Also, engineering solutions such as flood control projects have so far had limited achievements in this flat yet complex river system. On the other hand, nonstructural measures like timely flood forecasting and warning dissemination are viewed as a cost-effective way to reduce flood damages and loss of life (Akanda, 2012; Chowdhury, 2000). To help residents evacuate and secure their property ahead of time, the warning provided by a reliable forecasting system is thus critical (Chowdhury, 2004). The Center for Environmental and Geographic Information Services (CEGIS, 2006) estimated that a 2-day lead time could only reduce postflood households and agricultural costs by 2%, while a 7-day forecast could reduce costs by 20% for the flood vulnerable areas in Bangladesh. The Asian Disaster Preparedness Center (ADPC, 2002) noted that the minimum forecast horizon that allows farming communities to take effective actions against flood/drought is about 10 days, and optimal lead time is in excess of 3 weeks (Webster & Hoyos, 2004).

At present, the Flood Forecasting and Warning Centre (FFWC) of Bangladesh Water Development Board (BWDB) disseminates two types of forecasting data for the major river points in the flood-prone areas of the country: first, 1–5 days' lead time deterministic flood forecast; and second, 1–10 days' lead time probabilistic flood forecast with the collaboration of Climate Forecast Application Network, CFAN (http://www.ffwc.gov.bd, http://www.cfanclimate.com). However, there is scope for improvement in the existing operational forecasting capability that could ensure further improvement in the flood disaster management of Bangladesh. The main challenge that FFWC faces in producing more reliable flood forecasting information inside Bangladesh is the prediction of accurate upstream inflow through the Ganges, Brahmaputra, and Meghna rivers at the border points with India. The essential idea is, if upstream inflows through these three rivers are successfully predicted with greater reliability, FFWC's current operational flood forecasting model (ie, a one-dimensional river model based on MIKE 11) can provide skilled water level forecast inside Bangladesh. The aim of this study, therefore, is to produce forecast discharge at the Hardinge Bridge point on the Ganges River (Fig. 1), which is one of the two most important boundary points of FFWC's flood forecasting model. The other important point is Bahadurabad on the Brahmaputra River, but the present chapter will not cover that.

This chapter presents a simple statistical streamflow forecasting approach for the Ganges at Hardinge Bridge point as a proof-of-concept of linear model application for a large monsoonal river system. The proposed linear model uses observed streamflow and gridded precipitation data for the upstream basin areas to provide satisfactory forecasts for up to 10-day lead time. It appears that the improved Ganges flow forecasting up to 10-day lead time will improve FFWC's operational flood forecast, and will contribute to enhancing Bangladesh's resilience against floods in coming years.

1.2 Current and Prior Efforts in Bangladesh Flood Forecasting

The FFWC was established in 1972 by the BWDB to assist national preparedness for floods (Chowdhury, 2000). As the drainage basins of GBM system mainly lie outside of Bangladesh (only 8% of the combined drainage area is within Bangladesh), traditional methods of flood forecasting would require continuous cooperation from its upstream neighbor India to provide flow data. Currently, only the water level data from a few upstream gauging stations inside India are available through the Joint River Commission (JRC) of India and Bangladesh during the flood season, though sometimes these data are not available for real-time forecasting purposes. In summary, the available upstream data is not adequate to provide longer lead time forecasting for the flood-prone areas of Bangladesh.

Consequently, a numerical hydrodynamic model based on MIKE 11, developed by the Danish Hydraulic Institute (DHI), has been incorporated into the flood forecasting system since 1992 (Jakobsen, Hoque, Paudyal, & Bhuiyan, 2005). FFWC predicts water level at the entry points of the Ganges and the Brahmaputra River from India to Bangladesh, as shown in Fig. 1, during the flood season. These estimates are then used as the boundary conditions of MIKE 11 flood model to forecast water level at downstream points. Hence, the subjective nature of this boundary condition estimation — due to lack of information about upstream conditions inside India — for the Ganges and the Brahmaputra puts a major constraint to the FFWC flood forecasting system. FFWC issued flood forecasts up to 3-day until 2013; from 2014, FFWC increased their deterministic flood forecasting lead time to 5-day using the same method.

A related initiative named the CFAN, an international consortium led by the Georgia Institute of Technology, USA came forward with the ensemble-based 1- to 10-day discharge forecast of the Ganges and Brahmaputra Rivers inside Bangladesh to overcome the limitation of FFWC in generating its forecast beyond 3-day forecast horizon in 2003. Hopson and Webster (2010) described this forecasting scheme for 1- to 10-day lead time flood forecasting at Hardinge Bridge and Bahadurabad for 2003–07. The European Center for Medium-Range Weather Forecasts (ECMWF) ensemble prediction schemes were adopted, and the meteorological model output was calibrated using the precipitation inputs from National Aeronautics and Space Administration (NASA) and National Oceanic and Atmospheric Administration (NOAA) to reduce system errors. Results suggested that except for moderate forecast ability in 2003, all major floods were well forecasted at a 10-day lead time during 2003–07 for the Brahmaputra River. The timing of flood (the onset and end of the flood period) and forecast flood risk probabilities were skillful, despite the considerable overestimation or underestimation of peak magnitudes. For the Ganges, forecasts were less skillful after 5-day lead time; the authors suggested that inappropriate incorporation of the Ganges water management scenario in the upstream might be the reason for that. Webster et al. (2010) extended the forecast to 2009 for both rivers. At present, FFWC receives this ensemble flow prediction on a daily basis from CFAN, uses them as the boundary condition for their flood model, and thus provides 1- to 10-day lead time probabilistic flood forecast.

The advantage of their work is, up to 10-day lead time, the observed discharge is generally contained within the ensembles spread, the forecast scheme displays capability in capturing the onset and end of flood period, and high flood risk probabilities were forecasted most of the time during the flood period. It suggests that the forecast system has the capability of accounting for uncertainty in rainfall and flow forecasting. However, the challenge is how to interpret the ensemble forecasting results and issue effective warnings, how to utilize it to make predictions for other river stations within the country, and ultimately how to communicate these forecasts to the user community and provide appropriate guidance for action. To overcome this, FFWC uses the only ensemble mean, ensemble mean plus one standard deviation (upper-bound), and mean minus one standard deviation (lower-bound) of CFAN predictions in their operational flood forecast model.

Hirpa et al. (2013) employed passive microwave sensing technique to estimate the river width at more than 20 locations upstream of the Hardinge Bridge and Bahadurabad points to calculate the nowcast and forecast discharge at these two points. For a different lead time, the maximum correlation between satellite-derived flow (SDF) and gauge discharge observations was selected as the basis for forecasting. The SDF forecast tended to underestimate the peaks for both rivers, especially for the Brahmaputra, but showed capability in capturing the rising and recession limbs of the hydrograph. After combining with "persistence forecast," the forecast performance was greatly improved; for the Brahmaputra, the forecast had shown considerably better performance than autoregressive moving-average (ARMA) model beyond the 3-day horizon. The forecast for the Ganges was slightly worse than the ARMA model up to 10 days. However, Hirpa et al. argued that the weakness of using the ARMA model was that the continuous past observations must be available, and thus their approach still provided useful perspectives.

Hossain et al. (2014) outlined a method of using JASON-2 satellite altimetry for forecasting the Ganges and Brahmaputra flow by propagating forecasts from upstream (Indian) locations to downstream (Bangladesh) locations through a hydrodynamic river model. The forecast river levels were then compared with the 5-day later "nowcast" simulation by a river model developed in Hydrologic Engineering Centers River Analysis System (HEC-RAS) and using in situ river level at the upstream boundary points in Bangladesh. They analyzed their forecast accuracy in six river segments or stretches of the forecasting domain including rivers such as the Ganges, Brahmaputra, Padma, Surma, Upper Meghna, and Lower Meghna on seasonal separation. They reported 5-day lead time forecast with an average root mean square error (RMSE) ranging from 0.5 to 1.5 m and a mean bias (underestimation) of 0.25–1.25 m in river water level estimation relative to nowcast or HEC-RAS model output. Comparison of observed water levels at three river stations (Bahadurabad, Sirajganj on the Brahmaputra, and Hardinge Bridge on the Ganges) showed an average forecast error ranging from 0.4 to 0.5 m and RMSE from 0.2 to 0.7 m.

Biancamaria, Hossain, and Lettenmaier (2011) utilized Topex/Poseidon satellite altimetry measurements in India on the Ganges and Brahmaputra to estimate water elevation anomalies at Hardinge Bridge and Bahadurabad. They reported good quality forecast for 5-day lead time and suggested that although 10-day lead time forecast yielded higher errors, it could still be valuable in the operational forecast. However, several limitations exist for this approach. First, the historical long record of satellite altimeters data is not available, thus restricting long-term study within this framework (less than 2-year data was used in this study). Second, as the current generation satellite altimeters were primarily designed for oceanographic purposes, the low time resolution (10-day repeat period in this study) and relatively inaccurate river heights measurement remain an issue. In addition, to realize real-time forecasts, additional efforts are needed to shorten the repeat period, and the combined use of different satellite products might become promising to improve forecast performance. Biancamaria et al. also discussed that the future Surface Water and Ocean Topography (SWOT) wide swath altimeter would provide improved forecast coverage and accuracy, and would be able to address this shortcoming.

Siddique-E-Akbor, Hossain, Lee, and Shum (2011) compared water level estimates obtained from two methods, Envisat satellite altimetry, and hydrodynamic-hydrologic (HEC-RAS) model, for three rivers in Bangladesh. They reported that in general the average RMSE was larger than 2.0 m, and Brahmaputra had both the lowest RMSE (0.67 m) and highest correlation (1.0). As the HEC-RAS model simulated estimate had relatively lower RMSE (0.7–1.2 m), they suggested that the model estimates might be an alternative benchmarking inland altimetry data when in situ gauging observations were not available.

Akhtar, Corzo, van Andel, and Jonoski (2009) applied artificial neural network (ANN) in Ganges flow forecasting. The study period is 2001–05, with 2001–03 for training and 2004–05 for verification. The input variables consist of previous river flow, previous lagged area-average rainfall, sum of lagged rainfall, and actual evapotranspiration (AET), and they were grouped into four options in the study. As they noted that low flow was less dominated by the previous upstream rainfall but more by local groundwater dynamics, only high flow was selected for the analysis as the goal was to build flood forecast model. Experiments suggested that within 5-day forecast horizon, rainfall and AET failed to add value compared to the persistence forecast (only previous flow used as input), and sum of lagged rainfall seemed to produce better results than lagged area-average rainfall when combined with previous flow as inputs. Akhtar et al. conducted experiments up to 10-day lead time, and results showed that the value of rainfall (sum of lagged rainfall as input) started to become evident beyond 7-day. It is worth mentioning their use of flow length and travel time as an approach to incorporate spatial precipitation into the ANN model. As the Ganges runs through such a large basin, the spatial distribution of precipitation could be critical in predicting river flow at Hardinge Bridge. The disadvantage of this work is that the ANN model structure is unknown, and for statistical methods the training period is considered short.

1.3 A Simple Flood Forecasting Scheme: Hypothesis and Questions

Despite various attempts to use modeled, forecasted, and remotely sensed precipitation data products to increase the forecasting lead time, comparative utility of these computationally intensive methods and simple numerical approaches have not been fully explored in the context of the Ganges. We hypothesize that, for the Ganges and similar large monsoonal river basins — where availability and access to upstream basin data are limited and detailed forecasting model setup are prohibitively expensive — enhanced quantitative precipitation may not lead to enhanced streamflow forecasting accuracy. The reason could be explained as the limited accuracy of available precipitation forecast products and uncertainty associated with the detailed hydrologic modeling. We provide a simple linear regression modeling framework for the Ganges River basin to explore the utility of such forecasts to enhance the streamflow forecasting accuracy. With this premise, we plan to answer the following questions: What is the achievable accuracy of simple linear forecasting models for the Ganges that uses only lagged values of flow? What is the utility of adding retrospective observations or "perfect forecasts" of precipitation over the contributing area of the basin, in terms of forecasting skill and lead time? How much improvement in streamflow forecasting accuracy does the medium-range precipitation forecasts add to streamflow forecasts beyond 5-day lead time in large basins with seasonally dominated hydrology?

The idea of "perfect forecasts" of precipitation is derived from a premise that no forecast product could be as accurate as observed data. As far as observed data is concerned, APHRODITE is one of the gridded rainfall data that is prepared with gauge observations and is available in the public domain and easily accessible. Applying APHRODITE data thus allows us to test the validity of our hypothesis that even the most accurate rainfall forecast beyond 6- to 7-day may not improve the 10-day forecasting capability for the Ganges flow. Considering the limitation of APHRODITE's availability on real-time, we introduced the TRMM 3B42 rainfall estimates to our modeling scheme and compared the results derived from APHRODITE dataset.

1.4 Datasets Used in the Study

The historical records of daily discharge of the Ganges at Hardinge Bridge and daily precipitation over the Ganges basin are analyzed to characterize the pattern of the flood period and the relationship between rainfall and river discharge. The river discharge at Hardinge Bridge was collected from BWDB. This is the daily rated discharge, calculated from water level measurements by Bangladesh authority (Apr. 1, 1934–Mar. 31, 2009), with about a year missing records during the 1971–72 liberation war of Bangladesh. Two sets of gridded basin-wide daily rainfall data were used in the study: the APHRODITE V1101 (1951–2007) with 0.5 degree spatial resolution (http://www.chikyu.ac.jp/precip/) and TRMM 3B42V7 (1998–2013) with 0.25 degree resolution (http://gdata1.sci.gsfc.nasa.gov/daac-bin/G3/gui.cgi?instance_id=TRMM_3B42RT). The APHRODITE data was created primarily with data derived from raingauge-observation network, while TRMM is the satellite rainfall estimates.

2 THE GANGES FLOODS AND RAINFALL-RUNOFF RELATIONSHIPS

2.1 The Seasonality of the Ganges Basin

The southwest monsoon originating from the Bay of Bengal between Jun. and Oct. is the main source of precipitation for the Ganges River basin, being accompanied with the occasional tropical cyclones that also originates from the Bay of Bengal. On average, about 85–87% rainfall occurs during four monsoon months (Jun.-Sep.), and only a small amount of rainfall occurs in Nov.-Jan.

In the upper Gangetic Plain in Uttar Pradesh (India), annual rainfall averages 760–1020 mm, in the Middle Ganges Plain of Bihar (India) 1020–1520 mm, and in the delta region 1520–2540 mm (FAO, 2016). The seasonality of the basin rainfall is captured in the Ganges flow measurement at the Hardinge Bridge point, and is shown as historical monthly average rainfall and discharge in Fig. 2.

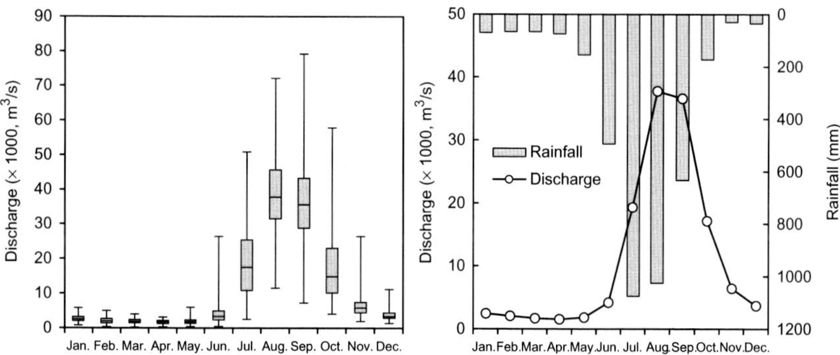

Fig. 2 The seasonality of the average monthly flow (1934–2009) (left panel) and rainfall-flow relationship (1951–2007) (right panel) of the Ganges basin at the Hardinge Bridge point. The box-and-whisker plot in the left panel shows the upper and lower quartiles deviated from the median as a box. The whiskers, the lines extending vertically from boxes, indicate variability (maximum and minimum) outside the upper and lower quartiles. In the right panel, monthly rainfall is plotted in secondary *y*-axis and values are in reverse order.

2.2 Rainfall-Runoff Relationship Over the Ganges Basin

The drainage area of the Ganges is $944 \times 10^3 \text{km}^2$ at Hardinge Bridge, and it involves a wide range of travel time for water to reach Hardinge Bridge from its original point of precipitation in the upstream regions. To add appropriate lags to rainfall over the Ganges basin, we calculated average runoff travel time from four large domains or zones of the Ganges to the Hardinge Bridge point (Fig. 3). The hydrological spatial analysis of ArcGIS and spatially distributed unit hydrograph (SDUH) concept has been applied to calculate the runoff travel time map or isochrones. The method involves using Shuttle Radar Topography Mission (SRTM) generated digital elevation data, U.S. Geological Survey (USGS) prepared land use, and the Food & Agriculture Organization (FAO) prepared soil coverage data in its hydrological analysis. The resulting isochrones map thus provides the flow travel time for each raster cell or grid of the basin that it takes to reach the basin outlet, and the domain average travel time is calculated from the isochrones map. The domain average travel time is found at 0–7, 8–13, 14–18, and 19–25 days for domains one to four, respectively.

To examine the rainfall spatial variability, the four domain average rainfall totals are compared to one another. The annual domain average rainfalls for domain one to four are 608, 845, 969, and 822 mm respectively. Compared to annual basin average rainfall at about 847 mm, domains two and four receives a similar range of rainfall while domain one is deficient, and domain three receives more than domain average rainfall. Webster et al. (2010) plotted 10-day average basin-wide precipitation over the Ganges for 2004 (Jun.-Oct.), shifted it forward 15 days, and found a good relationship between it and daily discharge at Hardinge Bridge. Similarly, Chowdhury and Ward (2004) found that the successive flows in Jul., Aug., and Sep. at Hardinge Bridge were highly correlated to rainfall received upstream in Jun., Jul., and Aug. with R^2 is 0.54, 0.48, and 0.53, respectively. They also noticed that the seasonal flow (Jun.-Sep., JJAS) was highly correlated with same period rainfall ($R^2 = 0.56$).

As the rainfall is scarce, and flow is low during most of the year for the Ganges basin, the annual relation between rainfall and river flow provides a misleading measure of predictability. Thus, it is important to explore the correlation between rainfall and discharge during Jul. to Oct., more particularly during Aug. to Sep. The historical data shows that the highest flood usually occurs during these 2 months in the Ganges. The basin average rainfall could explain 84% of the Ganges discharge for the entire year, but only 37% for Aug. and Sep.

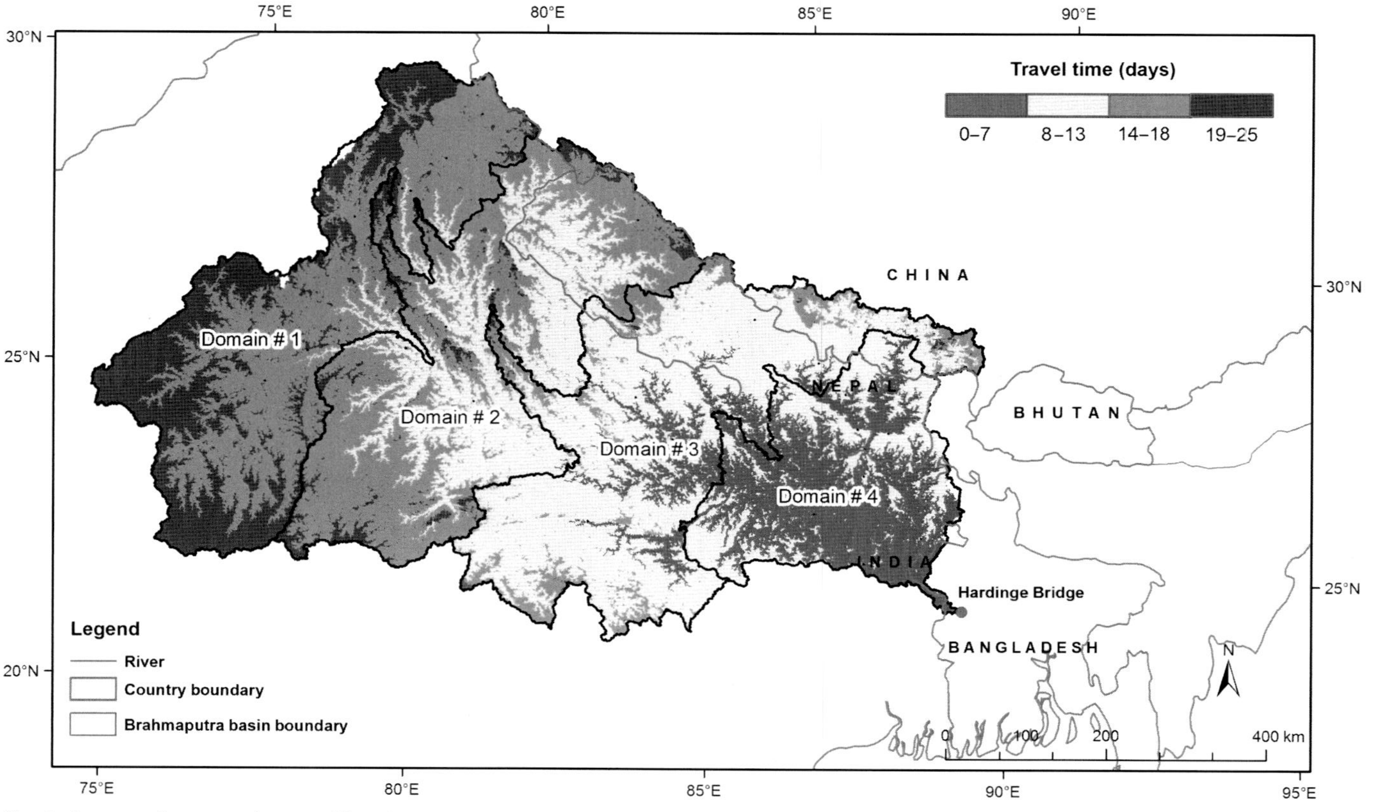

Fig. 3 Average flow travel time of four large domains of the Ganges basin.

3 A SIMPLE *Q-Q* AND *Q+P-Q* MODEL FOR THE GANGES

In this section, two statistical models are proposed to predict the Ganges discharge at the Hardinge Bridge point, and comparisons are made among different lead times. The first one is the Q-Q model, where "Q" denotes river discharge at Hardinge Bridge, the second one is *Q+P-Q* model, where "*P*" denotes upstream Ganges basin rainfall. Both models utilize flow persistence, and the only difference is that *Q+P-Q* model takes rainfall observation into account. The rainfall observation could be either retrospective observation or observation plus forecast rainfall. Hence, the *Q+P-Q* model is divided into two types: *Q+P-Q* "without forecast rainfall" and "with forecast rainfall" model. The domain average rainfall, described in Section 2.2, is used instead of basin average rainfall to incorporate rainfall spatial variability in the proposed linear model.

3.1 A Persistence-Based *Q-Q* Model

The autocorrelation analysis of the Ganges flow shows that there is a strong persistence of streamflow at the daily scale up to several days. Therefore, we introduce a Q-Q model based on daily streamflow or discharge values in Eq. (1), where Q denotes the streamflow. The experimentation suggests that today's and the previous day's flow of the Ganges could represent the persistence forecast potential, and the model structure is as follows:

$$Q_{t+n} = \alpha_n Q_t + \beta_n Q_{t-1} + \gamma_n \tag{1}$$

where Q_{t+n} is the forecasted flow at Hardinge Bridge on day $(t+n)$ with forecasts are made on current day t, Q_t is river discharge on day t, Q_{t-1} is river discharge on day $(t-1)$, α_n, β_n, and γ_n are regression model coefficients obtained for different lead time n during model calibration.

3.2 A Simple *Q+P-Q* Model

Introducing the rainfall observation into Q-Q model provides the structure of *Q+P-Q* model based on the understanding that the upstream rainfall has the potential to be a good predictor for flow at Hardinge Bridge. In this model, *Q* denotes river flow at Hardinge Bridge, and *P* denotes upstream rainfall over different domains of the Ganges basin as shown in Fig. 3. The calculation of the domain average rainfall and applied lags are described in Section 2.2. The corresponding lags are added to the domain average rainfall to account for the average travel time from each domain to the point of interest (eg, Hardinge Bridge). The model structure is as follows:

$$Q_{t+n} = a_n Q_t + b_n Q_{t-1} + c_n P_1 + d_n P_2 + e_n P_3 + f_n P_4 + g_n \tag{2}$$

where Q_{t+n}, Q_t, Q_{t-1} denote the same as Eq. (1); P_1, P_2, P_3, and P_4 are lagged domain average daily rainfall for domains one to four; and a_n, b_n, c_n, d_n, e_n, f_n, and g_n are model coefficients obtained for different lead time n from the calibration period.

3.3 Adding the "Perfect Rainfall Forecast"

We did not use any real rainfall forecast product in this study. Rather we considered APHRODITE or TRMM observed rainfall as the "perfect rainfall forecast" data, as if we produced forecasts with prior knowledge of the rainfall observations. The idea of introducing the "perfect rainfall forecast" into the Q+*P*-Q model is to examine the maximum level of forecast skill that the proposed linear model could achieve if we had the most accurate forecasted rainfall data.

Since the APHRODITE rainfall data is available for 1951–2007, the flow records and rainfall data are divided into 1951–90 as calibration and 1991–2007 as a validation period to calibrate and validate the proposed linear model. For the TRMM dataset that is available from 1998, the calibration and validation periods are divided into 1998–2001 and 2002–07, respectively.

4 RESULTS

To assess the performance of different models, we used the coefficient of determination (R^2), adjusted coefficient of determination (adj R^2), RMSE and Nash-Sutcliffe Efficiency (NSE) metrics, which are defined as follows:

$$R^2 = \left(\frac{\sum_{i=1}^{m} \left(y_{\text{obs}} - y'_{\text{obs}}\right)\left(y_{\text{mod}} - y'_{\text{mod}}\right) /}{\sqrt{\left(\sum_{i=1}^{m} \left(y_{\text{obs}} - y'_{\text{obs}}\right)^2\right)} \sqrt{\left(\sum_{i=1}^{m} (y_{\text{mod}} - y'_{\text{mod}}\right)}} \right)^2 \quad (3)$$

$$R^2\left(\text{adj}\right) = 1 - \left(1 - R^2\right)\left(m-1\right) / \left(m-p-1\right) \quad (4)$$

$$\text{RMSE} = \sqrt{\left(1/m \sum_{i=1}^{m} \left(y_{\text{obs}} - y_{\text{mod}}\right)^2\right)} \quad (5)$$

$$\text{NSE} = 1 - \sum_{i=1}^{m} \left(y_{\text{obs}} - y_{\text{mod}}\right)^2 / \sum_{i=1}^{m} \left(y_{\text{obs}} - y'_{\text{obs}}\right)^2 \quad (6)$$

where R^2 measures how well the model fits the data, R^2 (adj) is adjusted R^2 takes the number of predictors into account, y_{obs} is observation, y_{mod} is the model estimate of y_{obs}, y'_{obs} is average value of y_{obs}, y'_{mod} is average value of y_{mod}, m is the number of observations, and p is the number of independent regressors, that is, the number of variables in the model. The NSE can range from negative infinity to one; when it is negative, the observed mean provides a better predictor than the modeled value; a value of zero indicates that predictions are as good as the observed mean; while the closer NSE value is to one, the more accurate the model is, and a value of one means a perfect model.

Our results of the utility of simple linear model in enhancing the accuracy of 5-, 7-, and 10-day lead time streamflow forecast of the Ganges River are summarized in Table 1 and Figs. 4–7. As shown in Fig. 4A and Table 1, the simple lagged streamflow model (called *Q-Q* model) has the comparable forecasting ability to a lagged streamflow model combined with precipitation observation (called *Q+P-Q* without forecast rainfall or just *Q+P-Q* model) for up to 5-day forecasting lead time. For 5-day lead time forecast, adjusted R^2 is 0.81 and RMSE is 5451 m^3/s for the *Q-Q* model — that uses only lagged discharge value of the Ganges for its regression — during monsoon (Jul.-Oct.) of 2002–07 (Table 1). The contribution of adding observed upstream rainfall to the *Q-Q* model is limited within 5-day lead time; R^2 (adj) only improves from 0.81 to 0.87 for APHRODITE and 0.88 for TRMM rainfall dataset. The RMSE improves from 5451 to 4458 m^3/s for APHRODITE and 4289 m^3/s for TRMM dataset, respectively (Table 1). With forecasting lead time beyond 5-day, enhancement of forecasting accuracy with the addition of precipitation information is noticeable. The R^2 (adj) value of *Q-Q* model for 7- and 10-day forecast is 0.64 and 0.47 that improves up to 0.80 and 0.63 for APHRODITE and 0.82 and 0.68 for TRMM rainfall data use in *Q+P-Q* model, respectively. The RMSE improves from 7337 m^3/s of *Q-Q* model to 5559 m^3/s for APHRODITE and 5179 m^3/s for TRMM data use in 7-day flow forecast. The corresponding improvement for the 10-day Ganges

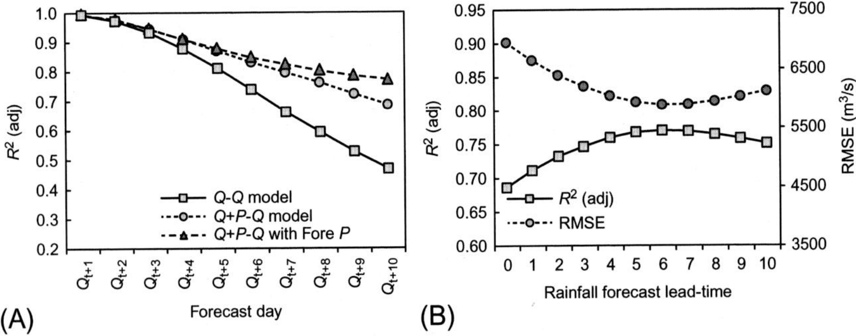

Fig. 4 (A) Flow prediction capability of *Q-Q* and *Q+P-Q* model (without and with "perfect forecast rainfall") during Jul.–Oct., 2002–07; (B) 10-day flow prediction capability of *Q+P-Q* model with the addition of 1- to 10-day lead time "perfect forecast rainfall." The R^2 (adj) is shown in primary *y*-axis and RMSE in secondary *y*-axis.

Table 1 Performance of linear model for 5-, 7-, and 10-day lead time forecast during monsoon (Jul.–Oct.) over the period 2002–07

	5-day			**7-day**					**10-day**				
	Q-Q	***Q+P-Q* w/o FR**		***Q-Q***	***Q+P-Q* w/o FR**		***Q+P-Q* with 3-day FR**		***Q-Q***	***Q+P-Q* w/o FR**		***Q+P-Q* with 6-day FR**	
Stats	–	**APH**	**TR**	–	**APH**	**TR**	**APH**	**TR**	–	**APH**	**TR**	**APH**	**TR**
R^2 (adj)	0.81	0.87	0.88	0.66	0.80	0.82	0.82	0.84	0.47	0.68	0.73	0.77	0.80
NSE	0.80	0.87	0.88	0.64	0.80	0.82	0.82	0.84	0.42	0.68	0.73	0.77	0.80
RMSE	5451	4458	4289	7337	5559	5179	5138	4910	9350	6944	6376	5882	5535

Note: *APH*, APHRODITE; *TR*, TRMM; *w/o*, without; *FR*, forecast rainfall.

Q+P-Q with forecast rainfall was not applied for 5-day lead time.

forecast is from 9350 m³/s of Q-Q model to 6944 m³/s for APHRODITE and 6376 m³/s for TRMM data use in *Q*+*P*-*Q* model.

The improvement of flow prediction capability of Q-Q, Q+*P*-Q without forecast rainfall and Q+*P*-Q with "perfect forecast rainfall" model are shown in Fig. 4A. Fig. 4B shows the 10-day flow prediction capability of Q+*P*-Q model by introducing "perfect forecast rainfall" along with observed rainfall into its regression. When rainfall forecast lead time increases from 0 to 6 days, R^2 (adj) increases from 0.68 to 0.77 for both APHRODITE and TRMM dataset. The model performance remains more or less unchanged or even deteriorates after that. It indicates that 6-day rainfall forecast is adequate for 10-day lead time flow forecast at Hardinge Bridge. Also it suggests that the upstream basin rainfall over the last 4 days does not contribute much to the current day's Ganges flow measured at the Hardinge Bridge point.

Fig. 5 shows the long-term performance of *Q*+*P*-*Q* model (without forecast rainfall) for 10-day lead time forecast over the period 2002–07. The model demonstrates the capability to forecast the rising and recession limbs of the peak in all years and captures the peaks for average to medium flood years. However, the model misses the peaks for large flood years (ie, 2003 and 2004) and thus loses the capability to predict high flows during an extreme flood event. This is probably because long-term calibration tends to maximize the average model performance, thus generating somewhat unsatisfactory results for extreme flood years.

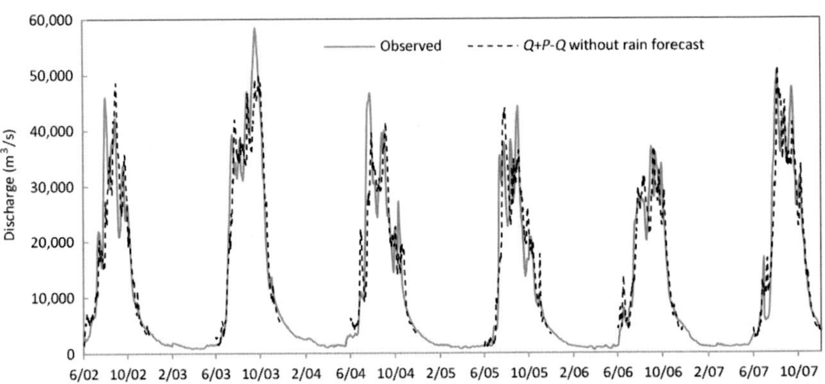

Fig. 5 The Ganges observed and 10-day forecast flow of *Q*+*P*-*Q* without forecast rainfall model for the period 2002–07.

The performance of Q-Q, Q+*P*-Q without forecast rainfall and "with rainfall forecast" model in predicting the Ganges flow has been shown in Figs. 6 and 7 as hydrographs for two relatively large flooding years in recent

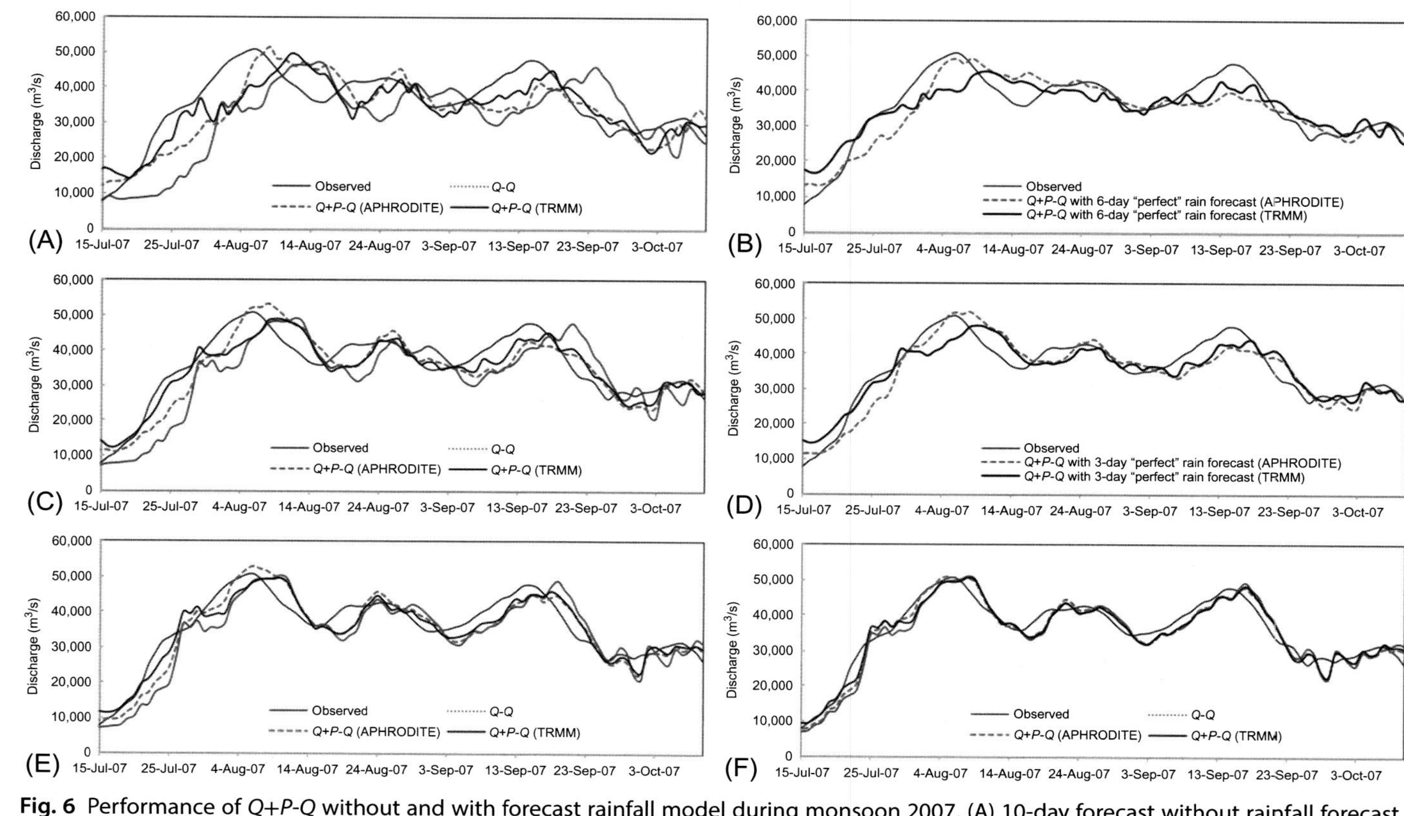

Fig. 6 Performance of *Q+P-Q* without and with forecast rainfall model during monsoon 2007. (A) 10-day forecast without rainfall forecast, (B) 10-day forecast with rainfall forecast, (C) 7-day forecast without rainfall forecast, (D) 7-day forecast with rainfall forecast, (E) 5-day forecast without rainfall forecast, and (F) 3-day forecast without rainfall forecast.

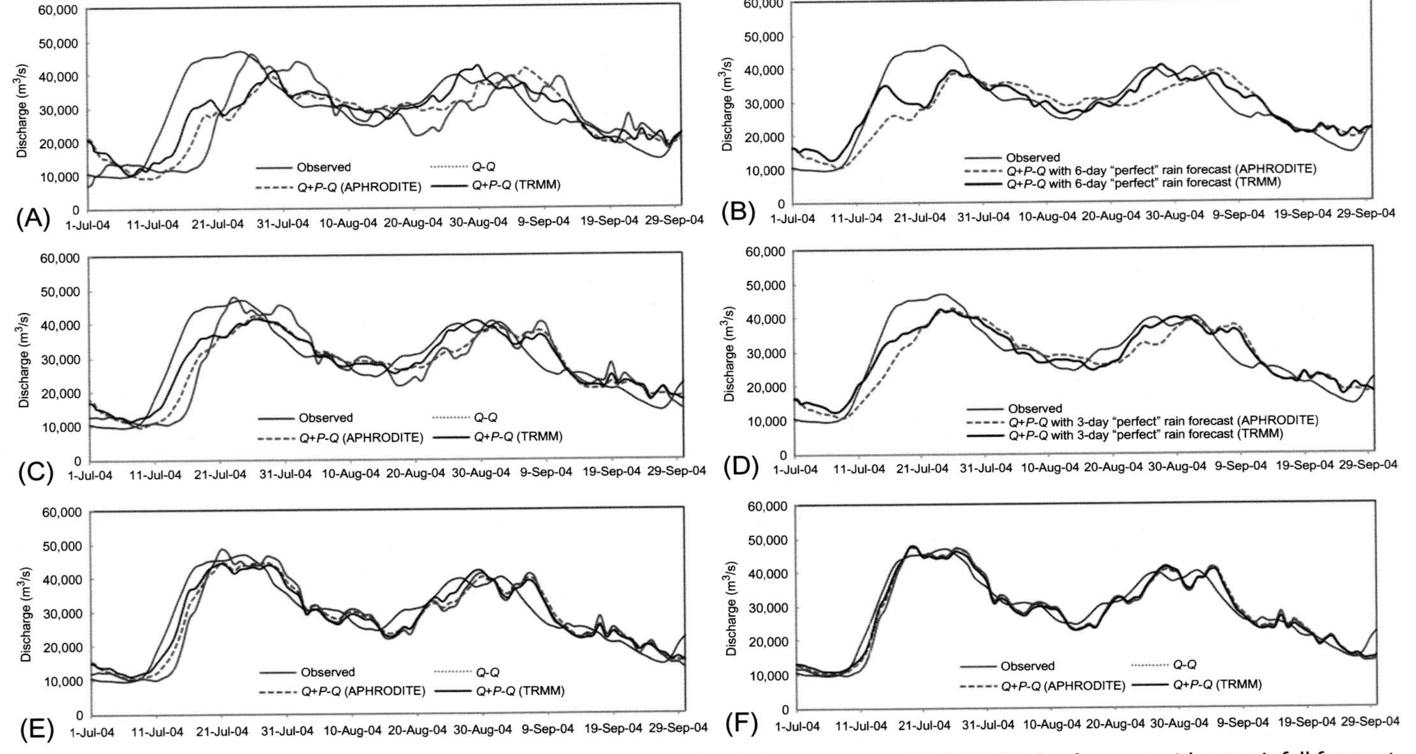

Fig. 7 Performance of *Q+P-Q* without and with forecast rainfall model during monsoon 2004. (A) 10-day forecast without rainfall forecast, (B) 10-day forecast with rainfall forecast, (C) 7-day forecast without rainfall forecast, (D) 7-day forecast with rainfall forecast, (E) 5-day forecast without rainfall forecast, and (F) 3-day forecast without rainfall forecast.

history of Bangladesh — floods in 2004 and 2007. Up to 10-day lead time, the performance of $Q+P$-Q models (without and with forecast rainfall) are exceptionally good during the 2007 monsoon; they successfully capture the Ganges flow rise and fall during this high flood season with reasonable accuracy. The overall performance of the model during monsoon 2002–07 is shown in Table 1. The model performance is equally good up to the 7-day lead time for 2004 flood season, but does not capture the early flood peak (from 15 to 30 July 2004) well in 10-day flow forecast. However, the performances in predicting the Ganges flow for the remaining period of the 2004 flood season is as good as it is for the 2007 flood period.

5 DISCUSSION

This chapter explored the utility of using a linear input-output model to provide streamflow forecasts up to 10 days. The contribution of adding upstream observed rainfall to the model that used only lagged values of Ganges flow (ie, Q-Q model) is not significant within 5-day lead time. A plausible explanation of why the inclusion of rainfall hardly improves the model performance within 5-day lead time is that the current and previous flows already contain most of the basin response in an aggregated form in determining the future flow within the stipulated 5-day period. It also suggests that for such a large basin, day-to-day persistence is adequate for forecasting lead time up to 5 days, and detailed modeling of the basin response may not add much to enhance forecasting accuracy.

However, the model improvement becomes noticeable beyond 5-day lead time. Similar observations could be drawn from the study of Akhtar et al. (2009); they applied ANN models to flow forecasting at Hardinge Bridge, and demonstrated little added value in introducing rainfall as input until 5-day lead time.

The need for extended forecasts — beyond 5 days — has inspired recent efforts utilizing quantitative precipitation forecasts from various numerical weather prediction (NWP) models. The hydrological community has also shown interest in ensemble flood forecasting, which incorporates ensembles of NWPs instead of single deterministic forecasts to make predictions (Thielen et al., 2009). Cloke and Pappenberger (2009) reviewed a number of case studies on ensemble flood forecasting, and found that the majority are based on single hindcasted events, while only a few have evaluated multiple events over several years. It is also worthwhile to note that most of these studies are for much smaller basins except for the Ganges and the Brahmaputra (Webster et al., 2010).

Cloke and Pappenberger (2009) summarized key challenges in using ensemble flood forecasts. First, the NWPs may not be very useful because of their accuracy of precipitation forecasts. Second, as extreme floods are rare, there are not enough instances to calibrate the models adequately or evaluate their performances, plus most studies tend to avoid reporting failures (false alarms) of their particular systems. Third, the intensive demand for computer power might restrict the use of ensemble methods in operational systems, especially for the case of the recently proposed grand-ensemble forecast (He et al., 2009) which further burdens computation by increasing the number of ensemble members. Fourth, it remains a challenge in terms of how to interpret the forecasting results, how to incorporate them into decision-making, and how to best disseminate the results to the end-users. In addition, for hindcasts or real-time forecasts, the model performance evaluation tends to be quite subjective; there is no generally agreed-upon scoring system for assessment and basin-wide comparison between these studies.

To sum up, a linear modeling framework that uses only lagged values of flow is able to provide skilled forecasts for 3- to 5-day lead time. At the same time, introducing upstream rainfall observation into the linear model can provide skilled forecast for 7- to 10-day lead time, necessary for improved disaster planning and management. The introduction of forecast rainfall improves the *Q+P-Q* model noticeably compared to flow forecast without rainfall forecast. However, it appears that adjusted R^2 and RMSE do not improve beyond rainfall forecasting lead time of 5–6 days for 10-day flow forecast. Consequently, we suggest that for large basins, it may not be necessary to match the rainfall forecast lead time with the streamflow forecast lead time. A shorter lead time with more accurate forecast rainfall (eg, 5–6 days forecast rainfall for the Ganges basin) could be found effective to enhance the forecasting accuracy of streamflow up to 10-day lead time.

We tested our model using APHRODITE and TRMM rainfall dataset for the upstream basin areas. Both datasets provide almost identical results when they are applied to the proposed linear models. The TRMM rainfall data are easily available on a near real-time basis. It also provides a perfect opportunity to apply this dataset in the proposed linear model for predicting the 1- to 10-day Ganges flow inside Bangladesh in a daily operational flood forecasting activities.

Findings from this study are expected to be applicable to other large river basins in South and Southeast Asia with strongly seasonal hydrology such as the Brahmaputra, the Mekong, or the Irrawaddy. It will be worthwhile to examine the adequacy of the proposed method for other large basins from this region. The location of the stream gauge in our study is located at Hardinge Bridge, 132 km downstream of the point where the Ganges enters Bangladesh from India. Over 90% of the basin area lies upstream of this point; thus, the rainfall-runoff process is controlled by the heterogeneous variability of rainfall, land use, topography, and nature of soil type, as well as groundwater interactions. However, our findings suggest that detailed heterogeneous representation may not be necessary for operational forecasting of flow at Hardinge Bridge with a 10-day lead time. More importantly, the sociopolitical reality of lack of real-time data sharing and cooperation between India and Bangladesh strengthen the utility of our model. Our findings suggest that a careful utilization of available streamflow and rainfall data into a simple linear model can be employed with minimal cost for the Ganges basin flood forecasting operations in Bangladesh, and potentially for other large river basins for the region.

ACKNOWLEDGMENT

This work was supported, in part, by grants from the US National Science Foundation (RCN-SEES 1140163 and NSF-IGERT 0966093).

REFERENCES

ADPC. (2002). *Application of climate forecasts in the agriculture sector.* Climate forecasting applications in Bangladesh project report 3, Bangkok: Asian Disaster Preparedness Center. 29 pp.

Akanda, A. S. (2012). South Asia's water conundrum: Hydroclimatic and geopolitical asymmetry, and brewing conflicts in the eastern Himalayas. *International Journal of River Basin Management, 10*(4), 307–315.

Akanda, A. S., Jutla, A. S., Gute, D. M., Sack, B. R., Alam, M., Huq, A., et al. (2013). Population vulnerability to biannual cholera peaks and associated macro-scale drivers in the Bengal delta. *American Journal of Tropical Medicine and Hygiene, 89*(5), 950–959.

Akhtar, M. K., Corzo, G. A., van Andel, S. J., & Jonoski, A. (2009). River flow forecasting with artificial neural networks using satellite observed precipitation pre-processed with flow length and travel time information: Case study of the Ganges river basin. *Hydrology and Earth System Sciences, 13*(9), 1607–1618.

Biancamaria, S., Hossain, F., & Lettenmaier, D. P. (2011). Forecasting transboundary river water elevations from space. *Geophysical Research Letters, 38*, L11401.

CEGIS. (2006). *Centre for Environmental and Geographical Service (CEGIS) early warning system.* Final report to the Asian Development Bank. 263 pp. Available at: http://cfan.eas.gatech.edu/PJWebster/Floods/.

Chowdhury, M. R. (2000). An assessment of flood forecasting in Bangladesh: The experience of the 1998 flood. *Natural Hazards, 22*(2), 139–163.

Chowdhury, R., & Ward, N. (2004). Hydro-meteorological variability in the greater Ganges-Brahmaputra-Meghna basins. *International Journal of Climatology, 24*(12), 1495–1508.

Chowdhury, M. R., & Ward, M. N. (2007). Seasonal flooding in Bangladesh—Variability and predictability. *Hydrological Processes, 21*(3), 335–347.

Cloke, H. L., & Pappenberger, F. (2009). Ensemble flood forecasting: A review. *Journal of Hydrology, 375*(3–4), 613–626.

FAO. (2016). AQUASTAT website. Food and Agriculture Organization of the United Nations (FAO). http://www.fao.org/nr/water/aquastat/basins/gbm/index.stm. Website accessed on [2015/06/15].

He, Y., Wetterhall, F., Cloke, H. L., Pappenberger, F., Wilson, M., Freer, J., et al. (2009). Tracking the uncertainty in flood alerts driven by grand ensemble weather predictions. *Meteorological Applications, 16*(1), 91–101.

Hirpa, F. A., Hopson, T. M., De Groeve, T., Brakenridge, G. R., Gebremichael, M., & Restrepo, P. J. (2013). Upstream satellite remote sensing for river discharge forecasting: Application to major rivers in south Asia. *Remote Sensing of Environment, 131*, 140–151.

Hopson, T. M., & Webster, P. J. (2010). A 1–10-day ensemble forecasting scheme for the major river basins of Bangladesh: Forecasting severe floods of 2003–07. *Journal of Hydrometeorology, 11*(3), 618–641.

Hossain, F., Siddique-E-Akbor, A. H., Mazumder, L. C., ShahNewaz, S. M., Biancamaria, S., Lee, H., et al. (2014). Proof of Concept of an Altimeter-Based River Forecasting System for Transboundary Flow Inside Bangladesh. *IEE Journal of Selected Topics in Applied Earth Observations and Remote Sensing,* 7(2), 587–601.

Jakobsen, F., Hoque, A.K.M.Z., Paudyal, G. N., & Bhuiyan, M. S. (2005). Evaluation of the short-term processes forcing the monsoon river floods in Bangladesh. *Water International, 30*(3), 389–399.

Monirul Qader Mirza, M., Warrick, R. A., Ericksen, N. J., & Kenny, G. J. (2001). Are floods getting worse in the Ganges, Brahmaputra and Meghna basins? *Environmental Hazards, 3*(2), 37–48.

Rasid, H., & Paul, B. K. (1987). Flood problems in Bangladesh: Is there an indigenous solution? *Environmental Management, 11*(2), 155–173.

Siddique-E-Akbor, A. H. M., Hossain, F., Lee, H., & Shum, C. K. (2011). Inter-comparison study of water level estimates derived from hydrodynamic–hydrologic model and satellite altimetry for a complex deltaic environment. *Remote Sensing of Environment, 115*(6), 1522–1531.

Thielen, J., Bogner, K., Pappenberger, F., Kalas, M., del Medico, M., & de Roo, A. (2009). Monthly-, medium-, and short-range flood warning: Testing the limits of predictability. *Meteorological Applications, 16*(1), 77–90.

Webster, P. J., & Hoyos, C. (2004). Prediction of monsoon rainfall and river discharge on 15–30-day time scales. *Bulletin of the American Meteorological Society, 85*(11), 1745–1765.

Webster, P. J., Jian, J., Hopson, T. M., Hoyos, C. D., Agudelo, P. A., Chang, H., et al. (2010). Extended-range probabilistic forecasts of Ganges and Brahmaputra floods in Bangladesh. *Bulletin of the American Meteorological Society, 91*(11), 1493–1514.

INDEX

Note: Page numbers followed by *b* indicate boxes, *f* indicate figures and *t* indicate tables.

A

Acoustic Doppler current profilers (ADCPs) technology
- in China
  - Doppler current-meter, 72, 74*f*
  - electric wave current-meter, 72, 74*f*
  - evaporation observation site, 72, 73*f*
  - flood peak measuring, 72, 75*f*
  - hydrometric boat, 72, 73*f*
  - measurement from bridge, 72, 74*f*
  - online sediment measurement, 72, 75*f*
  - rainfall observation site, 72, 73*f*
- streamflow data, 379, 381, 394

Acoustic Doppler velocity meter (ADVM), 381

Acoustic velocity meter (AVM), 381

Actual evapotranspiration (AET), 405

ADCPs technology. *See* Acoustic Doppler current profilers (ADCPs) technology

Advanced Hydrologic Prediction Services (AHPS), 252–255, 289–290

Advanced Weather Interactive Processing System (AWIPS)
- API strategy, 289
- flash flood watch and warning program, 301
- HADS, 262
- infrastructure, 288–289
- MRMS system, 296
- NEXRAD system, 290
- NWS, 252–255
- QPF, 274
- RFC operational workflow, 284*f*
- temperature estimation, 274
- WAN, 285–287

Agence Française de Presse (AFP), 107

AHPS. *See* Advanced Hydrologic Prediction Services (AHPS)

Amazon River basin
- Ensemble Kalman Filter technique, 59–60
- ESP approach, 59–60
- forecast system, 59
- IPH/UFRGS, 58–59
- location of, 58–59
- MGB-IPH, 59
- Solimoes/Amazon main stem, 60, 61*f*

Andalusian Environmental Information Network (REDIAM), 317

Antecedent precipitation index (API) model, 257–259

ANTILOPE radar, 133

ANZEMC. *See* Australian New Zealand Emergency Management Committee (ANZEMC)

APHRODITE, 406–407, 411–414, 418

ArcGIS, 408

Artificial neural network (ANN), 405

Asian Disaster Preparedness Center (ADPC), 401

Atmospheric jets, 95

Australian Bureau of Meteorology
- climate, 4–7, 6*f*
- coastal flooding, 7
- demographics, 3–4, 4–5*f*
- drought, 9
- emergency agencies, 36
- flash floods, 7–8
- flood forecasting and warning services, 15–16
- flood warning system, 13–15, 15*f*, 18
- flood watch, 16–18, 16–17*b*
- hazards, warnings, and forecasts division, 13, 14*f*
- history of, 10–13
- hydroclimatic networks, 33
- hydrological modeling, 34–35
- institutional factors, 36
- monitoring networks, 33–34
- National Operations Unit, 37
- operational forecasting and systems
  - data systems and quality control, 19–21, 20–21*f*
  - expertise role, 32

Australian Bureau of Meteorology *(Continued)*
hydrologic simulation models (*see* Hydrologic simulation models)
HyFS, 29–32, 30–31*f*
HYMODEL, 29, 31, 31*f*
production and dissemination of, 18, 19*f*
statistical models, 22–23, 22*f*
quality control and processing, 34
riverine floods, 8
warning and communication, 35
Australian New Zealand Emergency Management Committee (ANZEMC), 38
Automated flood warning systems (AFWS), 302
Automatic Parameter Optimization Program (OPT), 268
Automation of Field Operations and Services (AFOS), 254
Autoregressive moving-average (ARMA) model, 214, 403–404
AWIPS. *See* Advanced Weather Interactive Processing System (AWIPS)
Aylon river headwaters, 159–162

B

Baden-Württemberg Flood Forecasting Centre (HVZ BW), 129–130
Balkan floods, 341, 341–342*f*
BALTEX research project, 138–139
Bangladesh flood forecasts
AET, 405
ANN, 405
BWDB, 401–402
ECMWF, 402–403
FFWC, 401–402
flatness nature, 399
flood-prone areas in, 401
HEC-RAS, 404–405
JASON-2 satellite, 404
MIKE 11, 402
passive microwave sensing technique, 403–404
Topex/Poseidon satellite altimetry measurements, 404–405
Bangladesh Water Development Board (BWDB), 401–402
Bayesian Model Averaging (BMA), 329
Brahmaputra river (Bahadurabad), 401–402
Brazil
Amazon River basin (*see* Amazon River basin)
CEMADEN, 45
climate, 41–42, 43*f*
Doce river basin, CPRM, 46–47
flood forecast, 42–43
geography, 41, 42*f*
HPP dams, 44–45
hydropower reservoirs and power generation, 43–44
MGB-IPH model, 44–45
ONS, 44
operational forecast system, 45–46
PREVIVAZH model, 44
Sao Francisco forecasting system, 51–53, 52–53*f*
Tocantins River (*see* Tocantins River basin forecasting system)
Upper Uruguay River forecasting system, 48–51, 48*f*, 50–51*f*
Brazilian Geological Service (CPRM), 46–47, 47*f*
Bureau of Hydrology of the Ministry of Water Resources (MWR-BoH), 71, 75
Bureau of Meteorology (Bureau). *See* Australian Bureau of Meteorology
Bureau of Reclamation, 256
BWDB. *See* Bangladesh Water Development Board (BWDB)
Bye Report, 205–206

C

Center for Environmental and Geographic Information Services (CEGIS), 401
Center for Natural Disaster Monitoring and Alert (CEMADEN), 45
Central African jets, 95
Central European Floods, 339, 340*f*
Centre for Ecology and Hydrology's (CEH), 212
Centro de Previsao de Tempo e Estudos Climáticos (CPTEC), 49
Cerrado biome, 41

China
CNFFS (*see* China National Flood Forecasting System (CNFFS))
flood control and management
flood defenses, 68
flood diversion and storage, 68
harmonious coexistence, humans and water, 69
headquarters, 69
heavy rainstorm and dam-failures, 69–70
risk concept, 70
hydrological forecasting and prediction
data transmission, 72–76
DEM, 84–85
forecasting administration, 71
hydrological information, 71
hydrological monitoring, 71–72, 72*f*
IFP, 84
rainfall and flood joint forecast, 84
RS and GIS technologies, 84–85
hydrological forecasting methods, 76–77
China Meteorological Administration (CMA), 75
China Ministry of Water Resources, 70
China National Flood Forecasting System (CNFFS)
calibration system, 77–78, 78*f*
catchment boundary, 78, 80*f*
ensemble streamflow prediction component, 77–78, 84*f*
extended stream flood prediction system, 77–78, 78*f*
flood forecasting models, 77–78, 79*t*
forecasted result optimizing, 78, 81*f*
GIS functions, 78, 79*f*
man-machine alternative interface
discharge amending in, 78, 83*f*
rainfall amending in, 78, 82*f*
model parameter calibrating, 78, 80*f*
operational forecast system, 77–78, 78*f*
rainfall amending in, 78, 82*f*
real-time operational forecasting, 78, 81*f*
reservoir forecasting and operation, 78, 82*f*
Thiessen polygons creation, 77–78, 83*f*
Civil Contingencies Act, 231, 237
Climate Forecast Application Network (CFAN), 401–403
Climate Prediction Center (CPC) morphing technique (CMORPH), 186–187
Climate Research Unit (CRU), 92–93
CNFFS. *See* China National Flood Forecasting System (CNFFS)
Coastal flood forecasting
in Australia, 7
in Great Britain
CS3X, 222
ECMWF, 222–224, 224*f*
FFC, 222, 226
Liverpool gauge, 226, 227–228*f*
short-and medium-range deterministic wave forecasts, 222
Storm Tide Warning Service, 225
surge ensemble output, 222
Colombia (South America)
design and development
forecasting outputs and products, 360–362, 361*f*
forecasting processes and models, 358–360, 359*f*, 360*t*
forecasting system infrastructure, 356, 357*f*
monitoring data, 356–357
HEC-HMS hydrological event model, 353–354
HEC-RAS hydrodynamic model, 353–354
IDEAM, 350, 353–357, 358*t*, 360–362, 365–366
Medellin, 353–354
physiography and demographics
Andean regions, 351–353
Magdalena-Cauca basin, 351–353, 352*f*
pilot development
Bogotá River, 355
criteria for, 354
expansion of, 362, 363–364*f*
integration of the system, 365
interinstitutional collaboration, 365–366
Magdalena river, 354
Upper Cauca basin, 354–355, 362
Río Frío river, 353–354
Tunjuelo river, 353–354

Commission of Inquiry (COI), 12–13
Common AWIPS Visualization Environment (CAVE), 274
Commonwealth Scientific and Industrial Research Organisation (CSIRO), 25
Community Collaborative Rain, Hail and Snow Network (CoCoRaHS), 264
Community Hydrologic Prediction System (CHPS), 274
Concept of Operations (CONOPS), 257
Congo River Basin
 climate change, 106–107
 disaster management systems
  data management, 115, 115*f*
  flood management policy, 118–120
  ground-based monitoring, 112–113, 113*f*
  institutional arrangement, 118–120
  rainfall-runoff modeling, 115–118
  satellite-based observation, 114
 flood disasters, 87
 hydro climate processes (*see* Hydro climate processes)
 land use dynamics, 105–106
 map of, 87, 88*f*
 physiographic setting and physical characteristics
  land elevation areas, 91–92, 91*f*
  main rivers and sources, 89, 90*f*
  see-saw phenomenon, 89–91
  swells/rises, 89
 trends and socio-economic impacts
  AFP, 107
  Dartmouth Flood Observatory, 107
  EM-DAT, 107
  flash flood, 108, 110*f*
  hydro climate disasters, 108, 110*f*
  natural disasters, 107–108, 109*t*
  riverine flood, 108, 110*f*
  stem receives flow contribution, 111
Continental Shelf Model (CS3X), 222
Continuous ranked probability scores (CRPS), 338–339, 339*f*
Cooperative Observer Program (Coop), 261–262
Cooperative Water Program, 374
Corporación Autónoma Regional de Cundinamarca (CAR), 355
COSMO-LEPS forecasts, 330–331
COSMO-RU model, 175
CS3X. *See* Continental Shelf Model (CS3X)
Current-meter method, 72

D

Danish Hydraulic Institute (DHI), 402
Data Acquisition Processing System (DAPS), 262–263
Delft-FEWS, 29, 315, 355–356, 358–360, 365
Digital Elevation Model (DEM), 190
Digital Precipitation Array (DPA), 294
Diploma in Operational Hydrometeorology and Flood Forecasting, 210
Direct water balance methods, 177
Doce river flood forecasting system, 46–47, 47*f*

E

Early warning flood forecasting systems (EWFFS)
 Amur River basin, 179
 Kuban River basin, 179
 short-range forecast techniques, 179, 180*f*
 structure of, 179*f*
ECMWF. *See* European Centre for Medium-Range Weather Forecasts (ECMWF)
Economic Commission for Central African States (ECCAS), 119–120
EFAS. *See* European Flood Awareness System (EFAS)
EFAS Information System (EFAS-IS), 318–319, 332–334
El-Gera upstream gauge station, 187–188
El Nino Southern Oscillation (ENSO), 5–7, 92–93, 349
Emergency disaster database (EM-DAT), 107
Emergency Response Coordination Centre (ERCC/EC), 314
Ensemble Kalman Filter technique, 59–60
Ensemble prediction systems (EPSs), 52–53
Ensemble Streamflow Prediction Analysis and Display Program (ESPADP), 268–269
Ensemble streamflow prediction (ESP) approach, 59–60

Ensemble Verification System (EVS), 300
Environment Act of 1995, 205
Environment Agency (EA), 204–205, 209
Environment Agency's Lessons Learned report, 205–206
EU-FLOOD-GIS database, 322–323
European Centre for Medium-Range Weather Forecasts (ECMWF), 222–224, 281–283, 317, 402–403
European Corine Land Cover 2000, 322–323
European Flood Awareness System (EFAS), 187
  case studies
    Balkan floods in May 2014, 341, 341–342*f*
    Central European Floods in summer 2013, 339, 340*f*
    inter-annual variation, 337, 338*f*
    performance of, 338–339, 339*f*
  Copernicus EMS, 314–315
  data and information exchange, 313–314
  dissemination of forecasts
    EFAS-IS, 332–334
    email alerts and daily overview, 334–335, 334*t*
  flood risk management, 313
  forecast generation
    infrastructure, 325
    scheduling and execution, 326–327
  forecast products
    EFAS thresholds and return periods, 328, 328*t*
    flash flood alerts, 330, 331*f*
    post-processed forecasts, 329, 329*f*
    rainfall animation, 330–331, 332*f*
    soil moisture and snow anomaly maps, 331
  meteorological models, 320–321
  objectives of, 314
  operational performance
    monitoring, 335–336, 336*t*
    system performance, 336–337, 337*t*
  pan-European early warning systems, 315
  structure, 317–318
  trans-boundary forecasting system, 315
European Flood Risk Management Directive, 147–148
European Precipitation Index based on simulated Climatology (EPIC), 330
EWFFS. *See* Early warning flood forecasting systems (EWFFS)
Extended Streamflow Prediction (ESP) System, 268–270
Extreme Rainfall Alert (ERA) service, 217

F

False alarm rate (FAR), 235
Father of Systematic Streamgaging (Newell), 373
Federal Ministry of Transport and Digital Infrastructure, 141–142
Federal Office for the Environment (FOEN), 128
FEWS. *See* Flood early warning system (FEWS)
FGMOD river basin model, 138–139
FKIS-Hydro, 315
Flash flood guidance (FFG), 252–254, 270
Flash floods
  in Australia, 7–8
  in Congo River Basin, 108, 110*f*
  in Israel, 156
  in Scotland, 219*b*
Flood Action Plan, 129
Flood alerts, 159, 161*f*
Flood and Water Management Act of 2010, 209
Flood Control Act of 1936, 374
Flood Control Act of 1938, 251–252
Flood early warning system (FEWS)
  CPRM, 47
  in United States, 274–276
Flood Forecasting and Warning Centre (FFWC), 401–402
Flood Forecasting Centre (FFC), 209–210, 218, 222, 238
Flood Guidance Statement (FGS), 217, 231, 231*f*
Floodline Warnings Direct (FWD), 206
Flood Risk Management (Scotland) Act, 204
Flood Warning Centre Rhine (HMZ Rhein), 129

Flood warning systems (FWS), 302
Food and Agriculture Organization of the United Nations (FAO), 190

G

Ganges basin
- adjusted coefficient of determination (adj R^2), 411
- coefficient of determination (R^2), 411
- ensemble flood forecasting, 417–418
- perfect rainfall forecast, 411, 414
- persistence-based *Q*-*Q* model, 410
- rainfall-runoff relationship over, 408, 409*f*
- RMSE and NSE metrics, 411
- seasonality, 407, 407*f*
- simple *Q*+*P*-*Q* model, 410–411, 414–418, 414*f*
- time forecast, 412–414, 412*f*, 413*t*, 414–416*f*

Ganges-Brahmaputra-Meghna (GBM) river system
- Bangladesh (*see* Bangladesh flood forecasts)
- Ganges basin (*see* Ganges basin)
- geographical setting of, 399, 400*f*

Gash River Basin, Sudan
- flood event in 2007, 183, 184*f*
- flood hydrograph, 192–193
- geographical setting, 187–188, 188*f*
- geospatial datasets, 186–187, 186*f*
- hydrological parameter extraction, 191
- hydrometeorological model setup, 191–192, 192*f*
- Kassala Bridge stations, 195, 196–197*f*, 197*t*
- model calibration and validation, 193–194
- NASA TRMM, 195
- probabilistic forecasting system, 187
- prototype flood forecasting system, 187
- rainfall-runoff models, 184–185
- satellite-based rainfall monitoring, 185–186
- spatial and nonspatial databases, 190
- tele-meteorological rainfall data, 187
- terrain processing, 190–191
- topographic model setup, 189

Gash River Training Unit (GRTU), 184–185
Gash Spate Irrigation Scheme (GSIS), 187–188
Génie Rural a 4 Parametres (GR4J), 25
Geographic Information Systems (GIS), 254–255
Geostationary Operational Environmental Satellite (GOES), 260–263, 289, 389
German Weather Service (DWD), 125, 132–133
Germany
- DWD, 125
- flood partnerships, 147–148
- forecast dissemination, 145–147, 147*f*
- hydrological and meteorological data, 130–133, 131*f*
- national and international networking, 125, 126*f*, 127
- NWP, 133–136
- regional organization, 137–138
- Rhine basin, 127–130, 127–128*f*
- river forecasting models
  - LARSIM water balance model, 138–141
  - WAVOS Rhein, 143–145, 144*f*
  - WAVOS Water Level Forecasting System, 141–143, 142*f*
- snowmelt forecasts, 136–137
- transboundary data exchange, 137–138

G2G model. *See* Grid-to-Grid (G2G) model
GIS-based hydrological water balance model, 116–117
Global Ensemble Forecasting System (GEFS), 52–53, 306–307
Global Land Cover 2000 (GLC2000), 322–323
Global Precipitation Measurement (GPM), 186–187
Global Runoff Discharge Center (GRDC), 116–117
GLOBCARBON project, 322–323
GloFAS, 315
GOES. *See* Geostationary Operational Environmental Satellite (GOES)
Gravity Recovery and Climate Experiment (GRACE) satellite mission, 116–117

Great Britain
annual average rainfall, 201, 202*f*
aquifers and porous rock, 203–204
astronomical tides/spring tides, 203
convective systems, 240–241
countrywide flood forecasting approach
coastal, 222–224, 223–224*f*, 227–228*f*
drivers, history, and context, 211–212
G2G model (*see* Grid-to-Grid (G2G) model)
groundwater, 229
surface water, 217–218, 219*f*, 221*f*
estuaries and coastal modeling, 241–242
FFC, 238
flood Risk, England, Wales, and Scotland, 204
forecasting and warning dissemination, 229, 230*f*
integrated modeling and forecasting, 242–243
landscape, 205–206
lead time forecasting, 243
observations and instrumentation, 241
people, skills, interpretation, and engagement, 244
performance of, flood forecasting and warning service
formal national measure, 237–238
local river model performance, 235–237, 237*t*
quality of, 233
Pitt Review, 209–210
products and services
FGS, 231, 231*f*
Hydromet Service, 232
internet-published flood forecasts, 232
surface water flood forecasts, 232, 234*f*
three-day flood risk forecast, 232
UKCMF service, 232
River Thames, 201, 202–203*b*
SEPA strategic developments, 239, 239*t*
SFFS, 210–211, 238
summer 2007 floods, 206–207, 207–208*f*
Grid-to-Grid (G2G) model
advantages, 214
error correction, 214
gauged and ungauged locations, 215, 216*f*
observations and forecast data, 215
physical-conceptual distributed hydrological model, 212–213
runoff production scheme, 213, 213*f*
SFFS, 215–217, 216*f*

H

Hardinge Bridge point (Ganges river), 401–404, 406–408
Health and Safety Laboratory (HSL), 218
HFS. *See* Hydrologic forecasting system (HFS)
Horn of Africa. *See* Gash River Basin, Sudan
HSL. *See* Health and Safety Laboratory (HSL)
Hurricane Rita, 394
Hurricane Sandy, 394
HVZ BW. *See* Baden-Württemberg Flood Forecasting Centre (HVZ BW)
Hydro climate processes in Congo River Basin
annual maximum, mean, and minimum flows, 96–97, 97*f*
atmospheric jets, 95
daily trend in streamflow, 96–97, 96*f*
ITCZ, 93–95
mean annual rainfall, 92–93, 93*f*, 95
meso-scale convective systems
rainfall belt, 95, 96*f*
SSTs, 94–95
wavelet approach
average scale spectrum, 100–101, 101*f*
disaster type, 102*t*
global wavelet spectrum, 98
interannual cycle, 101
maximum annual flows, 99–100, 100*f*, 104*f*
mean monthly flows, 99, 99–100*f*, 101–103, 103*f*
Morlet wavelet function, 98–99
NAO index, 104–105, 104*f*
natural disasters, 103, 104*f*
origins, 101–103
reference wavelet function, 98
wavelet spectrum, 98

Hydroelectric power plants (HPPs), 44–45
Hydrological Ensemble Prediction Experiment (HEPEX), 297–298
Hydrological Forecast Center of the Ministry of Water Resources, 71
Hydrological monitoring stations in China
  current-meter method, 72
  evaporation observation site, 72, 73*f*
  flood peak measuring, 72, 75*f*
  flow measurement
    Doppler current-meter, 72, 74*f*
    electric wave current-meter, 72, 74*f*
  hydrometric boat, 72, 73*f*
  measurement from bridge, 72, 74*f*
  moving-boat method, 72
  online sediment measurement, 72, 75*f*
  rainfall observation site, 72, 73*f*
  special stations, 71–72
  state level basic hydrometric stations, 71–72
Hydrologic Ensemble Forecast System (HEFS), 297–298
Hydrologic forecasting system (HFS)
  in Australia, 13, 29–32, 30*f*
  Roshydromet, 171–173, 172*f*
Hydrologic Research Laboratory (HRL), 254
Hydrologic simulation models in Australia
  catchment delineation, 23–24, 23*f*
  data assimilation, 28–29
  event-based models, 24–25
  GR4J, 25
  parameters, 25–26
  QPF, 26–28, 27*f*, 28*t*
  rainfall depth, 24
  reservoir modeling, 26
Hydrology Laboratory Research Modeling System (HL-RMS), 276
Hydrometeorological Automated Data System (HADS), 262–263
Hydromet Service, 232
Hydropower dams in Brazil, 42, 43*f*
HyFS. *See* Hydrologic forecasting system (HFS)

I

IDEAM. *See* Instituto de Hidrología, Meteorología y Estudios Ambientales (IDEAM)
Index velocity, 381
Institut national de Recherche en Sciences et Technologies pour l'Environnement et l'Agriculture (IRSTEA), 25
Instituto de Hidrología, Meteorología y Estudios Ambientales (IDEAM), 350, 353–357, 358*t*, 360–362, 365–366
Instituto de Pesquisas Hidráulicas-Universidade Federal do Rio Grande do Sul (IPH/UFRGS), 49
Integrated Water Resources Science and Services (IWRSS), 303–304
Interactive Forecast Program (IFP), 84, 265
Interactive Verification Program (IVP), 300
International Commission for the Protection of the Rhine (ICPR), 129
International Commission of Congo-Oubangui-Sangha (CICOS), 118
Inter tropical convergence zone (ITCZ), 92–95
Inverse Distance Weighting (IDW), 271
Israel
  climate and hydrological characteristics, 154–156, 155*f*, 157–158*f*
  Mediterranean region, 153
  precipitation patterns, 153–154
  Tel Aviv Metropolis, HEC-HMS model, 159–165, 162–164*f*
  WRF-Hydro Model (*see* Israeli WRF-Hydro Model)
Israel Hydrological Service (IHS), 156
Israeli WRF-Hydro Model
  channel network, 159, 160*f*
  flood alerts, 159, 161*f*
  forecast maps for, 159, 161*f*
  map of, 158–159, 159*f*
  NCAR, 156–158
ITCZ. *See* Inter tropical convergence zone (ITCZ)

J

Japan Aerospace Exploration Agency (JAXA), 186–187

K

Kassala City, Sudan. *See* Gash River Basin, Sudan

L

Lake Tanganyika Water Authority (LTA), 119–120
Land use/land cover (LULC), 190
La Niña conditions, 349
Large Area Runoff Simulation Model (LARSIM), 138–141, 139–140*f*
Lead Local Flood Authorities (LLFAs), 217
LISFLOOD hydrological model, 320–328, 330–331, 338–339
Local Readout Ground Stations (LRGS), 260–261

M

Manual Calibration Program (MCP), 268
Mareb River, 187–188
Mean areal values of precipitation (MAP), 268, 271–273
Mean areal values of temperature (MAT), 268, 274
Meteorological Model-based Ensemble Forecast System (MMEFS), 297–298
Meteorology Act of 1906, 10–11
Meteorology Act of 1955, 19–20
Met Office, 209–210
MIKE 11 flood model, 402
MIKE HYDRO BASIN interface, 118
Ministry of Water Resources and Electricity (MoWRE), 184–185
MOGREPS-UK model, 217–218
Monitor AVançado de ENchentes (MAVEN) system, 46
Moselle, 137–138
Moving-boat method, 72
Multi-Radar, Multi-Sensor (MRMS) System, 296
Multisensor Precipitation Estimator (MPE), 293–295
MWR-BoH. *See* Bureau of Hydrology of the Ministry of Water Resources (MWR-BoH)

N

Nash Sutcliffe coefficient, 49, 51*f*, 194–195
Nash-Sutcliffe efficiency (NSE), 235, 323, 324*f*
National Aeronautics and Space Administration (NASA), 186–187
National Center for Atmospheric Research (NCAR), 156–158
National Centers for Environmental Prediction (NCEP), 306–307
National Climatic Data Center (NCDC), 261–262
National Environmental Satellite, Data and Information Service (NESDIS), 262–263
National Flood Emergency Framework for England, 229
National Meteorology Agency, Ethiopia, 191–192
National Oceanic and Atmospheric Administration/National Weather Service ((NOAA/NWS). *See* NOAA/NWS, United States
National Rivers Authority (NRA), 205
National Severe Storms Laboratory (NSSL), 296
National Streamflow Information Program (NSIP), 260, 375–376
National System Operator, 44
National Water Center (NWC), 304–307
National Water Information System (NWIS), 260–261
National Weather Service (NWS). *See* NOAA/NWS, United States
Natural Hazards Partnership, 218
Natural Resource Conservation Service (NRCS), 256
Natural Resources Wales models, 204, 226
NEMO model, 242
NESDIS Command and Data Acquisition (CDA) System, 262–263
NESDIS GOES Data Collection System (DCS), 262–263
NexGen products, 28
NEXRAD. *See* Next Generation Radar (NEXRAD)
Next Generation Radar (NEXRAD), 252–254, 256, 260, 262, 281, 281*f*, 290, 292*f*, 293–295, 297*f*, 301, 306*f*
NIKLAS program, 132

NOAA Global Ensemble Reforecast Data Set, 55–56
NOAA/NWS, United States, 249
advancements, 249–250
AFOS, 254
AFWS, 302
AHPS, 252–255, 289–290
API model, 257–259
AWIPS, 288–289
CHPS, 274–276
Congressional bill, 251–252
data acquisition and processing
CoCoRaHS, 264
Coop, 261–262
data needs for, 259–264
HADS, 262–263
U.S. Geological Survey stream gaging program, 260–261
event-based hydrologic modeling systems, 252–254
FFG, 252–254
flash flood watches and warnings, 301
Flood Control Act of 1938, 251–252
forecast evaluation
service system, 300–301
verification, 300
future developments
IWRSS, 303–304
NWC, 304–307
GIS, 254–255
HEFS, 297–298
HRL, 254
hydrologic forecasting constraints, 249
Kansas river flooding, 251–252
mainframe computer system-based NWSRFS, 252–254
mission statement of, 251
MPE, 293–295
MRMS system, 296
NEXRAD, 252–254, 256, 260, 262, 281, 281*f*, 290, 292*f*, 293–295, 297*f*, 301, 306*f*
NWSRFS, 249–250
calibration system, 268
ESP system, 268–270
FFG System, 270
IFP, 265
models, 271
OFS, 265
QPE, 273–274
QPF, 274
temperature estimation, 274
OH, 252, 254
OHD, 254
OHRFC, 252
Organic Act of 1890, 251
QPE, 290–296
QPF, 252–254
RDHM, 276–281
RFCs, 252–254, 253*t*
river and flood program origin, 251
river forecasting paradigm, 255–257
SAC-SMA model, 252–254
Weather Bureau, 251–252
WFO, 281–283, 282*f*
North Atlantic Oscillation (NAO), 98
NSIP. *See* National Streamflow Information Program (NSIP)
Numerical weather predictions (NWPs)
in Australia, 20–21, 37
in Brazil, 49
in Germany
COSMO DE EPS, 136
COSMO-LEPS, 134
DWD's forecasting chain, 133
ECMWF Ensemble Prediction System ENS, 134
EPS data, 136
misforecasts, 133–134
RNWP-PEPS, 136
runoff forecast, 133–134, 135*f*
in Great Britain, 212, 222–224, 232
in United States, 281–283
NWISWeb, 395
NWSChat, 391
NWSRFS Operational Forecast System, 265
NWS River Forecast System (NWSRFS), 249–250
NWS Telecommunications Gateway (NWSTG), 262–263

O

Office of Hydrologic Development (OHD), 254
Office of Hydrology (OH), 252, 254

Ohio River Forecast Center (OHRFC), 252
One-dimensional hydrodynamic-numerical model WAVOS-1D, 143
Operational forecast system
in Australia
data systems and quality control, 19–21, 20–21*f*
expertise role, 32
hydrologic simulation models (*see* Hydrologic simulation models)
HyFS, 29–32, 30–31*f*
HYMODEL, 29, 31, 31*f*
production and dissemination of, 18, 19*f*
statistical models, 22–23, 22*f*
in Brazil, 45–46
in China, 77–78, 78*f*
in Colombia, 350, 365
in Ganges basin, 401–403
in Great Britain, 211–212
in United States, 264–265, 266*b*, 268, 272*t*
OPT *see* Automatic Parameter Optimization Program (OPT)
Organic Act of 1890, 251

P

Panel on Climate Change, Assessment Report 4, 106
Physically based hydrologic models, 177
Physical-statistical methods, 177
Portable Batch System (PBS) software, 325
Precipitation Estimation from Remotely Sensed Imagery Using Artificial Neural Networks (PERSIANN), 186–187
Probability of detection (POD), 235

Q

Quantitative Precipitation Estimation (QPE), 273–274, 290–296
Quantitative Precipitation Forecast (QPF), 44–45, 252–254, 274, 330
Queensland floods, 12–13

R

RADOLAN data, 132–133
Real Time Control Tools (RTC-Tools), 52–53
REDIAM. *See* Andalusian Environmental Information Network (REDIAM)
Relative Volume Error *(RVE)*, 194–195
Research Distributed Hydrologic Model (RDHM), 276–281
Rhine basin, 127–130, 127–128*f*
Rhineland-Palatinate, 130, 132, 138, 145–147, 147*f*
Rijkswaterstaat Waterdienst (RWS), 129, 317
River Forecast Centers (RFCs), 249, 250*f*, 252–254, 253*t*
Riverine floods, 8
Rivers and Harbors Act of 1927, 374
River-Transfer Hydrological Model, 116–117
Russian Federal Service for Hydrometeorology and Environmental Monitoring (Roshydromet)
EWFFS (*see* Early warning flood forecasting systems (EWFFS))
forecasted phenomena and forecast range, 170
HFS, 171–173, 172*f*
hydrometeorological data, 173–176, 174*f*
operational hydrological forecasts, 170–171
runoff characteristics, 169
runoff formation models, 171
spring flood predictions, 176–177
RWS. *See* Rijkswaterstaat Waterdienst (RWS)

S

Sacramento Soil Moisture Accounting (SAC-SMA) model, 252–254
Salam-Alikum downstream gauge station, 187–188
Sao Francisco River forecasting system, 51–53, 52–53*f*
Satellite Rainfall Estimates (SRE), 191–192
Scottish Environment Protection Agency (SEPA), 204
Scottish Flood Forecasting Service (SFFS), 210–211
SCS-Unit Hydrograph techniques, 192
Sea surface temperatures (SST), 94–95
Service Central d'Hydrometeorologie et d'Appui a la Previsions des Inondations (SCHAPI) in France, 210
Service de Prévision des Crues Rhin-Sarre in Strasbourg (DREAL Alsace), 129
Service Level Agreements, 36

Short Range Numerical Weather Prediction Program (SRNWP), 136
Short-Term Ensemble Prediction System (STEPS), 20–21
Shuffle Complex Evolution (SCE) method, 268
Shuttle Radar Topography Mission (SRTM), 190, 322–323, 408
Slovak Hydrometeorological Institute (SHMU), 317
SNOW model, 136–137
SOBEK software, 143
Soil Conservation Service (SCS), 192
Solimoes/Amazon main stem, 60, 61*f*
Southern African Development Commission (SADC), 119–120
Southern Oscillation, 5–7
Spanish ELIMCO Sistemas, 317
Spatially distributed unit hydrograph (SDUH), 408
SPATSIM software, 118
Standard Hydrometeorological Exchange Format (SHEF), 285–287, 298
Standard Particle Swarm 2011 (SPSO-2011), 323
State Flood Forecasting or Warning Centre, 125
Station Duty Manual, 257
Storm-surge sensors (SSSs), 394
Storm Tide Forecasting Service, 206
Storm Tide Warning Service, 225
Stream burning technique, 191
Streamflow data
  delivery of streamflow information
    DCSs, 389
    GOES, 389
    landline and cellular telephone technology, 389
    radio transmission communications, 389
    streamflow forecast, 390–391
  Ganges basin (*see* Ganges-Brahmaputra-Meghna (GBM) river system)
  measuring stage
    gage datum, 377
    nonrecording gage, 378
    permanent gage datum, 378
    pressure measurement, 377–378
    radar-level sensor, 377–378, 378*f*
    streamgage structure, 376*f*, 377–378
  measuring streamflow
    ADCPs, 379, 381
    ADVM, 381
    AVM, 381
    cross-section discretization method, 379
    direct measurements, 379
    discharge measurements, 379
    index velocity, 381
    indirect measurements, 379
    midsection velocity-area method, 379
  stage and volumetric discharge, 376–377
  stage-discharge relation
    complex ratings, 381–383, 383*f*
    index-velocity method, 386–388
    rating controls, 383–384
    rating curve for the Platte River, 385–386, 386*f*
    rating extension, 384–385
    shifting-control method, 385–386
    simple stage-discharge ratings, 381–383, 382*f*
    temporary changes, 385–386
    uncertainty in ratings, 388
  streamflow defined, 371
  streamgaging (*see* Streamgaging United States)
Streamgaging United States
  history
    assessment programs, 373
    father of systematic streamgaging (Newell), 373
    NSIP, 375–376
    training camp, in Embudo, 373
    USACE, 373–374
    war years (1939–1947), 374
  opportunities for streamgage program
    ADCP, 394
    Hurricane Rita, 394
    Hurricane Sandy, 394
    James river, 391–393, 392*f*
    lake LaMoure, 392–393, 393*f*
    NWISWeb, 395
    Pipestem and Jamestown reservoirs, 392–393
    RDGs, 391–393
    real-time delivery, 395
    SSSs, 394
    WaterWatch, 395
  streamgage diagram, 376*f*

Supervisor Monitoring Scheduler (SMS) software, 326
Surface Water and Ocean Topography (SWOT), 404–405
Surface Water Decision Support Tool (SWFDST), 217–218
Swedish Meteorological and Hydrological Institute (SMHI), 317
Systematic flood forecasting, 11

T

Tel Aviv Metropolis, 159–165, 162–164*f*
Tocantins River basin forecasting system
  exceedance diagrams, 56, 57*f*
  forecasting system, 53–54
  hydropower production, 53, 54*f*
  limitations of, 58
  MapWindow GISr, 56
  MergeHQ, 55
  QPF, 55–56
  rain gauge network, 54–55
  sequence of, 56, 57*f*
  Visual Basic .NET programming language, 56
TOPKAPI model, 84–85
Três Marias reservoir operation, 53
Tropical Rainfall Measuring Mission (TRMM), 53–55, 186–187, 406–407, 411–414, 418

U

UK Coastal Monitoring and Forecasting Service, 206
Unified River Basin Simulator (URBS), 24–26
United Kingdom Coastal Monitoring Forecasting Service (UKCMF), 222, 232
United States
  NOAA/NWS (*see* NOAA/NWS, United States)
  streamgaging (*see* Streamgaging United States)
United States Army Corps of Engineers (USACE), 256
Upper Uruguay River forecasting system, 48–51, 48*f*, 50–51*f*
U.S. Army Corps of Engineers (USACE), 373–374
U.S. Geological Survey (USGS), 371
U.S. Geological Survey stream gaging program, 260–261

V

Volga river reservoir, 177

W

Water Evaluation and Planning, 118
Water Resources Planning tool, 118
WaterWatch, 395
Waterways and Shipping Administration (WSV), 129
WAVOS Rhein, 143–145, 144*f*
WAVOS Water Level Forecasting System, 141–143
Weather Bureau, 251–252
Weather Forecast Offices (WFOs), 249, 281–283, 282*f*
Weather Predictions (QPF), 26–28, 27*f*, 28*t*
Weather Regime Analysis tool, 224*f*
Weather Research and Forecasting (WRF) model, 153–154
Wide Area Network (WAN), 76
World Meteorological Organization's, 170
World Meteorological Situation, 315
WRF-Hydro model. *See* Israeli WRF-Hydro Model

Y

Yangtze River Flood Control Headquarters, 69
Yarqon basin in Israel, 159–162, 162*f*
Yarqon Drainage Authority, 159–162
Yellow River Flood Control Headquarters, 69